Advances in Quantitative Remote Sensing in China—In Memory of Prof. Xiaowen Li

Advances in Quantitative Remote Sensing in China—In Memory of Prof. Xiaowen Li

Volume 1

Special Issue Editors

Shunlin Liang
Guangjian Yan
Jiancheng Shi

MDPI • Basel • Beijing • Wuhan • Barcelona • Belgrade

MDPI

Special Issue Editors

Shunlin Liang
University of Maryland
USA

Jiancheng Shi
Institute of Remote Sensing and Digital Earth
China

Guangjian Yan
Beijing Normal University
China

Editorial Office
MDPI
St. Alban-Anlage 66
Basel, Switzerland

This is a reprint of articles from the Special Issue published online in the open access journal *Remote Sensing* (ISSN 2072-4292) from 2017 to 2018 (available at: https://www.mdpi.com/journal/remotesensing/special_issues/quantitative_rs)

For citation purposes, cite each article independently as indicated on the article page online and as indicated below:

LastName, A.A.; LastName, B.B.; LastName, C.C. Article Title. *Journal Name* **Year**, *Article Number, Page Range.*

Volume 1
ISBN 978-3-03897-270-9 (Pbk)
ISBN 978-3-03897-271-6 (PDF)

Volume 1–2
ISBN 978-3-03897-278-5 (Pbk)
ISBN 978-3-03897-279-2 (PDF)

Contents

About the Special Issue Editors

Shunlin Liang earned a Ph.D. from Boston University in 1993. His main research interests focus on the estimation of land surface variables from satellite data, earth energy balance, and the assessment of environmental changes. He authored the book "Quantitative Remote Sensing of Land Surfaces" (Wiley, 2004), co-authored the book "Global LAnd Surface Satellite (GLASS) Products: Algorithms, Validation and Analysis"(Springer, 2013), edited the book "Advances in Land Remote Sensing: System, Modeling, Inversion and Application" (Springer, 2008), and co-edited the books "Advanced Remote Sensing: Terrestrial Information Extraction and Applications" (Academic Press, 2012) and "Land Surface Observation, Modeling, Data Assimilation" (World Scientific, 2013). He was also the Editor-in-Chief of the nine-volume book series "Comprehensive Remote Sensing". He has published over 320 SCI-indexed papers and 32 book chapters. According to Google Scholar, his H-index is 64. He is an IEEE fellow.

Guangjian Yan is a professor and vice director of the State Key Laboratory of Remote Sensing Science. He received his B.A. and M.A. degrees from the Beijing Institute of Technology in 1993 and 1996, respectively. He earned a Ph.D. degree from the Institute of Remote Sensing Applications, Chinese Academy of Sciences in 1999. He was selected as the New Century Excellent Talent at the University of China in 2005. He has been the principle investigator of national programs (numbers 973 and 863) and the key program of NSFC. His main research interests include multi-angular remote sensing, vegetation remote sensing, and radiation budget. He invented a series of multi-angle platforms for ground-based and airborne remote sensing measurement. He proposed a path length distribution model which can improve the accuracy of LAI indirect measurement more than one time. He has published more than 200 papers.

Jiancheng Shi received his B.A. in Hydrogeology and Engineering Geology from the University of Lanzhou in China in 1982. He earned his M.A. and Ph.D. degrees in Geography from the University of California, Santa Barbara (UCSB) in 1987 and 1991, respectively. He then worked at the Institute for Computational Earth System Sciences (later the Earth Research Institute) at UCSB as a research professor. In 2010, he joined the Institute of Remote Sensing and Digital Earth, Chinese Academy of Sciences as director and worked as a senior research scientist at the State Key Laboratory of Remote Sensing Science in Beijing, China. His research interests include microwave remote sensing of water cycle-related components. He has published more than 300 journal and conference papers. He is a Fellow of IEEE and SPIE.

remote sensing

MDPI

Editorial

Recent Progress in Quantitative Land Remote Sensing in China

Shunlin Liang [1,2,3,*], Jiancheng Shi [1,4] and Guangjian Yan [1,4]

[1] State Key Laboratory of Remote Sensing Science, Beijing Normal University and Institute of Remote Sensing and Digital Earth, Beijing 100875, China; shijc@radi.ac.cn (J.S.); gjyan@bnu.edu.cn (G.Y.)

[2] Department of Geographical Sciences, University of Maryland, College Park, MD 20742, USA

[3] School of Remote Sensing Information Engineering, Wuhan University, Wuhan 430072, China

[4] School of Remote Sensing Science and Engineering, Faculty of Geographical Science, Beijing Normal University, Beijing 100875, China

* Correspondence: sliang@umd.edu; Tel.: +1-301-405-4556

Received: 12 September 2018; Accepted: 14 September 2018; Published: 18 September 2018

During the past forty years, since the first book with a title mentioning quantitative and remote sensing was published [1], quantitative land remote sensing has advanced dramatically, and numerous books have been published since then [2–6] although some of them did not use quantitative land remote sensing in their titles. Quantitative land remote sensing has not been explicitly defined in the literature, but we consider it as a sub-discipline of remote sensing including the following components (see Figure 1): radiometric preprocessing, inversion, high-level product generation, and applications. Many inversion algorithms rely on physical models of radiation regimes of landscapes, which link with remotely-sensed data. Generating high-level satellite products of land surface biophysical and biochemical variables create the key bridge between remote sensing science and applications. Conducting in situ measurements for validation of inversion algorithms and satellite products is also a critical component. Application of satellite products to address scientific and societal relevant issues will ultimately decide the future of quantitative land remote sensing.

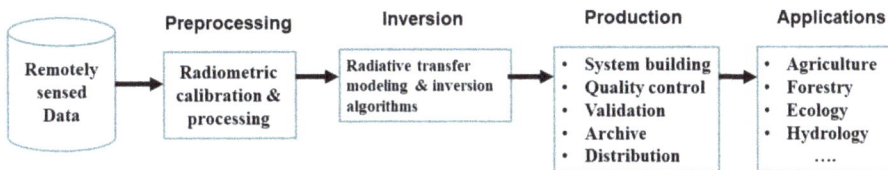

Figure 1. The scope of quantitative land remote sensing.

One of the major drivers of the rapid development of quantitative remote sensing in China is the availability of a huge amount of satellite data not only from the international space agencies but also from Chinese satellite sensors. Figure 2 shows the major Chinese satellite missions for land surface monitoring, such as the China-Brazil Earth resource satellites (CBERS), environment (Huang-Jing, HJ), resources (Zhi-Yuan, ZY), meteorological (Feng-Yun, FY), and high-resolution (Gao-Fen, GF) satellite series. Most of them are polar-orbiting satellites, but GF-5 and FY-4 are geostationary satellites. With the constellation of multiple satellites, both high spatial and temporal resolutions are being achieved.

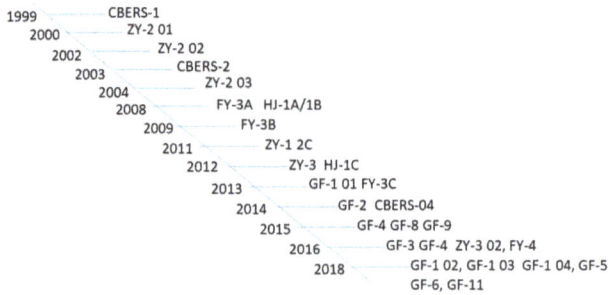

```
1999 ·········· CBERS-1
  2000 ········· ZY-2 01
    2002 ········ ZY-2 02
      2003 ········ CBERS-2
        2004 ······· ZY-2 03
          2008 ······ FY-3A  HJ-1A/1B
            2009 ······ FY-3B
              2011 ······ ZY-1 2C
                2012 ····· ZY-3  HJ-1C
                  2013 ···· GF-1 01 FY-3C
                    2014 ··· GF-2  CBERS-04
                      2015 ·· GF-4 GF-8 GF-9
                        2016 · GF-3 GF-4  ZY-3 02, FY-4
                          2018 · GF-1 02, GF-1 03  GF-1 04, GF-5
                                 GF-6, GF-11
```

Figure 2. Major Chinese satellites relevant to land remote sensing.

Because of the increased data volume and sophistication of information extraction, one of the trends in quantitative remote sensing is the production of high-level satellite products, mostly by the data centers with centralized facilities and specialized experts. It started from the NASA Earth Observing System (EOS) program in the 1990s. Since then, China has started to produce and distribute satellite products worldwide. One of the major product suites is the Global Land Surface Satellite (GLASS) products [7,8]. It has been expanded from the original 5 products into the present 12 products (see Table 1) that are being distributed free of charge through the China National Data Sharing Infrastructure of Earth System Science (http://www.geodata.cn/thematicView/GLASS.html) and the Global Land Cover Facility at the University of Maryland (http://glcf.umd.edu/data).

The GLASS products have some unique features, for example, long-time times series (several products span from 1981 to present), high-spatial resolution of the radiation products (5 km instead of the typical resolutions of ~100 km), and high quality and accuracy [9–11]. Efforts are being made in China [12] to develop more Climate Data Records (CDR) that are defined as the time series of measurements of sufficient length, consistency, and continuity to determine climate variability and change by the National Research Council [13].

Table 1. Overview of the Global Land Surface Satellite (GLASS) products and their characteristics.

No.	Product	Spatial Resolution	Temporal Resolution	Temporal Range	References
1	Leaf area index	1–5 km, 0.05°	8 days	1981–2017	[14,15]
2	Albedo	1–5 km, 0.05°	8 days	1981–2017	[16–18]
3	Emissivity	1–5 km, 0.05°	8 days	1981–2017	[19,20]
4	FAPAR	1–5 km, 0.05°	8 days	1981–2017	[21]
5	Downward shortwave radiation	0.05°	1 day	1983, 1993, 2000–2017	[22]
6	PAR	0.05°	1 day	1983, 1993 2000–2017	[22]
7	Longwave net radiation	0.05°	Instantaneous	1983, 1993, 2003, 2013	[23,24]
8	All-wave net radiation	0.05°	1 day	1983, 1993 2000–2017	[25]
9	Land Surface Temperature	1–5 km, 0.05°	Instantaneous	1983, 1993, 2003, 2013	[26]
10	Fraction of vegetation cover	500 m, 0.05°	8 days	1981–2017	[27]
11	Latent heat (ET)	1–5 km, 0.05°	8 days	1981–2017	[28]
12	Gross Primary Productivity	1–5 km, 0.05°	8 days	1981–2017	[29]

Many members of our community have made significant contributions to the development of quantitative land remote sensing. Professor Xiaowen Li was one of leading figures. Trained as an electrical engineer, Professor Li started to work on physical modeling of the vegetation radiation field in the early 1980s under the supervision of Professor Alan Strahler. He developed the well-known Li–Strahler geometric-optical vegetation reflectance model [30,31], and later coupled it with radiative transfer modeling [32,33]. He pioneered the simplified "kernels" to model land surface directional reflectance for developing the MODIS surface albedo products [34], These "kernels" have been widely used for analyzing various satellite observations. He also explored the angular behavior and scaling of the thermal-infrared remote sensing signatures [35], and proposed to constrain the remote sensing inversion using prior knowledge [36]. In the second half of his career, Professor Li devoted his time and energy to facilitate and promote quantitative land remote sensing research in China by leading several extensive research projects, directing the Research Institute on Remote Sensing under the Chinese Academy of Sciences, and helping establish the State Key Laboratory of Remote Sensing Science under the Chinese Ministry of Science and Technology. Those are just few examples of areas where Professor Li has made outstanding contributions. A comprehensive summary of his achievements has been provided by Liu et al. [37].

In memory of Professor Li, we organized the Third National Forum on Quantitative Remote Sensing at Beijing Normal University during 14–15 July 2017. There were 296 meeting participants from 65 research institutes and universities in China, and almost all aspects of quantitative land remote sensing were discussed.

The papers of this Special Issue are mainly from this forum. Although 40 articles cannot comprehensively characterize different aspects of quantitative land remote sensing in China, they clearly represent the current level of research in this area by Chinese scientists. These papers are related to various satellite data products, such as incident solar radiation [38–40], chlorophyll fluorescence [41], surface directional reflectance [42–44], aerosol optical depth [45], albedo [46,47], land surface temperature [48–50], upward longwave radiation [51], leaf area index [52–55], fractional vegetation cover [56], forest biomass [57], precipitation [58], evapotranspiration [59–61], freeze/thaw [62], snow cover [63], vegetation productivity [64–68], phenology [69,70], biodiversity indicators [71], drought monitoring [72], forest disturbance [55], air-quality monitoring [73], sensor design [74], and sampling strategy [75] for validation with in situ measurements. Most of these papers are based on optical-thermal remotely-sensed observations, but a few papers are also based on microwave [62,63] and Lidar [54,76] data.

Although these 40 papers do not represent a large sample, they demonstrate that few studies have been undertaken on physical modeling for understanding remotely-sensed signals and use of Chinese satellite data in their analysis. This latter shortcoming calls for the further improvement of Chinese satellite data quality.

Acknowledgments: This work was supported in part by the National Key Research and Development Program of China (No. 2016YFA0600101) and the National Natural Science Foundation of China (No. 41331173). We would like to thank the members of the Scientific Steering Committee and the Organizing Committee of the Third National Forum on Quantitative Remote Sensing at Beijing Normal University for their great contributions.

Conflicts of Interest: The authors declare no conflicts of interest

References

1. Swain, P.H.; Shirley, M.D. *Remote Sensing: The Quantitative Approach*; McGraw Hill: New York, NY, USA, 1978.
2. Liang, S. *Quantitative Remote Sensing of Land Surfaces*; John Wiley & Sons, Inc.: New York, NY, USA, 2004.
3. Liang, S. *Advances in Land Remote Sensing: System, Modeling, Inversion and Application*; Springer: New York, NY, USA, 2008.
4. Liang, S.; Li, X.; Wang, J. *Advanced Remote Sensing: Terrestrial Information Extraction and Applications*; Elsevier Science Bv: Amsterdam, The Netherlands, 2012.
5. Myneni, R.; Ross, J. *Photon-Vegetation Interactions: Applications in Optical Remote Sensing and Plant Ecology*; Springer: Berlin/Heidelberg, Germany, 1991.

6. Tang, H.; Li, Z.L. *Quantitative Remote Sensing in Thermal Infrared: Theory and Applications*; Springer: Berlin, Germany, 2014.

7. Liang, S.; Zhang, X.; Xiao, Z.; Cheng, J.; Liu, Q.; Zhao, X. *Global LAnd Surface Satellite (GLASS) Products: Algorithms, Validation and Analysis*; Springer: Berlin, Germany, 2013.

8. Liang, S.; Zhao, X.; Yuan, W.; Liu, S.; Cheng, X.; Xiao, Z.; Zhang, X.; Liu, Q.; Cheng, J.; Tang, H.; et al. A long-term global land surface satellite (GLASS) dataset for environmental studies. *Int. J. Digit. Earth* **2013**, *6*, 5–33. [CrossRef]

9. Xiao, Z.; Liang, S.; Jiang, B. Evaluation of four long time-series global leaf area index products. *Agric. For. Meteorol.* **2017**, *246*, 218–230. [CrossRef]

10. Xiao, Z.; Liang, S.; Sun, R. Evaluation of three long time series for global fraction of absorbed photosynthetically active radiation (fapar) products. *IEEE Trans. Geosci. Remote Sens.* **2018**, *56*, 5509–5524. [CrossRef]

11. Xu, B.; Li, J.; Park, T.; Liu, Q.; Zeng, Y.; Yin, G.; Zhao, J.; Fan, W.; Yang, L.; Knyazikhin, Y.; et al. An integrated method for validating long-term leaf area index products using global networks of site-based measurements. *Remote Sens. Environ.* **2018**, *209*, 134–151. [CrossRef]

12. Liang, S.; Tang, S.; Zhang, J.; Xu, B.; Cheng, J.; Cheng, X. Production of the global climate data records and applications to climate change studies. *J. Remote Sens.* **2016**, *20*, 1401–1499.

13. NRC. *Climate Data Records from Environmental Satellites: Interim Report*; The National Academies Press: Washington, DC, USA, 2004.

14. Xiao, Z.; Liang, S.; Wang, J.; Xiang, Y.; Zhao, X.; Song, J. Long-time-series global land surface satellite leaf area index product derived from MODIS and AVHRR surface reflectance. *IEEE Trans. Geosci. Remote Sens.* **2016**, *54*, 5301–5318. [CrossRef]

15. Xiao, Z.Q.; Liang, S.; Wang, J.D.; Chen, P.; Yin, X.J.; Zhang, L.Q.; Song, J.L. Use of general regression neural networks for generating the GLASS leaf area index product from time-series MODIS surface reflectance. *IEEE Trans. Geosci. Remote Sens.* **2014**, *52*, 209–223. [CrossRef]

16. Qu, Y.; Liu, Q.; Liang, S.; Wang, L.; Liu, N.; Liu, S. Improved direct-estimation algorithm for mapping daily land-surface broadband albedo from MODIS data. *IEEE Trans. Geosci. Remote Sens.* **2014**, *52*, 907–919. [CrossRef]

17. Liu, N.; Liu, Q.; Wang, L.; Liang, S.; Wen, J.; Qu, Y.; Liu, S.H. A statistics-based temporal filter algorithm to map spatiotemporally continuous shortwave albedo from MODIS data. *Hydrol. Earth Syst. Sci.* **2013**, *17*, 2121–2129. [CrossRef]

18. Liu, Q.; Wang, L.; Qu, Y.; Liu, N.; Liu, S.; Tang, H.; Liang, S. Primary evaluation of the long-term GLASS albedo product. *Int. J. Digit. Earth* **2013**, *6*, 69–95. [CrossRef]

19. Cheng, J.; Liang, S. A novel algorithm for estimating broadband emissivity of global bare soil using MODIS albedo product. *IEEE Trans. Geosci. Remote Sens.* **2013**, *51*, 2619–2631.

20. Cheng, J.; Liang, S.; Verhoef, W.; Shi, L.; Liu, Q. Estimating the hemispherical broadband longwave emissivity of global vegetated surfaces using a radiative transfer model. *IEEE Trans. Geosci. Remote Sens.* **2016**, *54*, 905–917. [CrossRef]

21. Xiao, Z.; Liang, S.; Sun, R.; Wang, J.; Jiang, B. Estimating the fraction of absorbed photosynthetically active radiation from the MODIS data based GLASS leaf area index product. *Remote Sens. Environ.* **2015**, *171*, 105–117. [CrossRef]

22. Zhang, X.; Liang, S.; Zhou, G.; Wu, H.; Zhao, X. Generating global land surface satellite incident shortwave radiation and photosynthetically active radiation products from multiple satellite data. *Remote Sens. Environ.* **2014**, *152*, 318–332. [CrossRef]

23. Cheng, J.; Liang, S. Global estimates for high-spatial-resolution clear-sky land surface upwelling longwave radiation from MODIS data. *IEEE Trans. Geosci. Remote Sens.* **2016**, *54*, 4115–4129. [CrossRef]

24. Cheng, J.; Liang, S.; Wang, W.; Guo, Y. An efficient hybrid method for estimating clear-sky surface downward longwave radiation from MODIS data. *J. Geophys. Res. Atmos.* **2017**, *122*, 2616–2630. [CrossRef]

25. Jiang, B.; Liang, S.; Ma, H.; Zhang, X.; Xiao, Z.; Zhao, X.; Jia, K.; Yao, Y.; Jia, A. GLASS daytime all-wave net radiation product: Algorithm development and preliminary validation. *Remote Sens.* **2016**, *8*, 222. [CrossRef]

26. Zhou, J.; Zhang, X.; Zhan, W.; Zhang, H. Land surface temperature retrieval from MODIS data by integrating regression models and the genetic algorithm in an arid region. *Remote Sens.* **2014**, *6*, 5344–5367. [CrossRef]

27. Jia, K.; Liang, S.; Liu, S.H.; Li, Y.W.; Xiao, Z.Q.; Yao, Y.J.; Jiang, B.; Zhao, X.; Wang, X.; Xu, S.; et al. Global land surface fractional vegetation cover estimation using general regression neural networks from MODIS surface reflectance. *IEEE Trans. Geosci. Remote Sens.* **2015**, *53*, 4787–4796. [CrossRef]

28. Yao, Y.; Liang, S.; Li, X.; Hong, Y.; Fisher, J.B.; Zhang, N.; Chen, J.; Cheng, J.; Zhao, S.; Zhang, X.; et al. Bayesian multimodel estimation of global terrestrial latent heat flux from eddy covariance, meteorological, and satellite observations. *J. Geophys. Res. Atmos.* **2014**, *119*, 4521–4545. [CrossRef]

29. Yuan, W.P.; Liu, S.; Zhou, G.S.; Zhou, G.Y.; Tieszen, L.L.; Baldocchi, D.; Bernhofer, C.; Gholz, H.; Goulden, M.L.; Hollinger, D.Y.; et al. Deriving a light use efficiency model from eddy covariance flux data for predicting daily gross primary production across biomes. *Agric. For. Meteorol.* **2007**, *143*, 189–207. [CrossRef]

30. Li, X.; Strahler, A. Geometric-optical modeling of a coniferous forest canopy. *IEEE Trans. Geosci. Remote Sens.* **1985**, *23*, 705–721. [CrossRef]

31. Li, X.; Strahler, A. Geometric-optical bi-directional reflectance modeling of a coniferous forest canopy. *IEEE Trans. Geosci. Remote Sens.* **1986**, *24*, 906–919. [CrossRef]

32. Li, X.; Strahler, A.H. Geometric-optical bidirectional reflectance modeling of the discrete crown vegetation canopy: Effect of crown shape and mutual shadowing. *IEEE Trans. Geosci. Remote Sens.* **1992**, *30*, 276–292. [CrossRef]

33. Li, X.; Strahler, A.H.; Woodcock, C.E. A hybrid geometric optical-radiative transfer approach for modeling albedo and directional reflectance of discontinuous canopies. *IEEE Trans. Geosci. Remote Sens.* **1995**, *33*, 466–480.

34. Wanner, W.; Li, X.; Strahler, A. On the derivation of kernels for kernel-driven models of bidirectional reflectance. *J. Geophys. Res. Atmos.* **1995**, *100*, 21077–21089. [CrossRef]

35. Li, X.; Strahler, A.H.; Friedl, M.A. A conceptual model for effective directional emissivity from nonisothermal surfaces. *IEEE Trans. Geosci. Remote Sens.* **1999**, *37*, 2508–2517.

36. Li, X.; Gao, F.; Wang, J.; Strahler, A. A priori knowledge accumulation and its application to linear BRDF model inversion. *J. Geophys. Res. Atmos.* **2001**, *106*, 11925–11935. [CrossRef]

37. Liu, Q.; Yan, G.; Jiao, Z.; Wen, J.; Liang, S.; Wang, J. From geometric-optical optical remote sensing modeling to quantitative remote sensing science—In memory of Academician Xiaowen Li. *Remote Sens.* **2018**. new submit.

38. Yang, L.; Zhang, X.; Liang, S.; Yao, Y.; Jia, K.; Jia, A. Estimating surface downward shortwave radiation over china based on the gradient boosting decision tree method. *Remote Sens.* **2018**, *10*, 185. [CrossRef]

39. Zhang, H.; Huang, C.; Yu, S.; Li, L.; Xin, X.; Liu, Q. A lookup-table-based approach to estimating surface solar irradiance from geostationary and polar-orbiting satellite data. *Remote Sens.* **2018**, *10*, 411. [CrossRef]

40. Zhou, Y.; Yan, G.; Zhao, J.; Chu, Q.; Liu, Y.; Yan, K.; Tong, Y.; Mu, X.; Xie, D.; Zhang, W. Estimation of daily average downward shortwave radiation over antarctica. *Remote Sens.* **2018**, *10*, 422. [CrossRef]

41. Hu, J.; Liu, X.; Liu, L.; Guan, L. Evaluating the performance of the SCOPE model in simulating canopy solar-induced chlorophyll fluorescence. *Remote Sens.* **2018**, *10*, 250. [CrossRef]

42. Wu, Q.; Song, C.; Song, J.; Wang, J.; Chen, S.; Yu, B. Impacts of leaf age on canopy spectral signature variation in evergreen Chinese fir forests. *Remote Sens.* **2018**, *10*, 262. [CrossRef]

43. Wen, J.; Liu, Q.; Xiao, Q.; Liu, Q.; You, D.; Hao, D.; Wu, S.; Lin, X. Characterizing land surface anisotropic reflectance over rugged terrain: A review of concepts and recent developments. *Remote Sens.* **2018**, *10*, 370. [CrossRef]

44. Zhang, C.; Ren, H.; Liang, Y.; Liu, S.; Qin, Q.; Ersoy, O. Advancing the prospect-5 model to simulate the spectral reflectance of copper-stressed leaves. *Remote Sens.* **2017**, *9*, 1191. [CrossRef]

45. Tian, X.; Liu, S.; Sun, L.; Liu, Q. Retrieval of aerosol optical depth in the arid or semiarid region of Northern Xinjiang, China. *Remote Sens.* **2018**, *10*, 197. [CrossRef]

46. Lin, X.; Wen, J.; Liu, Q.; Xiao, Q.; You, D.; Wu, S.; Hao, D.; Wu, X. A multi-scale validation strategy for albedo products over rugged terrain and preliminary application in Heihe River Basin, China. *Remote Sens.* **2018**, *10*, 156. [CrossRef]

47. Hao, D.; Wen, J.; Xiao, Q.; Wu, S.; Lin, X.; Dou, B.; You, D.; Tang, Y. Simulation and analysis of the topographic effects on snow-free albedo over rugged terrain. *Remote Sens.* **2018**, *10*, 278. [CrossRef]

48. Tang, B.; Zhao, X.; Zhao, W. Local effects of forests on temperatures across Europe. *Remote Sens.* **2018**, *10*, 529. [CrossRef]

49. Yu, W.; Ma, M.; Li, Z.; Tan, J.; Wu, A. New scheme for validating remote-sensing land surface temperature products with station observations. *Remote Sens.* **2017**, *9*, 1210. [CrossRef]
50. Meng, X.; Cheng, J.; Liang, S. Estimating land surface temperature from feng yun-3C/MERSI data using a new land surface emissivity scheme. *Remote Sens.* **2017**, *9*, 1247. [CrossRef]
51. Zhou, S.; Cheng, J. Estimation of high spatial-resolution clear-sky land surface-upwelling longwave radiation from VIIRS/S-NPP data. *Remote Sens.* **2018**, *10*, 253. [CrossRef]
52. Zhao, J.; Li, J.; Liu, Q.; Wang, H.; Chen, C.; Xu, B.; Wu, S. Comparative analysis of Chinese HJ-1 CCD, GF-1 WFV and ZY-3 MUX sensor data for leaf area index estimations for maize. *Remote Sens.* **2018**, *10*, 68. [CrossRef]
53. Zhou, J.; Zhang, S.; Yang, H.; Xiao, Z.; Gao, F. The retrieval of 30-m resolution LAI from landsat data by combining MODIS products. *Remote Sens.* **2018**, *10*, 1187. [CrossRef]
54. Li, S.; Dai, L.; Wang, H.; Wang, Y.; He, Z.; Lin, S. Estimating leaf area density of individual trees using the point cloud segmentation of terrestrial LiDAR data and a voxel-based model. *Remote Sens.* **2017**, *9*, 1202. [CrossRef]
55. Wang, J.; Wang, J.; Zhou, H.; Xiao, Z. Detecting forest disturbance in northeast China from GLASS LAI time series data using a dynamic model. *Remote Sens.* **2017**, *9*, 1293. [CrossRef]
56. Yang, L.; Jia, K.; Liang, S.; Liu, M.; Wei, X.; Yao, Y.; Zhang, X.; Liu, D. Spatio-temporal analysis and uncertainty of fractional vegetation cover change over northern China during 2001–2012 based on multiple vegetation data sets. *Remote Sens.* **2018**, *10*, 549. [CrossRef]
57. Wang, M.; Sun, R.; Xiao, Z. Estimation of forest canopy height and aboveground biomass from spaceborne LiDAR and landsat imageries in Maryland. *Remote Sens.* **2018**, *10*, 344. [CrossRef]
58. Zeng, Q.; Wang, Y.; Chen, L.; Wang, Z.; Zhu, H.; Li, B. Inter-comparison and evaluation of remote sensing precipitation products over China from 2005 to 2013. *Remote Sens.* **2018**, *10*, 168. [CrossRef]
59. Li, X.; Xin, X.; Peng, Z.; Zhang, H.; Yi, C.; Li, B. Analysis of the spatial variability of land surface variables for ET estimation: Case study in HiWATER Campaign. *Remote Sens.* **2018**, *10*, 91. [CrossRef]
60. Zhang, L.; Yao, Y.; Wang, Z.; Jia, K.; Zhang, X.; Zhang, Y.; Wang, X.; Xu, J.; Chen, X. Satellite-derived spatiotemporal variations in evapotranspiration over northeast China during 1982–2010. *Remote Sens.* **2017**, *9*, 1140. [CrossRef]
61. Wang, X.; Yao, Y.; Zhao, S.; Jia, K.; Zhang, X.; Zhang, Y.; Zhang, L.; Xu, J.; Chen, X. MODIS-based estimation of terrestrial latent heat flux over north america using three machine learning algorithms. *Remote Sens.* **2017**, *9*, 1326. [CrossRef]
62. Hu, T.; Zhao, T.; Shi, J.; Wu, S.; Liu, D.; Qin, H.; Zhao, K. High-resolution mapping of freeze/thaw status in china via fusion of MODIS and AMSR2 data. *Remote Sens.* **2017**, *9*, 1339. [CrossRef]
63. Liu, X.; Jiang, L.; Wu, S.; Hao, S.; Wang, G.; Yang, J. Assessment of methods for passive microwave snow cover mapping using FY-3C/MWRI data in China. *Remote Sens.* **2018**, *10*, 524. [CrossRef]
64. Yu, T.; Sun, R.; Xiao, Z.; Zhang, Q.; Liu, G.; Cui, T.; Wang, J. Estimation of global vegetation productivity from global land surface satellite data. *Remote Sens.* **2018**, *10*, 327. [CrossRef]
65. Hu, L.; Fan, W.; Ren, H.; Liu, S.; Cui, Y.; Zhao, P. Spatiotemporal dynamics in vegetation GPP over the great khingan mountains using GLASS products from 1982 to 2015. *Remote Sens.* **2018**, *10*, 488. [CrossRef]
66. Xie, X.; Li, A.; Jin, H.; Yin, G.; Bian, J. Spatial downscaling of gross primary productivity using topographic and vegetation heterogeneity information: A case study in the Gongga Mountain Region of China. *Remote Sens.* **2018**, *10*, 647. [CrossRef]
67. Lin, S.; Li, J.; Liu, Q.; Huete, A.; Li, L. Effects of forest canopy vertical stratification on the estimation of gross primary production by remote sensing. *Remote Sens.* **2018**, *10*, 1329. [CrossRef]
68. Cui, T.; Sun, R.; Qiao, C.; Zhang, Q.; Yu, T.; Liu, G.; Liu, Z. Estimating diurnal courses of gross primary production for maize: A comparison of sun-induced chlorophyll fluorescence, light-use efficiency and process-based models. *Remote Sens.* **2017**, *9*, 1267. [CrossRef]
69. He, Z.; Li, S.; Wang, Y.; Dai, L.; Lin, S. Monitoring rice phenology based on backscattering characteristics of multi-temporal RADARSAT-2 datasets. *Remote Sens.* **2018**, *10*, 340. [CrossRef]
70. Zheng, Z.; Zhu, W. Uncertainty of remote sensing data in monitoring vegetation phenology: A comparison of MODIS C5 and C6 vegetation index products on the Tibetan Plateau. *Remote Sens.* **2017**, *9*, 1288. [CrossRef]
71. Wu, J.; Liang, S. Developing an integrated remote sensing based biodiversity index for predicting animal species richness. *Remote Sens.* **2018**, *10*, 739. [CrossRef]

72. Xia, L.; Zhao, F.; Mao, K.; Yuan, Z.; Zuo, Z.; Xu, T. SPI-based analyses of drought changes over the past 60 years in China's major crop-growing areas. *Remote Sens.* **2018**, *10*, 171. [CrossRef]
73. Yun, G.; Zuo, S.; Dai, S.; Song, X.; Xu, C.; Liao, Y.; Zhao, P.; Chang, W.; Chen, Q.; Li, Y.; et al. Individual and interactive influences of anthropogenic and ecological factors on forest PM2.5 concentrations at an urban scale. *Remote Sens.* **2018**, *10*, 521. [CrossRef]
74. Zhou, H.; Wang, J.; Liang, S. Design of a novel spectral albedometer for validating the MODerate resolution imaging spectroradiometer spectral albedo product. *Remote Sens.* **2018**, *10*, 101. [CrossRef]
75. Yin, G.; Li, A.; Verger, A. Spatiotemporally representative and cost-efficient sampling design for validation activities in wanglang experimental site. *Remote Sens.* **2017**, *9*, 1217. [CrossRef]
76. Xie, D.; Wang, X.; Qi, J.; Chen, Y.; Mu, X.; Zhang, W.; Yan, G. Reconstruction of Single Tree with Leaves Based on Terrestrial LiDAR Point Cloud Data. *Remote Sens.* **2018**, *10*, 686. [CrossRef]

remote sensing

MDPI

Article

Evaluating the Performance of the SCOPE Model in Simulating Canopy Solar-Induced Chlorophyll Fluorescence

Jiaochan Hu [1,2], Xinjie Liu [1], Liangyun Liu [1,*] and Linlin Guan [1]

[1] Key Laboratory of Digital Earth Science, Institute of Remote Sensing and Digital Earth, Chinese Academy of Sciences, Beijing 100094, China; hujc@radi.ac.cn (J.H.); liuxj@radi.ac.cn (X.L.); guanll@radi.ac.cn (L.G.)

[2] College of Resources and Environment, University of Chinese Academy of Sciences, Beijing 100049, China

* Correspondence: liuly@radi.ac.cn; Tel.: +86-10-8217-8163

Received: 15 December 2017; Accepted: 4 February 2018; Published: 6 February 2018

Abstract: The SCOPE (soil canopy observation of photochemistry and energy fluxes) model has been widely used to interpret solar-induced chlorophyll fluorescence (SIF) and investigate the SIF-photosynthesis links at different temporal and spatial scales in recent years. In the SCOPE model, the fluorescence quantum efficiency in dark-adapted conditions (FQE) for Photosystem II (fqe2) and Photosystem I (fqe1) were two key parameters of SIF emission, which have always been parameterized as fixed values derived from laboratory measurements. To date, only a few studies have focused on evaluating the SCOPE model for SIF interpretation, and the variation of FQE values in the field remains controversial. In this study, the accuracy of the SCOPE model to simulate the canopy SIF was investigated using diurnal experiments on winter wheat. First, ten diurnal experiments were conducted on winter wheat, and the canopy SIF emissions and the SCOPE model's input parameters were directly measured or indirectly retrieved from the spectral radiances, gross primary productivity (GPP) data, and meteorological records. Second, the SCOPE-simulated SIF emissions with fixed FQE values were evaluated using the observed canopy SIF data. The results show that the SCOPE model can reliably interpret the diurnal cycles of SIF variation and provide acceptable results of SIF simulations at the O_2-B (SIF_B) and O_2-A (SIF_A) bands with RRMSEs of 24.35% and 23.67%, respectively. However, the SCOPE-simulated SIF_B and SIF_A still contained large systematical deviations at some growth stages of wheat, and the seasonal cycles of the ratio between SIF_B and SIF_A (SIF_A/SIF_B) cannot be credibly reproduced. Finally, the SCOPE-simulated SIF emissions with variable FQE values were evaluated using the observed canopy SIF data. The simulating accuracy of SIF_B and SIF_A can be improved greatly using variable FQE values, and the SCOPE simulations track well with the seasonal SIF_A/SIF_B values with an RRMSE of 20.63%. The results indicated a clear seasonal pattern of FQE values for unbiased SIF simulation: from the erecting to the flowering stage of wheat, the ratio of fqe1 to fqe2 (fqe1/fqe2) gradually increased from 0.05–0.1 to 0.3–0.5, while the fqe2 value decreased from 0.013 to 0.007. Our quantitative results of the model assessment and the FQE adjustment support the use of the SCOPE model as a powerful tool for interpreting the SIF emissions and can serve as a significant reference for future applications of the SCOPE model.

Keywords: solar-induced chlorophyll fluorescence; fluorescence quantum efficiency in dark-adapted conditions (FQE); SCOPE; Fraunhofer Line Discrimination (FLD); gross primary productivity (GPP)

---

1. Introduction

Solar-induced chlorophyll fluorescence (SIF) refers to the emission of red and far-red light from chlorophyll during the absorption of photosynthetically active radiation under natural sunlight.

The SIF spectrum is a continuous broadband spectrum that covers the approximately spectral range of 650–850 nm. Its spectral shape is characterized by one peak at around 685 nm and another at around 740 nm [1,2]. The emitted SIF is the sum of the chlorophyll fluorescence of photosystem I (PSI) and photosystem II (PSII). PSII contributes to the SIF emission in both the red and far-red spectral regions, whereas PSI contributes to the SIF emission only in the far-red region [3]. As a result, the intensity and shape of the SIF spectrum can reflect the amount of energy absorbed by PSII and PSI [4,5]. In addition, several studies have determined the physics-physiology mechanism connecting function of the photosynthetic apparatus with chlorophyll fluorescence from active florescence induction measurement and demonstrated that the fluorescence signal can be a reliable and observable indicator of the plant's photosynthetic status [4–6]. To date, extensive SIF-photosynthesis research has focused on investigating the empirical correlations between SIF and GPP, and demonstrated that SIF measurements can offer a promising approach for detecting the terrestrial vegetation's actual photosynthetic activity [7–11]. Since Plascyk introduced the Fraunhofer Line Discrimination (FLD) method [12] to extract SIF signals from the observed vegetation-reflected radiance, various studies have demonstrated the possibility of measuring SIF at Fraunhofer lines or atmospheric absorption bands (e.g., an O_2-B band at approximately 687 nm and an O_2-A band at approximately 760 nm) on the ground, from airborne platforms, and from satellites (for review, see [13]). Recently, global SIF maps derived from hyperspectral satellite data have become available [14–18]. Meanwhile, SIF's application for global monitoring of plant photosynthesis has become a hot research area [9,14,19].

The soil canopy observation, photochemistry, and energy fluxes (SCOPE) model [20] has become a virtual laboratory for interpreting SIF and investigating SIF-photosynthesis links on diurnal or seasonal scales. With a full physiological representation of photosynthesis and fluorescence, SCOPE has been regarded as a robust deterministic model for interpreting SIF and photosynthesis in various studies. For example, it has been used to provide training and test data sets for new SIF retrieval methods [21,22]; to derive empirical relationships between the seasonal maximum carboxylation rate (V_{cmax}) and SIF, which are used to retrieve the global photosynthetic capacity of crops [19,23]; to evaluate the predictive power of SIF to estimate gross primary productivity (GPP); to investigate the sensitivities of both GPP and SIF and their relationship to the biochemical parameters, as well as to the environmental conditions at different spatial-temporal scales [24–27]; and to assess the influence of confounding factors such as physiological and structural interferences or temporal scaling effects on SIF-GPP relationships [8].

Despite the SCOPE model-integrated existing theories of radiative transfer, energy balance, micrometeorology, and plant physiology, the model is analytical. Thus, it inevitably contains assumptions due to model abstractions for SIF representations and uncertainties in driving variables [28]. To date, only a few studies have involved the experimental validation of the SCOPE model for SIF interpretation. Verrelst et al. provided insight into the key variables that drive the reflectance and SIF emission simulations, based on a global sensitivity analysis (GSA) of the SCOPE model [29]. The results showed that leaf composition, leaf area index, leaf inclination, irradiance, and V_{cmax} are the most important factors affecting the SIF simulation and need to be accurately parameterized to produce unbiased SIF interpretations. By comparing the simulated SIF with corresponding field observations, Van der Tol et al. assessed the impacts of leaf pigment concentrations and canopy structures on simulated SIF, as well as the impacts of PQ and NPQ [28]. However, they focused on revealing information about the biochemical regulation of the energy pathways contained in the SIF signal, without offering the quantitative accuracy of SIF simulation, and only the simulations of far-red SIF at O_2-A band (SIF_A) have been investigated. Several studies have reported that the red SIF at the O_2-B band (SIF_B) is more closely connected to plant photosynthesis, possibly because SIF_B is located near the fluorescence peak emitted by PSII [3,30,31]. Additionally, the ratio of SIF_A to SIF_B can express SIF's spectral shape, which can provide important information regarding physiological and biochemical activities in vegetation [5,6,32]. Therefore, both the intensity and shape of the SCOPE-simulated SIF spectra must be evaluated quantitatively.

On the other hand, the fluorescence quantum efficiency in dark-adapted conditions (defined as F_0-level fluorescence yield) for PSII (fqe2) and PSI (fqe1) was always set as fixed values derived from laboratory measurements, which may be unsuitable for the accurate simulation of SIF. According to Van der Tol et al. [28], the suggested values of fqe2 and the ratio of fqe1 to fqe2 (fqe1/fqe2) for SCOPE were 0.01 and 0.2, respectively. While, for early versions of the SCOPE model (before version 1.53), the fqe2 value was suggested to be 0.01 by the model developer. This priori value can be obtained from the work by Genty et al., [33], but it is unknown whether the value is universal [20]. The measured fluorescence yields values at F_0-level have been reported as around 0.02 [34–36]. In the study by Trissl et al. [37], three different levels (0.01, 0.021, and 0.018) of the fluorescence yields at F_0-level were considered, and the contribution from PSI to the total fluorescence signal is reported as around 20%. As reported by Björkman and Barbara [38], the FQE values have been observed to change with different vegetation species, chlorophyll contents, and exposures of leave surfaces to the sun. Besides, all the laboratory-measured fqe2 values derived from active fluorescence measurements may have some uncertainties due to the contamination of PSI fluorescence and the PSII closure caused by measuring flashes during the measurements of PSII fluorescence at F_0 level under dark-adapted conditions [35–37,39]. Therefore, there are still a lot of uncertainties in the estimation of FQE and its variation remains controversial. In this context, two issues arise: (i) the accuracy of SCOPE-simulated SIF emissions with fixed FQE values needs to be evaluated using the observed SIF data and (ii) suitable FQE values should be determined using field experiment observations, if SIF simulated with fixed FQE was not sufficiently accurate.

Therefore, in this paper, we focused on two objectives: (i) quantitatively evaluating SCOPE's performance for modeling both SIF intensities at O_2-B and O_2-A bands, and the ratio between them (SIF_A/SIF_B); and (ii) determining the FQE values using observations of ten field experiments on winter wheat. After the input parameters of the SCOPE model were directly measured or indirectly retrieved with high accuracy, the model was implemented to simulate diurnal SIF emissions compared with the observations across wheat's growing season in 2015 and 2016. This paper is outlined as follows. Section 2 describes the experimental data sets, the parameter inversion methods, and the SIF simulation process. Section 3 shows the results of parameter retrieving and model evaluation with fixed and variable FQE values. Section 4 discusses the uncertainties and prospects of this study. Finally, the most important conclusions are given in Section 5.

2. Materials and Methods

2.1. SCOPE (Soil Canopy Observation of Photochemistry and Energy Fluxes)

2.1.1. SCOPE Model Description

SCOPE is a vertical (1-D), integrated, radiative transfer and energy balance model [19]. This model combines radiative transfer of solar radiation and radiation emitted by the vegetation (thermal and SIF) with the energy balance in which a biochemical module handles the fluorescence emission efficiency depending on the two de-excitation pathways: photochemical quenching of excitation energy via electron transport (PQ) and non-photochemical quenching of excitation energy via thermal energy dissipation (NPQ) [40]. It calculates directional top-of-canopy reflected radiation, emitted thermal radiation, and SIF signals together with energy, water, and CO_2 flux. In this work, we employed version 1.61 to interpret SIF and GPP. The model consists of several modules combined to simulate SIF and photosynthesis. The model's main features related to the SIF and GPP simulations are briefly described here (for more details, see [20]).

At the leaf level, two modules are used to simulate the SIF emission. One is the leaf radiative transfer module called Fluspect that handles the radiative transfer of incident light and SIF emission in the leaf. The other is the biochemical module that handles the emission efficiencies of photosystems depending on the PQ and NPQ at photosystem level. At the canopy level, the optical radiative transfer module (RTMo) governs the incident light on the individual leaves and the propagation of

SIF throughout the canopy based on the scattering of arbitrarily inclined leaves (SAIL) model [41]. It calculates radiation transfer in a multilayer canopy to obtain reflectance and fluorescence in the observation direction as a function of solar zenith angel and leaf inclination distribution. The spectral resolution of the modeled spectra is 1 nm in the range of 400–2500 nm for reflectance and 640–850 nm for fluorescence.

Fluspect is an extension of the PROSPECT model [42] that adds SIF radiative transfer within the leaf. Fluspect calculates the probability that excitation at a specific wavelength (400–750 nm) results in fluorescence at a longer wavelength (640–850 nm) at the illuminated and the shaded sides of the leaf. Furthermore, when implementing Fluspect, two photosystems (PSI and PSII) are responsible for fluorescence. As a result, Fluspect's output consists of leaf reflectance and transmittance, as well as four fluorescence excitation-emission probability matrices: one for each photosystem at the illuminated and shaded sides of the leaf [20,24,28]. Fluspect's input parameters consist of all the leaf composition parameters as described with the PROSPECT parameters and the fqe1 and fqe2.

The biochemical module is employed to scale the SIF emission efficiencies of PSII (i.e., PQ and NPQ) as a function of micrometeorological conditions (e.g., irradiance, temperature, relative humidity, and wind speed) and photosynthesis parameters (e.g., the V_{cmax} and the Ball-Berry stomatal conductance parameter m) relative to the efficiency in dark or pre-dawn conditions. For representation of photosynthesis (i.e., PQ), either the models proposed by Farquhar et al. (for C3 species) [43] and Von Caemmerer (for C4 species) [44] or the model presented by Collatz et al. [45,46] are/is adopted. In these photosynthesis models, V_{cmax} is an important biochemical variable for carbon assimilation, which describes the maximum carboxylation rate of RuBisCO. It is assumed to decrease exponentially with the depth in the canopy and is calibrated by the temperature correction parameters. When implementing the biochemical module adopted in [25], the response of SIF emission efficiency is empirically calibrated to a number of datasets collected in field and laboratory experiments of unstressed and drought-stressed vegetation, referred to hereafter as TB12 and TB12-D, respectively. The MD12 module [47] has a more explicit parameterization of fluorescence quenching mechanisms. Instead of the empirical calibration in the TB12 and TB12-D, this module can reproduce intermediate conditions using two additional variables: the rate constant of sustained thermal dissipation (kNPQs) and the fraction of functional reaction centers (qLs) [48].

The SCOPE model also simulates a diversity of fluxes, one of which is net photosynthesis of canopy (NPC). NPC represents the total gross photosynthesis less the flux of CO_2 associated with foliage respiration. Since photosynthesis is the exchange CO_2 flux between leaf and atmosphere, it is calculated by simply gathering the photosynthesis over the leaf region of the canopy in the SCOPE model [24]. Therefore, NPC from the SCOPE model can be used to compute GPP for approximate comparisons with the GPP observations over canopies by setting the respiration parameter to zero.

2.1.2. SCOPE Model Inputs

To simulate photosynthesis and fluorescence, the SCOPE model requires inputs related to meteorology, leaf optical properties and canopy structure, leaf biochemistry, and illumination/observation geometry (see Table 1 for details). These input parameters were derived from three sources: the field measurements, the related literatures, and the model inversion.

Table 1. Values or sources of the main input parameters of the SCOPE model used in our simulation

Parameters	Definition	Unit	Value/Source
Leaf biochemistry			
V_{cmo}	Maximum carboxylation capacity at 25 °C	$\mu mol\ m^{-2}\ s^{-1}$	Inversion
m	Ball-berry stomatal conductance parameter	—	9
Rdparam	Parameter for dark respiration (Rd = Rdparam × V_{cmo})	—	0
Leaf optical			
Cab	Chlorophyll content density	$\mu g/cm^2$	Measurement

Table 1. *Cont.*

Parameters	Definition	Unit	Value/Source
Cw	Leaf equivalent water thickness	cm	Measurement
Cdm	Dry matter content	g/cm^2	Measurement
N	Leaf thickness parameters	—	1.4
Canopy			
LAI	Leaf area index	m^2/m^2	Measurement
LIDFa	LIDF parameter a, which controls the average leaf scope	—	Inversion
LIDFb	LIDF parameter b, which controls the distribution's bimodality	—	−0.15
Fluorescence			
fqe2	Fluorescence efficiency for PSII in dark-adapted condition	—	0.01 or adjusted
fqe1/fqe2	Ratio of fqe1 to fqe2	—	0.2 or adjusted
Meteorology			
R_{in}	Broadband incoming shortwave radiation (0.4–2.5 μm)	W/m^2	Measurement
Ta	Air temperature	T	Measurement
p	Air pressure	hPa	Measurement
ea	Atmospheric vapor pressure	hPa	Measurement
u	Wind speed at measurement height	m/s	Measurement
Ca	Atmospheric CO_2 concentration	ppm	Measurement
Geometry			
LAT	Latitude	degree	Measurement
LON	Longitude	degree	Measurement
VZA	Observation zenith angle	degree	0

Most of the main input parameters needed in the SCOPE model were considered either known as their literature values or directly measured from the field experiments with sufficient confidence. A majority of leaf optical and canopy structural parameters—including Cab, Cw, Cdm, and LAI—were accurately measured from our field experiments. In addition, the meteorological variables were derived from our high-accuracy meteorological observations using the automatic weather station (AWS). These diurnal meteorological variables were imported into the model input files and loaded for the time series simulations with SCOPE. Meanwhile, the diurnal solar zenith angles were automatically calculated during the simulation using the inputs Julian day, time, and field site longitude and latitude. Thus, the SCOPE's parameterization can keep pace with the observed canopy's vegetation growth and environmental variation at both diurnal and seasonal time scales. On the other hand, some parameters' values were determined based on the related literatures. Following [49], the Ball-berry stomatal conductance parameter (m) should be set around 9 for well-watered C3 species, and the dark respiration parameter (Rdparam) value was set to zero regarding the output NPC as GPP. The LIDFa and LIDFb were two canopy structural parameters that determine the leaf inclination distribution function defined in [50]. LIDFa controls the average leaf scope, and LIDFb controls the distribution's bimodality. According to [29], LIDFa can largely affect both the simulated reflectance and fluorescence, and LIDFb has only a marginal impact on the simulated reflectance and fluorescence. Therefore, we consider only the variations in LIDFa in this study, and LIDFb was set to its default value of −0.15.

The V_{cmax} at 25 °C (denoted as 'V_{cmo}' in the SCOPE model) and LIDFa were two key variables driving the SIF simulation and were accurately retrieved from the in situ observations. According to the global sensitivity analysis of the SCOPE model in [29], for the TB12 module used in this work, the canopy-leaving SIF variability was determined mainly by four driving vegetation variables: Cab, LIDFa, LAI, and V_{cmo}. These key inputs need to be reliably confirmed to accurately interpret canopy SIF and photosynthesis. Cab and LAI were easily measured, while field measurements of leaf angle distribution and V_{cmo} consume lots of time and effort. Therefore, LIDFa and V_{cmo} were estimated using the model inversion method with measured reflectance spectra and GPP data.

The fqe2 and fqe1 were two key parameters that determined the simulated SIF intensity. They were two multiplicative factors added to the probability matrices of PSII and PSI fluorescence. Thus, in the

SCOPE model, the fqe2 and fqe1 values proportionally impacted the intensity of the PSII and PSI fluorescence spectra. The literature suggested fqe2 and fqe1/fqe2 values are approximately 0.01 and 0.2, respectively, which were determined from the active fluorescence measurements with PAM in the laboratory [25]. However, whether the fixed FQE values are suitable for vegetation in the field at different growth stages has not yet been sufficiently validated. In this study, we inspected the accuracy of SCOPE-simulated SIF with both literature-fixed FQE values and with variable FQE values. The variable FQE values were obtained by fitting the SIF simulations to the observed SIF data (as described in Section 2.4). Other parameters required by the SCOPE model were set to their default values (Table 1).

2.2. In Situ Measurements

To evaluate the SCOPE model performance at both diurnal and seasonal time scales, ten diurnal in situ experiments were conducted on winter wheat (*Triticum aestivum* L.) during the 2015~2016 vegetation growing season to measure the vegetation parameters, the diurnal flux and meteorological variables, and the canopy spectra (details listed in Table 2). Our selected field site was located in an open and flat area at the National Precision Agriculture Demonstration Base in the town of Xiaotangshan, Beijing, China (40.17°N, 116.39°E). Conventional fertilizer and irrigation management were used on the winter wheat in the sample plot, which had a uniform growth status.

2.2.1. Measurements of Vegetation Parameters

A destructive sampling method was used to measure the leaf optical and canopy structural parameters, including Cab, Cw, Cdm, and LAI. Near the spot of spectral and flux measurements, twenty tillers above the ground within a 1 m × 1 m sample area were cut and immediately sent to the laboratory. Meanwhile, the density of tillers in the sample area was investigated. The leaves of ten tillers were weighted and scanned with a Li-Cor 3100 area meter to calculate the LAI [51]. Several leaves of the other ten tillers were cut into pieces and uniformly mixed, and approximately 0.2 g of them was randomly picked to measure Cab using spectrophotometry [52]. All of these two-part leaves were over-dried at 60 °C until a constant weight was reached. The Cw and Cdm were then calculated using the measured fresh weight, dry weight, and leaf area.

The measured results for ten fieldwork days are listed in Table 2, along with the corresponding growth stages. The vegetation samples cover different growth stages, with various optical and structural parameters, which are suitable data sets for model validation. The measured LAI values aligned with their realistic patterns across the growing season, which to some degree verifies the measuring accuracy. With the growing of wheat, the LAI continually increased from the erecting to the booting stage. Meanwhile, there was an obvious decrease with the arrival of flowering.

Table 2. The growth stages and vegetation parameters of winter wheat at the time of ten diurnal experiments in 2015 and 2016.

	2015			2016		
	April 3	**April 13 & 14**	**April 24 & 25**	**April 8 & 9**	**April 18**	**May 3 & 4**
Growth stage	Erecting	Jointing	Booting	Erecting	Jointing	Flowering
LAI	1.5	2.1	2.4	2.5	2.9	1.9
Cab (μg/cm^2)	59.2	62.2	61.3	55.3	53.7	57.3
Cw (cm)	0.0138	0.0126	0.0158	0.0163	0.0199	0.0177
Cdm (g/cm^2)	0.0042	0.0040	0.0045	0.0048	0.0049	0.0043

2.2.2. Diurnal Flux and Meteorological Observations

The flux and meteorological variables were observed using an eddy covariance (EC) system and the AWS. The AWS was fixed on a stand at the center of our selected field site to collect the meteorological variables, including photosynthetically active radiation (PAR, μmol m^{-2} s^{-1}), air humidity (rH, %), vapor pressure deficit (VPD, hpa), soil temperature (T_{soil}, °C), and other input parameters of SCOPE (including R_{in}, Ta, p, ea, and u, as listed in Table 2) every 10 s. The AWS output was recorded at 10 min intervals using a CR1000 unit (Campbell Scientific Inc., Logan, UT, USA). Near the AWS, an EC system was installed on a stand to measure the exchange of energy, water vapor, and CO_2 across the canopy-atmosphere interface. The EC system included a 3D sonic anemometer (CSAT3, Campbell Scientific Inc., Logan, UT, USA) for measuring three-dimensional velocity and temperature and an open-path infrared gas analyzer (Li-7500, Li-Cor, Lincoln, NE, USA) that measured CO_2 and H_2O density. The sensors were installed at a height of 2.5 m above the ground. The main output parameters include the net ecosystem exchange of CO_2 flux (NEE, $mg/m^2/s$), latent heat flux (LE, W/m^2), sensible heat flux (H, W/m^2), friction velocity (u*, m/s), and the atmospheric CO_2 concentration for model input (i.e., Ca in Table 2). The data were stored in a CR3000 data logger (Campbell Scientific Inc., Logan, UT, USA) and processed with an average time of 30 min at a sampling frequency of 10 Hz.

With the obtained half-hour NEE data and the corresponding meteorological variables (including R_{in}, Ta, rH, LE, H, u*, and T_{soil}) as inputs, half-hour GPP data could be calculated using the online tool available at the Max Planck Institute for Biogeochemistry (MPI-BGC) website (http://www.bgc-jena.mpg.de/~MDIwork/eddyproc/). First, u* filtering was conducted to calculate the u* threshold for identifying conditions with insufficient turbulence and marking those conditions as data gaps to avoid biases in fluxes measured using eddy covariance. Subsequently, gap filling was carried out to fill the gaps in half-hourly eddy covariance data. The gap filling of the eddy covariance and meteorological data were performed with methods similar to [53] while also considering flux co-variation with meteorological variables and flux temporal auto-correlation [54]. Finally, flux partitioning was implemented for partitioning NEE into ecosystem respiration and GPP. Based on the night-time partitioning algorithm [54], respiration is estimated from the night-time and extrapolated to the daytime.

Figure 1 exhibits the measured date sets about half-hour Ta, VPD, and PAR observations from ten experiments in 2015 and 2016. It indicates that the weather was sunny and stable during ten of the experiments, except for observations at approximately 11:30 and 14:30 on 25 April 2015, at approximately 13:00 and 14:00 on April 18, and at approximately 15:00 on 4 May 2016, when it was cloudy. These GPP observations and corresponding spectral measurements were reserved for later statistical analysis, since they can validate the model's performance in different weather conditions.

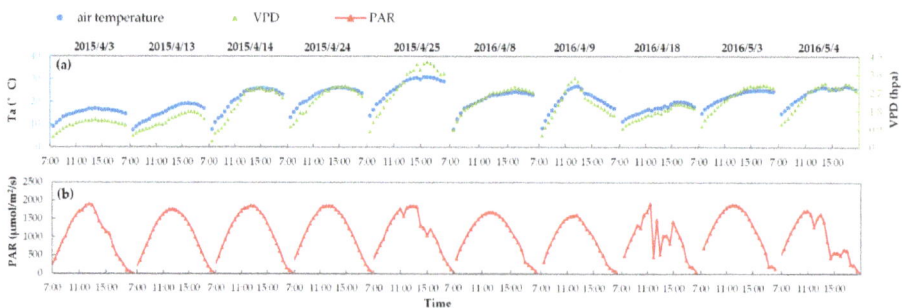

Figure 1. Diurnal observations of meteorological parameters between 7:00 to 19:00 made during ten experiments in 2015 and 2016: (**a**) half-hour air temperature and VPD observations and (**b**) half-hour PAR observations.

2.2.3. Diurnal Spectral Measurements

The diurnal measurements of the top-of-canopy spectra were taken using a customized Ocean Optics QE Pro spectrometer (Ocean Optics, Dunedin, FL, USA). This instrument recorded the solar irradiance and the canopy-reflected radiance spectra in the 645–805 nm spectral range with a spectral resolution of 0.31 nm, a sampling interval of 0.155 nm, and a signal-to-noise ratio (SNR) higher than 1000.

In this study, an automatic observation system was employed for continual spectral measurements. All spectral observations were acquired at nadir using the spectrometer's fiber optic (FOV: 25°), which was fixed on an erect turntable at a height of 2.8 m. A calibrated BaSO4 panel (of size 0.4 m × 0.4 m) was used as a reference to measure the solar irradiance spectrum. The measured canopy target was located in the south, and the reference panel was placed in the north. With the horizontal rotation of the observation stand, the solar irradiance spectrum could be automatically measured before and after each canopy spectrum measurement with a time lag of less than 30 s. Moreover, each measurement's field of view could remain the same. During each measurement, 5 single spectra were recorded, and each spectrum was produced by averaging 10 scans made at an optimized integration time with the application of dark current correction.

Measurements of all the canopy and panel radiance spectra were designed to be made every 0.5 h from 8:00 to 17:30, but several measurements failed due to instrumental problems. Thus, a total of 4, 7, 20, 3, 17, 8, 10, 14, 11, and 16 spectral measurements were made on 3, 13 & 14, and 24 & 25 April 2015, and on 8 & 9, 18 April and 3 & 4 May 2016, respectively. Besides, at 12:00 on 14 April 2015 and at 12:30 on 9 April and 3 & 4 May 2016, the shadow of a fiber optic probe on the reference panel disturbed the spectral measurements. Therefore, these spectral measurements were excluded in the later statistical analysis. As a result, a total of 4, 7, 20, 3, 17, 8, 9, 14, 10, and 15 (totally 106) valid spectral measurements were collected for our model validation on 3, 13 & 14, 24 & 25 April 2015 and on 8 & 9, 18 April and 3 & 4 May 2016, respectively.

2.2.4. SIF Retrievals from the Spectral Measurements

The Fraunhofer Line Discrimination (FLD) principle makes it possible to extract the weak SIF signal from the vegetation-reflected radiance at the Fraunhofer lines or the atmospheric absorption bands [12,55]. According to the accuracy assessment of the FLD-based SIF retrieval methods in [56,57], the 3FLD method [58] is most robust and can retrieve SIF with sufficient accuracy using spectral measurements by the QE pro spectrometer. Therefore, the 3FLD method was used for canopy SIF retrieval in this study. In the 3FLD method, the irradiance and radiance of a single reference channel used in the standard FLD method are replaced with the weighted averages for two channels at the left (for the shorter wavelength) and right (for the longer wavelength) shoulders of the absorption feature [58]. The weights of the two reference channels are defined as

$$w_{left} = \frac{\lambda_{right} - \lambda_{in}}{\lambda_{right} - \lambda_{left}}, w_{right} = \frac{\lambda_{in} - \lambda_{left}}{\lambda_{right} - \lambda_{left}} \tag{1}$$

in which λ is the wavelengths of the channels; and the subscripts 'in', 'left', and 'right' refer to the channels inside, at the left, and at the right shoulders of the absorption band, respectively. The SIF inside the absorption band can be calculated as Equation (2), in which I is the downwelling irradiance arriving at the top-of-canopy, and L is the total upwelling radiance at the TOC.

$$SIF_{in} = \frac{(I_{left}w_{left} + I_{right}w_{right})L_{in} - I_{in}(L_{left}w_{left} + L_{right}w_{right})}{(I_{left}w_{left} + I_{right}w_{right}) - I_{in}} \tag{2}$$

To sum up, from ten field experiments, we can synchronously derive the vegetation parameters, meteorological variables, and GPP at half-hour intervals, as well as diurnal reflectance and SIF datasets. These high-accuracy experimental datasets are sufficiently reliable as the SCOPE model inputs or for the validation of SIF simulations.

2.3. Inversion of LIDFa and V_{cmo} from In Situ Measurements

2.3.1. LIDFa Inversion from the Diurnal Canopy Reflectance Spectra

LIDFa is a leaf inclination distribution factor that will vary with the growth of wheat. In this study, the LIDFa values across the growing season were retrieved from the measured reflectance spectra by inverting the RTMo module with the look-up table (LUT) method. According to the GSA of the SCOPE model for reflectance simulation in [29], the LIDFa has a significant influence on the reflectance spectra in the region at around 500–1300 nm. Moreover, other key driving variables that govern the simulated reflectance spectra from 650 nm to 750 nm (including LAI, Cab, and Cdm) are all accurately measured from in situ experiments. Therefore, the LIDFa can be retrieved from our measured reflectance spectra from 645 nm to 805 nm. However, at different measurement times during the day, different LIDFa values will be retrieved due to the bi-directional reflectance characteristics of canopy and random measurement errors. So, the only and optimal LIDFa value must be determined for one day (hereafter denoted as the daily LIDFa). In addition, the adjacent two days were regarded as one day for LIDFa retrieval. The schematic overview of the LIDFa inversion is shown in Figure 2.

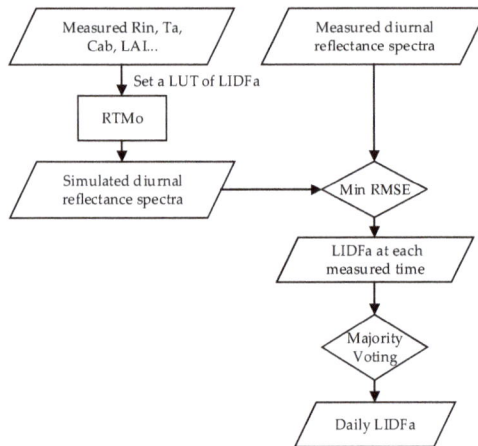

Figure 2. Schematic overview of the LIDFa inversion procedure.

First, the LIDFa at each measuring time was retrieved using the least root mean squared error (RMSE) method. Using the measured diurnal meteorological variables and other required inputs, the RTMo module was run with an LUT of different LIDFa values at half-hour time intervals for each fieldwork day. According to the six common kinds of leaf inclination distribution defined in [50], the LIDFa range was set to −1~1 with an interval of 0.05. Thus, for every half-hour, 41 canopy reflectance spectra under different LIDFa conditions could be collected from the SCOPE model outputs. Next, for the *i*-th LIDFa value of the *n*-th spectra measurement, we calculated the RMSE of simulated reflectance spectra R_{sim} to the measured reflectance spectra R_{mea}, as shown in Equation (3)

$$RMSE_R(i, n) = \frac{\sum\limits_{\lambda = 645}^{805} |R_{sim}(\lambda, i, n) - R_{mea}(\lambda, n)|}{N_\lambda} \tag{3}$$

in which λ represents the spectral wavelength and N_λ is the number of the spectral bands of the QE pro; *n* ranges from 1 to *N*, and *N* is the total number of the spectra measurements during one fieldwork day; and *i* ranges from 1 to 41. To avoid the influence of SIF emission, the apparent reflectance spectra around the absorption bands were smoothed by spline interpolation. The LIDFa, which produces the least RMSE, was selected as the retrieved LIDFa at the n-th spectra measurement (LIDFa$_n$). Thus, a vector $[$LIDFa$_1$ ... LIDFa$_n$ ... LIDFa$_N]$ can be calculated to represent the various LIDFa estimations at all the measuring times during one fieldwork day.

Secondly, the daily LIDFa was final determined by a majority voting method, as follows:

$$\text{daily LIDFa} = \text{mod} \begin{bmatrix} \text{LIDFa}_1^- & \cdots & \text{LIDFa}_n^- & \cdots & \text{LIDFa}_N^- \\ \text{LIDFa}_1 & \cdots & \text{LIDFa}_n & \cdots & \text{LIDFa}_N \\ \text{LIDFa}_1^+ & \cdots & \text{LIDFa}_n^+ & \cdots & \text{LIDFa}_N^+ \end{bmatrix} \tag{4}$$

in which LIDFa$_n^-$ and LIDFa$_n^+$ were calculated as LIDFa$_n$ minus and plus the LIDFa step in the LUT (0.05), respectively, and the mod is an operator that calculates a matrix's mode. Some strategies were applied to decrease the uncertainties. First, we adopted the mode instead of the mean value to represent the daily LIDFa to avoid the influence of the abnormal LIDFa values retrieved from incorrect measurements. Second, two vectors calculated as the originally retrieved vector minus and plus the LIDFa step in the LUT (0.05), respectively, were added into the matrix to avoid the 'pseudo mode' problem, which is likely to occur if we use only the mode on the originally retrieved vector. Take the vector $[-0.5, -0.55, -0.55, -0.6, -1, -1]$ as an example. This vector has two modes, -0.55 and -1, and -1 is a 'pseudo mode', because it is far away from the majority elements. If the elements of this vector plus and minus 0.05 are added into the matrix, the mode will be -0.55. Therefore, the majority voting approach can find the optimal daily LIDFa that is closest to the true LIDFa value.

2.3.2. V_{cmo} Inversion from Diurnal GPP Observations

The V_{cmo} values across the growing season were retrieved from the diurnal observations by inverting the SCOPE model with the LUT method. V_{cmo} is a crucial leaf biochemical parameter for calculating photosynthesis and fluorescence emission in the SCOPE model. It changes with different vegetation types [59,60], plant functional types [61], and different days of the year [62]. According to [25], V_{cmo} influences carbon assimilation of photosynthesis (i.e., PQ) and thus fluorescence emission efficiency. Therefore, V_{cmo} may be estimated from varying net CO_2 fluxes. Wolf et al. have successfully estimated the V_{cmo} by fitting a commonly used model to measured net CO_2 fluxes [63]. In our study, the simulated GPP can represent the net CO_2 fluxes, because the respiration rate is set to zero. Thus, the V_{cmo} values were retrieved by comparing simulated against observed GPP. The inversion of V_{cmo} intends to find the optimal V_{cmo} value for each fieldwork day (hereafter denoted as the daily V_{cmo}). The schematic overview of V_{cmo} inversion is shown in Figure 3.

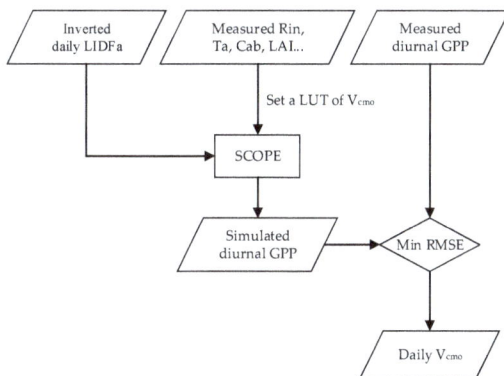

Figure 3. Schematic overview of the V_{cmo} inversion procedure.

The daily V_{cmo} was retrieved using the least RMSE method, which intends to find the diurnal GPP values that are most consistent with the measured ones. With the inverted daily LIDFa as inputs, SCOPE was run with an LUT with different V_{cmo} values at half-hour time intervals for each fieldwork day in the 2015~2016 period. Based on the literature [59,61], the range of V_{cmo} should be set to 10–200 µmol m^{-2} s^{-1} for C3 crops like wheat with a step 10 µmol m^{-2} s^{-1}. Thus, for every fieldwork day, 20 sets of half-hour GPP data under 20 V_{cmo} conditions could be collected from the SCOPE model outputs. Next, for the j-th V_{cmo} value, we calculated the RMSEs between the simulated half-hour GPP values GPP_{sim} with the measured ones GPP_{mea}, as shown in Equation (5)

$$RMSE_G(j) = \frac{\sum\limits_{m=1}^{M} |GPP_{sim}(j,m) - GPP_{mea}(m)|}{M} \tag{5}$$

in which m indicates the m-th measurement of GPP during one fieldwork day, and M is the total measurement times. Similar to the LIDFa inversion, the V_{cmo} that produces the least RMSE was selected as the inverted daily V_{cmo} for model input.

2.4. Settings of FQE Values

Two different FQE settings were adopted for the SIF simulations, as follows:

(1) Fixed FQE, simulations with fixed FQE values as the literature suggested values: 0.01 for fqe2 and 0.2 for fqe1/fqe2.
(2) Variable FQE, simulations with variable FQE values, which were estimated by fitting the SIF simulations to the observed SIF data and minimizing the systematic deviations in SIF$_A$ and SIF$_B$ simulations with the LUT of changing values of FQE.

As mentioned previously, the fqe2 and fqe1 are directly proportional to PSII and PSI fluorescence, respectively, and the PSII fluorescence spectra cover both the red and far-red bands, while the PSI fluorescence spectra cover only the far-red band. Thus, the fqe2 and fqe1/fqe2 are directly proportional to the simulated SIF$_B$ and SIF$_A$/SIF$_B$, respectively. In other words, if the simulated SIF$_B$ and SIF$_A$/SIF$_B$ values have systematic deviation, it is likely to be caused by the unsuitable FQE values. Adjusting FQE should find the optimal (fqe2, fqe1/fqe2) setting that makes the systematic deviation of diurnal SIF$_A$ and SIF$_B$ simulations minimum for each fieldwork day (hereafter denoted as the daily FQE values). According to the literature [28], the measured SIF normalized by PAR has a weak diurnal cycle for unstressed crops in a steady state. Therefore, the FQE values during one day are regarded as invariable in this study.

First, the SCOPE model was run with an LUT of different fqe2 and fqe1/fqe2 values at half-hour time steps corresponding to each diurnal experiment. The fqe2 was set from 0.005 to 0.02 with a step of 0.001, and the fqe1/fqe2 was set to 0.05~0.5 with a step of 0.05. Then, for every fieldwork day, 120 sets of half-hour SIF$_A$ and SIF$_B$ data under 120 different (fqe2, fqe1/fqe2) conditions were collected from the SCOPE model outputs. For each (fqe2, fqe1/fqe2) condition, the bias in daily averages of diurnal SIF$_A$ and SIF$_B$ simulations was adopted to describe the systematic deviation of diurnal SIF$_A$ and SIF$_B$ simulations, as defined in Equation (6)

$$bias_{SIF_B} = \frac{\overline{SIF_{B,sim}} - \overline{SIF_{B,mea}}}{\overline{SIF_{B,mea}}}, bias_{SIF_A} = \frac{\overline{SIF_{A,sim}} - \overline{SIF_{A,mea}}}{\overline{SIF_{A,mea}}} \tag{6}$$

in which $\overline{SIF}$ represents the daily average of diurnal SIF during one fieldwork day, the subscripts 'B' and 'A', respectively, represent the O$_2$-B and O$_2$-A bands, and the subscripts 'sim' and 'mea', respectively, represent the simulated and measured data. The computation of the daily average of diurnal SIF is necessary, because the diurnal variations in SIF are dominated by the diurnal cycles of irradiance, making it more difficult to reflect FQE effects. The absolute value of $bias_{SIF_B}$ and $bias_{SIF_A}$

can simultaneously reach their minimum value by adjusting the two FQE values, because fqe2 and fqe1/fqe2 separately governs the simulated SIF_B and SIF_A/SIF_B. Finally, the two FQE values that provide a minimum sum of the absolute value of $bias_{SIF_B}$ and $bias_{SIF_A}$ ($|bias_{SIF_B}|+|bias_{SIF_A}|$) were selected as the daily FQE values for each fieldwork day.

2.5. Experimental Process

Figure 4 displays schematically the process of SIF simulations and model evaluation. Using the inverted daily LIDFa and V_{cmo} (described in Section 2.3), the accurately measured vegetation parameters (described in Section 2.2.1), and meteorological variables (described in Section 2.2.2) as inputs, the SCOPE model first run with fixed FQE values at half-hour intervals for each fieldwork day. Other parameters required by the SCOPE model were set to their default or literature values described in Section 2.1.1. The SCOPE version 1.61 and biochemical module TB12 were adopted. Meanwhile, the daily FQE values were adjusted according to the systematic deviations of the SIF_A and SIF_B simulations. Overall, for each FQE setting (fixed and variable FQE), ten time series simulations were run along with ten diurnal experiments, and, eventually, from the SCOPE model outputs, ten simulated half-hour SIF spectra were collected and compared with the observed diurnal SIF data (described in Section 2.2.4). Finally, the accuracy of SIF_A, SIF_B, and SIF_B/SIF_A simulations with the two FQE settings was quantitatively evaluated.

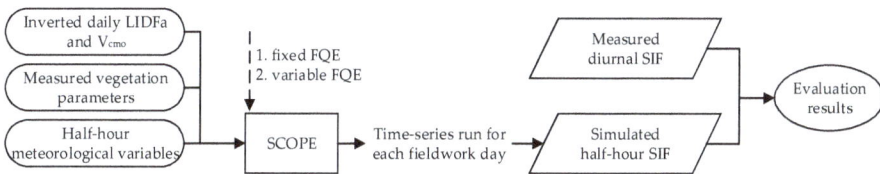

Figure 4. Schematic overview of SIF simulations and model evaluation.

3. Results

3.1. LIDFa and V_{cmo} Retrieved from In Situ Measurements

Using the inversion procedure as shown in Figures 2 and 3, daily LIDFa and V_{cmo} were retrieved on ten fieldwork days, as listed in Table 3. The seasonal changing patterns of our LIDFa retrievals were in accord with those realistic patterns across the growing season. As wheat grew from the erecting to the booting stage, the retrieved LIDFa continued to increase, which indicated that the leaves become flatter over time. In 2016, there was an obvious decrease following the increase due to the arrival of the flowering period. Note that the seasonal change pattern of LIDFa for two years is completely consistent with the seasonal variation of LAI listed in Table 1. This result verifies the reliability of the LIDFa inversion, because LAI and LIDFa are both canopy structural parameters and they should have similar change patterns across the growth season of wheat. The retrieved V_{cmo} varied from 45 to 110 μmol m^{-2} s^{-1}, which was in good accord with its well-known values for C3 crops like wheat or soybean. The ranges and seasonal change patterns of retrieved V_{cmo} values in 2015 and 2016 were consistent: the V_{cmo} values increased constantly from wheat's erecting to its flowering stage. In addition, the V_{cmo} values between two adjacent fieldwork days are almost unchanged. All of this evidence indicates that our retrieved V_{cmo} values are reliable.

Table 3. The retrieved daily LIDFa and V$_{cmo}$ values (unit: μmol m^{-2} s^{-1}) on ten fieldwork days in 2015 and 2016.

	2015					2016				
	April 3	April 13	April 14	April 24	April 25	April 8	April 9	April 18	May 3	May 4
LIDFa	−0.975	−0.875	−0.875	−0.625	−0.625	−0.875	−0.875	−0.75	−0.85	−0.85
V$_{cmo}$	50	80	80	110	110	45	55	65	95	100

3.2. Results of Reflectance and GPP Simulations

The simulated canopy reflectance and GPP separately contain information on two aspects: one is the leaf and canopy characteristic represented by RTMo module, and the other one is the plant physiological and photosynthesis state represented by the biochemical module. Both aspects impact the simulating accuracy of SIF. Therefore, the results of reflectance and GPP simulations should be inspected before the evaluation of SIF simulations.

Figure 5 displays the results of the simulated and measured reflectance spectra and their residuals at 10:00 for every fieldwork day. In general, the simulated reflectance spectra fit the measured ones well: the residual absolute values are less than 0.03 in the full region of 650 nm to 800 nm for all ten fieldwork days. Specifically, the simulated and measured reflectance spectra are approximately the same in the NIR from 750 nm to 800 nm (the residual absolute values in this region are less than 0.012), except for 25 April 2015. However, in the red and red edge region from 650 nm to 750 nm, the residuals between the two reflectance spectra are slightly higher. Figure 6 shows the RMSE statistics between the simulated and measured reflectance spectra for 106 spectral measurements in 2015 and 2016. As illustrated, the model reproduces the reflectance well: the RMSE values for 106 spectral measurements are between 0.0041 and 0.0437, and most of (67%) the RMSE values are lower than 0.02. According to [29], the leaf and canopy parameters—including LIDFa, LAI, Cab, and Cdm—together govern the variation in the reflectance spectra from 650 nm to 800 nm. So, reflectance simulation accuracy depends on the joint accuracy of these parameters. All these results indicate that using the measured and retrieved vegetation parameters as inputs, the SCOPE model can accurately model the leaf and canopy characteristic and reproduce the reflectance spectra.

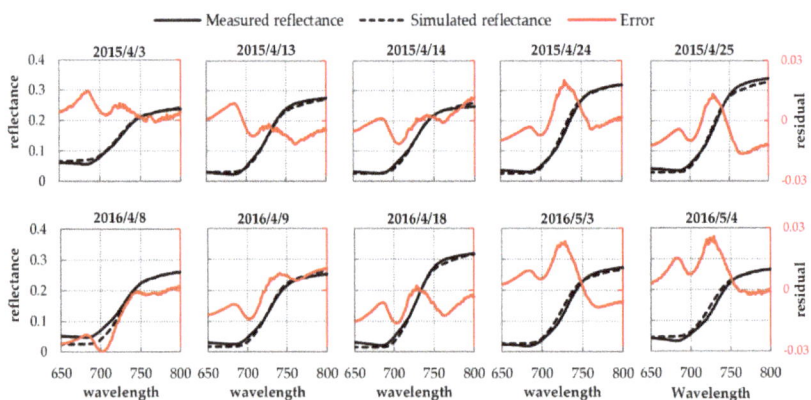

Figure 5. The simulated and measured reflectance spectra and their residuals at 10:00 for every fieldwork day in 2015 and 2016.

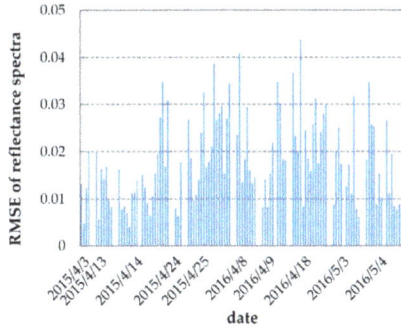

Figure 6. Root mean squared error (RMSE) values between the simulated and measured reflectance spectra for 106 spectral measurements in 2015 and 2016.

Figure 7 shows the diurnal cycles of the simulated and measured GPP on ten fieldwork days in 2015 and 2016. On one hand, the absolute intensities of the two GPP data sets (simulated and measured diurnal GPP) agree well: the RMSEs of two GPP data sets range from 1.514 µmol m^{-2} s^{-1} to 5.627 µmol m^{-2} s^{-1}, and the corresponding relative root mean square error (RRMSE) values range from 10.31% to 29.65% on the ten fieldwork days. On the other hand, the diurnal patterns of simulated GPP agree well with measured GPP on most fieldwork days, except for 18 April 2016, when the GPP observations were likely inaccurate and inconsistent with the PAR changes (see Figure 1b). This result indicates that the simulated diurnal GPP matches the PAR changes better than the measured GPP. This is because SCOPE's biochemical module can track the weather fluctuations via the meteorological inputs and thus accurately simulate the GPP, while the NEE observations are likely to be disturbed by changing air parameters like wind speed and direction. For further illustration, Figure 8 displays the correlation and RRMSE values between simulated and measured GPP for all half-hour flux observations in 2015 and 2016. The simulated GPP values are highly consistent with the measured ones: the scatters are close to the 1:1 line with a determination coefficient (R^2) of approximately 0.83 and an RRMSE of 20.69%. All these results of GPP simulations indicate that by using the directly measured or indirectly retrieved inputs from in situ observation, the SCOPE model can accurately represent the plant physiological state and interpret the vegetation photosynthesis.

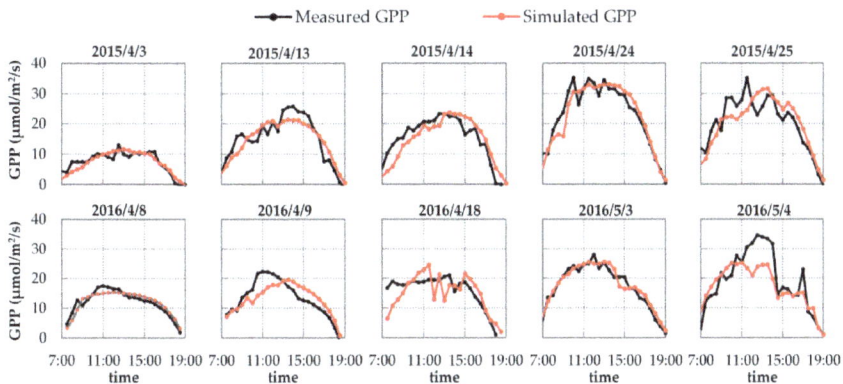

Figure 7. The diurnal cycles of simulated and measured GPP at half-hour intervals between 7:00 and 19:00 on ten fieldwork days in 2015 and 2016.

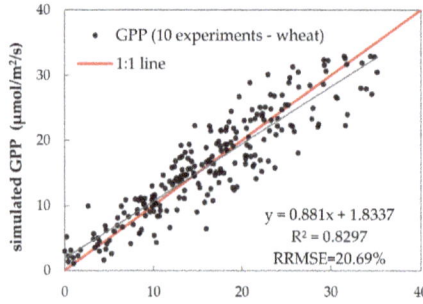

Figure 8. The correlation and relative root mean square error (RRMSE) value between simulated and measured GPP for all half-hour flux observations during ten experiments in 2015 and 2016.

3.3. Evaluation of SIF Simulations

3.3.1. Evaluation of SIF Simulations with Fixed FQE

First, we evaluated the simulating accuracy of SIF_B and SIF_A related to both their diurnal cycles and intensities. Figure 9 displays the diurnal cycles of the half-hourly simulated SIF with fixed FQE, compared to the measured SIF on ten fieldwork days in 2015 and 2016. As illustrated, the SCOPE model can well reproduce the diurnal cycles of the SIF variation for each fieldwork day. Many studies have proven that SIF is positively correlated with PAR and GPP at the canopy scale [7,8,57,64]. Similarly, here, the simulated SIF values increase until noon and then decrease over time (with the changing of SZA), which obviously agrees with the diurnal patterns of PAR and GPP observation. Like the diurnal PAR and GPP observations shown in Figures 1 and 7, the diurnal curves of SIF exhibit the same fluctuations due to unstable weather on 25 April 2015, and on 18 April and 4 May 2016. All these phenomena indicate that in the SCOPE model, the TB12 biochemical module can credibly regulate the diurnal variation of SIF emission efficiencies as a function of the dynamic micrometeorological variables. Nevertheless, on several fieldwork days—like for SIF_A on 24 April 2015 or for SIF_B on 3 May 2016—the systematic deviation in the diurnal SIF simulations is obvious: the simulated SIF_B or SIF_A values during the entire day collectively deviated from their measured points.

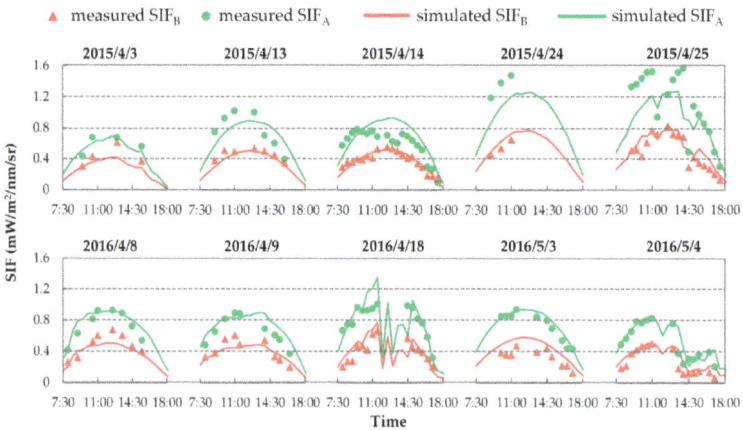

Figure 9. The diurnal cycles of simulated SIF with fixed FQE, compared to the measured SIF at O_2-B and O_2-A bands on ten fieldwork days in 2015 and 2016.

Table 4 concludes the quantitative assessment of the errors and deviations in SIF_B and SIF_A simulations with fixed FQE. This table lists the RRMSE of diurnal simulations for each fieldwork day (hereafter denoted as the daily RRMSE) and the RRMSE for all ten days of SIF_A and SIF_B simulations, as well as the $bias_{SIF_B}$ and $bias_{SIF_A}$ (as described in Section 2.4). The results show that the SCOPE model can provide acceptable accuracy for the SIF_B and SIF_A simulations with fixed FQE: the RRMSE values of SIF_B and SIF_A simulations for all 106 spectral measurements were 24.35% and 23.67%, respectively; and the daily RRMSEs were lower than 30% for each fieldwork day, except for the SIF_B on 3 & 4 May 2016. Nevertheless, the systematic deviations of diurnal SIF_A and SIF_B simulations were noteworthy on many fieldwork days, in particular for SIF_B in 3 April 2015 and 3 & 4 May 2016, and for SIF_A on 24 April 2015 the absolute bias values were greater than 20%. On these days, the error in diurnal SIF simulation was mainly caused by the systematic deviation, not the inconsistency of SIF change cycles. Note that the bias variation in SIF_B simulations shows a clear seasonal pattern that is coincident between 2015 and 2016. Specifically, at the erecting period (i.e., 3 April 2015 and 8 & 9 April 2016), the SIF_B was underestimated with the negative bias value; but at the booting or flowering period (i.e., 24 & 25 April 2015 and 3 & 4 May 2016), the SIF_B was largely overestimated with a positive bias value. These results indicated that the fixed FQE was not suitable for unbiased SIF_B and SIF_A simulations at all growth stages, and the unsuitable FQE values may be the major factor that caused the systematic deviations in simulated SIF. In addition, the bias values for SIF_B and SIF_A were mostly at different levels or even opposite in sign. For example, on 24 & 25 April 2015, the SIF_A was largely underestimated, while the SIF_B was overestimated; on 3 May 2016, the SIF_B was greatly overestimated while the simulated SIF_A values were just right. It implied that the SCOPE-simulated SIF_A/SIF_B with fixed FQE was probably unreliable, which should be investigated further.

Second, we evaluated the simulating accuracy of SIF_A/SIF_B to investigate whether the SCOPE model can reproduce the SIF spectral shape with fixed FQE. Figure 10 displays the correlation and RRMSE values between the diurnally simulated and measured SIF_A/SIF_B with fixed FQE for all 106 spectral measurements (Figure 10a) and the corresponding daily RRMSE for each single fieldwork day (Figure 10b). As illustrated, the SCOPE-simulated SIF_A/SIF_B values were not satisfying: for all 106 spectral measurements, most scatters were located far away from the 1:1 line with an R^2 of only 0.0286 and an RRMSE of nearly 30%, and most daily RRMSEs were larger than 25% (with a range from 23.57% to 33.29%) on ten fieldwork days. In addition, the measured SIF_A/SIF_B values changed with a range from 0.593 to 3.460, while the simulated SIF_A/SIF_B values were relatively stable with a range from 1.552 to 2.021 across wheat's growing season in 2015 and 2016.

Table 4. The relative root mean square error (RRMSE) and bias of diurnal SIF simulations with fixed FQE at O_2-B and O_2-A bands on ten fieldwork days.

Year	Date	O_2-B		O_2-A	
		RRMSE	bias	RRMSE	bias
	April 3	25.97%	−20.35%	11.14%	−1.74%
	April 13	12.47%	−10.78%	22.64%	−1.96%
2015	April 14	12.13%	0.71%	28.30%	18.29%
	April 24	13.43%	12.62%	22.30%	−22.24%
	April 25	24.54%	13.47%	26.43%	−13.85%
	April 8	20.23%	−13.70%	12.07%	6.93%
	April 9	19.53%	−5.83%	18.77%	12.43%
2016	April 18	29.54%	12.99%	21.69%	5.58%
	May 3	40.13%	38.16%	8.94%	6.43%
	May 4	38.67%	27.79%	23.51%	10.84%
all ten days		24.35%	—	23.67%	—

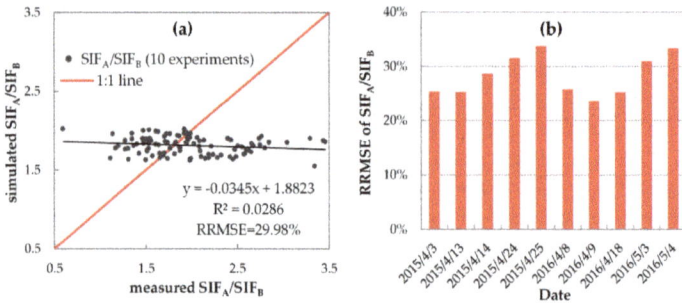

Figure 10. The correlation and relative root mean square error (RRMSE) values between simulated and measured SIF_A/SIF_B with fixed FQE: (**a**) the correlation and RRMSE for 106 spectral measurements and (**b**) the RRMSE values for each fieldwork day in 2015 and 2016.

Table 5 concludes the quantitative assessment of systematic deviation in the SIF_A/SIF_B simulations at all growth stages. The ratio of daily SIF_A and SIF_B averages $(\overline{SIF_A}/\overline{SIF_B})$ was calculated to express the systematic deviation, because the diurnal variations of SIF_A/SIF_B are dominated by the geometry of incidence and observation, making it more difficult to reflect seasonal variations of systematic deviation. As reported in Table 5, the simulated $\overline{SIF_A}/\overline{SIF_B}$ values were different from the measured ones at wheat's erecting and booting or flowering period, with RE's absolute value larger than 20%. Note that the RE variation in SIF_A/SIF_B simulations shows a consistent seasonal pattern between 2015 and 2016 that opposes the bias variation in SIF_B simulations (as mentioned previously). Specifically, at the erecting period, the SIF_A/SIF_B values were largely overestimated with positive RE values, but at the booting or flowering period, the SIF_A/SIF_B values were largely underestimated with negative RE values. In addition, with fixed FQE, the SCOPE model cannot credibly reproduce the seasonal cycles of SIF_A/SIF_B. As the wheat grows, the measured $\overline{SIF_A}/\overline{SIF_B}$ increased from the erecting to the flowering period with a wide range from 1.370 to 2.470, while the simulated $\overline{SIF_A}/\overline{SIF_B}$ remained in the range from 1.690 to 1.916, without clear seasonal changes. These results indicated that the fixed FQE was not suitable for unbiased SIF_A/SIF_B simulations at all growth stages. Considering that the fqe1/fqe2 value was directly proportional to the simulated SIF_A/SIF_B, the unregulated fqe1/fqe2 may be the major factor limiting the seasonal variation of simulated SIF_A/SIF_B.

Table 5. The simulated and measured ratio of daily SIFA and SIFB averages, as well as the relative error (RE) between them with fixed FQE for ten fieldwork days in 2015 and 2016.

	2015						2016			
	April 3	April 13	April 14	April 24	April 25	April 8	April 9	April 18	May 3	May 4
simulated $\overline{SIF_A}/\overline{SIF_B}$	1.690	1.853	1.857	1.705	1.698	1.885	1.916	1.859	1.705	1.732
measured $\overline{SIF_A}/\overline{SIF_B}$	1.370	1.686	1.590	2.470	2.237	1.522	1.605	1.990	2.213	1.997
RE	23.37%	9.88%	16.80%	−30.95%	−24.08%	23.90%	19.39%	−6.56%	−22.97%	−13.26%

3.3.2. Variable FQE Estimated from SIF Observations

Table 6 lists the variable FQE values estimated by minimizing the systematic deviations in SIF_A and SIF_B simulations for each fieldwork day. For further illustration, Figure 11 displays the seasonal cycle of fqe2 and fqe1/fqe2 from the erecting to wheat's flowering period in 2015 and 2016. As illustrated, from wheat's erecting to its flowering stage, the fqe2 value gradually decreased from 0.013 to 0.007, while the fqe1/fqe2 value exhibited an opposite trend and increased from 0.05 to 0.5.

The seasonal cycles of fqe2 and fqe1/fqe2 were consistent with variations of the systematic deviation in the SIF_B and SIF_A/SIF_B simulations (as mentioned previously), respectively. Specifically, the fqe2 value was close to its literature-fixed value of 0.01 only at the jointing period, while at the erecting period the value was larger (0.011–0.013), and at the booting or flowering period the value was lower (0.007–0.009). Also, the fqe1/fqe2 value was close to its literature-fixed value of 0.2 only at the jointing period, while at the erecting period the value was lower (0.05–0.1), and at the booting or flowering period the value was larger (0.3–0.5). These results confirm the seasonal FQE values, showing that the seasonal cycles of FQE values between 2015 and 2016 were greatly consistent.

Table 6. The variable fqe2 and fqe1/fqe2 values on ten fieldwork days in 2015 and 2016.

	2015					2016				
	April 3	April 13	April 14	April 24	April 25	April 8	April 9	April 18	May 3	May 4
fqe2	0.013	0.011	0.01	0.008	0.009	0.012	0.011	0.009	0.007	0.008
fqe1/fqe2	0.05	0.15	0.1	0.5	0.4	0.05	0.1	0.25	0.4	0.3

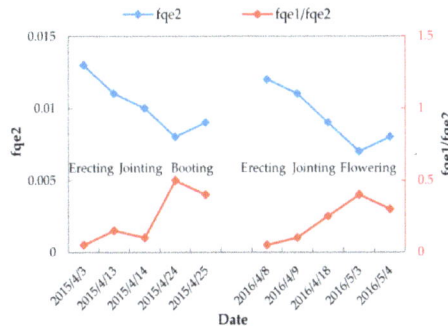

Figure 11. The seasonal cycles of fqe2 and fqe1/fqe2 on ten fieldwork days in 2015 and 2016.

3.3.3. Evaluation of SIF Simulations with Variable FQE

With the above variable FQE as inputs, the systematic deviations in SIF_B, SIF_A, and SIF_A/SIF_B simulations can be corrected; meanwhile, the limited range from SIF_A/SIF_B simulations can be extended due to the seasonal variations of fqe1/fqe2. Thus, the SCOPE model can provide more accurate SIF simulations related to both the individual bands (SIF_A and SIF_B) and the seasonal SIF_A/SIF_B values.

On the one hand, if simulated with variable FQE values, the simulation accuracy of SIF_B and SIF_A can be improved greatly. Like the quantitative assessments listed in Table 4, Figure 12 displays the correlation and RRMSE values between the simulated and measured SIF with variable FQE for all 106 spectral measurements (Figure 12a) and the corresponding daily RRMSEs for each single fieldwork day (Figure 12b). As illustrated, both the simulated SIF_B and SIF_A were consistent with the measured ones for all 106 spectral measurements: the scatters were located close to the 1:1 line, with the R^2 larger than 0.78 and an RRMSE of less than 20%. Additionally, the daily RRMSE values of SIF_B and SIF_A were lower than 30% (with a range from 5.97% to 29.80%) and 23% (with a range from 6.99% to 22.55%), respectively, during ten experiments in 2015 and 2016.

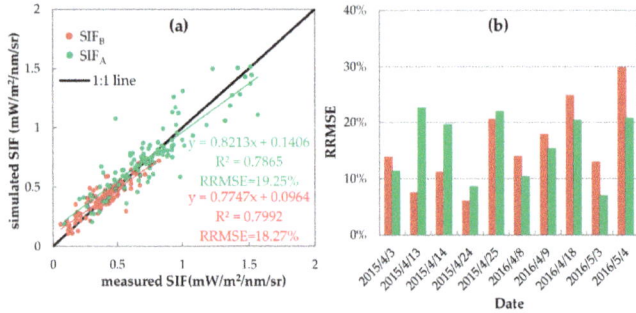

Figure 12. The correlation and relative root mean square error (RRMSE) values between simulated and measured SIF_A and SIF_B with variable FQE: (**a**) the correlation and RRMSE for 106 spectral measurements and (**b**) the RRMSE values for each fieldwork day in 2015 and 2016.

On the other hand, if simulated with variable FQE values, the SCOPE model can credibly reproduce the seasonal SIF_A/SIF_B values. Similar to Figure 12, Figure 13 shows the correlation and RRMSE values between the diurnally simulated and measured SIF_A/SIF_B with variable FQE. Unlike the unsatisfactory results shown in Figure 10, with variable FQE, both the range and values of simulated SIF_A/SIF_B were consistent with the measured ones: for all 106 spectral measurements, most scatters are close to the 1:1 line with an R^2 of 0.48 and an RRMSE of only 20.63%, and most daily RRMSE were lower than 20% (with a range from 4.98% to 24.20%) on the ten fieldwork days. All these results indicated that the variable FQE settings were more suitable than the fixed ones for acquiring unbiased SIF simulations.

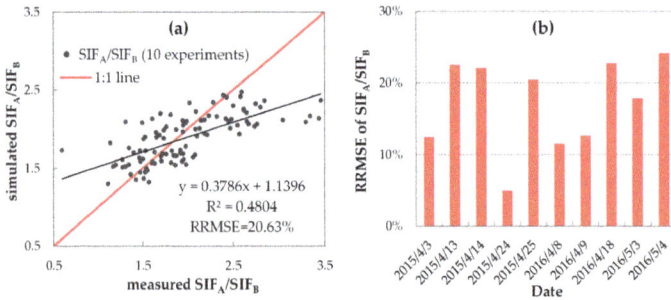

Figure 13. The correlation and relative root mean square error (RRMSE) values between simulated and measured SIF_A/SIF_B with variable FQE: (**a**) the correlation and RRMSE for 106 spectral measurements and (**b**) the RRMSE values for each fieldwork day in 2015 and 2016.

4. Discussion

In this study, SCOPE's driving input parameters were derived from ten field measurements in two ways: (i) direct measurements of vegetation and meteorological parameters and (ii) a model inversion approach to retrieve LIDFa and V_{cmo} from in situ reflectance and GPP observations. These accurate input parameters are vital for persuasive evaluation results of the SCOPE model. The determination of these parameters is one way to calibrate the main modules related to SIF simulation. In the SCOPE model, canopy SIF emissions are propagated using three modules: the Fluspect and biochemical modules at the leaf level and the RTMo module for canopy radiative transfer of SIF signal. Vegetation parameters (e.g., LIDFa, Cab, and LAI) determine the leaf and canopy characteristics represented with the RTMo module, and the meteorological and leaf biochemical parameters (e.g., V_{cmo}, R_{in}, and Ta) determine the plant physiological and photosynthesis state represented with the biochemical module. The simulated canopy reflectance and GPP outputs separately contain information on these two aspects. So, the accuracy of reflectance and GPP simulations can reflect the joint accuracy of input parameters, which has been verified in this study.

The V_{cmo} values across the growing season were inverted by fitting the SCOPE model to measured GPP data. The V_{cmo} is one of the key parameters in the biochemical module of SCOPE for GPP modeling [20,43]. However, Poolman et al. [65,66] pointed out that the Rubisco may not be the rate limiting factor of the rate of CO_2 assimilation, which means there may be some uncertainties of our method for the V_{cmo} estimation and new models reflecting this fact should be considered. Nevertheless, the retrieved V_{cmo} values match well with the literature values and show credible seasonal patterns. First, the retrieved V_{cmo} varied from 45 to 110 µmol m^{-2} s^{-1}, which was in good accord with those displayed in the literatures for C3 crops: V_{cmo} ranges from 35 to 83 µmol m^{-2} s^{-1} for wheat in [61], V_{cmo} ranges from 76 to 136 µmol m^{-2} s^{-1} for soybean in [67], and the common V_{cmo} value is 93 µmol m^{-2} s^{-1} for wheat in [68,69]. Second, the retrieved V_{cmo} values showed plausible seasonal change patterns, which agrees well with the seasonal changes of V_{cmo} retrieved from space-based SIF data in [23]. To date, the possibility of prescribing V_{cmo} from another information source is limited. V_{cmo} could be estimated using leaf nitrogen content [49,70], but this estimation relied only on few experimental results, and the validation is still needed. Zhang et al. [23] retrieved the V_{cmo} from changes in SIF, as observed by the GOME-2 satellite, but the satellite data were aggregated over space and time with the entire growth seasons containing drought and senescence, which is different from our focus on several diurnal cycles at the canopy scale. Therefore, our model-based inversion approach for V_{cmo} may provide an alternative way for the estimation of V_{cmo} on ground or global scales. As we focus on the evaluation the SCOPE model for SIF simulation in this study, further attempts on the more accurate estimation of V_{cmo} was not included.

However, there are some uncertainties within the process to derive FQE. Firstly, the SIF retrieved from the spectral data was regarded as the true value for accessing the accuracy of SIF simulation. Based on this assumption, SIF simulations could be validated. However, the SIF retrieval based on 3FLD method has some uncertainties: the accuracy is dependent on the spectral characteristics of the reflectance and irradiance spectra at the absorption band, and on the spectral resolution and SNR of the sensor used. Thus, the RRMSE of simulated SIF to measured SIF is not the real error of the SCOPE model's SIF simulation. Fortunately, the SIF observations derived from our experiments were sufficiently reliable thanks to the robust retrieval method with high spectral resolution and the SNR of the QE Pro spectrometer. According to the accuracy assessment using simulated data with the same SNR and SR of QE Pro spectrometer in [56,57], the RRMSE for the SIF retrieved using 3FLD methods is 13.2% at the O_2-B band and 9.5% at the O_2-A bands. Thus, the 3FLD method can retrieve SIF with sufficient accuracy and provide the reference value for evaluating SIF simulations. Therefore, the quantitative accuracy assessments made in this study can reflect the SCOPE model's reliability.

Secondly, the vegetation parameters and meteorological or flux variables derived from in situ observations were considered accurate in this study. They also suffer from some uncertainties due to instrumental or artificial errors in field measurements. These uncertainties affect the LIDFa and V_{cmo} retrieval accuracy and cause an additional error in SIF simulations. According to the error propagation presented in [28], the effects of Cdm and LIDFa or Cab and LIDFa on reflectance are opposite, implying that an overestimate (underestimate) in measured Cdm or Cab will lead to an overestimate (underestimate) in the LIDFa retrievals. Moreover, the effects of Cdm and LIDFa or Cab and LIDFa on fluorescence are also opposite. Thus, the effects of a simultaneous overestimate or underestimate can to some degree cancel out in the SIF simulation. Similarly, the effect of an overestimation or underestimate of the measured LAI can also be weakened in simulated SIF, because the effects of LAI and LIDFa on reflectance and fluorescence are both accordant. Therefore, the model calibration approach of reproducing the measured reflectance spectra can to some degree weaken the impacts of vegetation parametric uncertainties on SIF simulations.

These two sources of uncertainties will propagate to our FQE estimations and then affect their variation patterns. Nevertheless, the ranges and variation patterns of our FQE values were quite consistent between two growth seasons in 2015 and 2016. Moreover, the estimated fqe2 values were close to their literature values derived from some laboratory measurements. As reported in Van der Tol et al. [25,28], the fqe2 values was around 0.01 based on the laboratory measurements in Genty et al. [33]. Trissl et al. [37] considered three different levels of fqe2 values (0.01, 0.018, and 0.021) for the modeling of PSII photochemistry. Our estimated fqe2 values seem a little lower than the fqe2 values reported in literatures (around 0.02) [35,36]. This discrepancy may be caused by the uncertainties of parameter determinations in this study, or be caused by the errors of laboratory measurements in following two aspects. First, the measured PSII fluorescence signal inevitably included the contamination of PSI fluorescence, which would cause overestimation in fqe2 [35,37,39], while our estimated fqe2 values can avoid this error, as the simulated SIF for PSII and PSI was evaluated separately. Second, the measuring flashes used for the determination of the F_0-level fluorescence would cause some PSII closure, and thus the real F_0-level fluorescence would be overestimated to different degrees [36]. Besides, it has been reported in several literatures that the F_0-level fluorescence quantum efficiency obviously changes with different vegetation species, chlorophyll contents, and exposure of leave surfaces to the sun. Björkman and Barbara [38] measured the chlorophyll fluorescence characteristic at 77 K in 44 species of vascular plant, and found that the F_0-level fluorescence signal between two different C_3 species can be 2-fold variation. The variation of F_0-level fluorescence signal between the lower and upper leaf surfaces or the shaded and sun leaves were also remarkable. Björkman and Barbara [38] and Morales et al. [71] also observed declines in F_0-level fluorescence signal with increased chlorophyll content, probably due to an increase in the proportion of fluorescence that is reabsorbed. Moreover, as shown in Hussain et al. [72], the cinnamic acid stress will significantly reduce the efficiency of "open" PSII reaction centers in the dark-adapted state, and a tendency of increase in F_0-level fluorescence was observed during 2th and 4th days. These laboratory measurements can to some degree support and account for the variation of our FQE estimations with the growth of wheat. However, the influence of different factors on FQE variations is complicated and remains unsettled. Our results can provide a reference for the parameterization of FQE values with the winter wheat in the field. More control experiments with models and in the field need to be conducted for a better understanding of the variation of FQE values.

Further opportunities are available to investigate the simulated results of MD12 and SCOPE version 1.7 compared with this study's results. In this work, the TB12 biochemical module is generated in the SCOPE model to implement the simulations. For this module, SIF emission efficiency depends on the empirical calibration of a number of datasets collected in field and laboratory experiments. The MD12 module has a more explicit parameterization of fluorescence quenching mechanisms, while it needs two additional inputs (KNPQs and qLs) to express the intermediate conditions, which cannot be derived from our in situ measurements. Given the uncertainty in PSII-PSI fluorescence emission curves and corresponding fqe1 and fqe2 values, the PSI-PSII separation is explicitly avoided in the last version 1.7 of the SCOPE model. In the future, the SIF simulated results of this version can be evaluated and compared with this study's results.

More field observations and theoretical simulations are required to verify the results presented in this paper. At present, we conducted the experiments only on winter wheat, and the data sets are limited. Whether the seasonal patterns of FQE variation can be applied to other species has not been ascertained. In the future, the spectral measurements should be conducted across various species and plant functional types (PFT), with more frequent time series, at different locations, in multi-angle mode, along with the observations of vegetation parameters and flux exchanges. This research could become a reality with the achievement of our automatic fluorescence observation network. Moreover, active fqe1 and fqe2 measurements with PAM should be conducted in the field and on vegetation at different growth stages, which can further verify the seasonal cycles of our variable FQE values.

5. Conclusions

In this paper, we evaluated the SCOPE model's performance for SIF_A, SIF_B, and SIF_A/SIF_B simulations using high-accuracy spectral and flux observations in ten diurnal experiments on winter wheat. The SIF simulation accuracy with both fixed and variable FQE values was quantitatively assessed by comparison with the SIF retrieved from measurements using the 3FLD method with a QE pro spectrometer.

If simulated with fixed FQE values, the SCOPE model can reliably interpret SIF diurnal cycles and provide acceptable results for SIF_B and SIF_A simulations. The RRMSEs of SIF_A and SIF_B for all 106 spectral measurements in the ten diurnal experiments were 24.35% and 23.67%, respectively. Nevertheless, the fixed FQE values were not suitable for wheat at all growth stages. At wheat's erecting period and at its booting or flowering period, the systematical deviations in SIF_B and SIF_A/SIF_B simulations were noteworthy, and the seasonal cycles of SIF_A/SIF_B cannot be credibly reproduced, with a low determination coefficient (R^2) of 0.0286.

When the SIF simulation was conducted with variable FQE values, which vary with the growth of wheat, its accuracy was improved greatly. Specifically, the SCOPE model can accurately simulate the SIF_B and SIF_A with RRMSEs of 18.27% and 19.25%, respectively, and the SCOPE simulations track well with the seasonal SIF_A/SIF_B values with an RRMSE of 20.63% and a determination coefficient (R^2) of 0.48. The results indicated a clear seasonal pattern of suitable FQE values. When the growth stage changed from the erecting to the flowering stage, the fqe1/fqe2 increased from approximately 0.05–0.1 to approximately 0.3–0.5, while the fqe2 decreased from 0.013 to 0.07.

Therefore, although the SCOPE model can credibly simulate canopy SIF, the input FQE values should be carefully determined. Seasonal changes of the FQE values or the FQE values' dependence on plant physiological status cannot be ignored for accurate simulations of canopy SIF. Our quantitative results of the model assessment and FQE adjustment can serve as a significant reference for future application of the SCOPE model. However, the study is preliminary; more experiments are needed to determine FQE values in the SCOPE model.

Acknowledgments: This research was supported by the National Key Research and Development Program of China (2017YFA0603001), and the National Natural Science Foundation of China (41701396). The authors are thankful to Christiaan Van der Tol for providing the SCOPE model. We thank the anonymous reviewers for providing comments that helped to improve the quality of the original manuscript.

Author Contributions: Jiaochan Hu was primarily responsible for mathematical modeling and experimental design. Xinjie Liu improved the experimental analysis and revised the paper. Liangyun Liu contributed to the original idea for the research and to the experimental design. Linlin Guan provided support regarding the application of Diurnal flux and meteorological observations.

Conflicts of Interest: The authors declare no conflict of interest.

References

1. Pedrós, R.; Moya, I.; Goulas, Y.; Jacquemoud, S. Chlorophyll fluorescence emission spectrum inside a leaf. *Photochem. Photobiol. Sci.* **2008**, *7*, 498–502. [CrossRef] [PubMed]
2. Zarco-Tejada, P.J. Hyperspectral Remote Sensing of Closed Forest Canopies: Estimation of Chlorophyll Fluorescence and Pigment Content. Ph.D. Thesis, York University Toronto, Toronto, ON, Canada, December 2000.
3. Pfündel, E. Estimating the contribution of photosystem i to total leaf chlorophyll fluorescence. *Photosynth. Res.* **1998**, *56*, 185–195. [CrossRef]
4. Krause, G.H.; Weis, E. Chlorophyll fluorescence and photosynthesis—The basics. *Ann. Rev. Plant Biol.* **1991**, *42*, 313–349. [CrossRef]

5. Porcar-Castell, A.; Tyystjärvi, E.; Atherton, J.; van der Tol, C.; Flexas, J.; Pfündel, E.E.; Moreno, J.; Frankenberg, C.; Berry, J.A. Linking chlorophyll a fluorescence to photosynthesis for remote sensing applications: Mechanisms and challenges. *J. Exp. Bot.* **2014**, *65*, 4065–4095. [CrossRef] [PubMed]
6. Lichtenthaler, H.K.; Rinderle, U. The role of chlorophyll fluorescence in the detection of stress conditions in plants. *CRC Crit. Rev. Anal. Chem.* **1988**, *19*, S29–S85. [CrossRef]
7. Damm, A.; Elbers, J.A.N.; Erler, A.; Gioli, B.; Hamdi, K.; Hutjes, R.; Kosvancova, M.; Meroni, M.; Miglietta, F.; Moersch, A.; et al. Remote sensing of sun-induced fluorescence to improve modeling of diurnal courses of gross primary production (GPP). *Glob. Chang. Biol.* **2010**, *16*, 171–186. [CrossRef]
8. Damm, A.; Guanter, L.; Paul-Limoges, E.; van der Tol, C.; Hueni, A.; Buchmann, N.; Eugster, W.; Ammann, C.; Schaepman, M.E. Far-red sun-induced chlorophyll fluorescence shows ecosystem-specific relationships to gross primary production: An assessment based on observational and modeling approaches. *Remote Sens. Environ.* **2015**, *166*, 91–105. [CrossRef]
9. Guanter, L.; Zhang, Y.; Jung, M.; Joiner, J.; Voigt, M.; Berry, J.A.; Frankenberg, C.; Huete, A.R.; Zarco-Tejada, P.; Lee, J.-E. Global and time-resolved monitoring of crop photosynthesis with chlorophyll fluorescence. *Proc. Natl. Acad. Sci. USA* **2014**, *111*, 1327–1333. [CrossRef] [PubMed]
10. Liu, L.; Guan, L.; Liu, X. Directly estimating diurnal changes in GPP for C3 and C4 crops using far-red sun-induced chlorophyll fluorescence. *Agric. For. Meteorol.* **2017**, *232*, 1–9. [CrossRef]
11. Yang, X.; Tang, J.; Mustard, J.F.; Lee, J.E.; Rossini, M.; Joiner, J.; Munger, J.W.; Kornfeld, A.; Richardson, A.D. Solar-induced chlorophyll fluorescence that correlates with canopy photosynthesis on diurnal and seasonal scales in a temperate deciduous forest. *Geophys. Res. Lett.* **2015**, *42*, 2977–2987. [CrossRef]
12. Plascyk, J.A. The MK II fraunhofer line discriminator (FLD-II) for airborne and orbital remote sensing of solar-stimulated luminescence. *Opt. Eng.* **1975**, *14*, 339–346. [CrossRef]
13. Meroni, M.; Rossini, M.; Guanter, L.; Alonso, L.; Rascher, U.; Colombo, R.; Moreno, J. Remote sensing of solar-induced chlorophyll fluorescence: Review of methods and applications. *Remote Sens. Environ.* **2009**, *113*, 2037–2051. [CrossRef]
14. Frankenberg, C.; Fisher, J.B.; Worden, J.; Badgley, G.; Saatchi, S.S.; Lee, J.E.; Toon, G.C.; Butz, A.; Jung, M.; Kuze, A.; et al. New global observations of the terrestrial carbon cycle from GOSAT: Patterns of plant fluorescence with gross primary productivity. *Geophys. Res. Lett.* **2011**, *38*, 351–365. [CrossRef]
15. Frankenberg, C.; O'Dell, C.; Berry, J.; Guanter, L.; Joiner, J.; Köhler, P.; Pollock, R.; Taylor, T.E. Prospects for chlorophyll fluorescence remote sensing from the orbiting carbon observatory-2. *Remote Sens. Environ.* **2014**, *147*, 1–12. [CrossRef]
16. Guanter, L.; Frankenberg, C.; Dudhia, A.; Lewis, P.E.; Gómez-Dans, J.; Kuze, A.; Suto, H.; Grainger, R.G. Retrieval and global assessment of terrestrial chlorophyll fluorescence from gosat space measurements. *Remote Sens. Environ.* **2012**, *121*, 236–251. [CrossRef]
17. Joiner, J.; Yoshida, Y.; Vasilkov, A.P.; Corp, L.A.; Middleton, E.M. First observations of global and seasonal terrestrial chlorophyll fluorescence from space. *Biogeosciences* **2011**, *8*, 637–651. [CrossRef]
18. Joiner, J.; Guanter, L.; Lindstrot, R.; Voigt, M.; Vasilkov, A.P.; Middleton, E.M.; Huemmrich, K.F.; Yoshida, Y.; Frankenberg, C. Global monitoring of terrestrial chlorophyll fluorescence from moderate spectral resolution near-infrared satellite measurements: Methodology, simulations, and application to gome-2. *Atmos. Meas. Tech.* **2013**, *6*, 2803–2823. [CrossRef]
19. Guan, K.; Berry, J.A.; Zhang, Y.; Joiner, J.; Guanter, L.; Badgley, G.; Lobell, D.B. Improving the monitoring of crop productivity using spaceborne solar-induced fluorescence. *Glob. Chang. Biol.* **2016**, *22*, 716–726. [CrossRef] [PubMed]
20. Van der Tol, C.; Verhoef, W.; Timmermans, J.; Verhoef, A.; Su, Z. An integrated model of soil-canopy spectral radiances, photosynthesis, fluorescence, temperature and energy balance. *Biogeosciences* **2009**, *6*, 3109–3129. [CrossRef]
21. Liu, X.; Liu, L. Improving chlorophyll fluorescence retrieval using reflectance reconstruction based on principal components analysis. *IEEE Geosci. Remote Sens. Lett.* **2015**, *12*, 1645–1649.
22. Liu, X.; Liu, L.; Zhang, S.; Zhou, X. New spectral fitting method for full-spectrum solar-induced chlorophyll fluorescence retrieval based on principal components analysis. *Remote Sens.* **2015**, *7*, 10626–10645. [CrossRef]
23. Zhang, Y.; Guanter, L.; Berry, J.A.; Joiner, J.; van der Tol, C.; Huete, A.; Gitelson, A.; Voigt, M.; Kohler, P. Estimation of vegetation photosynthetic capacity from space-based measurements of chlorophyll fluorescence for terrestrial biosphere models. *Glob. Chang. Biol.* **2014**, *20*, 3727–3742. [CrossRef] [PubMed]

24. Verrelst, J.; van der Tol, C.; Magnani, F.; Sabater, N.; Rivera, J.P.; Mohammed, G.; Moreno, J. Evaluating the predictive power of sun-induced chlorophyll fluorescence to estimate net photosynthesis of vegetation canopies: A scope modeling study. *Remote Sens. Environ.* **2016**, *176*, 139–151. [CrossRef]

25. Van der Tol, C.; Berry, J.; Campbell, P.; Rascher, U. Models of fluorescence and photosynthesis for interpreting measurements of solar-induced chlorophyll fluorescence. *J. Geophys. Res. Biogeosci.* **2014**, *119*, 2312–2327. [CrossRef] [PubMed]

26. Koffi, E.; Rayner, P.; Norton, A.; Frankenberg, C.; Scholze, M. Investigating the usefulness of satellite derived fluorescence data in inferring gross primary productivity within the carbon cycle data assimilation system. *Biogeosci. Discuss.* **2015**, *12*, 707–749. [CrossRef]

27. Lee, J.E.; Berry, J.A.; Tol, C.; Yang, X.; Guanter, L.; Damm, A.; Baker, I.; Frankenberg, C. Simulations of chlorophyll fluorescence incorporated into the Community Land Model version 4. *Glob. Chang Biol.* **2015**, *21*, 3469–3477. [CrossRef] [PubMed]

28. Van der Tol, C.; Rossini, M.; Cogliati, S.; Verhoef, W.; Colombo, R.; Rascher, U.; Mohammed, G. A model and measurement comparison of diurnal cycles of sun-induced chlorophyll fluorescence of crops. *Remote Sens. Environ.* **2016**, *186*, 663–677. [CrossRef]

29. Verrelst, J.; Rivera, J.P.; van der Tol, C.; Magnani, F.; Mohammed, G.; Moreno, J. Global sensitivity analysis of the scope model: What drives simulated canopy-leaving sun-induced fluorescence? *Remote Sens. Environ.* **2015**, *166*, 8–21. [CrossRef]

30. Agati, G.; Mazzinghi, P.; Fusi, F.; Ambrosini, I. The F685/F730 chlorophyll fluorescence ratio as a tool in plant physiology: Response to physiological and environmental factors. *J. Plant Physiol.* **1995**, *145*, 228–238. [CrossRef]

31. Cheng, Y.-B.; Middleton, E.; Zhang, Q.; Huemmrich, K.; Campbell, P.; Corp, L.; Cook, B.; Kustas, W.; Daughtry, C. Integrating solar induced fluorescence and the photochemical reflectance index for estimating gross primary production in a cornfield. *Remote Sens.* **2013**, *5*, 6857–6879. [CrossRef]

32. Gitelson, A.A.; Buschmann, C.; Lichtenthaler, H.K. Leaf chlorophyll fluorescence corrected for re-absorption by means of absorption and reflectance measurements. *J. Plant Physiol.* **1998**, *152*, 283–296. [CrossRef]

33. Genty, B.; Briantais, J.M.; Baker, N.R. The relationship between the quantum yield of photosynthetic electron transport and quenching of chlorophyll fluorescence. *BBA—Gen. Subj.* **1989**, *990*, 87–92. [CrossRef]

34. Lazár, D. Chlorophyll a fluorescence induction. *Biochim. Biophys. Acta* **1999**, *1412*, 1–28. [CrossRef]

35. Lazár, D. Chlorophyll a fluorescence rise induced by high light illumination of dark-adapted plant tissue studied by means of a model of photosystem II and considering photosystem II heterogeneity. *J. Theor. Biol.* **2003**, *220*, 469–503. [CrossRef] [PubMed]

36. Lazár, D. Parameters of photosynthetic energy partitioning. *J. Plant Physiol.* **2015**, *175*, 131–147. [CrossRef] [PubMed]

37. Trissl, H.W.; Gao, Y.; Wulf, K. Theoretical fluorescence induction curves derived from coupled differential equations describing the primary photochemistry of photosystem II by an exciton-radical pair equilibrium. *Biophys. J.* **1993**, *64*, 974–988. [CrossRef]

38. Björkman, O.; Barbara, D. Photon yield of O_2 evolution and chlorophyll fluorescence characteristics at 77 K among vascular plants of diverse origins. *Planta* **1987**, *170*, 489–504. [CrossRef] [PubMed]

39. Lazár, D. Simulations show that a small part of variable chlorophyll a, fluorescence originates in photosystem I and contributes to overall fluorescence rise. *J. Theor. Biol.* **2013**, *335*, 249–264. [CrossRef] [PubMed]

40. Weis, E.; Berry, J.A. Quantum efficiency of photosystem II in relation to energy-dependent quenching of chlorophyll fluorescence. *Biochim. Biophys. Acta (BBA)-Bioenerg.* **1987**, *894*, 198–208. [CrossRef]

41. Verhoef, W. Light scattering by leaf layers with application to canopy reflectance modeling: The SAIL model. *Remote Sens. Environ.* **1984**, *16*, 125–141. [CrossRef]

42. Jacquemoud, S.; Baret, F. PROSPECT: A model of leaf optical properties spectra. *Remote Sens. Environ.* **1990**, *34*, 75–91. [CrossRef]

43. Farquhar, G.D.; von Caemmerer, S.; Berry, J.A. A biochemical model of photosynthetic CO_2 assimilation in leaves of C3 species. *Planta* **1980**, *149*, 78–90. [CrossRef] [PubMed]

44. Von Caemmerer, S. Steady-state models of photosynthesis. *Plant Cell Environ.* **2013**, *36*, 1617–1630. [CrossRef] [PubMed]

45. Collatz, G.J.; Ball, J.T.; Grivet, C.; Berry, J.A. Physiological and environmental regulation of stomatal conductance, photosynthesis and transpiration: A model that includes a laminar boundary layer. *Agric. For. Meteorol.* **1991**, *54*, 107–136. [CrossRef]

46. Collatz, G.J.; Ribas-Carbo, M.; Berry, J.A. Coupled photosynthesis–stomatal conductance model for leaves of C4 plants. *Aust. J. Plant Physiol.* **1992**, *19*, 519–538. [CrossRef]

47. Magnani, F.; Olioso, A.; Demarty, J.; Germain, V.; Verhoef, W.; Moya, I.; Van der Tol, C. Assessment of Vegetation Photosynthesis through Observation of Solar Induced Fluorescence from Space. In *Final Report for the European Space Agency under ESTEC Contract No. 20678/07/NL/HE*; ESA: Paris, France, 2009.

48. Porcar-Castell, A. A high-resolution portrait of the annual dynamics of photochemical and non-photochemical quenching in needles of Pinus sylvestris. *Physiol. Plant.* **2011**, *143*, 139–153. [CrossRef] [PubMed]

49. Xu, L.; Baldocchi, D.D. Seasonal trends in photosynthetic parameters and stomatal conductance of blue oak (*Quercus douglasii*) under prolonged summer drought and high temperature. *Tree Physiol.* **2003**, *23*, 865–877. [CrossRef] [PubMed]

50. Goel, N.S.; Strebel, D.E. Simple beta distribution representation of leaf orientation in vegetation canopies. *Agron. J.* **1984**, *76*, 800–802. [CrossRef]

51. Liu, L.; Wang, J.; Bao, Y.; Huang, W.; Ma, Z.; Zhao, C. Predicting winter wheat condition, grain yield and protein content using multi-temporal EnviSat-ASAR and Landsat TM satellite images. *Int. J. Remote Sens.* **2006**, *27*, 737–753. [CrossRef]

52. Ergun, E.; Demirata, B.; Gumus, G.; Apak, R. Simultaneous determination of chlorophyll a and chlorophyll b by derivative spectrophotometry. *Anal. Bioanal. Chem.* **2004**, *379*, 803–811. [CrossRef] [PubMed]

53. Falge, E.; Baldocchi, D.; Olson, R.; Anthoni, P.; Aubinet, M.; Bernhofer, C.; Burba, G.; Ceulemans, R.; Clement, R.; Dolman, H. Gap filling strategies for defensible annual sums of net ecosystem exchange. *Agric. For. Meteorol.* **2001**, *107*, 43–69. [CrossRef]

54. Reichstein, M.; Falge, E.; Baldocchi, D.; Papale, D.; Aubinet, M.; Berbigier, P.; Bernhofer, C.; Buchmann, N.; Gilmanov, T.; Granier, A. On the separation of net ecosystem exchange into assimilation and ecosystem respiration: Review and improved algorithm. *Glob. Chang. Biol.* **2005**, *11*, 1424–1439. [CrossRef]

55. Plascyk, J.A.; Gabriel, F.C. Fraunhofer line discriminator MK II—Airborne instrument for precise and standardized ecological luminescence measurement. *IEEE Trans. Instrum. Meas.* **1975**, *24*, 306–313. [CrossRef]

56. Liu, L.; Liu, X.; Hu, J. Effects of spectral resolution and SNR on the vegetation solar-induced fluorescence retrieval using FLD-based methods at canopy level. *Eur. J. Remote Sens.* **2015**, *48*, 743–762. [CrossRef]

57. Liu, L.; Liu, X.; Hu, J.; Guan, L. Assessing the wavelength-dependent ability of solar-induced chlorophyll fluorescence to estimate the GPP of winter wheat at the canopy level. *Int. J. Remote Sens.* **2017**, *38*, 4396–4417. [CrossRef]

58. Maier, S.W.; Günther, K.P.; Stellmes, M. Sun-induced fluorescence: A new tool for precision farming. In *Digital Imaging and Spectral Techniques: Applications to Precision Agriculture and Crop Physiology*; Kral, D.M., Barbarick, K.A., Volenec, J.J., Dick, W.A., Eds.; American Society of Agronomy Special Publication: Madison, WI, USA, 2003; pp. 209–222.

59. Wullschleger, S.D. Biochemical Limitations to Carbon Assimilation in C3 Plants—A Retrospective Analysis of the A/Ci Curves from 109 Species. *J. Exp. Bot.* **1993**, *44*, 907–920. [CrossRef]

60. Sellers, P.J.; Dickinson, R.E.; Randall, D.A.; Betts, A.K.; Hall, F.G.; Berry, J.A.; Collatz, G.J.; Denning, A.S.; Mooney, H.A.; Nobre, C.A.; et al. Modeling the exchanges of energy, water, and carbon between continents and the atmosphere. *Science* **1997**, *275*, 502–509. [CrossRef] [PubMed]

61. Kattge, J.; Knorr, W.; Raddatz, T.; Wirth, C. Quantifying photosynthetic capacity and its relationship to leaf nitrogen content for global scale terrestrial biosphere models. *Glob. Chang. Biol.* **2009**, *15*, 976–991. [CrossRef]

62. Mäkelä, A.; Hari, P.; Berninger, F.; Hänninen, H.; Nikinmaa, E. Acclimation of photosynthetic capacity in Scots pine to the annual cycle of temperature. *Tree Physiol.* **2004**, *24*, 369–376. [CrossRef] [PubMed]

63. Wolf, A.; Akshalov, K.; Saliendra, N.; Johnson, D.A.; Laca, E.A. Inverse estimation of Vcmax, leaf area index, and the Ball-Berry parameter from carbon and energy fluxes. *J. Geophys. Res. Atmos.* **2006**, *111*, 1003–1019. [CrossRef]

64. Liu, L.; Zhang, Y.; Wang, J.; Zhao, C. Detecting solar-induced chlorophyll fluorescence from field radiance spectra based on the fraunhofer line principle. *IEEE Trans. Geosci. Remote Sens.* **2005**, *43*, 827–832.

65. Poolman, M.G.; Fell, D.A.; Thomas, S. Modelling photosynthesis and its control. *J. Exp. Bot.* **2000**, *51*, 319–328. [CrossRef] [PubMed]

66. Poolman, M.G.; Olçer, H.; Lloyd, J.C.; Raines, C.A.; Fell, D.A. Computer modelling and experimental evidence for two steady states in the photosynthetic Calvin cycle. *Eur. J. Biochem.* **2001**, *268*, 2810–2816. [CrossRef] [PubMed]

67. Ainsworth, E.A.; Serbin, S.P.; Skoneczka, J.A.; Townsend, P.A. Using leaf optical properties to detect ozone effects on foliar biochemistry. *Photosynth. Res.* **2014**, *119*, 65–76. [CrossRef] [PubMed]

68. Kothavala, Z.; Arain, M.A.; Black, T.A.; Verseghy, D. The simulation of energy, water vapor and carbon dioxide fluxes over common crops by the Canadian Land Surface Scheme (CLASS). *Agric. For. Meteorol.* **2005**, *133*, 89–108. [CrossRef]

69. Lokupitiya, E.; Denning, S.; Paustian, K.; Baker, I.; Schaefer, K.; Verma, S.B.; Meyers, T.; Bernacchi, C.J.; Suyker, A.E.; Fischer, M. Incorporation of crop phenology in Simple Biosphere model (SiBcrop) to improve land-atmosphere carbon exchanges from croplands. *Biogeosciences* **2009**, *6*, 969–986. [CrossRef]

70. Wilson, K.B.; Baldocchi, D.D.; Hanson, P.J. Spatial and seasonal variability of photosynthetic parameters and their relationship to leaf nitrogen in a deciduous forest. *Tree Physiol.* **2000**, *20*, 565–578. [CrossRef] [PubMed]

71. Morales, F.; Abadía, A.; Abadía, J. Chlorophyll Fluorescence and Photon Yield of Oxygen Evolution in Iron-Deficient Sugar Beet (*Beta vulgaris* L.) Leaves. *Plant Physiol.* **1991**, *97*, 886–893. [CrossRef] [PubMed]

72. Hussain, M.I.; Reigosa, M.J. A chlorophyll fluorescence analysis of photosynthetic efficiency, quantum yield and photon energy dissipation in PSII antennae of *Lactuca sativa* L. leaves exposed to cinnamic acid. *Plant Physiol. Biochem.* **2011**, *49*, 1290–1298. [CrossRef] [PubMed]

remote sensing

MDPI

Article

Impacts of Leaf Age on Canopy Spectral Signature Variation in Evergreen Chinese Fir Forests

Qiaoli Wu [1,2], Conghe Song [3], Jinling Song [1,2,*], Jindi Wang [1,2], Shaoyuan Chen [1,2] and Bo Yu [1,2]

[1] State Key Laboratory of Remote Sensing Science, Jointly Sponsored by Beijing Normal University and Institute of Remote Sensing and Digital Earth of Chinese Academy of Sciences, Beijing 100875, China; wuql@mail.bnu.edu.cn (Q.W.); wangjd@bnu.edu.cn (J.W.); chenshaoyuan1020@gmail.com (S.C.); yuboray@163.com (B.Y.)

[2] Beijing Engineering Research Center for Global Land Remote sensing Products, Institute of Remote Sensing Science and Engineering, Faculty of Geographical Science, Beijing Normal University, Beijing 100875, China

[3] Department of Geography, University of North Carolina at Chapel Hill, Chapel Hill, NC 27599, USA; csong@email.unc.edu

* Correspondence: songjl@bnu.edu.cn; Tel.: +86-10-5880-5452; Fax: +86-10-5880-5274

Received: 5 January 2018; Accepted: 3 February 2018; Published: 8 February 2018

Abstract: Significant gaps exist in our knowledge of the impact of leaf aging on canopy signal variability, which limits our understanding of vegetation status based on remotely sensed data. To understand the effects of leaf aging at the leaf and canopy scales, a combination of field, remote-sensing and physical modeling techniques was adopted to assess the canopy spectral signals of evergreen *Cunninghamia* forests. We observed an approximately 10% increase in Near-Infrared (NIR) reflectance for new leaves and a 35% increase in NIR transmittance for mature leaves from May to October. When variations in leaf optical properties (LOPs) of only mature leaves, or both new and mature leaves were considered, the Geometric Optical and Radiative Transfer (GORT) model-simulated canopy reflectance trajectory was more consistent with Landsat observations (R^2 increased from 0.37 to 0.82~0.89 for NIR reflectance, and from 0.35 to 0.67~0.88 for EVI2, with a small RMSE (0.01 to 0.02)). This study highlights the importance of leaf age on leaf spectral signatures, and provides evidence of age-dependent LOPs that have important impacts on canopy reflectance in the NIR band and EVI2, which are used to monitor canopy dynamics and productivity, with important implications for RS and forest ecosystem ecology.

Keywords: leaf age; leaf spectral properties; leaf area index; *Cunninghamia*; Chinese fir; canopy reflectance; NIR; EVI2; geometric optical radiative transfer (GORT) model

1. Introduction

Forests cover approximately 30% of the Earth's land area (4.2×10^9 hectares). Globally, forests play critical roles in providing goods and services for terrestrial ecosystems, including filtering water, controlling water runoff, protecting soil, regulating the climate, and cycling and storing nutrients [1,2]. Many important biophysical processes in forests are conducted through leaves, including photosynthesis, transpiration, respiration, and light interception. Forests and other land vegetation currently remove approximately 30% of anthropogenic CO_2 emissions from the atmosphere through photosynthesis [3–5]. While high value has been placed on remote sensing (RS) for ecological research, management and modeling of forest canopy status at an ecosystem scale, a concomitant increase in understanding the factors that affect canopy reflectance has been only partially realized.

The interpretation of RS signals for forest canopies requires profound knowledge of the factors affecting their optical properties, which may be internal or external to the forest stand [6]. To extract useful information related to canopy growth using time-series data, anomalies in time-series data that are unrelated to real changes in canopies should be eliminated, such as clouds and aerosol contamination [7,8] as well as bidirectional reflectance distribution function (BRDF) effects caused by topography and sun-sensor geometry variation [9–14]. Abundant studies have confirmed the existence of significant seasonal patterns in the optical RS signal trajectory that remain after removing the impacts of these factors external to the forest stand [11,13,15,16]. Thus, variations in reflectance trajectories contain useful information related to factors internal to the stand. Based on previously reported experimental and modeling data, vegetation reflectance is primarily a function of tissue optical properties (reflectance and transmittance), canopy biophysical attributes (e.g., leaf area, foliage clumping) and background reflectance [17–21]. Leaf optical properties (LOPs) and leaf area index (LAI) are two of the main recognized internal factors involved in controlling canopy reflectance.

Seasonality in canopy spectral signals has been attributed to a varying LAI along with new leaf development and defoliation [22]. In deciduous forests, LAI shows high seasonality and might be the most significant factor affecting canopy reflectance, as all leaves are shed in winter. However, the situation is very different for mature evergreen forests, which show a relatively stable total leaf area every year. Thus, changes in canopy reflectance do not necessarily imply changes in LAI [23]. In evergreen forests, leaves have more than a one-year lifespan, and current-year new leaves remain throughout the winter. Although old leaves are shed every year, new leaves are also produced every year. The total leaf area in evergreen forests remains relatively stable throughout a year compared with deciduous forests. However, we still find significant seasonal variation in canopy reflectance in evergreen forests, which may not be caused by variations in LAI but rather by other internal factors. Further studies are needed to develop a more profound interpretation of the seasonality of optical RS signals for evergreen forests.

In addition to leaf area, LOPs are another significant factor that can change with time, which may strongly affect canopy reflectance [19,23,24], and leaf-age effects on canopy signals require more attention. Leaves affects the radiation field through its LOPs, including leaf reflectance and transmittance characteristics, which are wavelength-dependent [25,26]. A few supporting studies have confirmed the combined effects of LAI and LOPs on the seasonality of gross ecosystem CO_2 exchange (GEE), photosynthesis and NIR reflectance for Amazonian forests [23,27,28]. Many efforts aimed at accounting for effect of stand age on canopy reflectance with forest succession have improved the interpretation of canopy signals [29–33]. However, only a few studies have indicated that in addition to stand age, leaf age might also have a significant impact. Our ability to both interpret RS signals and develop new RS technologies for vegetation depends directly on our ability to resolve the multitude of factors controlling canopy and landscape reflectance signatures. This situation inspired us to further evaluate robust biophysical interpretations of the seasonality of optical RS signals, with a focus on evergreen forests, by examining the effects of age-dependent leaf properties on canopy reflectance.

Different age cohorts of leaves coexist in the canopy of evergreen forests, which can be classified to two main age groups: current-year new leaves ($\leq$1 a) and mature leaves (>1 a), which might exhibit different LOPs and, thus, different impacts on canopy reflectance. Mature leaves represent the majority and new leaves are the minority of leaves flushed every year. On the other hand, new leaves are mainly distributed at the top and in the outermost parts in tree crowns, while mature leaves are distributed in the lower and inner parts of the canopy. Thus, the seasonal variation of canopy reflectance might be strongly affected by changes in LOPs of both new leaves and mature leaves. This brings us to the crux of our study: the quantitative analysis and interpretation of the different leaf-aging effects of new leaves and mature leaves on seasonal variations in canopy reflectance. Hence, we addressed the following research questions in this study: (1) how do the LOPs of new leaves and mature leaves vary during the leaf maturation process? (2) How should the contributions of new leaves and mature leaves to canopy optical properties be quantified? (3) How does leaf aging affect the seasonal variation of the remotely sensed response spectral signals of evergreen forest canopies?

2. Materials

2.1. Study Sites

This study focused on two National Research Stations of Forest Ecosystems in Huitong county, Hunan province, China, as shown in Figure 1. The first station (Station 1, S1) is located at 26°40′N, 109°26′E, and the second station (Station 2, S2) is located at 26°50′N, 109°45′E. S1 was established in 1983, and S2 was established in 1996. Two main permanent sampling plots, ZH1 and WS3, were selected from S1 and S2, respectively, to measure and evaluate the effects of leaf aging on canopy optical properties. Two additional auxiliary plots, i.e., FZ1 and WS2, were also selected for comparison with the main plots. FZ1 exhibits the same plot conditions as ZH1 but is smaller, with a size of 30 × 40 m. WS3 is located next to WS2 and is eight years older than WS2.

Cunninghamia (Chinese fir) is a fast growing species, with its height increasing as much as 1 m per year [34], and usually matures approximately 20 a. The same *Cunninghamia* seedlings were planted in ZH1 and FZ1 after clear cutting. The stem density is the same in ZH1/FZ1, which exhibits half the density in WS2 (1920 trees ha^{-1}) and WS3 (1967 trees ha^{-1}). ZH1/FZ1 were planted with an initial stem density of 2500 trees ha^{-1} and thinned twice, once in 1997 and again in 2003, to a stable density of 1035 trees ha^{-1}.

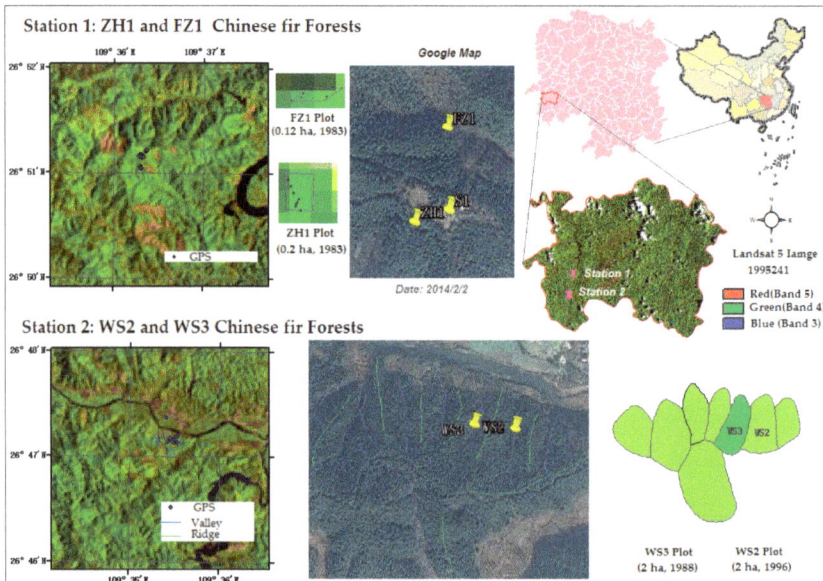

Figure 1. These four study plots are covered by *Cunninghamia lanceolata* (also known as Chinese fir) plantations, which were replanted after clear-cutting.

2.2. Field Data

2.2.1. Canopy Structural Parameter Measurements

Crown shape measurements were taken from a total of forty trees in S2. The size of the selected trees was evenly distributed in terms of height (*H*) (from 5.4 m to 20 m) and diameter at breast height (DBH) (from 7.7 cm to 37.8 cm). Among the 40 trees, seven trees were located in plots close to the study site, and 33 trees were located within sites ZH1 and FZ1. The parameters measured for the characterization of crown shape included the following: crown width in the north-south direction (*R*1) and the east-west direction (*R*2), tree height (*H*1) and height under the crown (*H*2), from which we can obtain the height of the crown center (*h*). A detailed description of the crown shape measurements can be found in Appendix A.

Only DBH and *H*1 were measured annually for each tree in the study plots, and other canopy structure parameters were only measured once for the 40 trees. Hence, to characterize the dominant stand canopy structure changes with time, we needed to build allometric relationships between other unknown canopy structural parameters with DBH or *H*1. The derived canopy structure parameters included the following:

- Crown radius (*R*): $R = (1.296 + 0.146 * DBH)/2$, $R^2 = 0.76$, RMSE = 0.72;
- Full tree height (*H*1): $H1 = 3.928 + 14.866 * (1 - \exp(-0.1865 * DBH^{8.285}))$, $R^2 = 0.94$, RMSE = 1.30;
- Crown center height (*h*): $h = -0.32 + 0.85 * H1$, $R^2 = 0.93$, RMSE = 0.98;
- Crown ellipticity (*b/R*), which was fixed at the mean value of the 40 field measurements: mean = 1.17, standard deviation (s.d.) = 0.46, since no significant relationship was found between *b/R* and DBH or H1.

Finally, the variations in canopy structure with stand development can be characterized by the means and standard deviations of the DBH and tree height (*H*1) for every individual tree.

2.2.2. LAI Measurements and Data Processing

Digital hemispherical photography (DHP) was the primary method for conducting regular monthly LAI measurements from 2005 until the present. From 2005 to 2006, photographs were taken by a worker every month using a CI-110 Plant Canopy Analyzer (Camas, WA, USA) to estimate the LAI. From 2007 onward; photographs were taken using a Canon EOS 40D digital camera equipped with a Nikon AF-DX 10.5 mm f/2.8 G ED fisheye lens. The camera was horizontally mounted at a fixed height of 0.2 m above the ground. The photographs were taken with automatic exposure under diffuse light conditions, typically soon before sunrise or after sunset.

In the ZH1 and FZ1 plots, measurements were taken on the 15th day of each month at five fixed locations per plot, facing in four cardinal directions. The images were processed using Gap Light Analyzer 2.0 software to calculate LAI_{DHP}. LAI_{DHP} measurements obtained with automatic exposure resulted in considerable underestimation because of underestimation of the ratio of green leaves to sky [35–39]. We used the effective LAI (LAI_e) measured by two LAI-2000 Plant Canopy Analyzers (LAI-2000, LI-COR) to correct the system bias in LAI_{DHP} data. One LAI-2000 unit was set at an open and flat area to measure diffuse sky light, and the other LAI-2000 unit was operated under the canopy at fixed locations as well as over the whole plots.

$$LAI_e = LAI_{DHP} + \varepsilon \tag{1}$$

where ε can be considered as a systematic bias in the DHP method due to automatic exposure problems, and ε was set to 1.83 for the DHP method using the average value.

The clumping index (Ω) was measured with the Tracing Radiation and Architecture of Canopies (TRAC, Natural Resources Canada) system to convert LAI_e to true LAI (LAI_t) using the following equation:

$$LAI_t = (1 - \alpha) * LAI_e * \gamma_e / \Omega \tag{2}$$

where the needle-to-shoot area ratio (γ_e) was set to 1.1 according to the results of the destructive sampling method described below, conducted in August 2015. The clumping index (Ω) was set to 0.8 and the value of the woody-to-total area ratio (α) was derived from the destructive samples for biomass estimation (0.2). A detailed description of the data preprocessing methods of LAI can be found in Appendix B.

2.2.3. Spectral Measurements: Leaf and Soil

Soil and leaf samples were collected from WS2 and WS3 at site 2. Current-year leaves were too short to be measured in April. Therefore, we collected current-year shoots every month during the leaf expansion period from May to August 2017 to measure the seasonal variations in LOPs. Trees were selected from WS2 and WS3 separately to obtain branches from different age cohorts. Twenty branches were randomly selected and destructively harvested for each age group to conduct spectral measurements. Young leaves and mature leaves of 0–3 a were collected from the same branches; the leaves were carefully stored in sealed plastic bags and measured within 48 h.

Spectral measurements over the full spectral range (350–2500 nm) were carried out in the laboratory using a portable spectroradiometer (SVC HR-1024) attached to an integrating sphere (Model 1800-12S, Licor). All spectra were standardized using a barium sulfate standard and the calibrated light source supplied by Licor with the integrating sphere. We arranged 8–10 flat needles closely together for each measurement, avoiding overlaps or gaps, to obtain leaf samples that were sufficiently wide to cover the gap on integrating sphere. Wide transparent tape was used to assemble the leaf leaves on the vacuum side of the blade, which was removed from the central region to avoid any impacts on LOPs. LOPs were measured on the upper surface of the *Cunninghamia lanceolata* leaf samples. A group of leaf sample is shown in Figure A4 in Appendix A.

2.3. Remote Sensing Observations

Landsat Observations

Landsat sensors have a long history and provide data with a fine spatial resolution (30 × 30 m) that can effectively capture forest stands in the landscape. In addition to their ideal spatial resolution, Landsat data provide certain other advantages, such as reducing the inconsistencies in observations by always viewing from almost the same direction (nadir view) and at the same time of day. However, one of the drawbacks of Landsat data (with a 16-day revisit period) is their relatively low temporal frequency, which is exacerbated by cloud-cover [40]. We collected all available L1T Landsat TM/ETM+ images for our study site (path/row: 125/41) from 1987 to 2016, and abstracted pixel reflectance values with no cloud contamination. Given the high frequency of cloud cover in forested areas, Landsat observations over one year are insufficient to describe seasonal patterns. Therefore, all available pixel reflectance values from all years were sorted by the day of year (DOY), based on which seasonal trajectories of canopy reflectance were constructed. Landsat red and NIR band surface reflectance data were used to calculate EVI2, which takes advantage of the auto-correlative properties of surface reflectance spectra between the red and blue bands:

$$EVI2 = 2.5 \frac{R_{NIR} - R_{RED}}{1 + R_{NIR} + 2.4R_{RED}} \tag{3}$$

where R_{NIR} is reflectance in the near infrared band and R_{RED} is reflectance in the red band.

3. Methods

3.1. Theoretical Foundation

Geometrical Optical Radiative Transfer (GORT) Model

The interaction between electromagnetic radiation and terrestrial plant canopies is a complex phenomenon and a key element in many RS applications. Among numerous methods for estimating the reflectance of forest canopies, the GORT model is based on the physical structure of the underlying scene. The GORT model is a hybrid of geometric optical (GO) and radiative transfer (RT) approaches for modeling canopy reflectance [41,42]. The GO model [43–45] quantifies single scattering in the canopy well and captures the fundamental properties of the canopy bidirectional reflectance distribution function (BRDF). The assumptions of the GO model are that the scene is composed of three-dimensional solid objects on a contrasting background and that the overall canopy reflectance can be modeled as a weighted sum of the spectral signatures of its individual scene components, based on their corresponding areal proportions within a pixel. The RT model is used to describe the multiple scattering within canopy elements in the GORT model, and the GO model and RT model are linked using canopy gap probabilities [46,47]. Due to the explicit consideration of crown gaps and mutual shadowing effects, the GORT model is suitable for simulating forest canopy reflectance with varying degrees of discontinuity. The GORT model has been successfully applied to predict the fundamental features of black spruce forest canopies [42], radiation penetrating the forest canopy to the forest floor [48] and spectral temporal manifestations of forest succession [49].

In the GO model [43–45], the reflectance of a pixel is modeled as the weighted sum of the spectral signatures of four scene components: sunlit ground, sunlit crown, shaded ground and shaded crown, as illustrated below:

$$S = K_g * G + K_c * C + K_z * Z + K_t * T \qquad (4)$$

where S is the average reflectance of the forest canopy inside a pixel; and K_g, K_c, K_z and K_t represent the areal proportions of the four components in the pixel; thus, K_g, K_c, K_z and K_t sum to unity. G, C, Z and T are the corresponding spectral signatures (reflectance in a given wavelength range) of the four components, as shown in Figure 2, which are functions of the proportions of incoming directional beams and diffuse solar radiation.

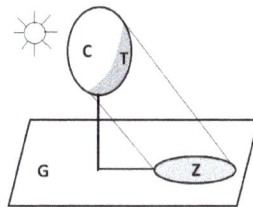

Figure 2. The four scene components used in the geometric optical (GO) model. C is the sunlit tree crown; T is the shaded tree crown; G is the sunlit background; and Z is the shaded background.

The canopy structure parameters listed in Table 1 for the stands are derived based on the field measurements of the DBH and tree height of every individual tree in the stands, and the allometric relationships between DBH or tree height. Parameter $h1$ and $h2$ must represent the structure of the dominant canopy layer. Therefore, we used the mean height minus s.d. to derive $h1$ and the mean height plus s.d. to derive $h2$. FAVD is calculated by dividing LAI by the crown volume, by treating the tree crown as an ellipsoid.

Table 1. Canopy structure parameters required for the GORT model to simulate canopy reflectance.

Symbols	Parameters	Source
	Canopy Structure Parameters	
$h1$	lower boundary of canopy center height	O [1] and CERN [2]
$h2$	upper boundary of canopy center height	O and CERN
R	horizon mean crown radius	O and CERN
b/R	crown spheroid ellipticity	1.17 (O)
λ	tree stem density (trees/ha)	CERN
FAVD	foliage area volume density (m^2/m^3)	O and CERN
k	leaf angle distribution factor	0.5 (random)
	Component Spectral Parameters	
r_L	leaf reflectance	O
t_L	leaf transmittance	O
r_G	soil/background reflectance	O
	Sun-Sensor Geometry	
SZN	sun zenith angle (°)	Time, Lon, Lat [3]
VZN	view zenith angle (°)	0
VAZ	view azimuth angle (°)	0

[1] O stands for observations from field experiments from 2015 to 2017. [2] CERN: National Ecosystem Research Network at Huitong National Research Station of Forest Ecosystem (HTF), China (http://htf.cern.ac.cn/meta/metaData). [3] Lon: Longitude; Lat: latitude.

3.2. Contribution of Component LOPs to Canopy LOPs

Different leaf-age cohorts coexist in an individual tree canopy in evergreen forests and can be classified into two major groups: new leaves and mature leaves. The LOPs of new leaves are quite different from those of mature leaves, thus, neither of them is representative of the LOPs of the whole canopy. Therefore, we need to consider the contributions of new leaves and mature leaves together to obtain the overall canopy LOPs.

3.2.1. Leaf Area Proportion of New and Mature Leaves

The first factor affecting the average canopy spectral properties is the proportion of the leaf area at different ages: the higher the proportion of the leaf area for a given age, the greater its impact on canopy optical properties. Thus, we need to consider the proportions of new leaves and mature leaves viewed by the sensor. L_{total} is the total LAI of all leaves, which consists of two parts:

$$L_{total} = L_{new} + L_{mature} \tag{5}$$

including the LAI of new leaves (L_{new}) and mature leaves (L_{mature}). The proportions of new and mature leaves were estimated based on destructive field measurements conducted by Zhongkun et al. [50], who studied the LAI of leaves at different leaf ages in Chinese fir, and we constructed the seasonally dynamic leaf area proportions by interpolating the data recorded using the phenology rule, assuming stable proportions to be reached at summer.

3.2.2. Spatial Organization of New and Mature Leaves

In addition to differences in the quantities of new and mature leaves, the spatial organization of new and mature leaves also has significant impacts on canopy component spectral signatures. Top crowns are mostly occupied by new leaves, and approximately 80% of the upper crown leaf area is occupied by new leaves in Chinese fir canopies [50]. Leaves distributed in upper crowns have significant impacts on canopy reflectance when viewed from the top of the crown. In contrast, old leaves (1 a, 2 a, 3 a) are mainly located in the inner and lower parts of tree crowns, which are less likely to be observed directly from the nadir view. Thus, contribution of new leaves located at the top of canopies cannot be ignored.

When we simulate canopy reflectance based on Landsat viewing geometry, new leaves occupy the majority of the field of view of the sensor, and solar radiation interacts with the leaves in the top layer first before reaching the lower canopy. Thus, new leaves in the upper canopy have a larger influence on canopy reflectance than mature leaves in the lower canopy. The influences of new and mature leaves on canopy LOPs are modeled as the areal-weighted averages of new and mature leaves observed by the sensors, as follows:

$$R_{ave} = w1 * R_{new} + w2 * R_{mature} \tag{6}$$

and

$$T_{ave} = w1 * T_{new} + w2 * T_{mature} \tag{7}$$

where R_{ave} and T_{ave} are the average leaf reflectance and transmittance at the canopy scale, respectively. R_{new} and R_{mature} are the reflectance of new and mature leaves, respectively. T_{new} and T_{mature} are the transmittance of new and mature leaves, respectively. Finally, the parameters of $w1$ and $w2$ are the areal weights for new and mature leaves, respectively.

Here, we model the areal weights, $w1$ and $w2$, considering both leaf-area and leaf-age impacts, as follows:

$$w1 = \frac{(1 - e^{-L_{new}})}{(1 - e^{-L_{total}})} \tag{8}$$

and

$$w2 = \frac{(1 - e^{-L_{total}}) - (1 - e^{-L_{new}})}{(1 - e^{-L_{total}})} = \frac{(e^{-L_{new}} - e^{-L_{total}})}{(1 - e^{-L_{total}})} \tag{9}$$

where $w1$ highlights the importance of new leaves in the crown in the top canopy, and the weight of mature leaves ($w2$) is estimated from the total contribution of all leaves after subtracting the occlusion effect of the new leaves.

3.3. LOPs at the Canopy Scale

LOPs at the canopy scale are a combination LOPs of both new leaves and mature leaves. At the leaf scale, LOPs are age-dependent and the LOPs of new leaves and mature leaves vary differentially. At the canopy scale, age-dependent LOPs provide a mechanism for producing seasonally varying forest albedo and changing NIR to red ratios, independent of changes in other canopy attributes. Our hypothesis is that component LOPs vary with leaf maturation, and contribute to the seasonality in canopy reflectance trajectories. In the following part, we describe methods for retrieving seasonal LOPs and examining the relationship between LOPs and canopy signals.

3.3.1. Model Sensitivity Analysis

To retrieve LOPs from Landsat observations using the GORT model, we need to obtain a comprehensive understanding of the sensitivity of model driven parameters. First, wide ranges are allowed for all parameters, and we used canopy structural measurements of young stands (stand age = 1 a) as the lower limit and canopy structural parameters of mature stands (stand age = 33 a) as the upper limit. Stem density and background reflectance were set to their true values. Since sensitivity analysis covered parameter ranges on a long temporal scale (1 to 33 a), insensitive parameters are eliminated. Then, the remaining sensitive parameters are involved in the new sensitivity analysis with a smaller range, taking the field measurements of each parameter as prior knowledge.

The global sensitivity of the model parameters is analyzed using the extension of the Fourier amplitude sensitivity testing (EFAST) method [51]. EFAST is a variance decomposition method determining what fraction of the variance in the model output can be explained by the variation in each input parameter (i.e., partial variance). The basis of the EFAST method is a parametric transformation that can reduce multidimensional integrals over the input parametric space to one-dimensional quadratures using a search curve that scans the whole input space [52]. Scanning is conducted so that each axis of the parametric space is explored at a different frequency. Then, Fourier decomposition is used to calculate both the first-order sensitivity (the contribution to the variance of the model output by each input, S_i) and total-order sensitivity (the first-order effect plus interactions with other inputs, S_{Ti}) of each input parameter, given as (Saltelli et al., 2008):

$$S_i = \frac{V_i}{V(Y)} = \frac{V[E(Y|X_i)]}{V(Y)} \tag{10}$$

and

$$S_{Ti} = S_i + \sum_{j \neq i} S_{ij} + \ldots = \frac{E[V(Y|X_{\sim i})]}{V(Y)} \tag{11}$$

where $X_{\sim i}$ denotes the variation in all input parameters except for X_i, and S_{ij} is the contribution to the total variance from the interactions between parameters.

Following Saltelli et al. [53], to compute S_i and S_{Ti}, we created a quasi-random sequence parameter sampling matrix, P, with dimensions of (m, n) for each SA test, where m is the sample size, and n is the number of input parameters. We set $m = 2^n$, which was sufficient to test the convergence of the sensitivity index and the stability of the rankings. Each row in matrix P represents a possible value set of X, and the quasi-random sequence helps to distribute the sampling points as uniformly as possible in the parameter space and avoid clustering, in addition to increasing the convergence rate. The global sensitivity analysis method is complex, and more details can be found in the work by Saltelli [53–55]. Fortunately, SimLab software [40] can help implement the EFAST method.

3.3.2. Model Inversion Strategy

In this section, we describe the GORT model inversion strategy used to retrieve LOPs from satellite observations. Landsat reflectance in the red and NIR bands and the derived EVI2 were used to retrieve sensitive parameters. The multi-stage inversion strategy proposed by Li et al. [56] was adopted for parameter retrieval. The main objective of the iterative inversion process is to adjust the model parameters so that the model output reflectance is as close as possible to the observed reflectance. The most sensitive parameter is retrieved first, and used as prior knowledge in the next inversion stage.

One of the most popular methods for solving the inversion problem is to minimize the cost function of control variables. In this study, the cost function [57] used to retrieve sensitive parameters from Landsat surface reflectance data is as follows:

$$J(x) = \frac{1}{2} \left(\sum_{n=1}^{N} \frac{\left[f_n(X) - y_n{}^{obs} \right]^2}{\delta_n{}^2} + \sum_{l=0}^{L} \frac{\left[X_l - X_l{}^{prior} \right]^2}{\delta_l{}^2} \right) \tag{12}$$

where $y_n{}^{obs}$ and $f_n(X)$ are the observed reflectance/EVI2 value and the corresponding modeled reflectance/EVI2 value, respectively. The variables $\delta_n{}^2$ and $\delta_l{}^2$ are the variances of the observational data and the prior distribution of parameters, respectively. The variables X_l and $X_l{}^{prior}$ are the parameter values and the initial values in the model, respectively. N is the number of observations, and L is the number of parameters. Sequential quadratic programming [58], an optimization algorithm for solving nonlinear programming problems, was adopted to search for the cost function minimum.

3.3.3. Validation: Direct and Indirect Methods

During the retrieval process, the LOPs of mature leaves in the red and NIR bands were set to fixed values, except for leaf transmittance in the NIR band, which was interpolated using field measurements; the LOPs of new leaves in the red band were also interpolated using SVC measurements for leaves. The retrieval results for new leaf LOPs in the NIR band were tested in direct and indirect ways, as shown in Figure 3.

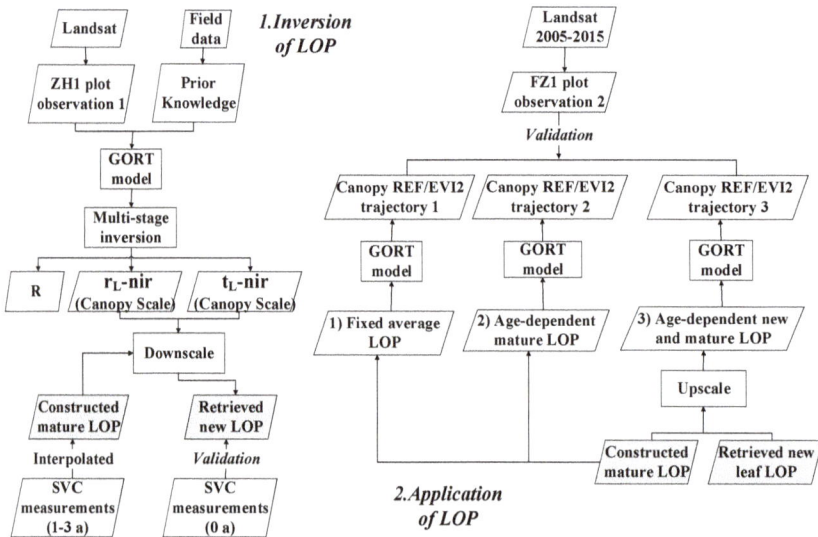

Figure 3. Flow chart of LOPs retrieval and application method.

We decomposed retrieved LOPs into new leaf (time variant) and mature leaf (known) components according to the calculated the contribution weights (w1 and w2) described in Section 3.2, and evaluated the decomposed LOPs transmittance using SVC field measurements directly. Additionally, we tested the applicability of the retrieved LOPs as a second-step validation. LOPs retrieved at one site (ZH1 plot) were applied to another site (FZ1 plot) to simulate canopy reflectance using the GORT forward simulating mode, and evaluated pixel reflectance using Landsat observations. FZ1 plot and ZH1 plot have similar growing conditions, but are located at different stands at Station 1. These sites exhibit different canopy structure properties but with the same tree species, stem density and stand age and close LAI values. In the simulation of canopy reflectance for FZ1 plot, the same LOPs, sun-sensor geometry and soil reflectance for the ZH1 plot were used, but a used different canopy structure was employed.

During the forward simulation process for canopy reflectance, three situations were considered and compared: (1) without taking leaf-age effects into consideration: setting LOPs parameters as fixed values using monthly average r_L-nir and t_L-nir of mature leaves; (2) considering the leaf-age effects for mature leaves: setting LOPs parameters using monthly varying r_L-nir and t_L-nir datasets of mature leaves; and (3) considering the leaf-age effects of both mature leaves and new leaves: setting LOPs parameters using monthly varying canopy-scale r_L-nir and t_L-nir datasets, which are up-scaled from the seasonal LOPs of new leaves and mature leaves. Finally, simulated canopy reflectance signatures were compared with pixel reflectance derived from Landsat time-series observations at FZ1 as an indirect validation.

4. Results

4.1. Sensitivity Analysis and Retrieval Results: GORT

4.1.1. Total-Order and Single-Order Sensitivity Analysis Results

Sensitivity analyses run two times for the GORT model for the red band and the NIR band. In the first sensitivity analysis, parameters with wide ranging variations were involved in the sensitivity analysis. In the second sensitivity analysis, we eliminated stem density (λ) and crown radius (r) in the analysis and focused on the remaining parameters, since stem density was known during our study time, and the average tree crown radius can be estimated based on the allometric relationship with DBH. In the second sensitivity analysis, the remaining parameters (LAI, h1, h2, λ, r_L, t_L, r_G) varied within the same range as in the first sensitivity analysis.

The sensitivity of the model input parameters was characterized using the total-order sensitivity index (Figure 4). For canopy reflectance in the red band, canopy reflectance is most sensitive to the crown radius and stem density in the first step of sensitivity analysis (Figure 4A). Given r and λ are known, LAI, h2, r_L and t_L become the most influential parameters in order (Figure 4B). For canopy reflectance in the NIR band, r_L, h2 and r are the main influential parameters over the long-term (Figure 4A). After removing the uncertainty of the crown radius and stem density, t_L becomes the third most influential parameter following r_L and h2.

To reveal the potential for retrieving uncertain parameters from canopy reflectance, we further analyzed the single-order sensitivity of each sensitive parameter in the GORT model outputs by fixing other parameters at stand average values. Taking the year of 2005 as an example, the single-order sensitivity results are shown in Figure 5. We found that during our study period, LAI and h2 were not influential regarding canopy spectral signatures in the red (Figure 5A) band, since LAI remains within a limited range (LAI > 3 m^2/m^2) for mature evergreen forests, and the crown radius remained the most influential parameter, especially in the red band (Figure 5A). For canopy reflectance in the NIR band, h2 was not influential within the range (18–23 m) for mature evergreen forests (Figure 5B). LOPs were sensitive parameters for both the red and NIR bands.

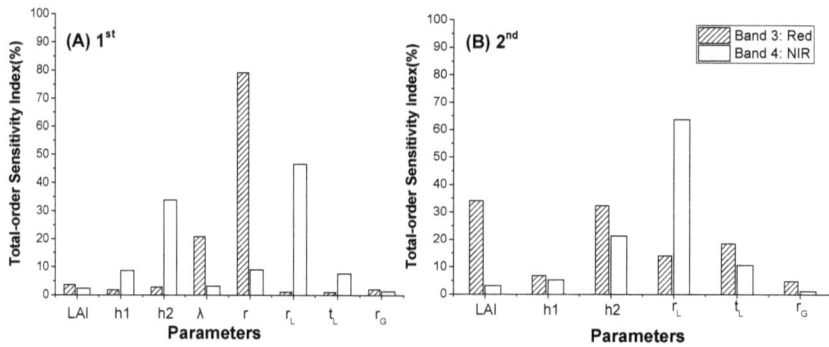

Figure 4. Results for the first (**A**) and second (**B**) global sensitivity analyses of the GORT model in the third (red) and fourth (NIR) Landsat bands for the following parameters: LAI (leaf area index), h1 (lower boundary of canopy center height), h2 (upper boundary of canopy center height), λ (tree stem density (trees/ha)), r (crown radius), r_L (leaf reflectance), t_L (leaf transmittance), and r_G (background reflectance).

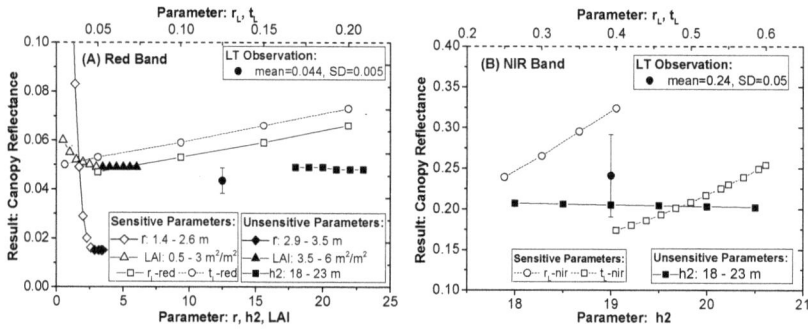

Figure 5. Single order sensitivity analysis of the impacts of sensitive parameters on GORT model outputs for the red band (**A**) and the NIR band (**B**). Only one of the model sensitive parameters was adjusted each time to study variations in the model output spectral signatures.

4.1.2. Prior Knowledge of Model Parameters

All prior knowledge of parameter values was obtained from field measurements and used for analyzing the sensitivity indexes of model parameters, as summarized in Table 2. According to total-order and first-order sensitivity analysis results, the influential parameters include: the crown radius (r), r_L and t_L, which were assumed to be adjustable during retrieval process. Non-influential parameters were set to values estimated using field measurements as described in Section 2.2. Multi-process model inversion was conducted in the following sequences: (1) Landsat red band reflectance $->$ crown radius (r); (2) Landsat NIR band reflectance/EVI2 $->$ leaf reflectance (r_L) for the NIR band; (3) Landsat NIR band reflectance/EVI2 $->$ leaf transmittance (t_L) for the NIR band; (4) Landsat red band reflectance/EVI2 $->$ leaf reflectance (r_L) for the red band; (5) Landsat red band reflectance/EVI2 $->$ leaf transmittance (t_L) for the red band. The results for the retrieved parameters are summarized in Table 2.

Table 2. Prior knowledge of input parameters and corresponding value ranges used in sensitivity analysis.

Parameter	Results	Prior Knowledge * (s.d.)	Lower Limit	Upper Limit
LAI	-	-	0.07	5.27
h1	-	-	1.49	13.5
h2	-	-	2.48	22.5
stem density (λ)	-	-	0.1035	0.252
crown radius (r) [1]	1.7	1.5	0.92	2.57
r_L-red		0.07(0.02)	0.05	0.12
t_L-red		0.04(0.03)	0.02	0.1
r_G-red	-	-	0.3	0.4
r_L-nir		0.52(0.02)	0.35	0.6
t_L-nir [1]		0.36(0.05)	0.25	0.4
r_G-nir [1]	-	-	0.35	0.45

[1] Adjustable parameters. * The prior knowledge is the initial value of parameters; the lower and upper limit values are the thresholds of the parameters.

4.2. Optical Properties of Individual New and Mature Leaves

4.2.1. Leaves at Different Ages

We first studied the leaf-age effect on leaf optical properties, and found that leaves in the canopy can be classified into two age groups: new leaves (0 a) and mature leaves (1–3 a). Field SVC measurements of the full spectra for all leaf samples were converted to Landsat-view using the Landsat relative spectral response (RSR) functions in the following three bands: green, red and NIR. Leaf samples were grouped into four age classes: 0 a, 1 a, 2 a and 3 a. Significant differences in LOPs were observed between 0 and 1 a leaves, while the differences between mature leaves at different ages (1–3 a) were not distinct (Figure 6). Thus, LOPs mainly varied within 0–1 a, i.e., during the maturation process of newly flushed leaves.

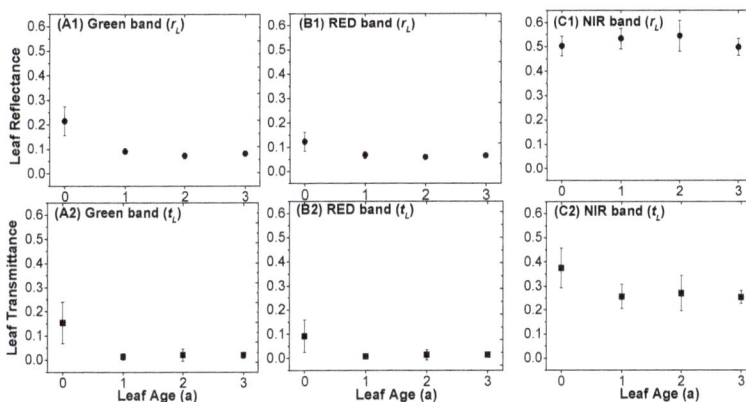

Figure 6. Leaf reflectance (r_L, **A1–C1**) and transmittance (t_L, **A2–C2**) for Chinese fir leaves at different ages (0–3 a) at three (green, red and NIR) short-wave bands. Leaves were collected on 5 May 2017.

Differences in leaf reflectance at different ages are observed in two visible bands and the NIR band (Figure 6A1–C1). New Leaves (0 a) exhibited a higher reflectance in the green and red bands compared with mature leaves (Figure 6A1–B1). In the NIR band, 1 a and 2 a leaves exhibited a slight increase in reflectance, but the reflectance of 3 a leaves was similar to that of 0 a leaves (Figure 6C1).

As can be observed in the Landsat-view transmittance signatures shown in Figure 6A2–C2, the trends in the changes in transmittance characteristics as a function of age were similar for these three bands. Significant differences were observed between 0 a and 1 a leaves, with the 0 a leaves showing a consistently higher level of transmittance compared with mature leaves. The decrease in transmittance as a function of increasing leaf age is due to the increase in absorption characteristics after a new leaf matures, which is also supported by the increasing area between the upper and lower set of curves shown in Figure 7A1–E1.

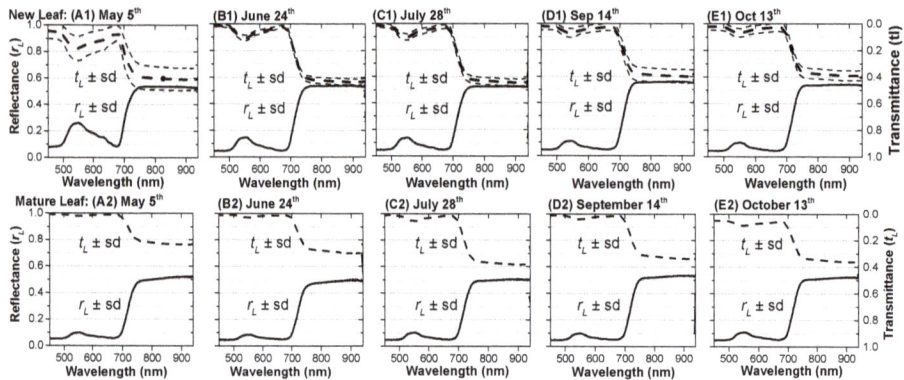

Figure 7. Field spectral curve data (mean ± 1 s.d.) in shortwave bands for new (**A1–E1**) and mature (**A2–E2**) needle samples (*n*) from Chinese fir collected on 5 May, 24 June, 28 July, 14 September and 13 October 2017. From May to October, *n* = 22, 15, 26, 30 and 39 respectively, and each sample includes 5 to 8 leaves. Reflectance data (r_L) for each month are presented as the lower set of curves within each plot, while transmittance data (t_L) are presented as the upper set of curves. Absorption characteristics are depicted based on the area between the upper and lower set of curves. SD values for new leaf t_L are depicted in the upper and lower dash lines, to avoid overlap with r_L, while SD values for all other r_L and t_L measurements are depicted in gray buffed areas.

4.2.2. Leaves in Different Seasons

The results of monthly LOPs measurements during the leaf expansion period (May to October) of 0 a leaves sampled from the Chinese fir trees are presented in Figure 7. As shown in the SVC spectral curve data for Chinese fir, significant differences were observed in the following spectral regions: the green peak (≈530–600 nm), the chlorophyll absorption well (≈660–690 nm) and along the NIR plateau (750–900 nm).

Differences in leaf reflectance are most pronounced in the green band. New leaves (0 a) exhibited a consistently lower green peak reflectance from May to October, while reflectance in the chlorophyll absorption region was shown to increase with age, especially during the first two months of the leaf expansion period, which is consistent with the increasing trend in NIR reflectance from 0 a to 1 a leaves as presented in Figure 6C1. Although this variation in NIR band reflectance is not as pronounced as in the green bands, it could have a greater impact at the canopy scale from the view of satellites, as illustrated in the following section.

With regard to leaf-level transmittance (Figure 7A1–E1), 0 a leaves showed a remarkable decrease in transmittance in the visible region from May to October, which was similar to that observed in the variations in leaf transmittance from 0 a to 1 a (Figure 7A2–B2). NIR transmittance characteristics first showed a slight increasing trend (from 1 May to 28 July) and presented a decreasing trend thereafter (from 28 July to 13 October), which coincided with the changing trend in new leaf NIR transmittance trajectory retrieved from Landsat observations (Figure 8A1). However, the observations from May to October could not fully explain the gap in the NIR transmittance of 0 a and 1 a leaves (Figure 8C2). One possible reason is that leaf transmittance decreases during winter (November to April), which may be supported by the new leaf transmittance retrieved from Landsat observations in winter time. As can be observed in Figure 8C1, NIR transmittance decreased significantly in November and December, but there is still a gap between the retrieved new leaf transmittance (0.2) and measured leaf transmittance (0.25) at the point of reaching 1 a in May of the following year after leaf production.

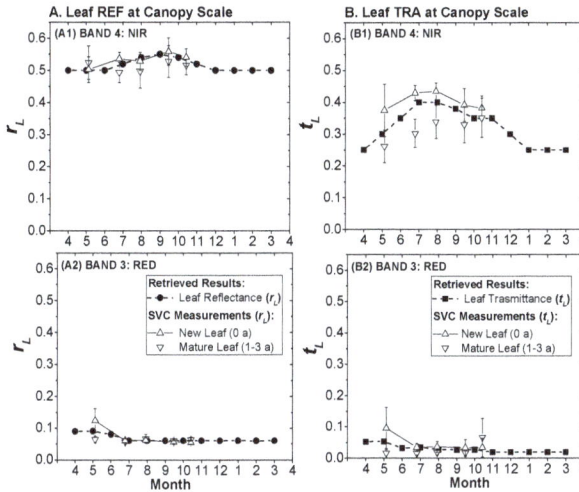

Figure 8. Retrieved results for leaf reflectance (r_L) and transmittance (t_L) at the canopy scale in the NIR band (**A1**,**A2**) and RED band (**B1**,**B2**).

4.3. Leaf-Age Effects on Variability in Landsat-Viewed Canopy Reflectance

Landsat records canopy LOPs from the sensor view and time-series data can be used to estimate seasonal variations in LOPs. First, we attempted to retrieve LOPs at the canopy scale from Landsat observations using the GORT model. Then, we estimated the new leaf component from the retrieved results by setting mature leaf component parameters as known parameters using field measurements. The results were evaluated via direct and indirect methods.

4.3.1. Leaf Optical Properties at the Canopy Scale

After considering the uncertainties and sensitivities of other parameters in the GORT model, seasonal LOPs were retrieved from Landsat observations using the proposed multi-stage inversion method by minimizing the cost function. Retrieved LOPs at the canopy scale are shown in Figure 8 and compared with field measurements of the spectral signatures of leaves. Changes in Landsat reflectance were associated with variations in leaf optical properties. As shown in Figure 8A1–B1, the most dramatic changes occurred in the NIR band, resulting in increased reflectance and transmittance during spring (April to September) and decreased reflectance and transmittance during winter (October to December). However, variations in NIR reflectance (0.5 to 0.55) were less notable than that of NIR transmittance (0.25 to 0.4). In the red band (Figure 8A2–B2), differences in LOPs mainly occurred in the first three months (May to July) and then became stable in the rest of the year. Both red reflectance and transmittance presented low values, and transmittance was slightly lower than reflectance. No LOPs changes in red band could be observed from Landsat alone, but including EVI2 data helped to some extent.

4.3.2. Seasonal Leaf Optical Properties

When viewed from the top of the crown, LOPs retrieved from Landsat observations are a combination of leaves of all ages. As analyzed in Section 3.3, we account for both the effects of leaf area and its spatial organization in the canopy to estimate the contribution of different leaves. Figure 9 shows the average seasonal variations in total leaf area (A) and leaf proportions (B) of new leaves and mature leaves in the canopy, and their contributions to canopy spectral properties (C). The time period starts at leaf expansion at age 0 (April) and ends with leaf maturity (March in the following year) at age 1, following the same twelve-month cycle with new leaf expansion each year. We can see that LAI does not change very dramatically for evergreen forests between seasons in our study period (Figure 9A). Thus, the satellite-observed seasonality of canopy reflectance is not determined by LAI, but rather is the result of variations in the LOPs of new leaves during the maturation process. Although new leaves only account for approximately 30% of the total leaf area (Figure 9B), new leaves made high contribution (approximately 80% at the peak time in summer) to canopy spectral properties from the crown top view (Figure 9C).

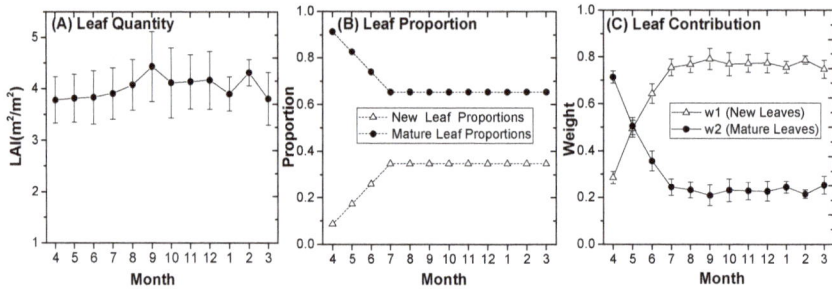

Figure 9. Seasonal variations in the leaf properties of the canopy, including (**A**) total leaf area: average LAI from stand age 22 to 33 (2005 to 2015); (**B**) leaf proportion: the percentage of new leaves and mature leaves in the canopy measured by Zhongkun et al. [50]; (**C**) weights for new leaves ($w1$) and mature leaves ($w2$) from stand age 22 to 33 (2005 to 2015).

The seasonal trajectories of mature leaf LOPs (Figure 10 B1,B2) were interpolated from field measurements, which were quite stable during growing season, with only one exception: NIR transmittance. Mature leaf transmittance in the NIR increased by 30% over the three months following the beginning of a new growing cycle in May (Figure 10B1), while mature leaves showed no significant difference in NIR reflectance during the whole growing season (May to November).

New leaf reflectance or transmittance can be estimated by deducting the contribution of mature leaves from the retrieved LOPs results, which are shown in Figure 10. It can be seen that the retrieved new leaf LOPs were in good agreement with the field observations (Figure 10A1,A2) from May to October. We can also see that new leaf LOPs in the red band decreased in the first three months (from April to June) and became stable thereafter (Figure 10A2). More importantly, new leaf NIR REF continued to increase from May to September after production (approximately 11%), while new leaf NIR TRA follow a nonlinear trajectory, increasing from May to August (approximately 16%) and decreasing thereafter (approximately 14%). A small increase in new leaf NIR REF and a small decrease in new leaf NIR TRA could both translate into a larger impact at the canopy scale, when the canopy is coated with new leaves, and the impact of the leaf-age effect at the canopy scale will be further studied in the following section.

A gap exists between the estimated new leaf NIR transmittance and SVC measurements in May, and the s.d. values of new leaf field measurements in May are greater than the measurements in other months (Figure 10A1,A2). This might be explained by the overestimation of measured leaf transmittance samples when leaves were small, which was caused by unavoidable small gaps because of minor misalignment in the arrangement of leaves. Large s.d. values for new leaf field measurements in May could also occur due to the differences in leaf expansion rate and maturity times in the early leaf flush period.

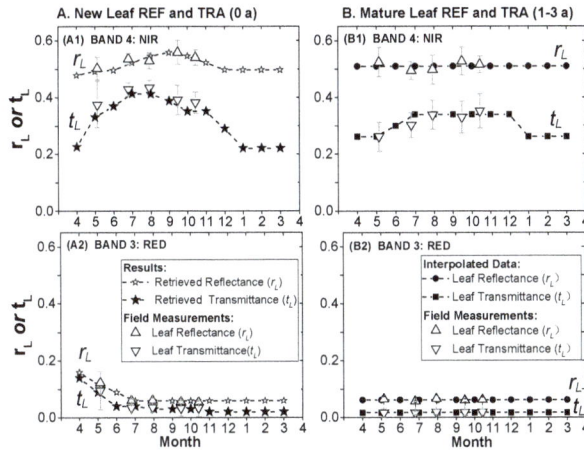

Figure 10. Validation of estimated new leaf LOPs in the NIR band (**A1**) and red band (**A2**) and the constructed mature leaf LOPs trajectories (**B1**,**B2**).

4.3.3. Leaf-Age Effects at Pixel Scale Based on Satellite Observations

LAI and sun-sensor geometry are widely recognized as the main factors contributing to seasonality in RS signals [12,19]. Thus, we need to distinguish leaf-age effects from the contributions of these two factors. To identify the contribution of leaf-age effects to the seasonality in Landsat signals, we compared the differences in the results with/without considering leaf-age effects. The forward-mode GORT model was used to simulate canopy reflectance at the pixel scale with Landsat-viewing geometry, and the model-driven parameters are listed in Table 1. We simulate canopy reflectance at NIR band and red band in the following three circumstances:

1. In the first circumstance, we ignore the leaf-age effects caused by aging mature leaves and growing new leaves, and only consider variations in LAI and sun geometry using field data. LOPs in the GORT model were fixed using the mean LOPs for mature leaves measured from May to September in 2017 (NIR band: $REF_{mature} = 0.51$, $TRA_{mature} = 0.26$ or 0.34; Red band: $REF_{mature} = 0.06$, $TRA_{mature} = 0.01$).

2. Based on circumstance 1, we include variations in LOPs caused by aging mature leaves. For comparative purposes, we first added variations in mature leaves using data shown in Figure 10B1 to drive the GORT model.

3. Based on circumstance 2, we further include variations in LOPs caused by production and expansion of new leaves. LOPs for the GORT model are shown in Figure 8B1. LOPs at the canopy scale retrieved at the ZH1 site are applied to the FZ1 site with different canopy structure parameters.

The overall GORT-simulated results obtained for the three circumstances are compared in Figure 11. After considering the mature leaf-age effects on canopy reflectance, the R^2 between the simulated canopy reflectance/EVI2 and the Landsat observations increased significantly (from 0.38 to 0.83 for canopy NIR reflectance and from 0.26 to 0.43 for canopy EVI2), as shown in Figure 11. However, the R^2 of both groups in the GORT simulation results was poor in the red band and RMSE remained low.

In circumstance 1, we can see that LAI is stable (Figure 9A) in these sub-tropical forests and is not the main factor contributing to the seasonality of canopy NIR REF. Seasonal variation in sun-sensor geometry from January to June to December (SZN varies from 56° to 25° to 56°) causes a small amount of seasonality of forest albedo, as shown based on the blue line in Figure 11A1. However, SZN alone does not sufficiently explain the seasonality of the Landsat canopy signals in our study sites. Assuming that the LOPs of mature leaves do not change, if we apply mature leaf LOPs measured at the peak time in summer (NIR TRA = 0.34) to other times, we might overestimate canopy NIR REF in winter; and if we apply mature leaf LOPs measured in early spring (NIR TRA = 0.26), we might underestimate canopy NIR REF in summer.

In circumstance 2, we consider the leaf-aging effects of the mature leaves, which constitute the majority (>60%) of the canopy LAI (Figure 9B). The LOPs of mature leaves are relatively stable in the red and NIR bands, therefore, mature leaves have no impact on canopy red REF. However, there is one notable exception: mature leaf NIR TRA is lower in winter and greater in summer and continued to increase from May to October (Figure 7A2–E2). The winter-summer differences in leaf NIR TRA contribute to a steeper NIR REF/EVI2 trajectory from April to July at the canopy scale, as shown based on the area between the blue and gray curves in Figure 11A1,C1.

In circumstance 3, we can observe how variations in the LOPs of new leaves contributed to the seasonal variation in canopy NIR reflectance and derived EVI2 (Figure 11A1,C1). As shown based on the area between the green and gray curves in Figure 11A1,C1, EVI2 trajectories follow the same pattern (Figure 11C1) as NIR reflectance. In general, new leaf production and expansion increases canopy NIR reflectance from May to October (Figure 11A1). An increased new leaf TRA (18%) also contributes to the increase in NIR REF at the canopy scale, but only before August. After September, an increase in NIR REF is caused by an increase in leaf REF. However, the changing trend of canopy NIR REF is nonlinear, which increases from May to August and decreases from August to October. As illustrated in Figure 10A1, canopy NIR REF increases from May to August partially due to increasing new leaf NIR REF during this period; conversely, canopy NIR REF decreases from August to October due to decreasing new leaf TRA at this interval.

The upper-layer-located new leaves provide a mechanism for producing greater seasonality of forest albedo in addition to mature leaves. A small increase in new leaf NIR REF (0.05 unit) has significant impacts at the canopy scale, since new leaves are generally located at the top of the canopy. New leaf expansion also contributes to negligible increase in canopy Red REF from March to July (Figure 11B1). EVI2 trajectories follow the same patterns (Figure 11C1) as NIR reflectance trajectories, arising from the linear dependence of EVI2 on NIR reflectance, as proved by a previous study [23]. We can conclude that both new and mature leaves contribute to the seasonality of forest albedo, which is independent of changes in other canopy attributes.

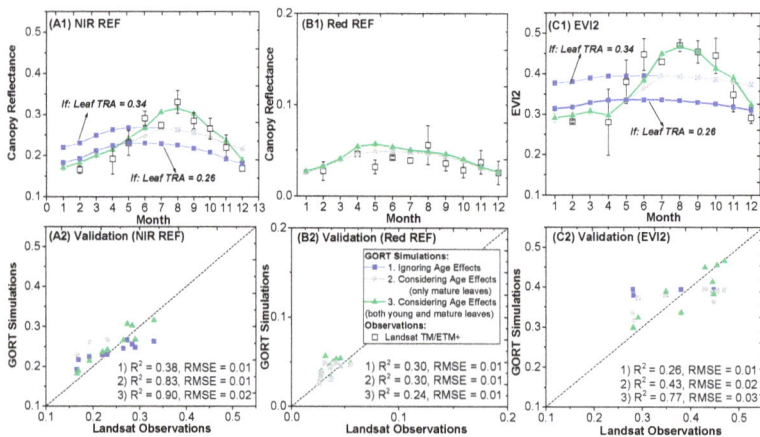

Figure 11. Aging effects of new leaves and mature leaves on seasonality of canopy signal trajectory in NIR band (**A1**), red band (**B1**) and EVI2 (**C1**). With the exception of the differences in leaf optical parameters, all other input parameters for the GORT model are the same in three circumstances. We evaluated the simulated canopy signature with Landsat observations (**A2–C2**). Both the simulated results and Landsat observations are at a monthly step with the mean value and s.d. (from 2005 to 2015).

5. Discussion

5.1. Spectral Changes and Leaf Aging

For Chinese fir, the leaf spectra mainly vary during the first year of leaf production (0–1 a), and then remain relatively stable from 1–3 a. During the new leaf expansion process, leaf water content (LWC) and the specific leaf area (SLA) decrease rapidly, resulting in variations in LOPs: including decreased transmittance and increased absorbance in the visible and NIR bands and decreased visible, but increased NIR reflectance. Leaf optical properties show strong age dependence and great seasonal difference for young and mature leaves. Spectral changes in the visible portion of the spectrum characterized the new leaves while increased NIR transmittance and decreased NIR absorbance characterized mature leaves. Changes observed in the visible spectrum matched similar observations for Carica [59] and Aldina [60]. In general, absorbance at the green peak (550 nm) increases, while reflectance and transmittance decrease at this band, which could be attributed to an increase in chlorophyll [59] and changes in leaf internal anatomy [61]. However, the changes of LOPs in the NIR band are complicated.

In agreement with previous findings [60], mature leaves were found to show a lower leaf albedo (leaf reflectance plus transmittance) at the NIR band than new leaves due to leaf aging. Early changes in the NIR band were noted during new leaf expansion, with a slight increase (11%) in leaf NIR REF being observed from May to October and a continuous increase (14%) in leaf TRA being recorded from May to August, followed by a slight decrease in NIR TRA during autumn and winter. Some studies [62] have found a dramatic increase in leaf NIR reflectance between July and August, which might be caused by differences in the leaf characteristics. Low-wax leaves display a continuous increase in NIR reflectance until maturity, while high-wax leaves show little increase in NIR reflectance [63]. Epicuticular waxes and thick cuticles could mask the effect of NIR increases caused by inter-cellular development in young high-wax leaves. These characteristics might explain why only a slight increase in NIR reflectance was found during leaf maturation in the present study, since Chinese fir is a high-wax species.

In contrast to previous studies, we observed a continuous increase in NIR transmittance for mature leaves (1–3 a). However, a significant gap in NIR transmittance existed between new leaves (measured from May to October) and mature leave (measured in May), and we only observed a small decrease in NIR transmittance for new leaves from September to October. However, a dramatic decrease in NIR transmittance might happen in winter time (November to December to January) according to the results retrieved from Landsat observations, which will be further studied in the future. Similar observations have been made by others [60], who found a consistent decrease in NIR transmittance during the last nine months of the leaf cycle.

5.2. Leaf-Aging Effects on Canopy Reflectance

At the canopy scale, canopy reflectance is a product of competing mechanisms of light absorption [60]. The interaction of photons with dense forests is characterized by strong scattering in the NIR and equally strong absorption in the shorter red and blue bands. The NIR reflectance of these forests is an order of magnitude greater than the reflectance at red (blue) band. On the other hand, aerosol scattering has greater impacts on reflectance in visible bands than the NIR band. In the NIR band, any mechanism that increases leaf absorption (such as decreased transmittance or reflectance) will have an enhanced effect on canopy reflectance [60]. Thus, we could expect leaf aging to have its largest impact in the NIR band.

Age-related LOPs are factors that require special attention but have been overlooked in previous studies focused on estimating vegetation status using optical signal trajectories. Previous studies also found changes in canopy NIR reflectance caused by the exchange of older leaves for newer leaves [23,24,27]. These studies considered the differences in LOPs between senesced old leaves and new leaves, but they ignored the simultaneous aging process of new leaves and mature leaves with a life span of more than three years. In this study, the observed seasonality of the canopy NIR reflectance of evergreen forests was due to age-dependent leaf optical properties at the NIR band, including both new leaf maturation and mature leaf aging, with little changes in total leaf area. New leaf maturation mainly increases NIR reflectance at the peak time in summer (May to October), and mature leaf aging mainly reduces NIR reflectance during early spring (January to June). We have already excluded differences caused by changes in the solar zenith angle [10], and our seasonally moist sub-tropical forests are free of drought impact [24,64–66] or the impact of epiphylls on mature leaves in tropical forests [23,60,67], which are highly debated factors causing changes in NIR reflectance in numerous studies, and the same seasonality pattern is observed in EVI because of its strong dependence on NIR reflectance.

It is important to note that the forest canopy is composed of mixed-age leaves. Mature leaves account for a major population, while a higher proportion of young leaves are observed by the Landsat satellite because of their distribution at the top of the canopy. Both new and mature leaves influence canopy NIR reflectance, and their impacts change with new leaf expansion. New leaf expansion changes the proportions of young and mature leaves within a forest canopy over the season, thus, further changes the canopy spectral signatures.

5.3. Potential Implications for Photosynthesis

Young leaves exhibit a higher photosynthetic capacity than older leaves, and it is therefore essential to track changes in the age-structure of leaves in the canopy, which could substantially improve the modeling of the seasonal dynamics of photosynthesis. Xiaoquan and Deying [68] studied the leaf-age effects on the photosynthetic characteristics of 18-year-old *Cunninghamia lanceolata* stands and found similar trends in leaf photosynthetic rate at different ages: the photosynthesis rate of newly flushed leaves increases during the maturation process and then decreases at the end of the growing season, while the photosynthesis rate of mature leaves (1 and 2 a) decreases as time progresses. This study suggested that considering the temporal variation in the LOPs of new leaves matters significantly in understanding the seasonal trend of canopy spectral signatures. Ignoring variations in leaf-scale properties may mislead our interpretations of seasonality of RS signals from evergreen forests where different-age leaf cohorts coexist in the canopy. This study provides new evidence of the importance of considering the phenology of LOPs at different ages and which could contribute to more accurately modeling of photosynthesis [69].

5.4. Implications of LOPs and Canopy Structure

At the canopy scale, seasonal variations in new leaf optical properties were shown to be the dominant factor producing seasonal variations in canopy reflectance and altering NIR to red ratios, independent of changes in other canopy attributes. However, our study explicitly focused on dense evergreen forests with stable LAI (LAI > 3). The parameters that drive the GORT model may vary on different temporal scales. For example, crown size and tree height vary annually, while LAI, leaf spectra, and sun geometry vary seasonally. We can explore the signal differences caused by LAI and canopy structure by comparing the annual differences in the GORT model outputs. We only applied the annual structural parameters from 2005 to 2015 (there were no data in 2009 and from 2012 to 2014) to drive the the GORT model. In addition to annual stable structural parameters, the SZN, LAI and canopy average LOPs were updated monthly. Although the LAI seasonality and structure parameters varied from year to year, NIR reflectance showed similar seasonal trajectories with little annual difference, as illustrated by the vertical error bar in Figure 11. There may be two explanations for this limited variation. First, there is a lack of LAI seasonality or the canopy is too dense with a high LAI; thus, small changes in leaf area are not as sensitive compared with the variations in LOPs. The results might be different for young stands, especially before canopy closure. Second, canopy structural parameters do not change significantly during the new leaf maturation process; thus, canopy structures have limited impacts on canopy reflectance during this period. As illustrated in other studies [23], LAI is also an important factor contributing to the seasonality of NIR reflectance.

This study quantified the effects of leaf age on canopy reflectance in mature evergreen forests to first explore the possible impacts on leaf quality when total leaf area does not exhibit significant seasonal changes. Given the findings of this case study, future studies in deciduous forests, where seasonal variations in LAI are more significant and leaf-age groups are less complicated, would allow further validating of our findings and apply our findings to better quantify the canopy spectral signatures to understand the role of forests in terrestrial ecosystems.

6. Conclusions

This study evaluated the effect of LOPs for leaves at different ages on the canopy spectral signature variation during the growing season for Chinese fir stands. For closed canopy evergreen stands with relatively stable LAI, new leaves exerted disproportional influence on the canopy season spectral signature due to the spatial distribution of the new leaves in the top and outer canopy. The most direct implications of our results are related to ecological or physiological studies that utilize remote sensing, and our findings provide a promising potential to improve the interpretation of RS signals. The most significant finding of this study at the leaf scale is the increase in the NIR transmittance of mature leaves and the increase in the NIR reflectance of new leaves. To date, most of the identified contributions of leaf age to the variation in leaf optical properties have involved changes in the differences between old leaves and new leaves. Simultaneous monitoring of new and mature LOPs with season, however, has not been previously documented. In this study, we observed an approximately 11% increase in NIR reflectance (0.05 unit) for new leaves and a 35% increase in NIR transmittance for mature leaves during the growing season.

Variations of LOPs at the leaf scale have significant impacts at the canopy scale, and contribute to seasonality of canopy NIR reflectance and EVI2. Due to the complexity of forest ecosystems, analyses based on field data alone cannot provide guidance in the interpretation of RS signals. Conversely, studies based on modeling alone, without proper ground measurements of the key factors driving canopy signal variation, can be misleading. This study combined field observations with the GORT model to elucidate the effects of leaf age on canopy-scale reflectance signals. We demonstrated that, in addition to sun-sensor geometry, the effects of leaf aging on LOPs were the major factor contributing to the seasonality of canopy reflectance for the Chinese fir stands:

- New leaf maturation is the main factor contributing to seasonality of canopy signals (NIR REF and EVI2), because of the distribution of these leaves in the top and outer canopy, as well as their increasing proportions with leaf growth. A small increase (0.05 unit) in new leaf NIR reflectance results in a significant increase in canopy NIR reflectance from spring to summer, while a decrease in new leaf NIR transmittance from August to October causes a decreasing trend in canopy NIR reflectance in autumn and winter.
- Mature leaf aging is another factor contributing to the seasonality of the canopy signals (NIR REF and EVI2) because of the significant proportion of mature leaves in the canopy. Mature leaf NIR transmittance is greater during the growing season than off the growing season. This difference in leaf TRA causes an increased difference in canopy reflectance between winter and summer.

Thus, the effects of leaf age cannot be ignored when conducting time series analyses using RS data for the evergreen needle leaf forests.

Acknowledgments: This research was supported in part by the National Key Research and Development Program of China (2016YFB0501502), the Special Funds for Major State Basic Research Project (2013CB733403), the National 863 Program (2013AA12A303, 2012AA12A301), the Special Funds for Scientific Foundation Project (2014FY210800-3) and the Fundamental Research Funds for the Central Universities. The authors like to express their great appreciation of the valued assistance from CERN and Huitong National Research Station of Forest Ecosystem (HTF), China. Datasets in our work are available upon request by contacting the corresponding author (songjl@bnu.edu.cn).

Author Contributions: Q.W. and J.S. conceived and designed the experiments; C.S. and J.W. contributed to the conception of the study; S.C. and B.Y. performed the experiments; Q.W. analyzed the data and wrote the paper. C.S., J.S. and J.W. have reviewed and edited the manuscript.

Conflicts of Interest: The authors declare no conflict of interest.

Appendix A

Appendix A.1 Data Description

Section 2.2.1 is supported with field data collected for Chinese fir forest stands located in the vicinity of the permanent ecosystem research stations in National Research Stations of Forest Ecosystems in Huitong county, Hunan province, southern China. This appendix provides details on data acquisition of canopy structure parameters.

Appendix A.2 Canopy Structural Parameter Measurements

Crown shape measurements were taken for a total of forty trees in study Site 2. Measured crown shape parameters are shown in Figure A1, among which, DBH was measured using diameter tapes, and crown radius and tree heights were measured using telemeter rods. The size of the selected trees was evenly distributed uniformly in terms of height (*H*) (from 5.4 m to 20 m) and diameter at breast height (DBH) (from 7.7 cm to 37.8 cm). Among the 40 trees, 7 are located at plots close to the study site, and 33 are located within the ZH1 and FZ1 sites. The parameters measured for characterizing crown shape are presented in Figure A1, including crown width in the north-south direction (*R1*) and the east-west direction (*R2*), tree height (*H1*) and height under crown (*H2*), from which we can obtain the height of crown center (*h*). We used the measurements of these 40 trees to build regression relationships for crown width, DBH, h and tree height. Figures A2 and A3 illustrated the field measurements of canopy structure parameters as well as their growth trajectories with DBH, which usually correlates well with tree ages.

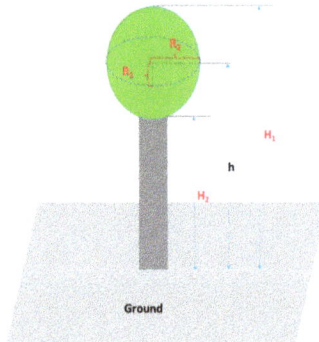

Figure A1. Measured parameters for sample trees selected to build regression relationships for the structural parameters of the tree crown and other measures, including tree height and DBH.

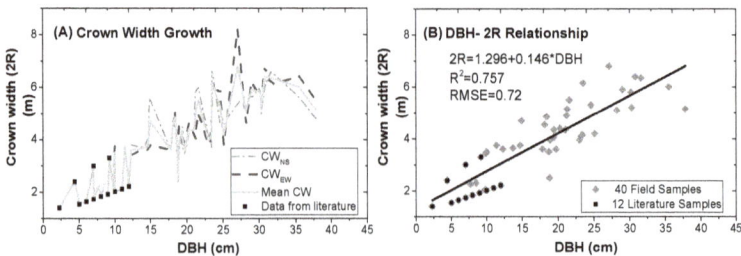

Figure A2. (**A**) Measured crown width (2*R*) for 40 sample trees with DBH in the north-south direction (CW$_{NS}$) and east-west direction (CW$_{EW}$); (**B**) Regression relationships between DBH and crown width (2*R*).

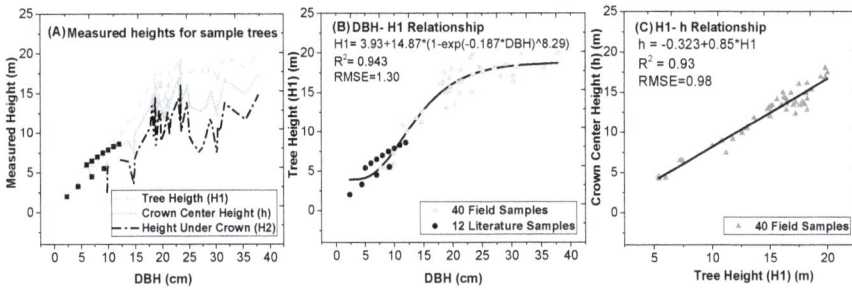

Figure A3. (**A**) Measured height values for 40 sample trees with increasing DBH. The measured data include tree height (*H1*), height of crown center (*h*) and height under crown (*H2*); Regression relationships between DBH and tree height (**B**) and between tree height and crown center height (**C**).

Appendix A.3 Leaf Area Proportion and Distribution in Crown

Table A1. Leaf area and proportion of leaves at different ages located at different crown positions for *Cunninghamia* trees in Hunan.

Leaf Age/Location	Bottom Crown	Central Crown	Upper Crown	Leaf Area $\sum$ (%)
0 a	1.16	1.41	3.24	5.8 (34.63%)
1 a	1.93	2.93	0.81	5.67 (33.85%)
2 a	1.80	1.59	0	3.39 (20.24%)
3 a	1.11	0.76	0	1.87 (11.64%)
Leaf Area $\sum$ (%)	6.00 (35.8%)	6.69 (40%)	4.05 (24.2%)	16.75 (100%)

Table origin: [50].

Appendix A.4 Leaf Sample for SVC Measurements

Figure A4. The vacuum side of one group of leaf leaves prepared for spectral measurement. Leaf samples were collected on 15 July 2016.

Appendix B

Appendix B.1 Data Description

Section 2.2.2 is supported with LAI field data collected for Chinese fir forest stands located close to the permanent ecosystem research stations in National Research Stations of Forest Ecosystems in Huitong county, Hunan province, southern China. Long-term LAI were measured monthly to drive the GORT model from year 2005 to 2015. In this time period, two different DHP methods were applied before and after year 2006. This appendix provides details on LAI data acquisition and processing to produce consistent LAI time-series.

Appendix B.2 Long-Term LAI_e Observations: DHP Methods

(1)   2005 and 2006: CI-110 Plant Canopy Analyzer

Plant canopy imaging was the primary method for conducting regular monthly LAI measurements from 2005 until present. From 2005 to 2006, pictures were taken every month by one worker using a CI-110 Plant Canopy Analyzer to estimate the LAI. From 2007 onward, photographs were taken by another fixed worker using a Canon EOS 40D digital camera equipped with a Nikon AF-DX 10.5 mm f/2.8 G ED fisheye convertor. The data measured before 2007 needed to be further preprocessed prior to further calculations to make them consistent with data measured after 2007. We have 30 pairs of data measured at the same positions and in the same directions in ZH1 and FZ1 to unify the original LAI time series data using data measured in December 2006 and January 2007. We employed two methods to unify these datasets: (1) Regression Method: The first method involves deriving two average seasonal LAI trajectories, one before 2007 and the other after 2007, and calculating their regression equations. (2) Lifting Method: The second method involves acquiring the mean absolute deviation (ΔLAI = 1.01) of the data before 2007 and those after 2007 using the mean difference between these 30 pairs of LAI data. The results of these two methods are different, and both are illustrated at the end of this section.

(2)   2007 to 2015: Fisheye Digital Camera

From 2007 afterwards, photographs were taken using a Canon EOS 40D digital camera equipped with a Nikon AF-DX 10.5 mm f/2.8G ED fisheye convertor. The camera was horizontally mounted at a fixed height of 0.2 m above the ground. The photographs were taken with automatic exposure under diffuse light conditions, typically soon after sunrise or immediately before sunset. In the ZH1 and FZ1 plots, measurements were taken for three layers: the herb layer, shrub layer and tree layer on the 15th day of each month at five fixed locations per plot and facing toward four cardinal directions. All measurements were made under diffuse light conditions to avoid introducing errors due to the presence of sunlit foliage. The images were processed by the Gap Light Analyzer 2.0 software to calculate LAI_{DHP}.

Incorrect exposure has been shown to cause significant underestimations in LAI_{DHP} measurements [35–39]. All DHP measurements were taken with automatic exposures, which resulted in considerable underestimation. The LAI measurements made using the DHP and LAI-2000 methods were shown to be significantly correlated in any zenith angle range. Therefore, we calibrated LAI_{DHP} by the LAI_e measured using an LAI-2000 plant analyzer, to quantify the systematic bias ε in DHP methods due to automatic exposure problems.

Appendix B.3 Converting LAI_e to LAI_t: LAI-2000 and TRAC Methods

LAI-2000 and TRAC measurements were used to convert LAI_{DHP} to true LAI. Tracing Radiation and Architecture of Canopies (TRAC, Natural Resources Canada, Canada Center for Remote Sensing, Saint-Hubert, QC, Canada) and an LAI-2000 Plant Canopy Analyzer (LAI-2000, LI-COR Inc, Lincoln, NE, USA) [41] were used to measure the LAI of each sample plot [42]. The LAI-2000 was used to measure the effective LAI (LAI_e), and TRAC was used to measure both the effective LAI (LAI_e) and the foliage clumping index (Ω).

The LAI-2000 method is based on the measurement of diffuse radiation attenuation in the blue band caused by the canopy, which is related to gap fraction. Further details on the theories and measurements behind LAI are given in Frazer et al. [70]. We employed two LAI-2000 units to measure the sky radiation and under-canopy radiation simultaneously. Additionally, we cross-calibrated these two LAI-2000 units before the field survey. One LAI-2000 unit is horizontally mounted on a rooftop facing toward our study sites and automatically records the above-canopy radiation every 5 min. Ninety-degree view caps were used on both units to avoid the influence of other objects on the sensors and made these measurements comparable with those from the DHP methods in four cardinal directions. The LAI-2000 measurements were taken under diffuse sky conditions in the early morning or after sunset. The TRAC method requires direct solar radiation, so we took TRAC measurements during midday at a constant walking pace along several parallel transects, which were perpendicular to the direction of tree stem shadows. Distance markers were registered every 5 m. The TRAC data were processed by TracWin software, which calculates LAI_e and the clumping index (Ω).

During August 2015, we used LAI-2000 instruments to measure the LAI in the WS2 and WS3 plots. We revisited these plots in June 2016, taking measurements with both the TRAC and LAI-2000 instruments. In June 2016, we surveyed the ZH1 and FZ1 sample plots and used the LAI-2000 instruments to measure the LAI at the same locations as measured by the long-term DHP methods. In comparison to the LAI-2000 instrument, the accuracy of DHP is affected by photograph exposure settings because these settings impact the ratio of green leaves to sky. An LAI-2000 unit was operated subsequently at the same locations for comparison with DHP, and then we convert LAI_e to true LAI values.

Additional correction parameters were required to convert LAI_e to LAI_t. The needle-to-shoot area ratio (γ_e) was measured using destructive sampling conducted in August 2015. The clumping index Ω was measured by the TRAC instrument. Additionally, the woody-to-total area ratio (α) for *Cunninghamia lanceolata* was derived from destructive sampling used to calculate biomass.

Appendix B.4 Unifying LAI Measurements Using Different Methods

LAI_{DHP} was corrected using LAI_e measurements from the LAI-2000 units to decrease the underestimation caused by the automatic exposure problem present in the DHP method. We made comparisons between the LAI as measured using DHP and LAI-2000 in the same locations in ZH1 and FZ1 to estimate the system bias (ε) of DHP. We applied a fixed bias instead of a regression relationship between DHP and LAI-2000 because our study site is pure *Cunninghamia lanceolata* plantations and the LAI values fall within a small range. Thus, it is not possible to build a robust regression relationship with limited LAI variations using the field measurements made on our study sites. Although previous studies have built and applied regression relationships between DHP and LAI-2000 LAI methods [71], we chose to estimate the system error of DHP with a fixed value in our study site. The reasons for this choice are listed as follows: first, regression relationships are usually site specific, lack universality and cannot be applied to other places. We used the existing regression relationship to our study sites, but the result was of poor quality. The regression relationship built by [71] resulted in a mean absolute error of 1.0 LAI when compared with field LAI-2000 measurements. Second, the accuracy of the regression relationship may vary in different ranges of values. For example, if the regression relationship fits well in the low-value range but fits poorly at high values, then when we apply it to other places with mainly high values, the relationship will fail. At this study site, the LAI is quite stable; as shown in Table A2, the standard deviation is quite small (approximately 0.25 as measured by LAI-2000), and the maximum and minimum LAI-2000 measurements within our study sites are quite close (3.19–3.36 for ZH1, and 3.18–3.3 for FZ1). Thus, correcting for the systematic error is a sufficient and reliable method for converting LAI values measured by the DHP method to LAI measured with LAI-2000.

Table A2. Comparison of effective LAI derived from digital hemispherical photography (DHP) and LAI-2000 methods.

Plot	ZH1			FZ1		
	DHP (5 pts * 4 Dirs)	LAI-2000 (5 pts * 4 Dirs)	LAI-2000 (26 pts * 2 Repeat)	DHP (5 pts * 4 Dirs)	LAI-2000 (5 pts * 4 dirs)	LAI-2000 (61 pts * 3 Repeat)
Maximum	1.68	3.52	3.36	1.56	3.85	3.3
Minimum	1.37	2.9	3.19	1.05	2.87	3.18
Mean	1.533	3.186	3.275	1.282	3.248	3.243
SD	0.12	0.23	0.12	0.2	0.39	0.06
System bias (ε) *	–	*1.653*	*1.724*	=	*1.966*	*1.962*

* The LAI were measured at 5 points, in four directions, and repeated two to three times for every plot. The system bias (ε) was calculated by (LAI-2000 LAI_e- DHP LAI_{DHP}).

According to Table A2, the system bias was set to 1.83 for the DHP method. The correction parameters required to convert LAI_e to LAI_t were set as follows: the needle-to-shoot area ratio (γ_e) was set to 1.1 according to the results of the destructive sampling conducted in August 2015. The clumping index Ω was produced by the TRAC instrument and set to 0.8. The value of the woody-to-total area ratio (α) for *Cunninghamia lanceolata* was derived from the destructive samples used to calculate biomass and was set to 0.2. An illustration of the data processing workflow may be useful for understanding how we unified the LAI field data measured by different instruments, as shown in Figure A5.

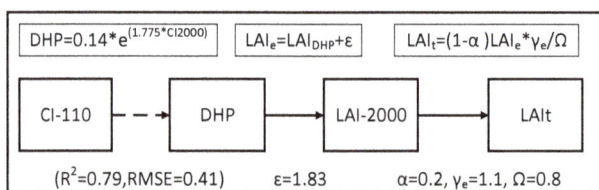

$$DHP=0.14*e^{(1.775*CI2000)} \qquad LAI_e=LAI_{DHP}+\varepsilon \qquad LAI_t=(1-\alpha)LAI_e*\gamma_e/\Omega$$

CI-110 - - → DHP → LAI-2000 → LAIt

$$(R^2=0.79, RMSE=0.41) \qquad \varepsilon=1.83 \qquad \alpha=0.2, \gamma_e=1.1, \Omega=0.8$$

Figure A5. Flowchart of LAI field data processing procedures. Before January 2007, monthly LAI was measured using a CI-110 instrument, and a DHP instrument was used afterwards.

References

1. Waring, R.H.; Schlesinger, W.H. *Forest Ecosystems: Concepts and Management*; Academic Press: Orlando, FL, USA, 1985; p. 75.
2. Dixon, R.K.; Brown, S.; Houghton, R.E.A.; Solomon, A.; Trexler, M.; Wisniewski, J. Carbon pools and flux of global forest ecosystems. *Science (Washington)* **1994**, *263*, 185–189. [CrossRef] [PubMed]
3. Schimel, D.; Stephens, B.B.; Fisher, J.B. Effect of increasing CO_2 on the terrestrial carbon cycle. *Proc. Natl. Acad. Sci. USA* **2015**, *112*, 436–441. [CrossRef] [PubMed]
4. Pan, Y.; Birdsey, R.A.; Fang, J.; Houghton, R.; Kauppi, P.E.; Kurz, W.A.; Phillips, O.L.; Shvidenko, A.; Lewis, S.L.; Canadell, J.G. A large and persistent carbon sink in the world's forests. *Science* **2014**, *333*, 988–993. [CrossRef] [PubMed]
5. Reich, P.B. Biogeochemistry: Taking stock of forest carbon. *Nat. Clim. Chang.* **2016**, *1*, 346–347. [CrossRef]
6. Guyot, G.; Guyon, D.; Riom, J. Factors affecting the spectral response of forest canopies: A review. *Geocarto Int.* **1989**, *4*, 3–18. [CrossRef]
7. Vermote, E.F.; Tanré, D.; Deuze, J.L.; Herman, M.; Morcette, J.-J. Second simulation of the satellite signal in the solar spectrum, 6s: An overview. *IEEE Trans. Geosci. Remote Sens.* **1997**, *35*, 675–686. [CrossRef]
8. Samanta, A.; Ganguly, S.; Vermote, E.; Nemani, R.R.; Myneni, R.B. Why is remote sensing of Amazon forest greenness so challenging? *Earth Interact.* **2012**, *16*, 1–14. [CrossRef]
9. Xiao, X.; Braswell, B.; Zhang, Q.; Boles, S.; Frolking, S.; Moore, B. Sensitivity of vegetation indices to atmospheric aerosols: Continental-scale observations in northern asia. *Remote Sens. Environ.* **2003**, *84*, 385–392. [CrossRef]

10. Galvão, L.S.; dos Santos, J.R.; Roberts, D.A.; Breunig, F.M.; Toomey, M.; de Moura, Y.M. On intra-annual evi variability in the dry season of tropical forest: A case study with MODIS and hyperspectral data. *Remote Sens. Environ.* **2011**, *115*, 2350–2359. [CrossRef]

11. Maeda, E.E.; Heiskanen, J.; Aragão, L.E.; Rinne, J. Can MODIS evi monitor ecosystem productivity in the Amazon rainforest? *Geophys. Res. Lett.* **2014**, *41*, 7176–7183. [CrossRef]

12. Maeda, E.E.; Galvão, L.S. Sun-sensor geometry effects on vegetation index anomalies in the Amazon rainforest. *GISci. Remote Sens.* **2015**, *52*, 332–343. [CrossRef]

13. Bi, J.; Knyazikhin, Y.; Choi, S.; Park, T.; Barichivich, J.; Ciais, P.; Fu, R.; Ganguly, S.; Hall, F.; Hilker, T. Sunlight mediated seasonality in canopy structure and photosynthetic activity of Amazonian rainforests. *Environ. Res. Lett.* **2015**, *10*, 064014. [CrossRef]

14. Moura, Y.M.; Galvão, L.S.; dos Santos, J.R.; Roberts, D.A.; Breunig, F.M. Use of MISR/Terra data to study intra- and inter-annual EVI variations in the dry season of tropical forest. *Remote Sens. Environ.* **2012**, *127*, 260–270. [CrossRef]

15. Lyapustin, A.I.; Wang, Y.; Laszlo, I.; Hilker, T.; Hall, F.G.; Sellers, P.J.; Tucker, C.J.; Korkin, S.V. Multi-angle implementation of atmospheric correction for MODIS (MAIAC): 3. Atmospheric correction. *Remote Sens. Environ.* **2012**, *127*, 385–393. [CrossRef]

16. Jones, M.O.; Kimball, J.S.; Nemani, R.R. Asynchronous Amazon forest canopy phenology indicates adaptation to both water and light availability. *Environ. Res. Lett.* **2014**, *9*, 124021. [CrossRef]

17. Goward, S.N.; Huemmrich, K.F. Vegetation canopy par absorptance and the normalized difference vegetation index: An assessment using the sail model. *Remote Sens. Environ.* **1992**, *39*, 119–140. [CrossRef]

18. Ollinger, S.V. Sources of variability in canopy reflectance and the convergent properties of plants. *New Phytol.* **2011**, *189*, 375–394. [CrossRef] [PubMed]

19. Asner, G.P. Biophysical and biochemical sources of variability in canopy reflectance. *Remote Sens. Environ.* **1998**, *64*, 234–253. [CrossRef]

20. Jacquemoud, S.; Verhoef, W.; Baret, F.; Bacour, C.; Zarco-Tejada, P.J.; Asner, G.P.; François, C.; Ustin, S.L. Prospect+ sail models: A review of use for vegetation characterization. *Remote Sens. Environ.* **2009**, *113*, S56–S66. [CrossRef]

21. Eriksson, H.M.; Eklundh, L.; Kuusk, A.; Nilson, T. Impact of understory vegetation on forest canopy reflectance and remotely sensed lai estimates. *Remote Sens. Environ.* **2006**, *103*, 408–418. [CrossRef]

22. Jordan, C.F. Derivation of leaf-area index from quality of light on the forest floor. *Ecology* **1969**, *50*, 663–666. [CrossRef]

23. Samanta, A.; Knyazikhin, Y.; Xu, L.; Dickinson, R.E.; Fu, R.; Costa, M.H.; Saatchi, S.S.; Nemani, R.R.; Myneni, R.B. Seasonal changes in leaf area of Amazon forests from leaf flushing and abscission. *J. Geophys. Res. Biogeosci.* **2012**, *117*. [CrossRef]

24. Asner, G.P.; Alencar, A. Drought impacts on the Amazon forest: The remote sensing perspective. *New Phytol.* **2010**, *187*, 569–578. [CrossRef] [PubMed]

25. Monteith, J.L.; Ross, J. *The Radiation Regime and Architecture of Plant Stands*; Springer: Dordrecht, The Netherlands, 1981; p. 344.

26. Asner, G.P.; Wessman, C.A.; Schimel, D.S.; Archer, S. Variability in leaf and litter optical properties: Implications for BRDF model inversions using AVHRR, MODIS, and MISR. *Remote Sens. Environ.* **1998**, *63*, 243–257. [CrossRef]

27. Doughty, C.E.; Goulden, M.L. Seasonal patterns of tropical forest leaf area index and CO_2 exchange. *J. Geophys. Res. Biogeosci.* **2008**, *113*. [CrossRef]

28. Wu, J.; Albert, L.P.; Lopes, A.P.; Restrepo-Coupe, N.; Hayek, M.; Wiedemann, K.T.; Guan, K.; Stark, S.C.; Christoffersen, B.; Prohaska, N. Leaf development and demography explain photosynthetic seasonality in Amazon evergreen forests. *Science* **2016**, *351*, 972–976. [CrossRef] [PubMed]

29. Horler, D.; Ahern, F. Forestry information content of thematic mapper data. *Int. J. Remote Sens.* **1986**, *7*, 405–428. [CrossRef]

30. Cohen, W.B.; Spies, T.A.; Fiorella, M. Estimating the age and structure of forests in a multi-ownership landscape of western Oregon, USA. *Int. J. Remote Sens.* **1995**, *16*, 721–746. [CrossRef]

31. Jakubauskas, M.E. Thematic Mapper characterization of lodgepole pine seral stages in Yellowstone National Park, USA. *Remote Sens. Environ.* **1996**, *56*, 118–132. [CrossRef]

32. Song, C.; Schroeder, T.A.; Cohen, W.B. Predicting temperate conifer forest successional stage distributions with multitemporal landsat thematic mapper imagery. *Remote Sens. Environ.* **2007**, *106*, 228–237. [CrossRef]

33. Nilson, T.; Peterson, U. Age dependence of forest reflectance: Analysis of main driving factors. *Remote Sens. Environ.* **1994**, *48*, 319–331. [CrossRef]

34. Eckenwalder, J.E. *Conifers of the World: The Complete Reference*; Timber Press: Portland, OR, USA, 2009.

35. Chen, J.; Black, T.; Adams, R. Evaluation of hemispherical photography for determining plant area index and geometry of a forest stand. *Agric. For. Meteorol.* **1991**, *56*, 129–143. [CrossRef]

36. Wagner, S. Calibration of grey values of hemispherical photographs for image analysis. *Agric. For. Meteorol.* **1998**, *90*, 103–117. [CrossRef]

37. Englund, S.R.; O'brien, J.J.; Clark, D.B. Evaluation of digital and film hemispherical photography and spherical densiometry for measuring forest light environments. *Can. J. For. Res.* **2000**, *30*, 1999–2005. [CrossRef]

38. Song, G.-Z.M.; Doley, D.; Yates, D.; Chao, K.-J.; Hsieh, C.-F. Improving accuracy of canopy hemispherical photography by a constant threshold value derived from an unobscured overcast sky. *Can. J. For. Res.* **2013**, *44*, 17–27. [CrossRef]

39. Liu, Z.; Jin, G.; Chen, J.M.; Qi, Y. Evaluating optical measurements of leaf area index against litter collection in a mixed broadleaved-Korean pine forest in China. *Trees* **2015**, *29*, 59–73. [CrossRef]

40. Tarantola, S.; Becker, W. Simlab software for uncertainty and sensitivity analysis. In *Handbook of Uncertainty Quantification*; Ghanem, R., Higdon, D., Owhadi, H., Eds.; Springer: Cham, Switzerland, 2016; pp. 1–21.

41. Li, X.; Strahler, A.H.; Woodcock, C.E. Hybrid geometric optical-radiative transfer approach for modeling albedo and directional reflectance of discontinuous canopies. *IEEE Trans. Geosci. Remote Sens.* **1995**, *33*, 466–480.

42. Ni, W.; Li, X.; Woodcock, C.E.; Caetano, M.R.; Strahler, A.H. An analytical hybrid gort model for bidirectional reflectance over discontinuous plant canopies. *IEEE Trans. Geosci. Remote Sens.* **1999**, *37*, 987–999.

43. Li, X.; Strahler, A.H. Geometric-optical modeling of a conifer forest canopy. *IEEE Trans. Geosci. Remote Sens.* **1985**, *23*, 705–721. [CrossRef]

44. Li, X.; Strahler, A.H. Geometric-optical bidirectional reflectance modeling of a conifer forest canopy. *IEEE Trans. Geosci. Remote Sens.* **1986**, *24*, 906–919. [CrossRef]

45. Li, X.; Strahler, A.H. Geometric-optical bidirectional reflectance modeling of the discrete crown vegetation canopy: Effect of crown shape and mutual shadowing. *IEEE Trans. Geosci. Remote Sens.* **1992**, *30*, 276–292. [CrossRef]

46. Serra, J. The boolean model and random sets. *Comput. Graph. Image Process.* **1980**, *12*, 99–126. [CrossRef]

47. Li, X.; Strahler, A.H. Modeling the gap probability of a discontinuous vegetation canopy. *IEEE Trans. Geosci. Remote Sens.* **1988**, *26*, 161–170. [CrossRef]

48. Ni, W.; Li, X.; Woodcock, C.E.; Roujean, J.-L.; Davis, R.E. Transmission of solar radiation in boreal conifer forests: Measurements and models. *J. Geophys. Res. All Ser.* **1997**, *103*, 29555–29566. [CrossRef]

49. Song, C.; Woodcock, C.E.; Li, X. The spectral/temporal manifestation of forest succession in optical imagery: The potential of multitemporal imagery. *Remote Sens. Environ.* **2002**, *82*, 285–302. [CrossRef]

50. Zhongkun, X.U.; Qingqian, X.U.; Rong, J. The characters of leaf area and net photosynthesis efficiency of needle at different parts and in different leaf age of *Cunninghamia lanceolata*. *Hunan For. Sci. Technol.* **2008**, *41*, 1–4.

51. Saltelli, A.; Bolado, R. An alternative way to compute fourier amplitude sensitivity test (FAST). *Comput. Stat. Data Anal.* **1998**, *26*, 445–460. [CrossRef]

52. Shi, H.; Xiao, Z.; Liang, S.; Zhang, X. Consistent estimation of multiple parameters from MODIS top of atmosphere reflectance data using a coupled soil-canopy-atmosphere radiative transfer model. *Remote Sens. Environ.* **2016**, *184*, 40–57. [CrossRef]

53. Saltelli, A.; Annoni, P.; Azzini, I.; Campolongo, F.; Ratto, M.; Tarantola, S. Variance based sensitivity analysis of model output. Design and estimator for the total sensitivity index. *Comput. Phys. Commun.* **2010**, *181*, 259–270. [CrossRef]

54. Saltelli, A.; Tarantola, S.; Chan, K.P.-S. A quantitative model-independent method for global sensitivity analysis of model output. *Technometrics* **1999**, *41*, 39–56. [CrossRef]

55. Saltelli, A.; Tarantola, S.; Campolongo, F.; Ratto, M. Sensitivity analysis in practice. *J. Am. Stat. Assoc.* **2004**, *101*, 398–399.

56. Li, X.; Gao, F.; Wang, J.; Zhu, Q. Uncertainty and sensitivity matrix of parameters in inversion of physical BRDF model. *J. Remote Sens.* **1997**, *1*, 161–172.

57. Tarantola, A. *Inverse Problem Theory and Methods for Model Parameter Estimation*; Society for Industrial and Applied Mathematics (SIAM): Philadelphia, PA, USA, 2005.

58. Nocedal, J.; Wright, S.J. *Sequential Quadratic Programming*; Springer: New York, NY, USA, 2006.

59. Lin, Z.F.; Ehleringer, J. Changes in spectral properties of leaves as related to chlorophyll content and age of papaya. *Photosynthetica* **1982**, *16*, 520–525.

60. Roberts, D.A.; Nelson, B.W.; Adams, J.B.; Palmer, F. Spectral changes with leaf aging in Amazon caatinga. *Trees* **1998**, *12*, 315–325. [CrossRef]

61. Rock, B.N.; Williams, D.L.; Moss, D.M.; Lauten, G.N.; Kim, M. High-spectral resolution field and laboratory optical reflectance measurements of red spruce and eastern hemlock needles and branches. *Remote Sens. Environ.* **1994**, *47*, 176–189. [CrossRef]

62. Gausman, H.W.; Allen, W.A.; Cardenas, R.; Richardson, A.J. Reflectance discrimination of cotton and corn at four growth stages. *Agron. J.* **1973**, *65*, 194–198. [CrossRef]

63. Chavanabryant, C.; Malhi, Y.; Wu, J.; Asner, G.P.; Anastasiou, A.; Enquist, B.J.; Cosio Caravasi, E.G.; Doughty, C.E.; Saleska, S.R.; Martin, R.E. Leaf aging of Amazonian canopy trees as revealed by spectral and physiochemical measurements. *New Phytol.* **2017**, *214*, 1049–1063. [CrossRef] [PubMed]

64. Samanta, A.; Ganguly, S.; Myneni, R.B. MODIS enhanced vegetation index data do not show greening of Amazon forests during the 2005 drought. *New Phytol.* **2011**, *189*, 11–15. [CrossRef] [PubMed]

65. Samanta, A.; Costa, M.H.; Nunes, E.L.; Vieira, S.A.; Xu, L.; Myneni, R.B. Comment on drought-induced reduction in global terrestrial net primary production from 2000 through 2009. *Science* **2011**, *333*, 1093. [CrossRef] [PubMed]

66. Xu, L.; Samanta, A.; Costa, M.H.; Ganguly, S.; Nemani, R.R.; Myneni, R.B. Widespread decline in greenness of Amazonian vegetation due to the 2010 drought. *Geophys. Res. Lett.* **2011**, *38*, 490–500. [CrossRef]

67. Toomey, M.; Roberts, D.; Nelson, B. The influence of epiphylls on remote sensing of humid forests. *Remote Sens. Environ.* **2009**, *113*, 1787–1798. [CrossRef]

68. Xiaoquan, Z.; Deying, X. Seasonal changes and daily courses of photosynthetic characteristics of 18-year-old chinese fir shoots in relation to shoot ages and positions within tree crown. *Sci. Silvae Sin.* **2000**, *36*, 19–26.

69. Wu, J.; Serbin, S.P.; Xu, X.; Albert, L.P.; Chen, M.; Meng, R.; Saleska, S.R.; Rogers, A. The phenology of leaf quality and its within-canopy variation are essential for accurate modeling of photosynthesis in tropical evergreen forests. *Glob. Chang. Biol.* **2017**, *23*, 4814–4827. [CrossRef] [PubMed]

70. Frazer, G.W.; Fournier, R.A.; Trofymow, J.A.; Hall, R.J. A comparison of digital and film fisheye photography for analysis of forest canopy structure and gap light transmission. *Agric. For. Meteorol.* **2001**, *109*, 249–263. [CrossRef]

71. Zhang, Y.; Chen, J.M.; Miller, J.R. Determining digital hemispherical photograph exposure for leaf area index estimation. *Agric. For. Meteorol.* **2005**, *133*, 166–181. [CrossRef]

remote sensing

MDPI

Review

Characterizing Land Surface Anisotropic Reflectance over Rugged Terrain: A Review of Concepts and Recent Developments

Jianguang Wen [1,2,3,*], Qiang Liu [2,4], Qing Xiao [1,3], Qinhuo Liu [1,2,3], Dongqin You [1,2], Dalei Hao [1,3], Shengbiao Wu [1,3] and Xingwen Lin [1,3]

[1]   State Key Laboratory of Remote Sensing Science, Institute of Remote Sensing and Digital Earth, Chinese Academy of Sciences, Beijing 100101, China; xiaoqing@radi.ac.cn (Q.X.); liuqh@radi.ac.cn (Q.L.); youdq@radi.ac.cn (D.Y.); haodl@radi.ac.cn (D.H.); wusb@radi.ac.cn (S.W.); linxw@radi.ac.cn (X.L.)
[2]   Joint Center for Global Change Studies (JCGCS), Beijing 100875, China; toliuqiang@bnu.edu.cn
[3]   University of Chinese Academy of Sciences, Beijing 100049, China
[4]   College of Global Change and Earth System Science, Beijing Normal University, Beijing 100875, China
[*]   Correspondence: wenjg@radi.ac.cn; Tel.: +86-010-64806255

Received: 24 December 2017; Accepted: 21 February 2018; Published: 27 February 2018

Abstract: Rugged terrain, including mountains, hills, and some high lands are typical land surfaces around the world. As a physical parameter for characterizing the anisotropic reflectance of the land surface, the importance of the bidirectional reflectance distribution function (BRDF) has been gradually recognized in the remote sensing community, and great efforts have been dedicated to build BRDF models over various terrain types. However, on rugged terrain, the topography intensely affects the shape and magnitude of the BRDF and creates challenges in modeling the BRDF. In this paper, after a brief introduction of the theoretical background of the BRDF over rugged terrain, the status of estimating land surface BRDF properties over rugged terrain is comprehensively reviewed from a historical perspective and summarized in two categories: BRDFs describing solo slopes and those describing composite slopes. The discussion focuses on land surface reflectance retrieval over mountainous areas, the difference in solo slope and composite slope BRDF models, and suggested future research to improve the accuracy of BRDFs derived with remote sensing satellites.

Keywords: anisotropic reflectance; BRDF; rugged terrain; solo slope; composite slope

1. Introduction

Rugged terrain covers approximately 24% of the Earth's land surface, plays an important role in the complex earth system, and forms a unique mountainous climate and ecosystem. Accurately estimating land surface variables over mountainous areas is of great importance for global hydrological and meteorological forecasts, as well as global ecological and environmental monitoring. Taking the benefits of advanced satellite instrumentation and accurate remote sensing modeling, topographic effects can be considered and neutralized in applications of surface parameter retrievals, such as improved land cover and land type mapping [1,2], key parameters of mountain radiation, and energy budget (albedo, land surface temperature (LST), and solar radiation) [3–5] and vegetation structure parameters (normalized difference vegetation index(NDVI), leaf area index (LAI), and fractional photosynthetically active radiation (FPAR)) [6,7]. Therefore, remote sensing satellite technique development over rugged terrain is crucial for extending remote sensing applications from flat surfaces to mountainous areas.

Remote sensing of anisotropic reflectance relates the land surface scattering behavior to its optics and structure, which is described with the bidirectional reflectance distribution function (BRDF) [8]

and can be observed in optical remote sensing. To retrieve BRDF properties of the land surface is the basis for land surface's physical parameter inversions [9–12], due to its structural information observed from different angles. In contrast, angular effects on surface reflectance should be removed by BRDF correction to normalize the reflectance [13–15] to implement classification and dynamic monitoring [16]. Thus, modeling the land surface BRDF and developing a corresponding remote sensing product is important for scientific research and remote sensing applications. However, rugged terrain complicates BRDF modeling by changing the amount of radiation detected by the sensor [17–20]. Generally, topography alters the illumination and viewing geometry and generates a relief shadow, observation masking, and multiple scattering, this results in the intense topographic dependence on total incident and reflectance radiance, which distorts BRDF characteristics [4,21]. The BRDF characteristics varies with wavelength and shows a BRDF wavelength dependence in surfaces of greater roughness [22,23]. Without considering the topographic effects on the land surface BRDF, the anisotropic reflectance estimation relative errors could be larger than 58% [24]. Therefore, the current remote sensing BRDF concept and modeling over rugged terrain should consider the topographic effects.

Previous studies have focused on the influence of geometric characteristics on the radiative transfer process over rugged terrain. Thus, the topographic correction models have first attempted to reduce topographic effects on reflectance [18,19,25,26]. Recently, there have been concerns about BRDF modeling over rugged terrain [17,20,21,27]. However, the operational BRDF product algorithm over rugged terrain is still being researched because it involves a complex process, which includes multi-angular reflectance dataset processing, an operational BRDF algorithm and its inversion method coupled with topography. From this perspective, high quality atmospheric correction that is used to obtain land surface reflectance is also the basic task to derive the rugged surface BRDF. Investigations have proven that the atmospheric correction of different resolution remote sensing data needs to consider the influence of topography [28,29]. However, with the decrease in spatial resolution, the influence of topography on pixel scale reflectance will gradually decrease. This causes a lot of the low resolution land surface reflectance to ignore the effects of topography during atmospheric correction (e.g., moderate resolution imaging spectroradiometer (MODIS) reflectance).The reflectance that does consider the topographic influence focuses on only the relatively high resolution remote sensing data (e.g., Landsat Thematic Mapper (Landsat/TM). Consequently, dominant BRDF modeling over rugged terrain occurs on an infinite slope surface [17,20,21,27]. However, the focus of low resolution remote sensing data is not the topography influence at the pixel level, but the topography influence at the sub-pixel level [29–31]. There should be a robust relationship between these two different anisotropic reflectance resolution. Thus, the characterization of land surface anisotropic reflectance over rugged terrain should be implemented from a systematic perspective by analyzing the critical scientific problems and reviewing current algorithms, which will benefit algorithm developers and broaden the interests of surface BRDF users.

In this paper, remote sensing BRDF modeling over rugged terrain according to the presented research chain is comprehensively reviewed, and the aim is to find an operational BRDF product potential solution for rugged terrain. This is important for quantitative remote sensing applications in mountainous areas. The paper is organized as follows: first, we analyzed the topographic effects on the BRDF and its scientific problems in Section 2. Second, the methods to solve the atmospheric correction and obtain the multi-angular reflectance are briefed in Section 3. The two kinds of BRDF modeling are described based on evolution histories in Sections 4 and 5, according to the spatial resolution between the digital elevation model (DEM) and remote sensing pixel. Then, we analyzed the challenges and opportunities for BRDF product generation over rugged terrain in Section 6. Finally, we summarize this paper.

2. BRDF in Rugged Terrain

2.1. Literatures Review

A review of the current literature is one of the best ways to clearly understand the timeline and milestones in BRDF research over rugged terrain. In this paper, we searched for articles titles matching the relevant key words about BRDF and rugged terrain on the Web of Science website, Thomson Reuter. BRDF has seven relevant query words: "reflect*", "reflectance*", "non*Lambertian", "BRDF", "BRF", "bi-directional reflect*", and "bidirectional reflect", "multiangle", "multi-angle", "multiangular" and "multi-angular". Rugged terrain has ten relevant query words: "mountain*", "hill*", "rugged terrain*", "complex terrain*", "topograph*", "slope*", "sloping terrain*", "rough* surface*", "random surface*", and "roughness". To exclude the irrelevant articles, we further screened the search results by subject and keywords. Finally, the articles about BRDF modeling over rugged terrain were collected from the Web of Science platform for citation statistics and analysis. Figure 1 shows the articles that were contributed by different research fields in recent decades. The results show that BRDF modeling over rugged terrain is of great importance in many scientific and engineering fields, including physics, engineering, optics, materials science, geology, remote sensing, geophysics, instrumentation, image science, chemistry, and spectroscopy. Judging by the number of articles, BRDF modeling over rugged terrain ranks sixth in remote sensing field.

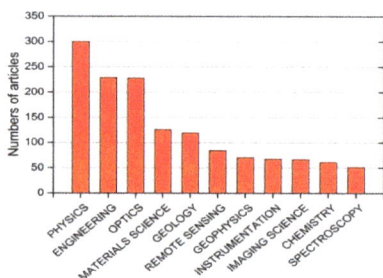

Figure 1. Literature statistics for bidirectional reflectance distribution function (BRDF) modeling over rugged terrain contributed by different research field in recent decades.

Figure 2a,b shows the annual publications and citations from 1983 to 2017 in the field of remote sensing. Since 1993, there have been several articles published every year. The variation in the number of published papers on this subject have followed a periodical pattern. In some characteristic years, the number of published paper is as many as six. In comparison, the citations of these published papers grow smoothly by year, especially since 2006, which is when the numbers of citations showed a rapid increase. This demonstrates that the subject of BRDF modeling over rugged terrain has received increasing attention and has gradually become a hotspot in studies of quantitative remote sensing.

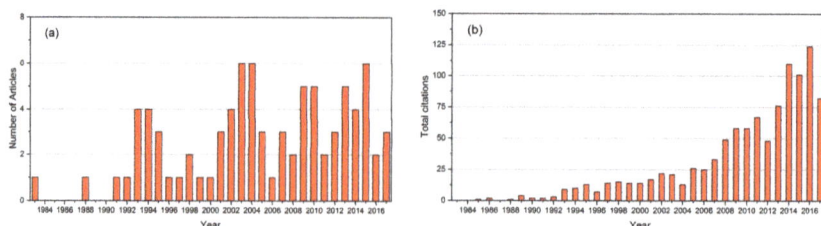

Figure 2. Literature statistics for BRDF modeling over rugged terrain from 1983 to 2017. (**a**) The numbers of published articles, and (**b**) total citations.

2.2. BRDF Definition and Its Topographic Effects

BRDF is defined mathematically as the ratio of the radiance L_r which is reflected by the target surface into a specific direction, to the collimated solar incident irradiance E_s on the same surface [32]. Two basic assumptions are implied in the current remote sensing BRDF quantity, which is shown in Equation (1) [33,34], and the assumptions are as follows: (1) the target surface is assumed to be a horizontal plane, and (2) the target surface is homogeneous with a uniform incident irradiance and outgoing irradiance. Namely, the radiative flux interaction and exchange is equivalent at every point over the target surface.

$$BRDF(\Omega_s, \Omega_v) = \frac{dL_r(\Omega_s, \Omega_v)}{dE_s(\Omega_s)} \tag{1}$$

where Ω_s and Ω_v indicate the illumination and viewing geometry, respectively.

However, steep rugged terrain changes local illumination and viewing geometry because the normal of the slope surface is not consistent with the vertical direction (shown in Figure 3), which is results in heterogeneous incident irradiances due to the three-dimensional (3-D) structure configuration of vegetation and topography [34]. Specifically, the slopes facing toward the sun will receive more illuminated irradiance than the slopes that are away from the sun [18]. The diffuse irradiance reflected from adjacent slopes will increase the global incident irradiance of the slope surface, and the distribution of shadowing and observation masking have a notable effect on the pixel reflectance [29,35]. Therefore, the question of how to describe these topographic effects and the radiation redistribution is a core concern in the BRDF model over rugged terrain. It is concluded that to accurately model BRDF over rugged terrain, it should be coupled with the terrain information supplied by a DEM.

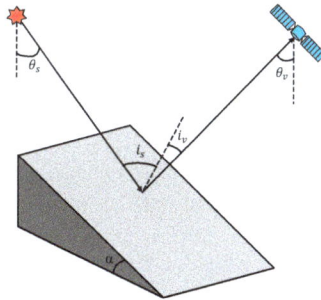

Figure 3. Configuration of solar illumination and sensor over a slope surface.

A spatial scale match between the DEM and remote sensing image pixel should be emphasized prior to coupling. According to the relationship between the spatial resolution of the available DEM and remote sensing pixel, the modeled topographies on the remote sensing pixel are classified into solo slope and composite slope (Figure 4). Under the assumption of the current DEM having a 30 m spatial resolution, the solo slope means that those remote sensing pixels have a spatial resolution comparable to the DEM dataset, and there is only a single slope surface, such as the Landsat/TM images, which have a 30 m spatial resolution. The composite slope refers to the situation when the remote sensing instrument with a large instantaneous field of view (IFOV) covers an area of few kilometers, which is result in a spatial resolution lower than the DEM dataset. There are numerous solo slopes contained within a remote sensing pixel. For example, the MODIS and advanced very high-resolution radiometer (AVHRR) sensors have a km-scale spatial resolution. However, if the remote sensing pixels have higher spatial resolution than the DEM, resampling the DEM to the same spatial resolution of the remote sensing image pixel is necessary. Otherwise, a higher resolution DEM should be provided.

The largest difference between topographic effects for these two types lies in the complex heterogeneous incident irradiance and topographic shadowing [4,29]. The pixel level topographic effect dominates the solo slope, including local geometry alteration, shadow cast, and diffuse irradiance reflected from adjacent slopes. In addition, with the lower spatial resolution, the topographic effects on BRDF become weaker due to a smooth overall slope [36]. However, in these cases, the sub-pixel level topographic effects are significant. Specifically, the sub-pixel level topographies affect the distributions of incident and reflected radiance by the distribution of sub-slope, sub-aspect, amount of shadow, and masking, which is leads to distorted BRDF characteristics.

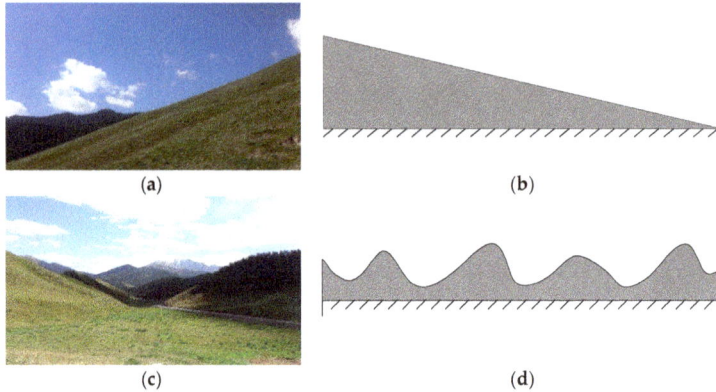

Figure 4. Graphics of topography relief: (**a**) nature surface of solo slope, (**b**) the topographic model of solo slope, (**c**) nature of composite slope, and (**d**) topographic model of composite slope.

Because the BRDF quantity is defined as the ratio of two infinitesimal surfaces, it is a theoretical concept and cannot be directly measured [13]. In remote sensing, the bidirectional reflectance factor (BRF) is adopted as a substitute for BRDF to describe the surface anisotropic scattering property [8,37]. BRF is defined as the ratio of reflected radiance from a target surface to that from an ideal and diffuse reference plane under identical illumination and viewing conditions. Thus, a key issue is how to define the reference plane when modeling the surface BRF. The horizontal reference plane at the highest point is widely favored in BRF models over a composite slope, which is shown in Figure 5 [29,30,35]. However, different opinions exist about a reference plane for the solo slope BRF model. Some physical-based analytical models adopt the reflectance of a slope-parallel white plane to calibrate surface BRF [17,21,27]. The derived reflectance is known as the slope BRF. The other analytical and most of 3-D computation BRF models are based on the horizontal reference plane, regardless of the underlying topography [20,38,39].

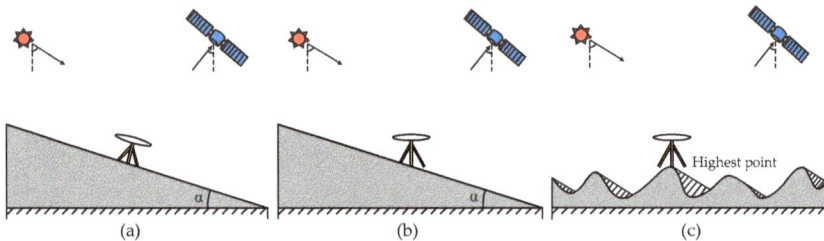

Figure 5. Reference plane configuration over solo slope ((**a**) slope-parallel white plane and (**b**) horizontal reference plane) and composite slope ((**c**) horizontal reference plane at the highest point).

2.3. Model Building Procedures and Scientific Problems

Many efforts have been devoted to investigating the topographic effects on surface BRDF. The comprehensive and systematic investigation of sloped surface BRDF estimation is crucial to understanding the basic theory and the scientific problems that are involved with the BRDF over rugged terrain. According to the relationship between the DEM scale and remote sensing pixel spatial resolution, two cases of BRDF modeling over rugged terrain are presented in Figure 6: solo slope BRDF modeling and composite slope BRDF modeling. A self-consistent solution and a validation technique for these two BRDFs should be emphasized in this procedure.

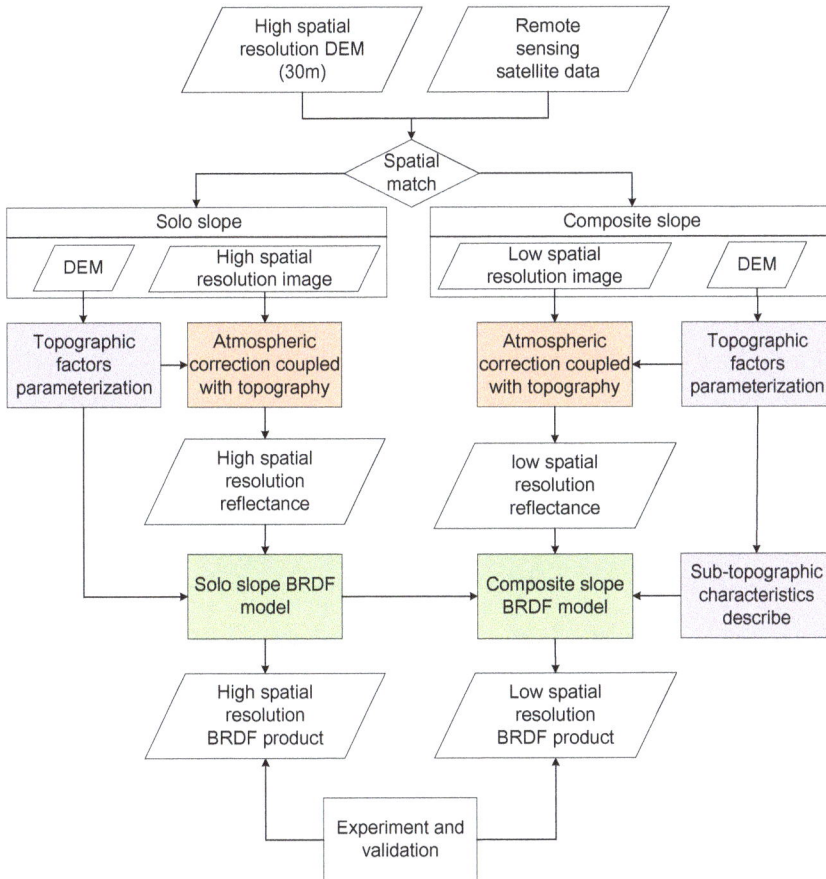

Figure 6. Key procedures of BRDF modeling over rugged terrain.

Thus, to derive a desired BRDF product, which describes the BRDF properties of the slope surface, a suitable and robust BRDF model is the basis for retrieval of the anisotropic reflectance distribution with multi angle satellite reflectance data. The following three key issues are needed to be addressed when modeling surface BRDF over rugged terrain.

(1) How should the solo slope and composite slope anisotropic reflectance properties be described?

Solo slope and composite slope surfaces have different topographic effect mechanisms in the BRDF, which leads to different core sensitivity factors considered in remote sensing BRDF modeling. Accurately building a BRDF model depends on how the topographic characteristics are parameterized. The pixel-level topographic effect dominates the reflectance of the solo slope through the alteration of local incidence and viewing geometry, as well as the distribution of the illuminated irradiance [20,29]. However, this effect is relatively small for the composite slope BRDF, and the sub-pixel level topographic effects become the primary factors, which include the shadow, distribution of sub-topographic slope and aspect statistic, and the redistribution of radiation within a remote sensing pixel [4,31].

(2) How can we implement the solo slope or composite slope atmospheric correction to correctly derive reflectance?

Currently, the surface anisotropy reflectance, coupled with atmospheric parameters, is derived from the airborne and satellite remote sensing observations. The second simulation of the satellite signal in the solar spectrum radiative transfer process (6S) [40] model and the MODerate resolution atmospheric TRANs mission (MODTRAN) [41] model are the two methods to describe the atmospheric effects and can be used to retrieve the land surface reflectance from the top of the atmosphere radiance. The topographic effect is always neglected in these algorithms. Some efforts have been made to extend the 6S atmospheric correction algorithm, coupled with a topography consideration for a solo slope [18,19]. For a composite slope, the topographic effects on the atmospheric correction process are not included, because the slope of the pixel level is considered as flat. Otherwise, if the orientation of sub-topography is statistically dependent on the overall slope and aspect, the atmospheric correction should be coupled with topography.

(3) How can we develop the BRDF (BRF) validation technique and conduct experiments over rugged terrain?

When compared with flat surfaces, the spatial and temporal heterogeneity of surface BRDF appears more obvious over mountainous areas. The validation of the mountainous surface BRDF is an important issue, which includes sampling strategy and respective experiment conduction. With current observations covering several scales from ground measurements with ground-based instruments, meter-scale with unmanned aerial vehicles and km-scale with satellites, the multi-scale validation technique may potentially address this issue.

3. Remote Sensing Atmospheric Correction over Rugged Terrain

The BRDF properties of land surfaces are commonly retrieved against multiple directional reflectance to sufficiently sample anisotropy. Thus, it is necessary that the atmospheric correction is completed prior to the BRDF retrieval. However, the topography intensely affects the atmospheric correction, and consequently, also affects the land surface reflectance. Without DEM consideration in the atmospheric correction, the reflectance is varied in its response to similar topographic features, where the solar and sensor geometries correspond to a flat surface. According to Figure 5, the corrected reflectance without a DEM is not the reflectance referenced to the slope BRF defined a slope-parallel reference plane or the remote sensing BRF defined a horizontal reference plane, where the sensor view angle is the local angle corresponding to the slope surface. Although topographic correction can normalize the reflectance to that of flat surface, such as C correction [42], sun-canopy-sensor (SCS) correction [43], and their integrated method [44], the reflectance is still not applicable to the slope surface BRDF model retrieval due to its contradiction with the geometry defined slope of the BRDF model. To obtain an accurate reflectance over rugged terrain, the atmospheric correction should be coupled with topography, which is based on the mountain radiation transfer prototype model and DEM.

3.1. Lambertian-Based Atmospheric Correction

The mountain radiation transfer algorithms describe the radiance received by the sensor over a mountainous area. On rugged terrain, the irradiance at the target surface is composed of solar direct irradiance E_s, sky diffuse irradiance E_d, and adjacent terrain reflected irradiance E_a, which are shown in Figure 7.

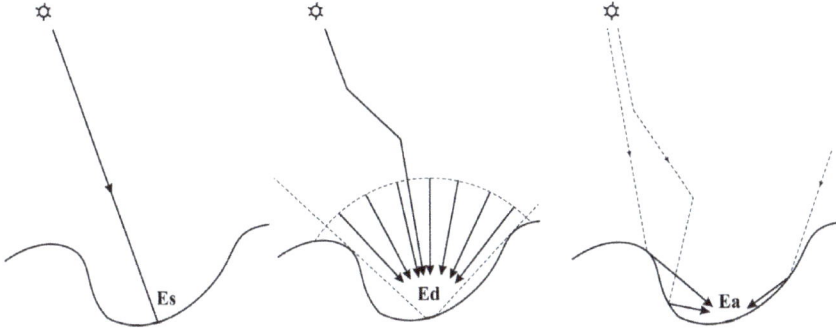

Figure 7. Irradiance at the land surface.

Direct solar radiation over rugged terrain is the most important component of the total surface radiation that reaches to the surface. When compared with flat surfaces, this depends on the relative local incident angle between the sun and the normal to slope surface. Therefore, the direct solar radiation changes with different slope surfaces. The sky diffuse irradiance will be reduced when the sky dome overlying a surface is not an integrated hemisphere of a horizontal surface. However, the adjacent terrain reflected irradiance increases the total radiation reaching the slope surface. Thus, the total radiation is the function of the DEM and atmospheric parameters.

If the slope surface reflectance is ρ and the solar and sensor geometry are (θ_s, ϕ_s) and (θ_v, ϕ_v), respectively, the sensor received radiance L can be expressed as follows [35]:

$$L = L_p + (E_s + E_d + E_a)\rho e^{-\tau/\cos\theta_v} / \pi(1-s\rho) \tag{2}$$

where L_p is the path radiance, τ is the atmospheric aerosol optical depth, and s is the atmospheric diffuse albedo. E_s can be expressed as follows [45]:

$$E_s = \Theta E_0 \cos(i_s) e^{-\tau/\cos(\theta_s)} \tag{3}$$

where E_0 is the exo-atmospheric solar irradiance and Θ is a binary coefficient, which is set to zero to show whether a pixel is shadowed and set to one otherwise.

The sky diffuse irradiance E_d is as follows [43]:

$$E_d = E_h \times (k\frac{\cos(i_s)}{\cos(\theta_s)} + (1-k)V_d) \tag{4}$$

$$V_d = \frac{1}{2\pi}\int_0^{2\pi} \left[\cos\alpha \sin^2 H_\varphi + \sin\alpha \cos(\varphi-\beta)(H_\varphi - \sin H_\varphi \cos H_\varphi)\right]d\varphi \tag{5}$$

where E_h is the sky diffuse radiance on a horizontal surface, k is an anisotropy index related to the atmospheric transmittance for direct irradiance and values between 0 and 1, and V_d [45] is sky view factor defined as the unobstructed portion of the sky at any given point. H_φ is the horizontal angle from the zenith downward to the local horizon for direction φ. α and β are the slope and aspect angles, respectively.

The adjacent terrain reflected irradiance E_a can be expressed as follows [20]:

$$E_a = \sum_N \frac{L_N \cos T_M \cos T_N dS_N}{r_{MN}^2} \tag{6}$$

where L_N is the radiance reflected from the land surface at point N and can be received by point M, dS_N is the area of N, T_M and T_N are the angles between the normal to the surface and the line of M and N, respectively, and r_{MN} is the distance between M and N.

The solution to Equation (2) is the reflectance ρ when the sensor radiance and the atmospheric variables affected by topography are accurately estimated, which plays an important role in early remote sensing topographic correction and land surface reflectance estimation over mountainous areas [25,46,47]. However, the literatures show that the mountain radiation transfer prototype model, which is based on Lambertian land surface assumptions, often leads to an overcorrected reflectance [18,19,26]. Therefore, the radiation between the earth's surface and atmosphere and the anisotropic land surface reflectance are the two core issues, where scientists are greatly concerned about improving the reflectance quality over mountainous areas.

3.2. Non-Lambertian-Based Atmospheric Correction

Recent literature shows that without considering the surface BRDF effects in atmospheric correction, the result will be errors of up to 10–20% in the worst cases [48]. Surface BRDF retrieval and atmospheric correction can be coupled in a converging iteration, which is operationally employed to achieve MODIS land surface reflectance retrievals [49]. It is distinct from atmospheric correction and BRDF correction, where the BRDF normalizes the reflectance to a certain geometry [13–15,50]. However, if the BRDF cannot be deduced from the remote sensing data themselves through inversion and iterative coupling, an alternative solution is to apply the BRDF prior knowledge from different remote sensing data [50], where the BRDF shape, rather than its magnitude, is required to resolve the effects. Previous research reported that land surface BRDF prior knowledge can improve the atmospheric and topographic correction quality and reduce the uncertainties in reflectance over rugged terrain [18,19,26].

According to the 6S atmospheric model, the target radiance to the sensor is described as the sum of four terms: (1) the photons directly transmitted from the sun to the target and directly reflected back to the sensor, (2) the photons scattered by the atmosphere, reflected by the surrounding target and directly transmitted to the sensor, (3) the photons directly transmitted to the target but scattered by the atmosphere on their way to the sensor, and, finally, (4) the photons that have at least two interactions with the atmosphere and one with the target. Thus, Equation (2) can be rewritten as follows [15,16]:

$$\begin{aligned} L = L_p + \frac{1}{\pi}(E_s\rho_{dd}(i_s, \varphi_s, i_v, \varphi_v)e^{-\tau/\cos(\theta_v)} + (E_d + E_a)\rho_{hd}(i_v, \varphi_v)e^{-\tau/\cos(\theta_v)} \\ + E_s\rho_{dh}(i_s, \varphi_s)t_d(\theta_v) + (E_d + E_a)\rho_{hh}t_d(\theta_v) + (E_s + E_a + E_d)\frac{(e^{-\tau/\cos(\theta_v)} + t_d(\theta_v)s\rho_{hh}^2)}{1 - s\rho_{hh}}) \end{aligned} \tag{7}$$

where $\rho_{dd}(i_s, \varphi_s, i_v, \varphi_v)$, $\rho_{hd}(i_v, \varphi_v)$, $\rho_{dh}(i_s, \varphi_s)$, and ρ_{hh} are the slope surface directional-directional reflectance, hemispheric-directional reflectance, directional-hemispheric reflectance, and bi-hemispheric reflectance, respectively. Similar to the MODIS atmospheric correction method [48], $\rho_{dd}(i_s, \varphi_s, i_v, \varphi_v)$ is resolved by introducing BRDF prior knowledge [16]. One of the assumptions for the BRDF coupled mountain radiation transfer model is that the BRDF shape depends on the land cover, and the BRDF effect for the slope is the BRDF for the rotated angles. An image dependent BRDF shape was first developed from a regression method or a regionally averaged BRDF shape using an image scene [51]. Another method is that a statistics-based MODIS BRDF prior knowledge look-up table (LUT) was proposed as the BRDF shape and used in the BRDF-based atmospheric correction (BRATC) [19].

4. Solo Slope BRDF Model

4.1. Physical Basis

The topographic effects on the solo slope BRDF depend on the slope and aspect angle, and the structural and optical properties of the vegetation. The surface's slope and aspect change the local solar incidence and the sensor observation directions. Although vegetation shows crown geotropism, regardless of the terrain slope, its projection on the slope surface varies with the slope and aspect, and the direct and diffuse irradiance are redistributed. This changes the incident radiation received by the slope surface and the sensor observed radiation. Thus, the slope surface BRDF characteristics are distorted by the topography.

The solo slope BRDF model focuses on development of the slope changed radiation and accurate estimation of the 3D structure of vegetation and soil over the slope surface. Many physical BRDF models, such as the scattering by arbitrarily inclined leaves (SAIL) model [52], geometric-optical and radiative transfer (GORT) model [53], and forest reflectance and transmittance (FRT) model [54], as well as the computer simulation model of discrete anisotropic radiative transfer (DART) [38] and radiosity-graphics combined model (RGM) [55], have addressed the problems associated with solar radiation and complex 3-D canopy structure and background. The simulated BRF varied widely between these models under the activity provided by the radiative transfer model intercomparison (RAMI) [56], where a complex 3-D vegetation scenario is used as the benchmarking. However, the topography will introduce difficulty in the vegetation canopy anisotropic reflectance characterization. The reflectance anisotropy of the solo slope is the following function:

$$BRF = f(\Omega_s, \Omega_v, DEM, para, ref, E) \tag{8}$$

where Ω_s, Ω_v, and DEM are the geometric parameters used to describe the anisotropic reflectance; *para* indicates the canopy structure and land biophysical parameters, including crown height, crown density, crown shape, LAI, etc.; *ref* represents the optical reflectance properties, such as leaf and background reflectance; and E is the incident direct and diffuse irradiance.

Specifically, the topography alters the local solar incidence angle and the sensor observation geometry, as Figure 8 shows, which induces area changes in the canopy shadows cast on the background, as well as the mutual shadowing relationship between discrete heterogeneous tree crowns [21,57], photon path length within the homogeneous vegetation layer [20,27], and effective local illuminated slope irradiance [58]. Therefore, differences in terrain configuration significantly vary in reflected signals of surfaces with similar land cover, structural, and optical properties. A rotational transition matrix between horizontal coordinates and local slope coordinates is adopted to correct the geometric relationship.

$$\begin{bmatrix} \sin\theta_{is(v)}\cos\phi_{is(v)} \\ \sin\theta_{is(v)}\sin\phi_{is(v)} \\ \cos\theta_{is(v)} \end{bmatrix} = \begin{bmatrix} \cos\alpha & 0 & -\sin\alpha \\ 0 & 1 & 0 \\ \sin\alpha & 0 & \cos\alpha \end{bmatrix} \begin{bmatrix} \sin\theta_{s(v)}\cos(\phi_{s(v)} - \beta) \\ \sin\theta_{s(v)}\sin(\phi_{s(v)} - \beta) \\ \cos\theta_{s(v)} \end{bmatrix} \tag{9}$$

where θ and φ are the zenith and azimuth angles, respectively, the subscripts *is* and *iv* represent the local incident and observation geometries, and the subscripts *s* and *v* represent the incident and observation geometries.

However, the rotational geometric correction leads to crown inclination. This contradicts with the geotropic nature of tree crowns where the trees grow vertically and orient with the gravitational field, regardless of the slope. The crown structural and optical properties over the solo slope remain the same as those on the flat surface. For example, the leaf angle distribution (LAD) and leaf reflectance and transmittance reflectance are not affected by the terrain [27]. However, for the discrete forest stands, the crown shadowing projections on the background are influenced by the geotropic nature of tree crowns (as shown in Figure 9). The geometry correction using Equation (9) without negative geotropism consideration will lead to an incorrect crown shadowing (Figure 9b), as well as an inappropriate canopy reflectance. However, the tree crowns are virtually inclined after the geometric correction (Figure 9c). The importance of the geotropic nature of tree crowns has been stressed in the topographic correction of forested terrain [43,59].

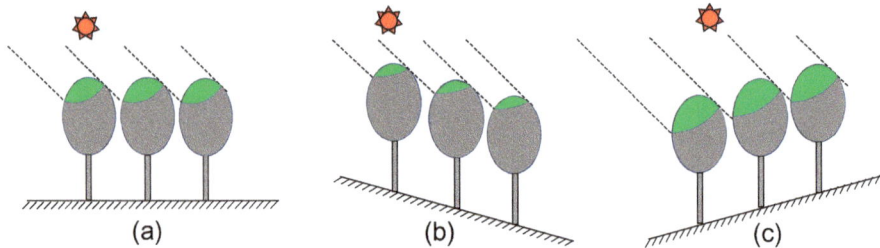

Figure 8. Canopy shadow cast on flat and sloped forest. (**a**) Flat forest. (**b**,**c**) Sloped forest. The dotted lines represent the incident solar beam.

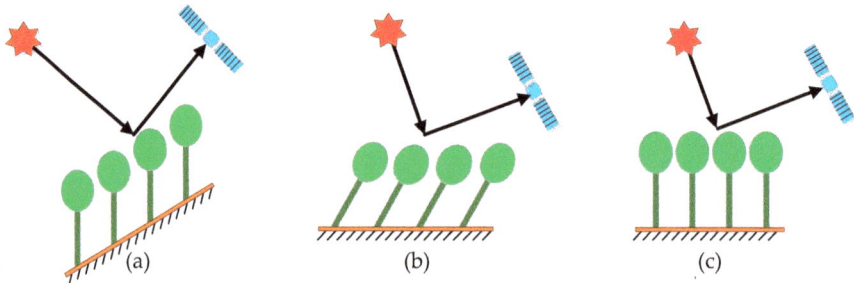

Figure 9. Topographic effects on crown sun-canopy-sensor geometry. (**a**) Forest stand on solo slope surface, (**b**) geometry correction without negative geotropism consideration, and (**c**) geometry correction with negative geotropism consideration.

Another topographic effect on the canopy reflectance is through the redistribution of surface global incident solar radiation, which modifies the upper boundary condition during the radiative transfer process. When compared to the flat terrain, the topography redistributes the direct solar irradiance by changing the surface's local illumination and viewing geometry. The terrain restricts the diffuse skylight and enhances the diffuse irradiance that is reflected from adjacent slopes [25,59–61]. The last two diffuse components can account for 40% of the global radiation of a sunlit slope when the solar zenith angle is high, and this can even approach 100% when the slope is obstructed by adjacent terrains [46,62]. Moreover, varied elevation will induce rapid changes in concentrations of aerosol, water vapor, and cloud properties, which give rise to significant variations in the amount of direct and diffuse incident irradiance [63].

In the past decades, physical solo slope BRDF models have been proposed to account for the topographic effects on pixel reflectance. These are the BRDF models based on radiative transfer, geometric-optical, and hybrid methods (Table 1). Three key scientific issues include the geometry correction, negative geotropism of trees, and the irradiance redistribution, which affect the topographic BRDF. Since vegetation cover, such as forest and grassland, dominates the complex mountainous land ecosystem, current physical solo slope BRDF models mainly focus on the vegetation.

Table 1. List of physical solo slope surface BRDF models.

Model Category	Model Name	Geometry Correction	Tree's Negative Geotropism	Diffuse Irradiance	Typical Reference
Radiative transfer	ROSST	√	√	×	Combal et al. [27]
	PLC	√	√	√	Yin et al. [20]
Geometric optical	GOMST	√	×	×	Schaaf et al. [21]
	GOST1	√	√	×	Fan et al. [17]
Hybrid	SLCT	√	×	√	Mousivand et al. [64]
	GOST2	√	√	×	Fan et al. [24]
	GOSAILT	√	√	√	Wu et al. [65]

The symbols √ and × indicate that with and without consideration in the BRDF model, respectively. ROSST is the improved ROSS model for solo slope terrain; PLC is A Physical Solo Slope Canopy Reflectance Model Based On The Path Length Correction; GOMST is the geometric-optical mutual shadowing model for solo slope terrain; GOST is the 4-scale geometric-optical model for solo slope terrain; SLCT is the soil–leaf-canopy model for solo slope terrain; GOSAILT is the hybrid model of GOMS and scattering by arbitrarily inclined leaves (SAIL) coupled topography.

4.2. Model Development

4.2.1. Radiative Transfer Model

The radiative transfer process captures the vegetation canopy reflectance and further retrieves the vegetation physical and biophysical parameters, which treats the forest canopy as an homogeneous, turbid medium with discrete leaf elements [66]. The descriptions of leaf structural and optical characteristics, such as leaf size and shape, LAI, LAD, scattering phase function, and single scattering albedo, are key to this approach [53].

The importance of the topographic effect in the radiative transfer was first proposed by Combal et al. [27] who successfully extended the Ross radiative transfer theory [67] for flat plant canopies to solo slope with a vertical plant stand. The topographic effect on local geometry relationship has been considered in the ROSST model [27] through the transformation between horizontal and slope coordinate systems. The geotropic nature of tree crowns was accounted for by using the leaf structural and the optical characteristics defined in the horizontal coordinate to solving the one-dimensional (1-D) analytical radiative transfer equation. However, only the direct solar radiation was considered, the effect of diffuse skylight, path radiance, and diffuse irradiance from adjacent slopes are neglected. Recently, the topographic effect on the bidirectional gap probability has been regarded as the primarily factor affecting canopy reflectance [20,53,68]. A physical solo slope canopy reflectance model based on the path length correction (PLC) [20] was proposed to account for the topographic effect on the canopy photon path length and its BRDF. The geometry correction, geotropic nature of the tree crown, and the diffuse skylight are coupled in this model. However, the diffuse irradiance from neighboring terrains is still neglected.

4.2.2. Geometric-Optical Model

The geometric-optical model has an advantage in understanding the 3-D complex crown structure's effects on the canopy reflectance [69,70], in which the pattern of sunlit and shaded crowns and backgrounds seen in a particular direction were considered to be the key factor. According to the geometric-optical theory, the canopy reflectance is assumed to be composed of four scene components: sunlit crown, sunlit background, shaded crown, and shaded background with their respective areal proportions.

The topographic effect on the canopy reflectance in the geometric-optical model was firstly evaluated by Schaaf et al. [21], who extended the Li-Strahler geometric-optical mutual shadowing model for a solo slope surface (GOMST) through a simple geometry correction, while retaining other structural and optical properties the same as those in the horizontal forest. The accurate estimation of the topographic effect on the crown cast shadow for the background is critical for canopy reflectance.

When a slope faces toward the sun, less crown shadows are projected on the background, and more shadows are cast on the background when it is away from the sun [18,57]. The trees are assumed to be perpendicular to the solo slope without negative geotropism in the GOMST model, which will lead to an underestimation in the red reflectance. This is because the areal proportion of the background will be underestimated, while the areal proportion of the crown will be overestimated in this case. However, the canopy reflectance and albedo also appear to be significantly affected by terrain even without consideration of the trees' negative geotropism [21]. Fan et al. [17] incorporated the topographic effect into the 4-scale geometric-optical model for solo slope terrain (GOST1), which acknowledged the geotropic nature of tree crowns. However, the diffuse irradiance components are neglected in the current geometric-optical models, which will cause an underestimation of global incident irradiance and surface reflected signals.

4.2.3. Hybrid Model

The radiative transfer model is accurate in estimating the canopy reflectance with micro-scale leaf reflectance, especially for high orders of scattering and diffuse irradiance effects, and the geometric-optical model is accurate in describing the single scattering results from 3-D forest crown structure. Thus, hybrid models that combine the two approaches can capture discontinuous canopy reflectance at the landscape scale [52,53,71].

In mountainous regions, the GOST1 model was coupled with the recollision probability theory to parameterize the component reflectivity, which is called the GOST2 model [24]. The multiple scattering within canopy-background system is considered in the GOST2 model. However, GOST2 reflectance seems to be overestimated. The main reason might be that a fixed relationship between the canopy structural parameters, LAI and photon recollision probability was implemented in the GOST2 model. Like the GOST1 model, the diffuse irradiance components are neglected in the GOST2 model. The soil–leaf-canopy (SLC) model has successfully captured discontinuous canopy reflectance through incorporating the crown clumping effects, vertical leaf color gradient, and non-Lambertian soil background into the scattering by arbitrarily inclined leaves (SAIL) model [52]. It has been extended to the solo slope surface (SLCT) by simply applying the geometric correction without consideration of the geotropic nature of the tree crowns [64]. Therefore, the red and NIR reflectance of the SLCT model were underestimated and overestimated, respectively. Similar to SLCT, GOMST was recently extended by the SAIL model and coupled topography (GOSAILT), for sloping forest, where the effects of slope, aspect, geotropism of the tree crown, multiple scattering scheme, and diffuse skylight are considered [65]. This avoids the issues of reflectance simulation being underestimated and overestimated over rugged terrain.

The computation simulation model can be treated the same as the hybrid model since it can accurately simulate canopy reflectance for both continuous and discontinuous forest stands [38]. Currently, the Monte Carlo ray-tracing (MCRT) computational models have been modified for the solo slope through a coupling of the surface's complex topography by importing the digital elevation model (DEM) datasets or a bilinear surface interpolation based on some simple terrain parameters [38,72]. When compared with the flat terrain, the simulations for rugged mountainous regions face a greater burden of huge memory requirement and computational loading, especially for complex terrain with large maximum elevation differences or large scenes [73].

4.3. Topographic Effect on Solo Slope BRDF

According to the solo slope BRDF simulated results from previous studies, we can conclude that the hotspot still occurs in the solar direction, regardless of slope when the canopy is located on a solo slope terrain. However, the magnitude and shape of BRDF shows an almost random difference caused by shadowing patterns, and the local illumination angle varies with the slope elevation and aspect almost randomly [20,21,57]. For example, as shown in Figure 10, when compared to a flat canopy illuminated by the north solar angle, the slope increases the canopy red reflectance in the backward direction and decreases the red reflectance in the forward direction relative to the solar incidence, respectively. However, The canopy NIR BRDF shape over a steep slope (60°) is distorted, especially in the forward direction, which is where the reflectance is higher than that in backward direction (Figure 10f) [21]. However, the slope seems to have no effect on the canopy reflectance in the direction perpendicular to the aspect of the solo slope terrain because the path length remains constant in the nadir direction due to the geotropic nature [20]. The skewed BRDF in the hemisphere leads to a distinct variation in the albedo values. When compared to the slope angle, the surface albedo is more sensitive to the aspect over the steep slopes. In particular, a larger albedo occurs for the slope facing away from the sun than the sunward facing slope due to the increasing local solar zenith angle and mutual shadowing.

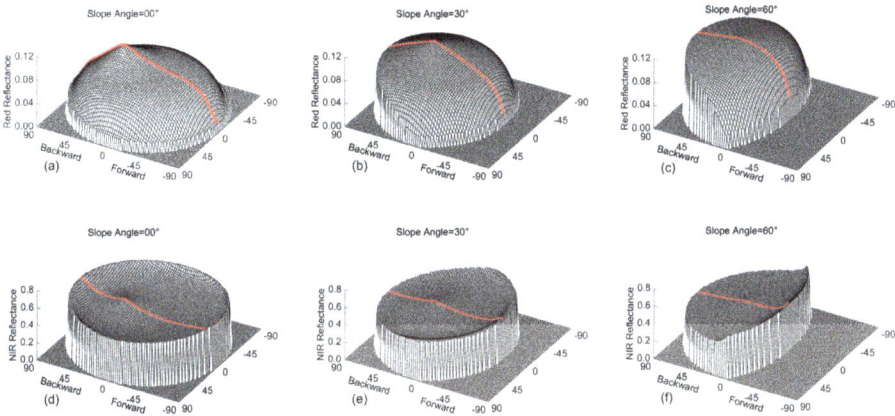

Figure 10. Solo slope reflectance simulated by the GOMST extended by the SAIL model and coupled topography (GOSAILT) model, where (**a**–**c**) are the red reflectance and the (**d**–**f**) are the NIR reflectance. The solar zenith is 30° and azimuth is 0°. The slopes aspect are also 0°; (**a**,**d**) are the flat terrain; (**b, e**) are the 30° slope; and (**c**,**f**) are the 60° slope; Red lines indicate the BRFs along the PP. The radial distance and polar angle of polar coordinate system are view zenith angle and the view azimuth angle, respectively.

5. Composite Slope BRDF Model

5.1. Physical Basis

The composite slope terrain, which is composed of many micro-sloping terrains within one pixel, is shown in Figure 11. For the composite sloping terrain, the topographic effects on reflectance are generally focused on the integrated effects of the micro-slopes within one remote sensing pixel, and they ignore the effects at the pixel level [29,74]. The micro-slope terrain variabilities lead to the shadows coming from both self-shadowing and shadows from the surrounding topography, and this alters the distribution of the composite slope incident radiation. Different spatial distribution characteristics of the micro-topography lead to different spatial geometric configurations of sun-sub-terrain-sensor, multi-scattering, and obstructing effects within the pixel. Characterizing and parameterizing the spatial distributions of the micro-slope topographic features are the key to modeling the BRDF over the composite slope terrain.

The anisotropy reflectance BRF_{coarse} of the composite slope terrain has the following functional form:

$$BRF_{coarse} = f(\Omega_s, \Omega_v, DEM_{fine}, BRF_{fine}) \tag{10}$$

where BRF_{fine} is the bidirectional reflectance of the micro-slope, which can be calculated with the BRDF models of the solo slope. DEM_{fine} represents the fine scale digital elevation models compared with the composite terrain.

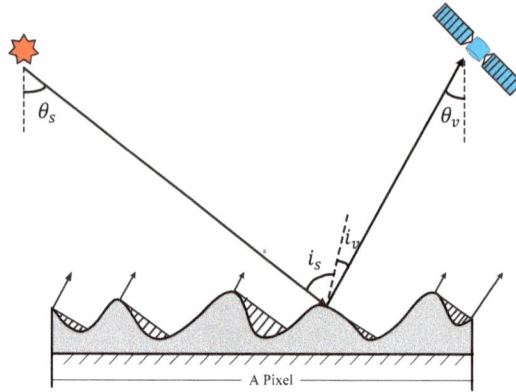

Figure 11. Radiative transfer process over the composite slope terrain.

Specifically, according to the principle of geometric optics and radiosity [75], under the assumption of a horizontal composite slope surface, its directional reflectance is as follows:

$$BRF_{coarse} = \frac{\int\limits_{A(s,v)} \cos i_{sj} BRF_{fine}(i_{sj}, i_{vj}, \varphi_{sj}, \varphi_{vj}) \cos i_{vj} dA_{tj}}{\cos \theta_s \int\limits_{A(v)} \cos i_{vj} dA_{tj}} \qquad (11)$$

where the subscript tj is the jth micro-slope; A_{tj} denotes the incremental surface area of the micro-slope; i_{sj}, i_{vj}, φ_{sj}, and φ_{vj} are the relative solar zenith angle, relative sensor zenith angle, relative solar azimuth angle, and relative sensor azimuth angle, respectively, which correspond to the micro-slope; θ_s and θ_v are the solar zenith angle and sensor zenith angle, which correspond to the horizontal plane; $A(s,v)$ denotes the micro-slopes that are both illuminated and visible; and, $A(v)$ denotes the visible micro-slopes.

From Equation (11), it can be concluded that BRDF modeling over the composite slope terrain is an inherently upscaling procedure. In addition to the effects of the micro-surface slope and aspect, the shadowing distribution within the composite slope is also identified as an essential factor to account for the topographic effects on BRDF. The amount and distribution characteristics of the shadow have great effects on the surface BRDF [76]. The shadowing function (also called the geometric attenuation factor) is built to describe the shadowing and masking effect [30,77–80]. Models are used to describe the complex upscaling process by combining the shadowing function S and an equivalent reflectance BRF_{eq}, which neglect the shadowing effect. In this case, the BRDF over the composite slope terrain can be written as follows:

$$BRF_{coarse} = BRF_{eq} \times S(\Omega_s, \Omega_v, DEM_{fine}) \qquad (12)$$

Essentially, the composite slope BRDF depends on the distribution characteristics of the interior topography of the remote sensing pixels. From the description of the topographic characteristics, BRDF models over the composite slope terrain can be divided into the special-shape based model, random field based model, and real DEM based model (Table 2).

Table 2. Overview of composite slope BRDF models.

Type	Terrain Description	Interior Topography Characteristics	Typical Reference
Special-shape	V-cavity	The surface consists of small symmetrical or non-symmetrical V-cavities	Torrance et al. [79]; Liu et al. [81]; Blinn et al. [82]
	Sphere-cavity	The surface consists of periodical positive sphere-cavities or negative sphere-cavities.	Buhlet al. [83]; Poulin et al. [84]; Koenderink et al. [85]
Random field	Random distribution	The height or the slope conforms to the Gaussian normal distribution, the exponential distribution, or other random distributions	Despan et al. [77]; Hapke [80]; Brockelman et al. [86]; Smith [87]
	Fractal	Describes the dependence of surface roughness on scale by a power law	Barsky et al. [78]; Shepard et al. [88]
DEM	DEM	The terrain is described by high spatial resolution digital elevation models	Wen et al. [29]; Roupioz et al. [31]

5.2. Model Development

5.2.1. Special-Shape Based Model

A simple and effective method used to characterize the topographic effects on BRDF is to take the terrain surface as a special-shape to model the surface anisotropic reflectance. Specifically, the composite slope surface is assumed to be composed of several elements with a repeated primitive shape, such as the V-cavities and spherical-cavities, which are shown in Figure 12. For the V-cavities, the distribution of the slope angle with respect to the horizontal plane is used to describe the surface roughness [78]. Positive sphere-cavities and negative sphere-cavities are the two types of terrain configurations, as shown in Figure 12b,c. The depth-to-diameter ratio of each spherical-cavity [79] and the distance between the centers of the two adjacent spherical-cavities [81] are the two main parameters used to describe the surface roughness. Although the actual terrain shape is probably much more complex, because it consists of various oriented micro-slopes, to describe the topographic relief and its shadow, a spherical or V geometry represents a reasonable physical approximation and mathematical treatment.

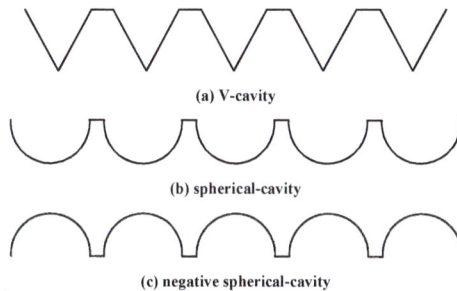

(a) V-cavity

(b) spherical-cavity

(c) negative spherical-cavity

Figure 12. Modeled surfaces with different spherical shape hypotheses.

For V-cavities, when considering the effects of shadowing and masking of facets by adjacent facets, the T-S model [79] first took this terrain configuration and further introduced a simplified, piecewise trigonometric function [82] to improve the qualification of shadowing effects. Although V-cavities have a simple configuration, the model assumes that adjacent micro-slopes have the same slope angle with opposite orientations, which will result in sharp, abnormal turning points in BRDF curves [81]. Therefore, by assuming that adjacent micro-slopes are oriented in opposing directions, but not symmetrical and the slope angle of each micro-slope follows a semi-Gaussian distribution, an improved anisotropic reflection model was developed, which is closer to the natural configuration of the topography. The results showed that this model reached better physical rationality and improved the accuracy of the BRDF model [81]. However, it would be difficult to elaborate an analytical model coupled with the multi-scattering component over the composite slope terrain. Although a simple and physically plausible construction is proposed for the multi-scattering effect and is coupled with the single reflection in the research by Kelemen et al. [89], the multi-scattering effect is usually neglected or handled as a constant diffuse factor, which is against practical observations [90].

The positive spherical cavities allow for large grooves with relatively sharp edges that are better than the V-cavities [83]. The negative spherical cavities can approximately depict a rough surface, like the small craters on the moon [84]. Buhl et al. (1968) [83] first assumed the rough surface to be a collection of positive spherical cavities, but did not present a complete mechanism that accounts for masking and shadowing effects for arbitrary incidence and exit. Following this, analytical anisotropic reflection models for positive sphere-cavities and negative sphere-cavities were derived [84], which take into account hiding and shadowing between spherical cavities and can be easily implemented. An algebraic solution for multi-scattering in the spherical cavity was further provided and confirmed that the exact (numerical) calculations for a spherical geometrical surface, coupled with the multi-scattering effect are physically realizable [85].

5.2.2. Random Field Based Model

While the hypothesis of V-cavities or sphere-cavities is appropriate mathematically, it is not physically plausible. The terrain surface is far from the V-cavities or spherical cavities. More complex and realistic configurations for the topography description are the random field, which include the fractal characteristics [91,92], exponential distributions [45,93,94], Laplace distribution [94], and Gaussian normal distribution [3,95]. These are widely applied in early quantitative mathematic reflectance models over the random rough surface [96] in optical engineering, radiophysics, metrology, computer graphics, and machine vision fields. Remote sensing scientists have developed a series of anisotropic reflectance models based on the random field, especially when it was assumed that the micro-slope distribution follows a random Gaussian distribution and self-affine fractals. For the surface with a Gaussian random distribution, the root mean square (RMS) slope [87], correlation length [87], and mean slope angle [30] are adopted to parameterize the characteristics of the random surface. For the surface with self-affine fractal characteristics, the Hurst exponent is the key to determining the roughness features. Figure 13 shows the rough surface with a typical, random normal distribution.

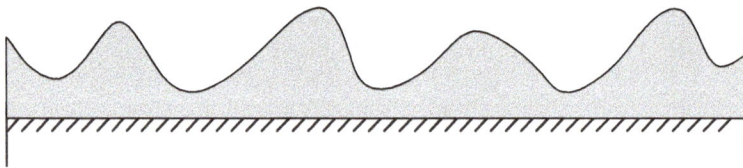

Figure 13. Random surface with normal distribution.

Initially, a series of shadowing functions have been developed over one-dimensional surfaces with a Gaussian random distribution to describe the interior topographic effects on the surface anisotropic reflectance. The first attempt of analytical derivation of the shadowing function can be traced back to the work of Beckmann [97] who derived the Beckmann shadowing function, which uses a Gaussian correlation function to characterize the random surface. However, it does not agree well with the numerical simulation [86] because of a mathematical error in the model derivation [98]. A rigorous shadowing function expression based on the surface with a Gaussian height field was presented by Wagner [99]. The Wagner shadowing function adopted integral approximations and ignored the correlation between the height of the rough surface and its slope, which leads to large analytical complexity. The Smith shadowing function further improved the Wagner shadowing function by introducing a normalization function and simplified the calculation complexity [87].

Heitz et al. (2016) [100] extended the Smith model to include micro-surface multiple scattering at rough material interfaces, which agree well with the computer simulation. However, these shadowing functions assume that the joint probability density of the random surface heights and slopes is uncorrelated. The effects on the Wagner and Smith shadowing functions were quantified and showed that they have large errors when the incidence angle is large because the real-world continuous surfaces have wider autocorrelation functions [101].

Extending the one-dimensional model to a two-dimension model is another important way to derive a shadow function [102], such as the Hapke shadowing function [80] and the Despan shadowing function [77]. The Hapke shadowing function is derived from the radiative transfer process and the raw equivalent slope principle. This function accounts for any general configuration of incidence, emission, and azimuth angles on the two-dimensional (2-D) surface [80], which involves only one arbitrary parameter of the mean slope angle. The Despan shadowing function was derived by using rigorous probabilistic techniques, including a conditional probabilistic distribution and total expectation formula [77], where the random rough surface is described as an isotropic 2-D Gaussian stochastic process with a Gaussian autocorrelation function.

Self-affine fractals are another way to describe natural surfaces [88]. They describe the dependence of surface roughness on scale by a power law. Shepard and Campbell (1999) first proposed an empirical formula for the fractal shadowing function, but no rigorous analytical models of self-shadowing on a fractal surface currently exist [88]. Further, an analytical integral form of the shadowing function for fractal surfaces with different fractal and roughness parameters [78] was offered and opens possibilities for further exploration of fractal surfaces behavior. An agreement was demonstrated between the fractal model and experimentally calculated shadowing functions using the Monte Carlo method.

5.2.3. DEM-Based Model

Random field models are usually constructed based on rigorous probabilistic techniques. Thus, in general, the random field of Gaussian normal distribution or other statistical distribution regarding the terrain is acknowledged at a large scale (more than 10 km). However, the real terrain surfaces often exhibit non-random spatial distribution characteristics at a relatively small scale (such as a 1 km remote sensing pixel). Investigations have shown the sub-terrain slope distributions within each kilometric pixel of more than 50% over the Tibetan Plateau is not normal [31]. It is questionable to directly apply the random field hypothesis to remote sensing BRDF modeling at a kilometer scale. Recently, accurate digital elevation models (DEM), such as GDEM2 [103] have become available at a global scale with a 30-m spatial resolution. These DEMs offer an efficient way to describe the surface topography. Rapid and accurate methods for calculating slope and azimuth, solar illumination angle, shadow factor, and sky view factors have been proposed [104]. These methods prompt us to model the BRDF by considering the micro-topographic effects based on the DEM.

According to the radiosity theory [75], a physical BRDF upscaling model based on Equation (11) and the multi-scattering consideration for sub-topographic effects [31] was developed. However, this procedure requires the micro-slope surface BRDF and topographic factors, which leads to complexity and low computational efficiency. Similarly to mean slope derived from the random field, the equivalent slope model (ESM) was proposed to derive the equivalent slope and aspect [29] to simplify this process. This model assumes that there is a virtual smooth slope where the incoming and outgoing radiation are the same as that of the composite sloping terrain and this is related to the position of the sun, sensor and micro-surface slope and aspect of the DEM, which is shown in Figure 14. Thus, ESM can account for the effects of the sub-topography and shadow distribution. Similar to the shadow function, a sub-topographic impact factor T can be derived from the equivalent slope. Therefore, BRF_{coarse} can be expressed as a function of anisotropic reflectance BRF_{eq} of the equivalent slope and the sub-topographic impact factor T:

$$BRF_{coarse} = BRF_{eq} \times T(\Omega_s, \Omega_v, DEM_{fine}) \tag{13}$$

where BRF_{eq} can be obtained by the solo slope BRDF models, and T is written as follows:

$$T(\Omega_s, \Omega_v, DEM_{fine}) = \frac{\cos i_v^e \cos i_s^e \int\limits_{A(s,v)} dA_{tj}}{\cos(\theta_s) \int\limits_{A(v)} \cos iv_j dA_{tj}} \tag{14}$$

where i_s^e, i_v^e, φ_s^e, and φ_v^e are the relative solar zenith angle, relative sensor zenith angle, relative solar azimuth angle, and relative sensor azimuth angle, respectively, which correspond to the equivalent slope.

When compared with the Hapke shadowing function S, which represents the amount of shadow on the composite slope, the sub-topographic impact factor T represents the effects of the tilted micro-slope distribution and mutual shadowing. For example, when the solar zenith angle is $0°$, S is one regardless of the sensor zenith angles and mean slopes. This is because there is no shadow when the sun is at nadir. However, T smaller than 1 when the mean slope is large and changes slightly with the view zenith angle when the solar zenith angle is $0°$. When the solar zenith angle is $45°$, both of these values are less than one [29].

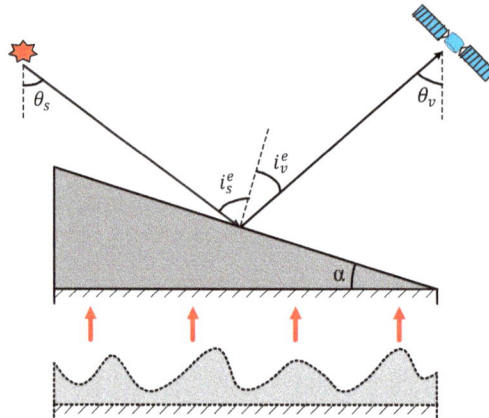

Figure 14. Equivalent slope: a virtual smooth surface.

5.3. Topographic Effect on Composite Slope BRDF

The magnitude and shape of BRDF over composite sloping terrain shows significant changes in response to the altered topography, which is because of the nonlinear dependence of BRDF on the micro-slope spatial configuration [29,76,105]. An asymmetric distributions of BRDF can be observed because of the shadow effects and the adaption of the sun-terrain-sensor geometry [29,72,105]. However, the BRDF peak location over the composite slope terrain is identical to that over flat terrain [72]. With the same sun and sensor directions, the deviation between the reflectance over composite slope terrain and that over flat terrain is sensitive to the mean slope. The deviation cannot be neglected even when the mean slope is small [36]. The deviation increases gradually with an increased mean slope, and it reaches 80% at about $37°$ of the mean slope when the SZA and VZA are around

60° [29]. The lowest deviations are found for the sun at zenith, and they increase considerably with an increase in SZA. Generally, there is a larger reflectance value when the VZA is larger. It also depends on the VAA and decreases when the viewing direction changes from the principal plane to the perpendicular principal plane [36].

6. Future Development and Perspective on BRDF Products Generation

Accurately generating a BRDF product over rugged terrain at the global scale is crucial for vegetation monitoring, as well as the energy budget for global climate change research. Various BRDF forward models have been developed to fit and explore the topographic effects of BRDF over rugged terrain. Because of a lack of operational algorithms for the BRDF product over rugged terrain, the current BRDF products do not account for the topographic effects, and thus, they show large uncertainties. According to the key steps of BRDF modeling, users should choose a high quality DEM and fast parameterization methods to obtain topographic factors. To extend the forward BRDF model to the operational one, it is necessary to adopt new and innovative ideas to overcome the limitations of BRDF applicability over rugged terrain, and to develop an effective method for validating the BRDF algorithm and satellite product performance.

6.1. High Quality DEM

The availability of high quality DEMs promotes the development of BRDF modeling over rugged terrain. In recent years, stereo photogrammetry and interferometric synthetic aperture radar techniques have been widely used to generate DEMs, as well as providing fast, reliable, and accurate solutions. Accurate DEM products are available at a global scale with resolutions of up to 30 m, which include GTOPO30 DEM, SRTM3DEM, SRTM CGIAR-CSI DEM, TanDEM-X DEM, ASTER GDEM1, and ASTERGDEM2.

The BRDF models of solo slope, composite slope, and the sloping reflectance retrieval show that their accuracies depends on the quality of the calculated topographic terms from the DEM. The issues of elevation accuracy, spatial resolution, and the co-registration accuracy between the DEM and satellite image are easy to identify but difficult to assess. Geo-location errors, elevation aberrations, and even blunders in the DEM base data can result in significant local errors in BRDF modeling. A high quality DEM is the basic necessity prior to successful BRDF modeling over rugged terrain. An external validation showed that the elevation accuracy of ASTERGDEM2 is 8.5 m and that of SRTM3 USGS is 6 m [103]. The precision of the DEM enables us to apply the observed correlation between shading and images [106] to improve the co-registration accuracy. DEM spatial resolution is another factor that should be considered in modeling the BRDF over rugged terrain. A low resolution DEM when compared to the remote sensing pixel will not provide detailed topographic information because the DEM resolution changes [107–109], mean slopes and curvatures decrease, and terrain details disappear. Thus, a scale appropriate for the satellite data is of great importance in BRDF modeling. We suggest that a high quality and high spatial resolution DEM is necessary for BRDF modeling based on both high resolution and low resolution remote sensing images, which correspond to the solo slope BRDF modeling and composite slope modeling, respectively.

6.2. Topographic Factor Parameterization

Because the DEM describes 3-D surface, several parameters to characterize landforms and surface-received solar radiation can be extracted from DEM datasets, and consequently, these can be used in the BRDF models. These include the slope and aspect describing the topography gradient and orientation, the topographic shadow mask indicating whether the target surfaces are sunlit, the sky view factor representing the proportion of the sky visible to the target surface, and the topographic configuration factor, illustrating the proportion of target slopes that are visible to the surrounding slopes [104]. All of these parameters are regional properties because their calculations depend on a suitable neighboring area. Although a formidable computational problem occurs during the calculation of these parameters, they require calculation only once, and thus, they can be stored as look up tables (LUT) before the actual model execution due to their stationary property, especially for deriving global products.

Specifically, the calculation of the slope and aspect of the target slope depends on the gradients in the east-west and north-south directions [110]. A moving 3 × 3 window is commonly used to derive the gradients. According to the pixels in the window adopted to calculate the gradients and their respective weights, six slope and aspect algorithms are frequently used. These algorithms include second-order finite difference [111,112], third-order finite difference [113,114], third-order finite difference weighted by reciprocal of squared distance [113], third-order finite difference weighted by reciprocal of distance [115], frame finite difference [116], and simple difference [117]. The algorithm for the topographic shadow mask indicates whether the slope is shadowed by the neighboring slope from the solar direction. The global decision tree [31], minimum radius search [118], and indirect horizontal angle [104] algorithms are the three frequently used methods. The sky view factor is defined as the ratio of the diffuse skylight that is received by the target slope to an unobstructed horizontal surface [104]. This value is restricted between 0 and 1. Values close to 1 indicate that the point is located at the top of the topography, and values that reach 0 indicate the point lies in the low part of a deep valley. This factor accounts for the slope and aspect, the obstruction from neighboring slopes, and the anisotropy of diffuse skylight. However, the anisotropic diffuse skylight is always neglected in the current sky view factor algorithm. Several algorithms have been proposed depending on whether the neighboring topographic obstruction effect is considered or the horizontal or inclined slope are taken as a reference [104,119–121]. Similar to the sky view factor, the topographic configuration factor is defined as the ratio of the nearby slope reflected irradiance received by the target slope on an unobstructed horizontal surface [104]. Directly calculating the topographic configuration factor is difficult, since the reflected irradiance of every slope facet visible to the target slope needs to be known. The alternative solution for the topographic configuration factor is that it can be expressed as the difference between the sky proportion for an infinite slope and the actual slope sky view factor [104,120].

6.3. Potential Method to Derive the BRDF Product over Rugged Terrain

Since the solo slope and composite slope BRDF models in this study are forward BRDF models, it is not possible to use these models as the BRDF operational algorithm to generate the BRDFs. The current linear kernel-driven model has been successfully adopted to derive the satellite BRDF/albedo products because of its simplicity, feasibility, and physical basis [122,123]. There has been no significant further progress in the kernel-driven model to fit the BRDF with multi-spectral and multi-angular reflectance. However, the effects of topography are rarely coupled in the kernel-driven models, which leads to uncertainties in the BRDF/albedo retrieval over mountainous areas [9,19]. Thus, a suitable forward BRDF model to derive the kernel function coupled with topographic effects is necessary to generate the BRDF product at the global scale.

As for the solo slope surface, most of the forward BRDF models cannot be directly derived with the kernel functions due to its complicated physical processes. Therefore, the current kernel driven models that are applied in mountainous areas are modified by a geometric relationships transformation between the horizontal surface and sloping plane. Specifically, this transform changes the solar and sensor view angles corresponding to the slope surface in the kernel function to reflect the sloping surface effects. Because of the shadow that is caused by the topography obstruction, the diffuse skylight serves as the dominant illumination energy source. Then, the kernel function derived from the BRDF model may be a constant according to a geometric-optical model, where the component spectral contrasts are neglected when the pure diffuse surface scatters. However, the spectral signatures of the scene components still show distinct difference even under the pure diffuse skylight illumination [71]. This may also be described by the hemispherical directional reflectance factor (HDRF, [8]) and the pixel HDRF displays an angular heterogeneity [8,124,125]. For the composite slope surface, the coarse-scale pixel directional reflectance is affected by the micro-slope internal to the pixel, in addition to the 3-D object structure itself, which includes the proportion of shadow, and micro-surface slope and aspect distribution. However, the current kernel driven model is applied directly to mountainous areas under the assumption that the topographic relief is a special 3-D object. The accuracy and uncertainties of this model are not yet credible and should be further evaluated. One of the deficiencies in the current kernel driven function is that it neglects topographic effects, which results in no topographic factors being included in the kernel function.

Therefore, a possible solution is to derive a united kernel driven function suitable for both solo-slope and composite-slope surfaces. According to the current BRDF model developed over rugged terrain, GOSAILT can be further derived as the solo slope kernel function, which is then is coupled with a sub-topography impact factor of the equivalent slope model (ESM) to form the composite slope kernel function. The ESM can extend the solo slope BRDF to the composite slope BRDF. The most important feature of these two models is that they are forwarded based on a real DEM, which will enable us to further promote the possibility of a united kernel function. Specifically, GOSAILT can be implemented to derive the linear semi-empirical kernel-driven model that is suitable for sloping terrains under both clear and overcast skies; this model has a similar framework to the RossThick-LiSparse Reciprocal (RTLSR) BRDF model [9], which includes sloped geometric-optical and volume scattering kernels. The ESM, an anisotropic reflectance model over the composite sloping terrain, was developed based on the equivalent slope principle. It extends the directional reflectance model for the solo sloping terrain to the reflectance model for the composite sloping terrain by a sub-topography impact factor, which describes the topographic influence. Similar to BRDF extension, by coupling with the sub-topography impact factor, the GOSAILT kernel function can be easily applied to the BRDF retrieval over rugged terrain.

Progress is also expected in the retrieval method development, which uses data from combined multi-sensors in mountainous areas. Low quality and cloud occlusion causes remote sensing data unavailability in mountainous areas to be more severe. Thus, the significant merit of combining multi-sensor reflectance is that it can provide additional multiple angular information, and then, this can improve the inversion accuracy of the BRDF on mountainous surfaces. For example, the multi-angular and multi-spectral kernel function (ASK) model [126,127] and multi-sensors combined BRDF inversion (MCBI) model [128] are proposed from the improvement of the BRDF kernel function, as well as the need to retrieve the BRDF synthetically by combined multiple sensor reflectance, which has a continuous spatial distribution and shorter-time scale of BRDFs.

Lastly, the remaining difficulties include the fast extraction of the DEM topographic factor and the support of the kernel function to fit the BRDF over rugged terrain. The look up table (LUT) might be a practical method to store all of the topographic factors, including slope, aspect, and shadow, as well as the sub-terrain impact factor with the different solar SZA, SAA, and DEM longitude and latitude. When the sloped kernel driven model is implemented, the global shadow, observing mast, sky view factor, topographic configuration factor, and ESM LUT provide the essential parameters that are needed to produce global BRDFs. For example, the SZA is from 0° to 65° with an interval 5°, and the SAA is from 0° to 330° with an interval 30° in the global topographic shadow mask (TSM) LUT. Figure 15 is the global topographic shadow mask, where the SZA is 40° and SAA is 0°.

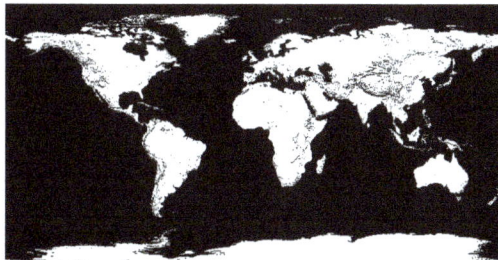

Figure 15. Global topographic shadow mask (TSM).

6.4. Validation Methods for the BRDF over Rugged Terrain

In situ multi-angular reflectance data, which is measured at sites with typical, homogeneous surfaces, is the ground reference truth for the land surface BRDF validation. However, the BRDF validation dataset is still far from sufficient to support the global BRDF validation, which is mainly due to the limitations of multi angle observation instruments and technique developments.

In mountainous areas, the BRDF measurement can be more difficult than those on flat surfaces due to the slope effects. First, there is still some controversy regarding the measurement method of the sloped BRDF. For example, whether the observation instrument should be parallel to the slope surface or the horizontal surface is a question that needs to be answered according the definition of the modeled BRDF. Second, mountainous area generally belongs to the forest land type, where the tree height might be too high for users to reach. It is difficult to carry out multi-angular forest canopy reflectance measurements with ground based instruments. One of the alternative methods is to use a tower to implement the multi-angular measurements, where the height of the tower should be higher than that of the forest canopy. Another alternative is the use of unmanned aerial vehicle (UAV) technology. A UAV can carry the optical CCD and provide multi-angular reflectance measurements for a small-scale area. The tower-based or UAV-based measurements only represent the scale of the solo slope. To validate the BRDF over the composite slope surface, a multi-scale validation strategy is important to solve the spatial scale mismatch between the ground-based and the satellite-based BRDF. However, the technique for BRDF upscaling over rugged terrain is still an ongoing subject of research, which has limited the applicability of the multi-scale validation strategy.

A current alternative validation method, especially for the BRDF validation of the composite-slope, is generally based on computer simulation technology [38,73,129], or the use of a miniature terrain sandbox to simulate the BRDF under the influence of topography as the reference truth. Computer simulation technology, such as discrete anisotropic radiative transfer (DART) [38], can set up different DEMs and different types of trees to build a real scene of solo slope or composite slope forest over rugged terrain. With the flux tracing technique, the multi-angular reflectance reference dataset is simulated. The terrain sandbox can simulate different typical composite terrains, as well as the different vegetation above the terrain. With the help of existing imaging spectroscopy technology, the multi-angular observation can be implemented to obtain the reference reflectance dataset.

7. Conclusions

In this paper, the model of bidirectional reflectance distribution function (BRDF) over rugged terrain has been comprehensively reviewed. The results of the literature analysis demonstrate that the subject of BRDF modeling over rugged terrain has been intensively addressed by remote sensing scientists over the past ten years. Referencing the BRDF definition, we proposed two kinds of BRDF over rugged terrain, according to the relationship between the spatial resolution of the DEM and remote sensing image pixel. These are the solo slope BRDF and the composite slope BRDF. Their scale difference and their self-consistencies should be emphasized.

The dominant factors of the BRDF over the solo slope and composite slope are different. The surface slope and aspect of the pixel level, which change the sun-terrain-sensor geometry, as well as the radiation distribution, are the factors controlling the solo slope BRDF. However, with the composite slope BRDF, besides the influence of the micro-slope within the pixel, the influencing factors are also the shadow distribution of the terrain occlusion, overall distribution of the micro-terrain, and the multiple scattering between micro-slopes. These sensitive factors should be concentrated on when modeling the BRDF over these two kinds of slopes.

Specifically, an accurate description of the interaction between the 3-D vegetation structure, soil, and atmosphere is of great importance for solo slope BRDF modeling. Radiative transfer, geometric-optical, and the hybrid theory are the three basic theories that are used to mathematically solve the interaction process. The geotropic nature of tree crowns and accurate parameterization of the components radiation signal of vegetation and soil in the solo slope BRDF modeling is important.

However, the description of the sub-topographic effects is the critical step for composite slope BRDF modeling, which is where the virtual random distributed topography and real DEM are the two solutions for the description of sub-topography inside the coarse scale pixels. The shadow function and the sub-topography impact factor are the two parameterizations. The sub-topography impact factor can be linked to the solo slope BRDF and the composite slope BRDF, where they both are based on the real DEM. When compared with the solo slope BRDF model, the development of a composite slope BRDF model should be further researched based on the simulated data and to achieve better accuracy.

According to the current BRDF model over rugged terrain, it is concluded that the topography can intensely affect both the shape and magnitude of the land surface BRDF. Generally, when a slope increases of the solo slope surface, the canopy reflectance opposite to the solar direction also increased, but the reflectance in the forward to solar direction decreased, which resulted in a skewed BRDF in the hemisphere. Thus, this consequently led to a distinct variation in the albedo values, as well as other parameters that are derived from the land surface BRDF. However, the hotspot direction and the solar orthogonal plane to the aspect of the solo slope terrain seem to have less topographic effects on the canopy reflectance. The composite slope surface, shadow, and sub-terrain slope result in asymmetric distributions of BRDF. The BRDF shape and magnitude depends on the mean slope, the dominated aspect of sub-terrain, the SZA, and SAA. With the mean slope increased, the topographic effects of BRDF are more intense under the same sun and sensor location. Even in the case of a relatively smaller mean slope, the deviation between the reflectance over composite sloping terrain and that over flat terrain is still significant.

Although relatively high quality DEMs are available, and the topographic factors can be parameterized quickly, it seems that the operational BRDF model used to fit the remote sensing satellite multi-angular reflectance does not show significant progress. Similar to the kernel driven model used in the MODISBRDF/albedo product, the kernel functions derived from the current forward BRDF model over rugged terrain are still a subject for ongoing research. Although GOSAILT seems to be able to derive the kernel function of the solo slope and composite slope, more efforts should be put toward operational BRDF model development and its validation over rugged terrain.

Acknowledgments: This work was supported by the Chinese Natural Science Foundation Project (NO: 41671363, 41331171).

Author Contributions: Jianguang Wen conceived the structure of this paper. Jianguang Wen, Shengbiao Wu, Dalei Hao and Xingwen Lin developed the BRDF model over rugged terrain. Qiang Liu, Qing Xiao, Qinhuo Liu, Dongqin You contributed to the discussion of scientific problems and the analysis of the results. All the authors conducted the manuscript revision.

Conflicts of Interest: The authors declare no conflict of interest.

References

1. Li, A.; Jiang, J.; Bian, J.; Deng, W. Combining the matter element model with the associated function of probability transformation for multi-source remote sensing data classification in mountainous regions. *ISPRS J. Photogramm. Remote Sens.* **2012**, *67*, 80–92. [CrossRef]
2. Vanonckelen, S.; Lhermitte, S.; Rompaey, A.V. The effect of atmospheric and topographic correction methods on land cover classification accuracy. *Int. J. Appl. Earth Obs. Geoinf.* **2013**, *24*, 9–21. [CrossRef]
3. Helbig, N.; Löwe, H. Shortwave radiation parameterization scheme for subgrid topography. *J. Geophys. Res. Atmos.* **2012**, *117*, 812–819. [CrossRef]
4. Wen, J.; Zhao, X.; Liu, Q.; Tang, Y.; Dou, B. An improved land-surface albedo algorithm with DEM in rugged terrain. *IEEE Geosci. Remote Sens. Lett.* **2013**, *11*, 883–887.
5. Yan, G.; Wang, T.; Jiao, Z.; Mu, X.; Zhao, J.; Chen, L. Topographic radiation modeling and spatial scaling of clear-sky land surface longwave radiation over rugged terrain. *Remote Sens. Environ.* **2016**, *172*, 15–27. [CrossRef]

6.   Pasolli, L.; Asam, S.; Castelli, M.; Bruzzone, L.; Wohlfahrt, G.; Zebisch, M.; Notarnicola, C. Retrieval of leaf area index in mountain grasslands in the alps from MODIS satellite imagery. *Remote Sens. Environ.* **2015**, *165*, 159–174. [CrossRef]

7.   Zhao, P.; Fan, W.; Liu, Y.; Mu, X.; Xu, X.; Peng, J. Study of the remote sensing model of FAPAR over rugged terrains. *Remote Sens.* **2016**, *8*, 309. [CrossRef]

8.   Schaepman-Strub, G.; Schaepman, M.E.; Painter, T.H.; Dangel, S.; Martonchik, J.V. Reflectance quantities in optical remote sensing—Definitions and case studies. *Remote Sens. Environ.* **2006**, *103*, 27–42. [CrossRef]

9.   Gao, F.; Schaaf, C.B.; Strahler, A.H.; Jin, Y.; Li, X. Detecting vegetation structure using a kernel-based BRDF model. *Remote Sens. Environ.* **2003**, *86*, 198–205. [CrossRef]

10. Schaaf, C.B.; Gao, F.; Strahler, A.H.; Lucht, W.; Li, X.; Tsang, T.; Strugnell, N.C.; Zhang, X.; Jin, Y.; Muller, J.-P. First operational BRDF, albedo nadir reflectance products from MODIS. *Remote Sens. Environ.* **2002**, *83*, 135–148. [CrossRef]

11. Xiao, Z.; Liang, S.; Wang, J.; Jiang, B.; Li, X. Real-time retrieval of leaf area index from MODIS time series data. *Remote Sens. Environ.* **2011**, *115*, 97–106. [CrossRef]

12. Zege, E.P.; Katsev, I.L.; Malinka, A.V.; Prikhach, A.S.; Heygster, G.; Wiebe, H. Algorithm for retrieval of the effective snow grain size and pollution amount from satellite measurements. *Remote Sens. Environ.* **2011**, *115*, 2674–2685. [CrossRef]

13. Roy, D.P.; Li, J.; Zhang, H.K.; Yan, L.; Huang, H.; Li, Z. Examination of Sentinel-2a multi-spectral instrument (MSI) reflectance anisotropy and the suitability of a general method to normalize MSI reflectance to nadir BRDFadjusted reflectance. *Remote Sens. Environ.* **2017**, *199*, 25–38. [CrossRef]

14. Roy, D.P.; Zhang, H.K.; Ju, J.; Gomez-Dans, J.L.; Lewis, P.E.; Schaaf, C.B.; Sun, Q.; Li, J.; Huang, H.; Kovalskyy, V. A general method to normalize Landsat reflectance data to nadir BRDF adjusted reflectance. *Remote Sens. Environ.* **2016**, *176*, 255–271. [CrossRef]

15. Schläpfer, D.; Richter, R.; Feingersh, T. Operational BRDF effects correction for wide-field-of-view optical scanners (BREFCOR). *IEEE Trans. Geosci. Remote Sens.* **2014**, *53*, 1855–1864. [CrossRef]

16. Bacour, C.; Bréon, F.-M.; Maignan, F. Normalization of the directional effects in NOAA–AVHRR reflectance measurements for an improved monitoring of vegetation cycles. *Remote Sens. Environ.* **2006**, *102*, 402–413. [CrossRef]

17. Fan, W.; Chen, J.M.; Ju, W.; Zhu, G. GOST: A geometric-optical model for sloping terrains. *IEEE Trans. Geosci. Remote Sens.* **2014**, *52*, 5469–5482.

18. Li, F.; Jupp, D.L.B.; Thankappan, M.; Lymburner, L.; Mueller, N.; Lewis, A.; Held, A. A physics-based atmospheric and BRDF correction for Landsat data over mountainous terrain. *Remote Sens. Environ.* **2012**, *124*, 756–770. [CrossRef]

19. Wen, J.; Liu, Q.; Tang, Y.; Dou, B.; You, D.; Xiao, Q.; Liu, Q.; Li, X. Modeling land surface reflectance coupled BRDF for HJ-1/CCD data of rugged terrain in Heihe river basin, China. *IEEE J. Sel. Top. Appl. Earth Obs. Remote Sens.* **2015**, *8*, 1506–1518. [CrossRef]

20. Yin, G.; Li, A.; Zhao, W.; Jin, H.; Bian, J.; Wu, S. Modeling canopy reflectance over sloping terrain based on path length correction. *IEEE Trans. Geosci. Remote Sens.* **2017**, *55*, 4597–4609. [CrossRef]

21. Schaaf, C.; Li, X.; Strahler, A. Topographic effects on bidirectional and hemispherical reflectances calculated with a geometric-optical canopy model. *IEEE Trans. Geosci. Remote Sens.* **1994**, *32*, 1186–1193. [CrossRef]

22. Croft, H.; Anderson, K.; Kuhn, N.J. Reflectance anisotropy for measuring soil surface roughness of multiple soil types. *Catena* **2012**, *93*, 87–96. [CrossRef]

23. Wang, Z.; Coburn, C.A.; Ren, X.; Teillet, P.M. Effect of surface roughness, wavelength, illumination, and viewing zenith angles on soil surface BRDF using an imaging BRDF approach. *Int. J. Remote Sens.* **2014**, *35*, 6894–6913. [CrossRef]

24. Fan, W.; Chen, J.M.; Ju, W.; Nesbitt, N. Hybrid geometric optical–radiative transfer model suitable for forests on slopes. *IEEE Trans. Geosci. Remote Sens.* **2014**, *52*, 5579–5586.

25. Proy, C.; Tanre, D.; Deschamps, P.Y. Evaluation of topographic effects in remotely sensed data. *Remote Sens. Environ.* **1989**, *30*, 21–32. [CrossRef]

26. Shepherd, J.D.; Dymond, J.R. Correcting satellite imagery for the variance of reflectance and illumination with topography. *Int. J. Remote Sens.* **2003**, *24*, 3503–3514. [CrossRef]

27. Combal, B.; Isaka, H.; Trotter, C. Extending a turbid medium BRDF model to allow sloping terrain with a vertical plant stand. *IEEE Trans. Geosci. Remote Sens.* **2000**, *38*, 798–810. [CrossRef]

28. Wang, K.; Zhou, X.; Liu, J.; Sparrow, M. Estimating surface solar radiation over complex terrain using moderate-resolution satellite sensor data. *Int. J. Remote Sens.* **2005**, *26*, 47–58. [CrossRef]

29. Wen, J.G.; Qiang, L.; Liu, Q.H.; Xiao, Q.; Li, X.W. Scale effect and scale correction of land-surface albedo in rugged terrain. *Int. J. Remote Sens.* **2009**, *30*, 5397–5420. [CrossRef]

30. Hapke, B. Bidirectional reflectance spectroscopy: 3. Correction for macroscopic roughness. *Icarus* **1984**, *59*, 41–59. [CrossRef]

31. Roupioz, L.; Nerry, F.; Jia, L.; Menenti, M. Improved surface reflectance from remote sensing data with sub-pixel topographic information. *Remote Sens.* **2014**, *6*, 10356–10374. [CrossRef]

32. NicoDEMus, F.E.; Richmond, J.C.; Hsia, J.J.; Ginsberg, I.W.; Limperis, T. *Geometrical Considerations and Nomenclature for Reflectance*; Radiometry; Jones and Bartlett Publishers, Inc.: Burlington, MA, USA, 1977; pp. 94–145.

33. Girolamo, L.D. Generalizing the definition of the bi-directional reflectance distribution function. *Remote Sens. Environ.* **2003**, *88*, 479–482. [CrossRef]

34. Snyder, W.C. Definition and invariance properties of structured surface BRDF. *IEEE Trans. Geosci. Remote Sens.* **2002**, *40*, 1032–1037. [CrossRef]

35. Parviainen, H.; Muinonen, K. Rough-surface shadowing of self-affine random rough surfaces. *J. Quant. Spectrosc. Radiat. Transf.* **2007**, *106*, 398–416. [CrossRef]

36. Combal, B.; Isaka, H. The effect of small topographic variations on reflectance. *IEEE Trans. Geosci. Remote Sens.* **2002**, *40*, 663–670. [CrossRef]

37. Martonchik, J.V.; Bruegge, C.J.; Strahler, A.H. A review of reflectance nomenclature used in remote sensing. *Remote Sens. Rev.* **2000**, *19*, 9–20. [CrossRef]

38. Gastellu-Etchegorry, J.-P.; Yin, T.; Lauret, N.; Cajgfinger, T.; Gregoire, T.; Grau, E.; Feret, J.-B.; Lopes, M.; Guilleux, J.; Dedieu, G.; et al. Discrete anisotropic radiative transfer (DART 5) for modeling airborne and satellite spectroradiometer and LIDAR acquisitions of natural and urban landscapes. *Remote Sens.* **2015**, *7*, 1667–1701. [CrossRef]

39. North, P. Three-dimensional forest light interaction model using a Monte Carlo method. *IEEE Trans. Geosci. Remote Sens.* **1996**, *34*, 946–956. [CrossRef]

40. Vermote, E.F.; Tanre, D.; Deuze, J.L.; Herman, M.; Morcette, J.J. Second simulation of the satellite signal in the solar spectrum, 6S: An overview. *IEEE Trans. Geosci. Remote Sens.* **1997**, *35*, 675–686. [CrossRef]

41. Berk, A.; Conforti, P.; Kennett, R.; Perkins, T.; Hawes, F.; Van Den Bosch, J. MODTRAN6: A major upgrade of the MODTRANradiative transfer code. *Proc. SPIE* **2014**. [CrossRef]

42. Reese, H.; Olsson, H. C-correction of optical satellite data over alpine vegetation areas: A comparison of sampling strategies for determining the empirical *c*-parameter. *Remote Sens. Environ.* **2011**, *115*, 1387–1400. [CrossRef]

43. Soenen, S.A.; Peddle, D.R.; Coburn, C.A. SCS + C: A modified sun-canopy-sensor topographic correction in forested terrain. *IEEE Trans. Geosci. Remote Sens.* **2005**, *43*, 2148–2159. [CrossRef]

44. Fan, Y.; Koukal, T.; Weisberg, P.J. A sun–crown–sensor model and adapted *c*-correction logic for topographic correction of high resolution forest imagery. *ISPRS J. Photogramm. Remote Sens.* **2014**, *96*, 94–105. [CrossRef]

45. Essery, R. Statistical representation of mountain shading. *Hydrol. Earth Syst. Sci.* **2004**, *8*, 1047–1052. [CrossRef]

46. Chen, Y.; Hall, A.; Liou, K.N. Application of three-dimensional solar radiative transfer to mountains. *J. Geophys. Res. Atmos.* **2006**, *111*, 5143–5162. [CrossRef]

47. Richter, R.; Schlaepfer, D. Geo-atmospheric processing of airborne imaging spectrometry data. Part 2: Atmospheric/topographic correction. *Int. J. Remote Sens.* **2002**, *23*, 2631–2649. [CrossRef]

48. Hu, B.; Lucht, W.; Strahler, A.H. The interrelationship of atmospheric correction of reflectances and surface BRDF retrieval: A sensitivity study. *IEEE Trans. Geosci. Remote Sens.* **1999**, *37*, 724–738.

49. Vermote, E.F.; Saleous, N.E.; Justice, C.O.; Kaufman, Y.J.; Privette, J.L.; Remer, L.; Roger, J.C.; Tanré, D. Atmospheric correction of visible to middle-infrared EOS-MODIS data over land surfaces: Background, operational algorithm and validation. *J. Geophys. Res. Atmos.* **1997**, *102*, 17131–17141. [CrossRef]

50. Li, F.; Jupp, D.L.B.; Reddy, S.; Lymburner, L.; Mueller, N.; Tan, P.; Islam, A. An evaluation of the use of atmospheric and BRDF correction to standardize Landsat data. *IEEE J. Sel. Top. Appl. Earth Obs. Remote Sens.* **2010**, *3*, 257–270. [CrossRef]

51. Flood, N. Testing the local applicability of MODIS BRDF parameters for correcting Landsat TM imagery. *Remote Sens. Lett.* **2013**, *4*, 793–802. [CrossRef]

52. Verhoef, W.; Bach, H. Coupled soil–leaf-canopy and atmosphere radiative transfer modeling to simulate hyperspectral multi-angular surface reflectance and TOA radiance data. *Remote Sens. Environ.* **2007**, *109*, 166–182. [CrossRef]

53. Li, X.; Strahler, A.H.; Woodcock, C.E. A hybrid geometric optical radiative transfer approach for modeling albedo and directional reflectance of discontinuous canopies. *IEEE Trans. Geosci. Remote Sens.* **1995**, *33*, 466–480.

54. Kuusk, A.; Nilson, T. A directional multispectral forest reflectance model. *Remote Sens. Environ.* **2000**, *72*, 244–252. [CrossRef]

55. Huang, H.G.; Chen, M.; Liu, Q.H.; Liu, Q.A.; Zhang, Y.; Zhao, L.Q.; Qin, W.H.; Zhang, J.; Foody, G.M. A realistic structure model for large-scale surface leaving radiance simulation of forest canopy and accuracy assessment. *Int. J. Remote Sens.* **2009**, *30*, 5421–5439. [CrossRef]

56. Widlowski, J.-L.; Mio, C.; Disney, M.; Adams, J.; Andredakis, I.; Atzberger, C.; Brennan, J.; Busetto, L.; Chelle, M.; Ceccherini, G. The fourth phase of the radiative transfer model intercomparison (RAMI) exercise: Actual canopy scenarios and conformity testing. *Remote Sens. Environ.* **2015**, *169*, 418–437. [CrossRef]

57. Gemmell, F. An investigation of terrain effects on the inversion of a forest reflectance model. *Remote Sens. Environ.* **1998**, *65*, 155–169. [CrossRef]

58. Smith, J.A.; Lin, T.L.; Ranson, K.L. The Lambertian assumption and Landsat data. *Photogramm. Eng. Remote Sens.* **1980**, *46*, 1183–1189.

59. Gu, D.; Gillespie, A. Topographic normalization of Landsat TM images of forest based on subpixel sun–canopy–sensor geometry. *Remote Sens. Environ.* **1998**, *64*, 166–175. [CrossRef]

60. Sandmeier, S.; Itten, K.I. A physically-based model to correct atmospheric and illumination effects in optical satellite data of rugged terrain. *IEEE Trans. Geosci. Remote Sens.* **1997**, *35*, 708–717. [CrossRef]

61. Wen, J.; Liu, Q.; Liu, Q.; Xiao, Q.; Li, X. Parametrized BRDF for atmospheric and topographic correction and albedo estimation in Jiangxi rugged terrain, China. *Int. J. Remote Sens.* **2009**, *30*, 2875–2896. [CrossRef]

62. Lee, W.L.; Liou, K.N.; Hall, A. Parameterization of solar fluxes over mountain surfaces for application to climate models. *J. Geophys. Res. Atmos.* **2011**, *116*, 94–104. [CrossRef]

63. Lefèvre, M.; Oumbe, A.; Blanc, P.; Espinar, B. Mcclear: A new model estimating downwelling solar radiation at ground level in clear-sky conditions. *Atmos. Meas. Tech.* **2013**, *6*, 2403–2418. [CrossRef]

64. Mousivand, A.; Verhoef, W.; Menenti, M.; Gorte, B. Modeling top of atmosphere radiance over heterogeneous non-Lambertian rugged terrain. *Remote Sens.* **2015**, *7*, 8019–8044. [CrossRef]

65. Wu, S.; Wen, J.; Tang, Y.; Zhao, J. Modeling anisotropic bidirectional reflectance of sloping forest. In Proceedings of the 2017 IEEE International Geoscience and Remote Sensing Symposium (IGARSS), Fort Worth, TX, USA, 23–28 July 2017; pp. 3874–3877.

66. Strahler, A.H. Vegetation canopy reflectance modeling—Recent developments and remote sensing perspectives. *Remote Sens. Rev.* **1997**, *15*, 179–194. [CrossRef]

67. Ross, J. Radiative transfer in plant communities. *Veg. Atmos.* **1975**, *1*, 13–55.

68. Cao, B.; Du, Y.; Li, J.; Li, H.; Li, L.; Zhang, Y.; Zou, J.; Liu, Q. Comparison of five slope correction methods for leaf area index estimation from hemispherical photography. *IEEE Geosci. Remote Sens. Lett.* **2015**, *12*, 1958–1962. [CrossRef]

69. Chen, J.M.; Leblanc, S.G. A four-scale bidirectional reflectance model based on canopy architecture. *IEEE Trans. Geosci. Remote Sens.* **1997**, *35*, 1316–1337. [CrossRef]

70. Li, X.; Strahler, A.H. Geometric-optical bidirectional reflectance modeling of the discrete crown vegetation canopy: Effect of crown shape and mutual shadowing. *IEEE Trans. Geosci. Remote Sens.* **1992**, *30*, 276–292. [CrossRef]

71. Ni, W.; Li, X.; Woodcock, C.E.; Caetano, M.R.; Strahler, A.H. An analytical hybrid GORT model for bidirectional reflectance over discontinuous plant canopies. *IEEE Trans. Geosci. Remote Sens.* **2002**, *37*, 987–999.

72. Jin, S.Y.; Susaki, J. A 3-D topographic-relief-correlated Monte Carlo radiative transfer simulator for forest bidirectional reflectance estimation. *IEEE Geosci. Remote Sens. Lett.* **2017**, *14*, 964–968. [CrossRef]

73. Huang, H.; Qin, W.; Liu, Q. RAPID: A radiosity applicable to porous individual objects for directional reflectance over complex vegetated scenes. *Remote Sens. Environ.* **2013**, *132*, 221–237. [CrossRef]

74. Gao, B.; Jia, L.; Menenti, M. An improved method for retrieving land surface albedo over rugged terrain. *IEEE Geosci. Remote Sens. Lett.* **2014**, *11*, 554–558. [CrossRef]

75. Ashdown, I. *Radiosity: A Programmer's Perspective*; John Wiley & Sons, Inc.: Hoboken, NJ, USA, 1994; pp. 7–10.

76. Parviainen, H.; Muinonen, K. Bidirectional reflectance of rough particulate media: Ray-tracing solution. *J. Quant. Spectrosc. Radiat. Transf.* **2009**, *110*, 1418–1440. [CrossRef]

77. Despan, D.; Bedidi, A.; Cervelle, B.; Rudant, J.P. Bidirectional reflectance of Gaussian random surfaces and its scaling properties. *Math. Geosci.* **1998**, *30*, 873–888.

78. Barsky, S.; Petrou, M. *The Shadow Function for Rough Surfaces*; Kluwer AcaDEMic Publishers: Dordrecht, The Netherlands, 2005; pp. 281–295.

79. Torrance, K.E.; Sparrow, E.M. Theory for off-specular reflection from roughened surfaces. *J. Opt. Soc. Am.* **1967**, *57*, 1105–1114. [CrossRef]

80. Hapke, B. *Theory of Reflectance and Emittance Spectroscopy: Photometric Effects of Large-Scale Roughness*; Cambridge University Press: Cambridge, UK, 1993.

81. Liu, H.; Zhu, J.; Wang, K. Modified polarized geometrical attenuation model for bidirectional reflection distribution function based on random surface microfacet theory. *Opt. Express* **2015**, *23*, 22788–22799. [CrossRef] [PubMed]

82. Blinn, J.F. Models of light reflection for computer synthesized pictures. *ACM SIGGRAPH Comput. Graph.* **1977**, *11*, 192–198. [CrossRef]

83. Buhl, D.; Welch, W.J.; Rea, D.G. Reradiation and thermal emission from illuminated craters on the lunar surface. *J. Geophys. Res.* **1968**, *73*, 5281–5295. [CrossRef]

84. Poulin, P.; Fournier, A. A model for anisotropic reflection. *ACM SIGGRAPH Comput. Graph.* **1990**, *24*, 273–282. [CrossRef]

85. Koenderink, J.J.; Doorn, A.J.V.; Dana, K.J.; Nayar, S. Bidirectional reflection distribution function of thoroughly pitted surfaces. *Int. J. Comput. Vis.* **1999**, *31*, 129–144. [CrossRef]

86. Brockelman, R.; Hagfors, T. Note on the effect of shadowing on the backscattering of waves from a random rough surface. *IEEE Trans. Antennas Propag.* **1966**, *14*, 621–626. [CrossRef]

87. Smith, B. Geometrical shadowing of a random rough surface. *IEEE Trans. Antennas Propag.* **1967**, *15*, 668–671. [CrossRef]

88. Shepard, M.K.; Campbell, B.A. Radar scattering from a self-affine fractal surface: Near-nadir regime. *Icarus* **1999**, *141*, 156–171. [CrossRef]

89. Kelemen, C.; Szirmay-Kalos, L. A microfacet based coupled specular-matte BRDF model with importance sampling. In Proceedings of the Eurographics 2011, Manchester, UK, 3–7 September 2001; pp. 25–34.

90. Shirley, P.; Smits, B.; Hu, H.; Lafortune, E. A practitioners' assessment of light reflection models. In Proceedings of the 1997 IEEE the Fifth Pacific Conference on Computer Graphics and Applications, Seoul, Korea, 13–16 October 1997; pp. 40–49.

91. Deems, J.S.; Fassnacht, S.R.; Elder, K.J. Fractal distribution of snow depth from LIDAR data. *J. Hydrometeorol.* **2006**, *7*, 285–297. [CrossRef]

92. Zhang, X.; Drake, N.A.; Wainwright, J.; Mulligan, M. Comparison of slope estimates from low resolution DEMs: Scaling issues and a fractal method for their solution. *Earth Surf. Processes Landf.* **2015**, *24*, 763–779. [CrossRef]

93. Essery, R. Spatial statistics of wind flow and blowing-snow fluxes over complex topography. *Bound.-Layer Meteorol.* **2001**, *100*, 131–147. [CrossRef]

94. Essery, R.; Marks, D. Scaling and parametrization of clear-sky solar radiation over complex topography. *J. Geophys. Res.* **2007**, *112*, 10122. [CrossRef]

95. Vico, G.; Porporato, A. Probabilistic description of topographic slope and aspect. *J. Geophys. Res. Earth Surf.* **2009**, *114*, 441–451. [CrossRef]

96. Sun, Y. Statistical ray method for deriving reflection models of rough surfaces. *J. Opt. Soc. Am. Opt. Image Sci. Vis.* **2007**, *24*, 724–744. [CrossRef]

97. Beckmann, P. Shadowing of random rough surfaces. *IEEE Trans. Antennas Propag.* **1965**, *13*, 384–388. [CrossRef]

98. Shaw, L.; Beckmann, P. Comments on "shadowing of random surfaces". *IEEE Trans. Antennas Propag.* **1966**, *14*, 253. [CrossRef]

99. Wagner, R.J. Shadowing of randomly rough surfaces. *J. Acoust. Soc. Am.* **1967**, *41*, 138–147. [CrossRef]

100. Heitz, E.; D'Eon, E.; D'Eon, E.; Dachsbacher, C. Multiple-scattering microfacet BSDFs with the Smith model. *ACM Trans. Graph.* **2016**, *35*, 58. [CrossRef]

101. Bourlier, C.; Saillard, J.; Berginc, G. Effect of correlation between shadowing and shadowed points on the Wagner and Smith monostatic one-dimensional shadowing functions. *IEEE Trans. Antennas Propag.* **2000**, *48*, 437–446. [CrossRef]

102. Bourlier, C.; Berginc, G.; Saillard, J. One- and two-dimensional shadowing functions for any height and slope stationary uncorrelated surface in the monostatic and bistatic configurations. *IEEE Trans. Antennas Propag.* **2002**, *50*, 312–324. [CrossRef]

103. Rexer, M.; Hirt, C. Comparison of free high resolution digital elevation data sets (ASTERGDEM2, SRTM v2.1/v4.1) and validation against accurate heights from the Australian national gravity database. *Aust. J. Earth Sci.* **2014**, *61*, 213–226. [CrossRef]

104. Dozier, J.; Frew, J. Rapid calculation of terrain parameters for radiation modeling from digital elevation data. *IEEE Trans. Geosci. Remote Sens.* **1990**, *28*, 963–969. [CrossRef]

105. Heitz, E. Understanding the masking-shadowing function in microfacet-based BRDFs. *J. Comput. Graph. Tech.* **2014**, *3*, 48–107.

106. Li, F.; Jupp, D.L.B.; Thankappan, M. Issues in the application of digital surface model data to correct the terrain illumination effects in Landsat images. *Int. J. Digit. Earth* **2015**, *8*, 235–257. [CrossRef]

107. Wang, X.; Yin, Z.Y. A comparison of drainage networks derived from digital elevation models at two scales. *J. Hydrol.* **1998**, *210*, 221–241. [CrossRef]

108. Wolock, D.M.; McCabe, G.J. Differences in topographic characteristics computed from 100- and 1000-m resolution digital elevation model data. *Hydrol. Processes* **2015**, *14*, 987–1002. [CrossRef]

109. Yin, Z.Y.; Wang, X. A cross-scale comparison of drainage basin characteristics derived from digital elevation models. *Earth Surf. Processes Landf.* **2015**, *24*, 557–562. [CrossRef]

110. Zhou, Q.; Liu, X. Analysis of errors of derived slope and aspect related to DEM data properties. *Comput. Geosci.* **2004**, *30*, 369–378. [CrossRef]

111. Fleming, M.D.; Hoffer, R.M. *Machine Processing of Landsat MSS Data and DMA Topographic Data for Forest Cover Type Mapping*; Purdue University: West Lafayette, IN, USA, 1979.

112. Ritter, P. Vector-based slope and aspect generation algorithm. *Am. Soc. Photogramm. Remote Sens.* **1987**, *53*, 1109–1111.

113. Horn, B.K.P. Hill shading and the reflectance map. *IEEE Proc.* **1981**, *69*, 14–47. [CrossRef]

114. Wood, J. The geomorphological characterisation of digital elevation models. *Diss. Theses-Gradworks* **1996**, *13*, 834–848.

115. Unwin, D.J.; Doomkamp, J.C. Introductory spatial analysis. In Proceedings of the Parallel Problem Solving from Nature—PPSN IV, Berlin, Germany, 22-26 September 2010; Volume 36, pp. 1307–1319.

116. Chu, T.H.; Tsai, T.H. Comparison of accuracy and algorithms of slope and aspect measures from DEM. In Proceedings of the GIS AM/FM ASIA'95, Bangkok, Thailand, 21–24 August 1995; pp. 21–24.

117. Jones, K.H. A comparison of algorithms used to compute hill slope as a property of the DEM. *Comput. Geosci.* **1998**, *24*, 315–323. [CrossRef]

118. Zakšek, K.; Podobnikar, T.; Oštir, K. Solar radiation modelling. *Comput. Geosci.* **2005**, *31*, 233–240. [CrossRef]

119. Helbig, N.; Löwe, H. Parameterization of the spatially averaged sky view factor in complex topography. *J. Geophys. Res. Atmos.* **2014**, *119*, 4616–4625. [CrossRef]

120. Richter, R. Correction of satellite imagery over mountainous terrain. *Appl. Opt.* **1998**, *37*, 4004–4015. [CrossRef] [PubMed]

121. Xin, L.; Koike, T.; Guodong, C. Retrieval of snow reflectance from Landsat data in rugged terrain. *Ann. Glaciol.* **2002**, *34*, 31–37. [CrossRef]

122. Roujean, J.L.; Leroy, M.; Deschamps, P.Y. A bidirectional reflectance model of the earth's surface for the correction of remote sensing data. *J. Geophys. Res. Atmos.* **1992**, *97*, 20455–20468. [CrossRef]

123. Wanner, W.; Li, X.; Strahler, A.H. On the derivation of kernels for kernel-driven models of bidirectional reflectance. *J. Geophys. Res. Atmos.* **1995**, *100*, 21077–21089. [CrossRef]

124. Bruegge, C.; Chrien, N.; Haner, D. A spectralon BRF data base for MISR calibration applications. *Remote Sens. Environ.* **2001**, *77*, 354–366. [CrossRef]

125. Gastellu-Etchegorry, J.P.; Guillevic, P.; Zagolski, F.; DEMarez, V.; Trichon, V.; Deering, D.; Leroy, M. Modeling BRF and radiation regime of boreal and tropical forests: I. BRF. *Remote Sens. Environ.* **1999**, *68*, 281–316. [CrossRef]

126. Liu, S.; Liu, Q.; Liu, Q.; Wen, J.; Li, X.; Xiao, Q.; Xin, X. The angular & spectral kernel model for BRDF and albedo retrieval. *IEEE J. Sel. Top. Appl. Earth Obs. Remote Sens.* **2010**, *3*, 241–256.

127. You, D.; Wen, J.; Liu, Q.; Liu, Q.; Tang, Y. The angular and spectral kernel-driven model: Assessment and application. *IEEE J. Sel. Top. Appl. Earth Obs. Remote Sens.* **2014**, *7*, 1331–1345.

128. Wen, J.; Dou, B.; You, D.; Tang, Y.; Xiao, Q.; Liu, Q.; Liu, Q. Forward a small-timescale BRDF/albedo by multisensor combined BRDF inversion model. *IEEE Trans. Geosci. Remote Sens.* **2017**, *55*, 683–697. [CrossRef]

129. Huang, H.; Qin, W.; Spurr, R.J.D.; Liu, Q. Evaluation of atmospheric effects on land-surface directional reflectance with the coupled RAPID and VLIDORT models. *IEEE Geosci. Remote Sens. Lett.* **2017**, *14*, 916–920. [CrossRef]

remote sensing

MDPI

Article

Advancing the PROSPECT-5 Model to Simulate the Spectral Reflectance of Copper-Stressed Leaves

Chengye Zhang [1,2,3,4], Huazhong Ren [1,2,3], Yanzhen Liang [5], Suhong Liu [6], Qiming Qin [1,2,3,*] and Okan K. Ersoy [4]

[1]   Institute of Remote Sensing and Geographic Information System, School of Earth and Space Sciences, Peking University, Beijing 100871, China; zhangchengye@pku.edu.cn (C.Z.); renhuazhong@pku.edu.cn (H.R.)

[2]   Beijing Key Lab of Spatial Information Integration and 3S Application, Peking University, Beijing 100871, China

[3]   Mapping and Geo-Information for Geographic Information Basic Softwares and Applications, Engineering Research Center of National Administration of Surveying, Beijing 100871, China

[4]   School of Electrical and Computer Engineering, Purdue University, West Lafayette, IN 47907, USA; ersoy@purdue.edu

[5]   Earth Observation System & Data Center, China National Space Administration, Beijing 100101, China; xueer3329@163.com

[6]   Faculty of Geographical Science, Beijing Normal University, Beijing 100875, China; liush@bnu.edu.cn

*   Correspondence: qmqinpku@163.com; Tel.: +86-10-6276-5715

Received: 26 September 2017; Accepted: 17 November 2017; Published: 20 November 2017

Abstract: This paper proposes a modified model based on the PROSPECT-5 model to simulate the spectral reflectance of copper-stressed leaves. Compared with PROSPECT-5, the modified model adds the copper content of leaves as one of input variables, and the specific absorption coefficient related to copper (K_{cu}) was estimated and fixed in the modified model. The specific absorption coefficients of other biochemical components (chlorophyll, carotenoid, water, dry matter) were the same as those in PROSPECT-5. Firstly, based on PROSPECT-5, we estimated the leaf structure parameters (N), using biochemical contents (chlorophyll, carotenoid, water, and dry matter) and the spectra of all the copper-stressed leaves (samples). Secondly, the specific absorption coefficient related to copper (K_{cu}) was estimated by fitting the simulated spectra to the measured spectra using 22 samples. Thirdly, other samples were used to verify the effectiveness of the modified model. The spectra with the new model are closer to the measured spectra when compared to that with PROSPECT-5. Moreover, for all the datasets used for validation and calibration, the root mean square errors (RMSEs) from the new model are less than that from PROSPECT-5. The differences between simulated reflectance and measured reflectance at key wavelengths with the new model are nearer to zero than those with the PROSPECT-5 model. This study demonstrated that the modified model could get more accurate spectral reflectance from copper-stressed leaves when compared with PROSPECT-5, and would provide theoretical support for monitoring the vegetation stressed by copper using remote sensing.

Keywords: vegetation remote sensing; reflectance model; spectra; leaf; copper; PROSPECT

---

1. Introduction

Remote sensing provides a rapid and large-scale tool for geobotanical prospecting [1–3] and environmental monitoring [4]. For the vegetation on copper deposits or the area polluted by industrial activities related to copper, excessive copper elements would be absorbed by root systems, and then stress the growth of plants and change the spectral reflectance of leaves. In addition, spectral reflectance is the vital foundation of vegetation remote sensing. Hence, the reflectance of leaves on copper-stressed vegetation is crucial for prospecting copper deposit and monitoring copper-pollution. Currently, most studies on remote sensing of vegetation stressed by heavy metal or other stress factors focus on the change of reflectance and vegetation indices using empirical statistical methods; many statistical models have been proposed, but little research has focused on the physical model [4–10]. However, for many users, statistical models cannot meet the requirement for understanding the action mechanism of copper (Cu) stress on leaf reflectance. Moreover, many parameters in statistical models are sensitive to the case study and have no physical meaning. Hence, a model that can accurately simulate the spectral reflectance of copper-stressed leaves should be developed, which would compensate for some of the deficiencies of statistical models.

The PROSPECT leaf optical properties model has been an important and most popular physical model to simulate leaf directional hemispherical reflectance and transmittance from 400 to 2500 nm [11,12]. Here we review the evolution over time of PROSPECT in order to illustrate the novelty of our study. To explain the interaction of isotropic light with a compact leaf, a theoretical model called "Plate Model" was proposed, which regards a compact leaf as an absorbing plate [13]. Plate Model was later generalized to the non-compact leaf [14]. Based on the generalized Plate Model, the PROSPECT model was proposed to simulate the spectral reflectance and transmittance of a leaf [11]. In this initial version, the reflectance and transmittance were calculated by the refractive index (n), a parameter describing the leaf structure (N), pigment concentration (C_{ab}), equivalent water thickness (C_w), and the corresponding specific absorption coefficients (K_{ab} and K_w), where n, K_{ab} and K_w, have been fitted by data with varying plant types and status. After this version, the content (C_m) and corresponding specific absorption coefficient (K_m) of the dry matter that influences the absorption features in shortwave infrared (SWIR), including cellulose, lignin, protein, hemicellulose, starch, and sugar, were introduced into the PROSPECT model [15–19]. The spectral resolution was improved from 5 nm to 1 nm in an unreleased version [20,21], and the model was modified to account for surface directional reflectance of a leaf [21,22]. In 2008, the PROSPECT model was further calibrated and two new versions (PROSPECT-4 and PROSPECT-5) were proposed [21]. The difference between PROSPECT-4 and PROSPECT-5 is that PROSPECT-5 separates total carotenoids from total pigments. In recent years, PROSPECT-5 has been popularly used in the remote sensing of vegetation instead of other versions.

However, the PROSPECT versions above were developed for healthy leaves. For copper-stressed leaves, excessive copper would change the spectral reflectance of leaves. Hence, to simulate the reflectance of copper-stressed leaves, the copper content and the specific absorption coefficient related to copper was added into the PROSPECT model [23]. Zhu et al. [23] initiated research on the physical reflectance model of copper-stressed leaves. However, in [23], there are still some problems needing to be addressed, whereas subsequent studies were not found in public literatures. These problems are shown as follows. (1) The structure parameter N of copper-treated leaf is determined using the reflectance at 800–1200 nm, and the absorption of copper ion was ignored in this wavelength range in the [23]. However, according to the theory of electron transition and experimental observation [24], there is a significant absorption of copper ion at 700–900 nm. (2) Carotenoid content has been treated as an input variable in the popular PROSPECT-5 model and has important influence on the reflectance of leaves. However, the carotenoid content and its specific absorption coefficient were ignored in the model developed in [23]. (3) The absorption characteristics of biochemical components (water, chlorophyll, dry matter, pigment) are inherent and should be remained unchanged in the advanced model for copper-stressed leaves.

Hence, based on above problems, the study on the reflectance model of copper-stressed leaves should be continued. In other words, the specific absorption coefficient related to copper needs to be improved, and a more accurate model for simulating the reflectance of copper-stressed leaf should be developed. Based on this motivation, we estimated the specific absorption coefficient related to copper and added it into the popular PROSPECT-5 model, and then developed a modified model to simulate the reflectance of copper-stressed leaf. The proposed model considers the carotenoid content, copper content and corresponding specific absorption coefficients. In addition, it avoids the coupling influence on the reflectance from structure parameter N and the absorption of copper ions.

2. Datasets

In this study, wheat (*Triticum aestivum* L., cultivar: "Xinchun-17") and pak choi (*Brassica chinensis* L., cultivar: "Shanghaiqing") were treated by copper with different levels using control experiments. Soil was collected from a vegetable garden without any contamination. Seeds were planted in the soil mixed with copper sulfate ($CuSO_4$) solutions in impermeable plastic pots (Figure 1). The copper contents in soil were controlled as 0, 25, 50, 100, 200, 400, 800, 1600, 3200, and 4800 mg/kg, respectively. The pot distribution of wheat and experimental scene of some samples are shown in Figure 1, and the pot distribution of pak choi is same as that of wheat. Therefore, the total number of pots is 60. For the group with 0 mg/kg copper content, the plants were regarded as healthy vegetation. Thus, the data from this group was not used for the calibration of the reflectance model for copper-stressed vegetation. This group was only used as a reference for observing the growth of copper-stressed vegetation. There are same characteristics (except copper content) of the experimental soil with different Cu levels, including nitrogen (N) content, phosphorus (P) content, potassium (K) content, water content, particle size, pH, and so forth. There are three pots for each copper-stress level experiment (three parallel experiments) to reduce accidental errors (Figure 1).

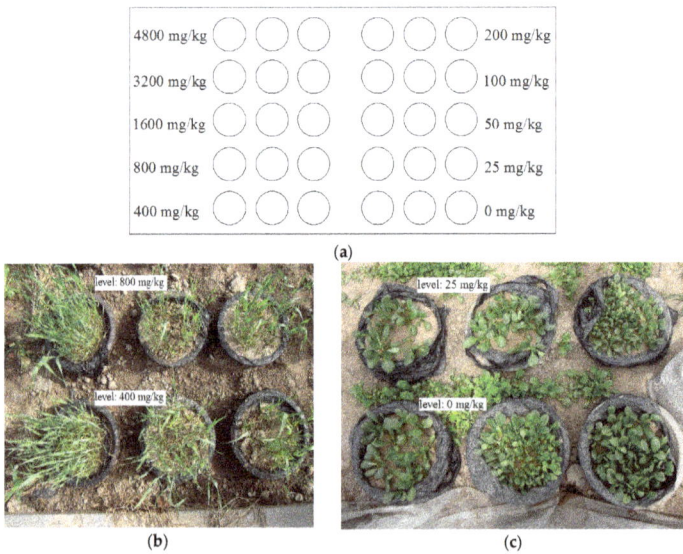

Figure 1. The experimental scene of some samples: (**a**) The distribution of plastic pots of wheat. (The pots distribution of pak choi is same with that of wheat. Circle: a pot); (**b**) Some samples of wheat (stress level: 400 mg/kg; 800 mg/kg); (**c**) Some samples of pak choi (stress level: 0 mg/kg; 25 mg/kg).

The spectral reflectance of leaves was measured by ASD (Analytical Spectral Devices) FieldSpec FR spectroradiometer (Boulder, CO, USA) with an Li-1800 integrating sphere (Li-Cor, Lincoln, NE, USA). The inside of the sphere is covered with $BaSO_4$. The diameter of sample port of the integrating sphere is about 14 mm, which could be covered by the measured leaves. The standard white reference was measured (the fiber faced the sample port) before each measurement of a sample (the fiber faced the reference). The light source is a halogen lamp. Due to the data quality from 1650 to 2500 nm from the integrating sphere, auxiliary spectra were also measured with a leaf clip, and these spectra from 1650 to 2500 nm were linearly scaled to replace the corresponding data from the integrating sphere.

In future studies, other integrating spheres with better data (e.g., RT-060-SF, Labsphere, NH, USA) are recommended to readers. The wavelength range of final measured spectra was from 350 to 2500 nm, and the spectral resolution was 1 nm, which is the same as that of the PROSPECT-5 model.

After the reflectance measurement in the laboratory, the leaves were immediately processed to measure biochemical components. Thus, the changing of biochemical components with different hours in a day almost has no impact in this study. The content (per surface area unit) of every biochemical component (i.e., chlorophyll, carotenoid, water, dry matter, and copper) in the leaves was measured using corresponding chemical methods. In detail, for each measurement, the leaf was sampled using a puncher with a hole of a known area. For the chlorophyll and carotenoid of a sample, the sample was triturated and then 80% acetone was used for extracting the chlorophyll and carotenoid. The absorbance of the solution was measured by a spectrophotometer at 470 nm, 646 nm, and 663 nm [23,25]. The concentration (mg/L) of chlorophyll and carotenoid was calculated based on the quantitative relationship between absorbance and pigment concentration (Lambert-Beer Law) [23,25]. According to the volume of the solution and the surface area of sample, the content (per surface area unit) of chlorophyll and carotenoid in a sample can be determined. For the measurement of copper content, the leaf sample was digested in concentrated nitric acid and perchloric acid. The solution was filtered, and the copper concentration was determined by atomic absorption spectrophotometry [23,26]. For the measurement of water content and dry matter content, the leaves were continuously heated at 105 °C for 30 min to kill the leaves to cease the metabolism and avoid some matter decomposition in the next step. The leaves were dried at 70 °C to get a constant weight, and then the leaves were weighed to determine the content of water and dry matter. Hence, for each sample, data on the content (per surface area unit) of copper, chlorophyll, carotenoid, water, and dry matter were collected as well as the corresponding spectral reflectance.

In addition, the images of internal structure of leaves with different stress levels were acquired by scanning electron microscopy (SEM). The samples were processed by freeze fracture technique and immediately put into the 2% glutaraldehyde solution (solvent: 0.1 mol/L potassium phosphate buffer, pH = 7.2). All the samples were preserved at constant 4 °C and observed with SEM [23]. The SEM used in this study was KYKY-EM 3200 with a resolution better than 6 nm.

In this study, the leaves were collected at different growth stages of the two vegetation (wheat: elongation stage, heading stage; pak choi: six leaves period, eight leaves period). The leaves sampled were the visually representative leaves in the plant. With the exception of the normal leaves (stress level = 0 mg/kg) and outliers, 33 groups of datasets of copper-stressed leaves were finally used for this study. The 33 groups of datasets of copper-stressed leaves were randomly divided into two parts (22 groups and 11 groups). Twenty-two groups of datasets were used to develop the new model, and the remaining 11 groups were used to perform validation of the new model.

In addition, this study also used a public dataset named LOPEX93 [27] for the comparison with the data of copper-stressed leaves on the leaf structure parameters. The LOPEX93 dataset includes the biochemical components and spectra of a variety of plants and has been widely used in the remote sensing of vegetation. In this study, the LOPEX93 dataset was downloaded from the website in [28].

3. Methods

In the PROSPECT model, a leaf is assumed to be composed of N homogeneous compact layers of biochemical components separated by $(N - 1)$ layers of air. N described the overall characteristics of the leaf structure and varies with different leaves. In the PROSPECT-5 version, the following parameters have been estimated and fixed in the model, and these parameters do not vary with different leaves: the angle of incidence of incoming radiation (α), refractive index ($n(\lambda)$), specific absorption coefficient of each biochemical components ($K_{ab}(\lambda)$, $K_{car}(\lambda)$, $K_w(\lambda)$, and $K_m(\lambda)$ represent the specific absorption coefficient of chlorophyll, carotenoid, water, and dry matter, respectively); $n(\lambda)$, $K_{ab}(\lambda)$, $K_{car}(\lambda)$, $K_w(\lambda)$, and $K_m(\lambda)$ are the functions of wavelength λ. Since above parameters were fixed, the simulations for the spectral reflectance of different leaves depend on the absorption coefficient of a compact layer ($K(\lambda)$).

The calculation method of $K(\lambda)$ is given by Equation (1) in PROSPECT-5 [21].

$$K(\lambda) = \frac{K_{ab}(\lambda) \cdot C_{ab} + K_{car}(\lambda) \cdot C_{car} + K_w(\lambda) \cdot C_w + K_m(\lambda) \cdot C_m}{N}, \tag{1}$$

where C_{ab}, C_{car}, C_w, C_m are the contents of chlorophyll, carotenoid, water (equivalent water thickness), and dry matter in leaf, respectively. Hence, to simulate the spectral reflectance of different leaves, following parameters which varies with different leaves are the five input variables of PROSPECT-5: N, C_{ab}, C_{car}, C_w, C_m. In other words, the PROSPECT-5 considers the absorption of four biochemical components: chlorophyll; carotenoid; water, and; dry matter.

However, several studies have indicated that heavy metal stress would damage the leaf structure, leading to a disorderly cell arrangement (e.g., [23,29]). Moreover, the copper content in copper-stressed leaf could be approximately 100 times more than that in normal leaf, so the absorption related to copper should not be ignored in the new model. Hence, we further analyzed the leaf structure parameter (N) and the specific absorption coefficient related to copper. Thus, α, $n(\lambda)$, $K_{ab}(\lambda)$, $K_{car}(\lambda)$, $K_w(\lambda)$, and $K_m(\lambda)$, which were analyzed and determined in PROSPECT-5 [21], were used in the new model with no change, because there are no existing studies or theories that link these parameters to copper stress.

The flowchart of the method used in this study is shown in Figure 2. (1) The leaf structure parameters of all the 33 samples were estimated by fitting the simulated spectra to the measured spectra in 400–510 nm using PROSPECT-5. The detailed method for calculating N could be found in Section 3.1; (2) 22 samples were used to estimate the specific absorption coefficient related to copper. The leaf structure parameter N and the specific absorption coefficient related to copper were determined independently in this study; (3) Other samples were used for the validation of the modified model. This procedure is explained in detail in Sections 3.1–3.3.

The modified model considers the absorption related to copper and adds the content of copper (C_{cu}) as an input variable when compared with PROSPECT-5. The input variables of both the modified model and the PROSPECT-5 model could be read from Table 1. In fact, only the content of copper (C_{cu}) was added to the modified model when it is compared with PROSPECT-5 (Table 1).

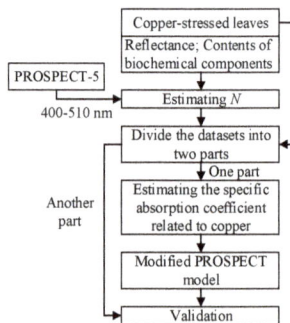

Figure 2. The steps of the method used in this study.

Table 1. The input variables of the modified model and the PROSPECT-5 model (for comparison).

Model	PROSPECT-5	Modified Model
Leaf structure parameter	N	N
Content of chlorophyll	C_{ab}	C_{ab}
Content of carotenoid	C_{car}	C_{car}
Equivalent water thickness	C_w	C_w
Content of dry matter	C_m	C_m
Content of copper	/	C_{cu}

3.1. Determination of N under Copper Stress

As previously mentioned, excessive copper in a plant would damage the leaf structure, which would result in a disorderly cell arrangement ([23,29]). To further illustrate the change of internal structure of copper-stressed leaves, some SEM images of leaves with different levels of copper stress are shown in Figure 3. It is clearly observed that more disordered internal structure was present in higher-stressed leaves, including atrophic mesophyll cells, disintegrating vascular bundle, and disorderly cell arrangement.

The PROSPECT model regards the leaf as N homogeneous compact layers of plates separated by $(N-1)$ layers of air. The leaf structure parameter (N) describes the leaf mesophyll structure and increases with a more disorderly cell arrangement. The value of N should be estimated accurately when the PROSPECT model is used for simulating the spectral reflectance of leaf.

The influence from the absorption by regular biochemical components at 800–1300 nm is the minimum in the whole wavelength. Moreover, the specific absorption coefficients of regular biochemical components is unknown and also needed to be estimated in the previous calibration on PROSPECT for healthy leaves ([21,23]). Hence, in previous studies on PROSPECT, the range of 800–1300 nm was selected for estimating N to avoid the influence from the absorption by regular biochemical components ([21,23]).

However, in this study, the specific absorption coefficients of above biochemical components in PROSPECT-5 remained unchanged in the modified model and did not need to be recalculated. So the influence from the absorption by regular biochemical components did not need to be considered. For estimating N in this study, the real influence may come from the absorption related to copper and the specific absorption coefficient related to copper was not yet determined. Hence, a wavelength range where the absorption of copper is at minimum should be selected to determine N of copper-stressed leaves to avoid the coupling influence. The absorption spectra of aqueous copper sulfate solutions were measured in [24], and the results demonstrated that the absorption of copper sulfate solutions almost equals zero in the range from 400 to 510 nm. As the wavelength λ ranges from 400 to 2500 nm in Equation (1), the reflectance in the whole wavelength would vary with different N. To verify this point and observe the detailed influence from N on the reflectance at 400–510 nm, the contents of biochemical components were fixed, and a series of reflectance spectra were generated with varying N. Figure 4 shows that changing N would cause the changes of reflectance in the whole wavelength range (400 to 2500 nm), including 400–510 nm. Therefore, we selected 400–510 nm, not 800–1300 nm, to estimate N.

In this study, we estimated the N of j-th leaf (N_j) by minimizing the merit function given by Equation (2) from 400 to 510 nm. In detail, for a leaf, the method is to find the minimum $J(N_j)$ and the corresponding N when N ranges from 1 to 7 with step = 0.01.

$$J(N_j) = (R_{mes,j}(\lambda_1) - R_{sim,j}(N_j, \lambda_1))^2 + (R_{mes,j}(\lambda_2) - R_{sim,j}(N, \lambda_2))^2, \tag{2}$$

where $R_{mes,j}$ and $R_{sim,j}$ are the j-th measured reflectance and corresponding modeled reflectance at the wavelength λ_1 and λ_2, respectively. The wavelengths λ_1 and λ_2 were where the measured reflectance reached maximum and minimum, respectively.

The leaf structure parameter N varies with different leaves and is a necessary input variable for both PROSPECT-5 and the modified model (Table 1). In this study, the leaf structure parameters N of 22 samples need to be estimated and then need to be used in the PROSPECT-5 model to calculate the specific absorption coefficient related to copper. The leaf structure parameters N of 11 samples for validation also need to be estimated to simulate the spectral reflectance of copper-stressed leaves using the modified model. Hence, the leaf structure parameters of all the samples need to be estimated. According to Equation (2), the calculation for N in a sample is independent and has nothing to do with other samples. Therefore, in this study, the leaf structure parameters N were estimated sample by sample, and the N of all the 33 copper-stressed samples were estimated.

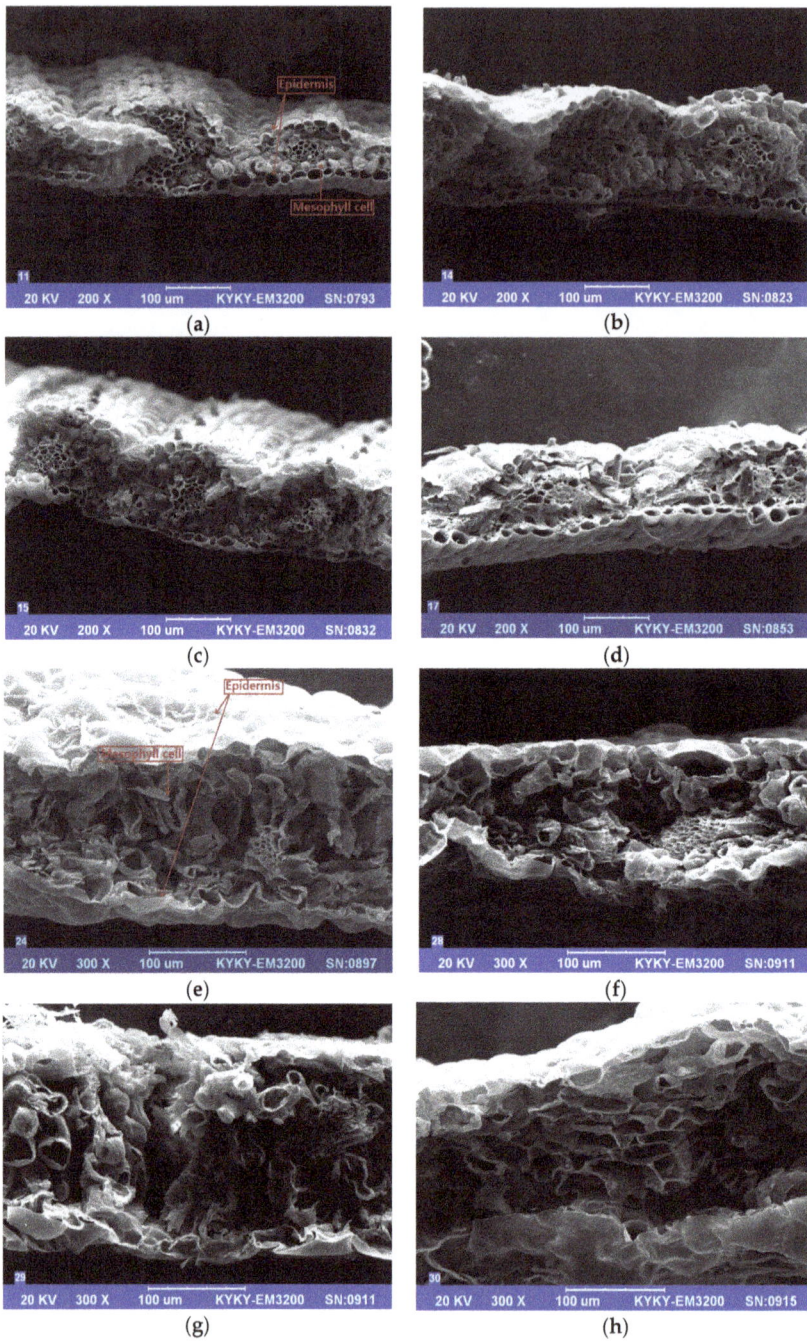

Figure 3. Scanning electron microscopy (SEM). images of leaves with different copper contents in soil: (**a**) Normal wheat; (**b**) Wheat with 200 mg/kg copper content in soil; (**c**) Wheat with 400 mg/kg copper content in soil; (**d**) Wheat with 1600 mg/kg copper content in soil; (**e**) Normal pak choi; (**f**) Pak choi with 200 mg/kg copper content in soil; (**g**) Pak choi with 400 mg/kg copper content in soil; (**h**) Pak choi with 800 mg/kg copper content in soil.

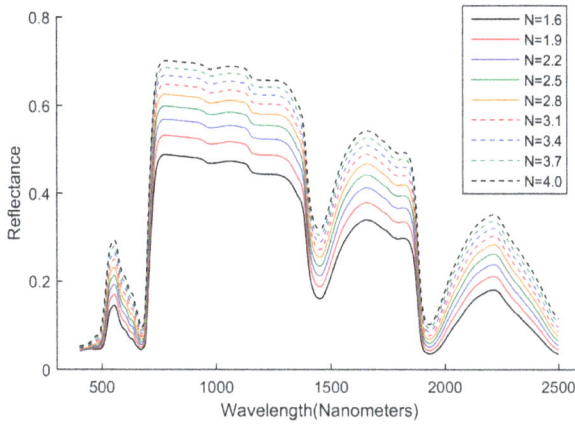

Figure 4. The modeled reflectance generated by PROSPECT-5 with different *N*. (C_{ab}, C_{car}, C_w, and C_m are fixed as 33 µg/cm^2, 8.6 µg/cm^2, 0.012 cm, 0.005 g/cm^2, respectively).

3.2. Determination of Specific Absorption Coefficient Related to Copper

Copper content in a healthy leaf is normally 5 to 30 mg/kg [4]. However, as previously mentioned, the copper content in a copper-stressed leaf would be approximately 100 times more than that in a normal leaf, which has been verified by the measured results from several previous studies (e.g., [4,6]). Hence, the absorption related to copper should be considered, and thus the calculation method of *K(λ)* was modified as Equation (3).

$$K_{new}(\lambda) = \frac{K_{ab}(\lambda) \cdot C_{ab} + K_{car}(\lambda) \cdot C_{car} + K_w(\lambda) \cdot C_w + K_m(\lambda) \cdot C_m + K_{cu}(\lambda) \cdot C_{cu}}{N},$$ (3)

where K_{cu} and C_{cu} are the specific absorption coefficient related to copper and the copper content in leaf, respectively. $K_{new}(\lambda)$ is the absorption coefficient in the new model.

The estimating method of K_{cu} in this study was similar with the methods for estimating specific absorption coefficients of other biochemical components in [21,23]. At each wavelength, we minimized the merit function given by Equation (4).

$$J(K_{cu}(\lambda)) = \sum_{j=1}^{Num} (R_{mes,j}(\lambda) - R_{sim,j}(K_{new}(\lambda), \lambda))^2,$$ (4)

where $R_{mes,j}(\lambda)$ and $R_{sim,j}(K_{cu}(\lambda), \lambda)$ are the *j*-th measured reflectance and corresponding modeled reflectance at the wavelength λ. *Num* is the number of samples for estimating K_{cu}. In this study, *Num* = 22.

An illustration on the modification strategy in this study should be presented here. The reflectance from 400 to 510 nm is controlled mainly by chlorophyll and carotenoids. Copper almost has no direct contribution to the reflectance from 400 to 510 nm. Although the copper-induced physiological implications (change of pigments) could affect the reflectance from 400 to 510 nm, this effectiveness should not be considered when the two pigments were measured correctly. Hence, the leaf structure parameter *N* was determined in the wavelengths from 400 to 510 nm since parameters of chlorophyll and carotenoids are known. In addition, the leaf structure parameter *N* and K_{cu} can be determined independently since *N* can be estimated by using reflectance spectra over the spectral range with no copper-related absorption.

3.3. Validation of the Modified Model

In this study, 11 groups of datasets on copper-stressed leaves were used for the validation of the new model. The simulated reflectance spectra were generated using PROSPECT-5 and the new model with the estimated N, respectively. The mean ($\pm$standard deviation, SD) of the simulated spectra and measured spectra of the 11 samples were calculated for comparison. The simulated spectra and measured spectra of two selected representative samples (No. 5 and No. 7) from 11 samples were also used for comparison. By observing the spectra, the modified model will be better than PROSPECT-5 if the spectra with the modified model is closer to the measured spectra. The differences between simulated reflectance and measured reflectance at seven key wavelengths, including 550 nm, 660 nm, 700 nm, 850 nm, 1400 nm, 1900 nm, and 2200 nm, were calculated. The modified model will be better than PROSPECT-5 if the differences with the modified model are nearer to zero than those with the PROSPECT-5 model. The root mean square error (RMSE) between the simulated spectra and measured spectra were calculated for both calibration and validation samples. The modified model will be better than PROSPECT-5 if the RMSEs with the modified model are less.

4. Results

The mean, standard deviation (SD), and coefficient of variation (CV) of datasets for calibration and validation were calculated (Table 2). Coefficient of variation characterizes the inter-difference degree in a dataset. In this study, all the coefficient of variations are significant high and the CVs with copper are the highest. Table 2 demonstrates that there are significant differences in leaf Cu and in other biochemical components for both calibration and validation datasets.

Table 2. Mean, standard deviation (SD), and coefficient of variation (CV) of datasets for calibration and validation (water: cm; dry matter: g/cm^2; chlorophyll, carotenoids, and copper: μg/cm^2).

	Calibration Datasets			Validation Datasets		
	Mean	**SD**	**CV**	**Mean**	**SD**	**CV**
water	0.0175	0.00771	44.04%	0.0213	0.00964	45.26%
dry matter	0.0041	0.00084	20.34%	0.0042	0.00094	22.08%
chlorophyll	32.64	7.853	24.06%	33.79	8.062	23.86%
carotenoids	13.36	5.942	44.48%	15.67	7.388	47.15%
copper	0.1220	0.09784	80.22%	0.1507	0.1571	104.22%

In this study, the leaf structure parameters N were estimated sample by sample, and the results of estimating N of all the 33 copper-stressed samples are presented here (Figure 5a). In addition, we randomly selected 33 groups of leaf structure parameters of normal leaves from the LOPEX93 dataset [27], which is also shown in Figure 5a. The box-and-whisker plots of the N of copper-stressed leaves and normal leaves are shown in Figure 5b. Figure 5a,b show that the leaf structure parameters of copper-stressed leaves are obviously more than those of normal leaves. Moreover, the distributed range of N with copper-stressed leaves is larger. The leaf structure parameters vs. leaf copper content with different symbols for each species were plotted in Figure 5c. The stress datasets were divided into three parts based on stress levels (low, medium, high). The averages of leaf structure parameters were plotted in Figure 5d for different classes and species. Figure 5c,d show that the leaf structure parameters tend to increase with stress levels.

The estimated specific absorption coefficient related to copper (K_{cu}) is shown in Figure 6. Following characteristics could be read from Figure 6: K_{cu} is near zero from 400 to 510 nm. High values of K_{cu} are presented at 590–710 nm with a valley near 680 nm, respectively. In addition, there are three major peaks near 1400 nm, 1900 nm, and 2400 nm, respectively.

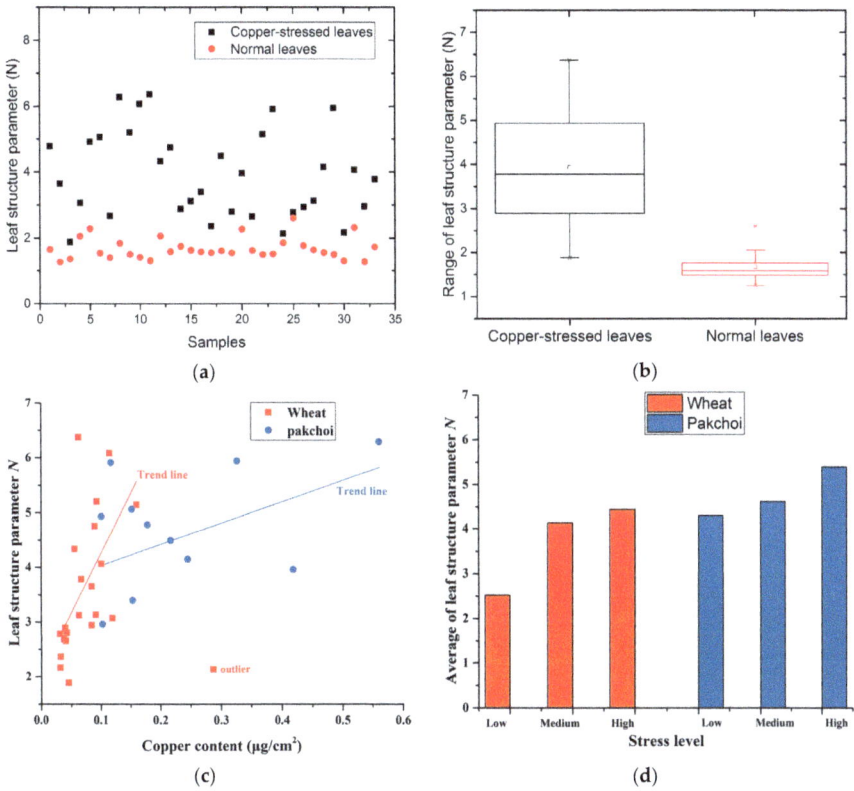

Figure 5. Leaf structure parameters *N* of copper-stressed leaves and normal leaves: (**a**) The value of leaf structure parameters; (**b**) The box-and-whisker plots of leaf structure parameters; (**c**) Leaf structure parameters vs. leaf copper content with different symbols for each species; (**d**) Bar graphs for the average of different stress levels (low, medium, high) and species.

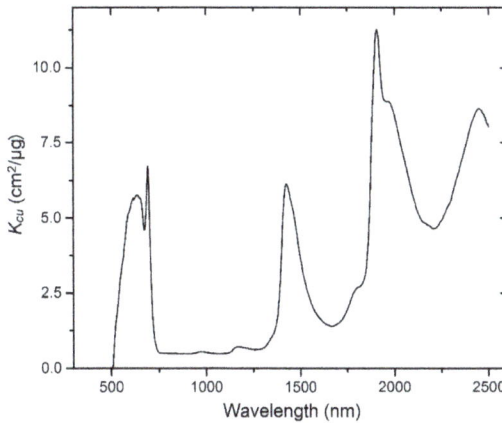

Figure 6. The specific absorption coefficient related to copper (K_{cu}).

The mean (±standard deviation, SD) of simulated spectra and measured spectra of the 11 samples is shown in Figure 7a. The simulated spectra and measured spectra of two selected representative samples (No. 5 and No. 7) from 11 samples are shown in Figure 7b,c. Samples of No. 5 and No. 7 are from wheat and pak choi, respectively. Figure 7 shows that, in the whole range of wavelength, the values of spectral reflectance with PROSPECT-5 are larger than the measured values, while the spectral reflectance with the new model is closer to the measured spectra. The 11 groups of RMSEs between the simulated spectra (PROSPECT-5 and the new model) and measured spectra are shown in Table 3. We also calculated the RMSE of each model for the calibration datasets (Table 4). For all the datasets used for validation and calibration, the RMSEs from the new model are less than that from PROSPECT-5.

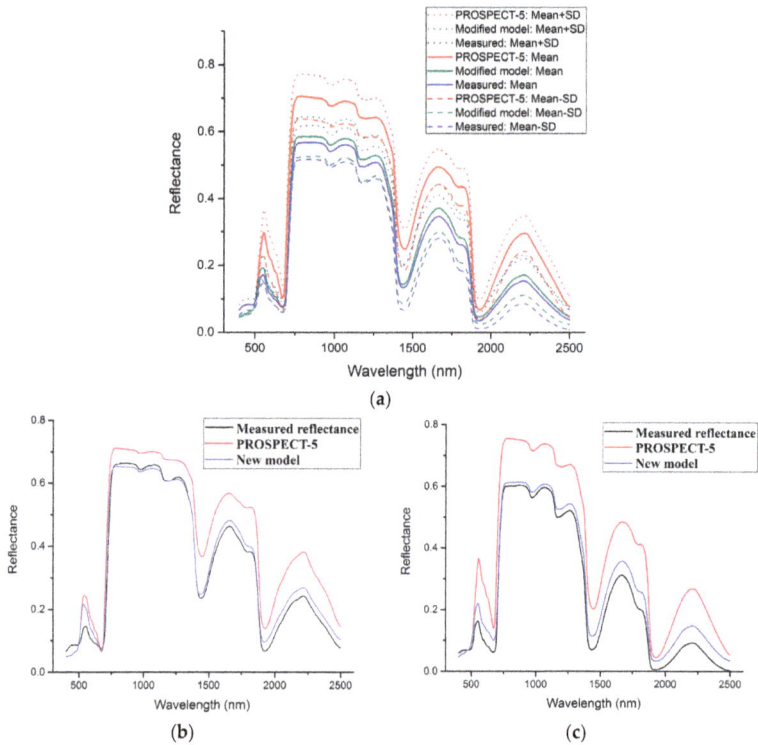

Figure 7. The simulated spectra (PROSPECT-5 and the new model) and measured spectra: (**a**) The mean (±standard deviation, SD) of simulated spectra and measured spectra of the 11 samples; (**b**) The selected representative sample: No. 5 sample (wheat: C_{ab} = 49.63 µg/cm^2; C_{car} = 8.94 µg/cm^2; C_w = 0.0096 cm; C_m = 0.0061 g/cm^2; C_{cu} = 0.0557 µg/cm^2); (**c**) The selected representative sample: No. 7 sample (pak choi: C_{ab} = 27.61 µg/cm^2; C_{car} = 19.32 µg/cm^2; C_w = 0.0320 cm; C_m = 0.0031 g/cm^2; C_{cu} = 0.1520 µg/cm^2).

To further illustrate the effectiveness of the new model, the reflectance values at several key wavelengths were selected for comparison where the reflectance curve of vegetation shows peaks, valleys, or obvious high values, including 550 nm, 660 nm, 700 nm, 850 nm, 1400 nm, 1900 nm, and 2200 nm. We calculated the differences between simulated reflectance and measured reflectance at these seven key wavelengths, and plotted these differences vs. wavelength and 11 samples, including wheat and pak choi (samples 1–6: wheat; samples 7–11: pak choi) (Figure 8). Figure 8 shows that the differences with the modified model are nearer to zero than those with the PROSPECT-5 model.

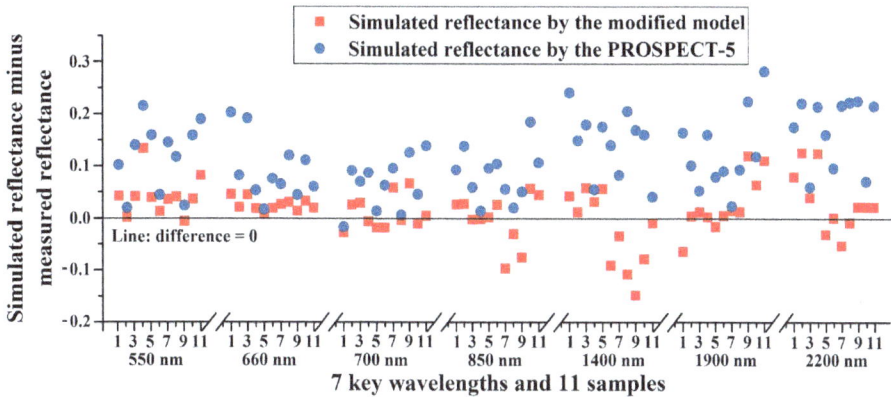

Figure 8. Differences between simulated reflectance and measured reflectance at key wavelengths.

Table 3. Root mean square errors (RMSEs) between the simulated spectra (PROSPECT-5 and the new model) and measured spectra of 11 groups samples used for validation.

Samples	No. 1	No. 2	No. 3	No. 4	No. 5	No. 6
PROSPECT-5	0.1567	0.1685	0.0810	0.0613	0.0988	0.0708
New model	0.0776	0.0576	0.0261	0.0225	0.0265	0.0486

Samples	No. 7	No. 8	No. 9	No. 10	No. 11	
PROSPECT-5	0.1489	0.1492	0.0781	0.1874	0.1924	
New model	0.0395	0.0942	0.0135	0.1038	0.0190	

Table 4. RMSEs between the simulated spectra (PROSPECT-5 and the new model) and measured spectra of 22 groups of samples used for calibration.

Samples	No. 1	No. 2	No. 3	No. 4	No. 5	No. 6
PROSPECT-5	0.1619	0.1943	0.1741	0.1444	0.1437	0.1489
New model	0.0943	0.1261	0.0496	0.0563	0.0491	0.0488

Samples	No. 7	No. 8	No. 9	No. 10	No. 11	No. 12
PROSPECT-5	0.1010	0.0433	0.0188	0.0791	0.0702	0.0377
New model	0.0312	0.0316	0.0173	0.0302	0.0210	0.0165

Samples	No. 13	No. 14	No. 15	No. 16	No. 17	No. 18
PROSPECT-5	0.0483	0.0543	0.1504	0.0254	0.1935	0.1214
New model	0.0175	0.0274	0.1043	0.0182	0.0949	0.0481

Samples	No. 19	No. 20	No. 21	No. 22		
PROSPECT-5	0.1328	01321	0.1347	0.1352		
New model	0.0438	0.0253	0.0112	0.0782		

5. Discussion

This paper advances the PROSPECT-5 model to simulate the reflectance of copper-stressed leaves and get accepted results of validation. There are some results needing further explanation and discussion.

The leaf structure parameters of copper-stressed leaves were estimated in the range from 400 to 510 nm, not from 800 to 1300 nm. Due to the low absorption of copper ion in this range, the estimation of the leaf structure parameters in this study avoided the influence from copper. The results of leaf structure parameters of copper-stressed leaves show three characteristics (Figure 5): they are generally larger than those of normal leaves; the distributed range is also larger; the leaf structure parameters tend to increase with stress levels. The most probable reason is that the structures of copper-stressed leaves were damaged due to excessive copper, and the cells were more disorderly with a higher level copper stress. This reason was also partially demonstrated by the SEM images (Figure 3). In addition, for estimating N in this study, we assumed the influence may come from the absorption related to copper, so we used a wavelength range where the absorption of copper ion is near zero, yet it might be that copper has other influences that we do not know about. Due to the indirect influence on reflectance from copper stress by changing the leaf structure, it is a good idea to establish a statistical relationship between N and copper content of a particular vegetation in the future. As there is a little information on the leaf structure parameter N considering monocotyledon and dicotyledon plants discrimination [11,15–21], this point may also deserve to be investigated in detail for copper-stressed vegetation in the future.

The simulated spectra from PROSPECT-5 are obviously higher than the measured spectra for copper-stressed leaves, while the simulated results from the new model are much nearer to the measured results (Figures 7 and 8). As previously mentioned, the copper content in normal leaf is relatively low (5 mg/kg to 30 mg/kg), but it is able to exceed 1300 mg/kg in a copper-stressed leaf. For normal leaves, the absorption by copper could be ignored and satisfying results could be acquired by PROSPECT-5. For copper-stressed leaves, excessive copper should be considered, and thus better results were acquired by the new model (Figures 7 and 8; Tables 3 and 4). Specifically, PROSPECT-5 was designed for healthy leaves, so it ignores the energy absorbed by copper-related matter, which leads to higher simulated reflectance for copper-stressed leaves. This is the reason why the simulated spectra from the modified model are much nearer to the measurements and the RMSEs with the modified model are less.

For the specific absorption coefficient related to copper (K_{cu}), Figure 6 suggests that it almost equals zero in the wavelength range of 400–510 nm, which is different from the results in [23]. However, the absorption spectra of aqueous copper sulfate solutions measured strictly in the laboratory [24], is consistent with the results in our study at 400–510 nm, which indicates that our results are more reasonable than that in [23].

The high values of K_{cu} at 590–710 nm, near 1400 nm, 1900 nm and 2400 nm, agree with the results in [23,30] well. For the explanation on the apparent characteristics near 1400 nm and 1900 nm, Zhu et al. [23] simply pointed that the coordinate bonds with hydroxide contribute to the observed features near 1400 nm and 1900 nm. Here, we try to give a more detailed explanation on all the observed features at 590–710 nm, 1400 nm, 1900 nm and 2400 nm. For copper-stressed vegetation, excessive copper in the soil was easily absorbed by plants via Cu^{2+} and $Cu(OH)^{+}$. In leaves, the Cu^{2+} and $Cu(OH)^{+}$ were transferred into chelate and fixed in organelle to hinder copper from diffusing and to protect other tissues in the plant [30,31]. According to the spectral theory, reflectance at 400–1300 nm is controlled by the electron transition of metal ions, while reflectance at 1300–2500 nm is determined by anionic group (e.g., hydroxyl, carbonate, sulfate) [32,33]. In this study, the high values of K_{cu} at 590–710 nm are similar with the absorption of copper sulfate and copper chloride [23,30], and the high values near 1400 nm and 1900 nm are similar with the absorption of water [22]. Hence, it can be inferred that Cu^{2+} contributes to the absorption at 590–710 nm, and hydroxyl contributes to the absorption near 1400 nm and 1900 nm. Sulfate ion shows absorption at 2400 nm [34]. The vegetation in this study was treated by copper sulfate ($CuSO_4$), so it was inferred that sulfate ions contributed to the high values of K_{cu} near 2400 nm. In addition, it appeared that the valley near 680 nm of K_{cu} could be observed. Vegetation stress always induces the enhancement of chlorophyll fluorescence [35–37]. According to the shape of chlorophyll fluorescence [35,38], it can be inferred that a high chlorophyll fluorescence (red fluorescence, RF) is also present near 680 nm. Thus, idealized K_{cu} needed to be reduced to enlarge the simulated reflectance to meet the enhancement from chlorophyll fluorescence. Therefore, RF is a very possible factor that makes contributions to this valley. Chlorophyll fluorescence also possesses high values called blue green fluorescence (BGF), ranging from approximately 350 nm to 520 nm [35,38], so some measured spectra are higher than simulated spectra generated by both PROSPECT-5 and the modified model in the visible range (Figure 7). In other words, this problem was also present in PROSPECT-5. K_{cu} is near zero in this wavelength range (Figure 6), so the enhancement on measured reflectance from BGF cannot be meted by reducing K_{cu}. Hence, chlorophyll fluorescence should be removed from the measured reflectance in the future to improve the ability to predict the spectra in the visible range. In total, K_{cu} reflects the overall influence from copper ion and its chelate, and other factors induced by copper, so we tentatively termed K_{cu} "specific absorption coefficient related to copper", not "specific absorption coefficient of copper ion". Moreover, copper stress impacts the leaf structure parameter N and we also changed the wavelength range for estimating N due to the copper presence. Hence, the modified model considers the physiological reaction of the leaf to copper presence, not just adding copper to the leaf spectra. In the future, what biochemical components that make contribution to the specific absorption coefficient related to copper (K_{cu}) and the mechanism of these absorptions should be further investigated and clarified. In addition, based the work in this paper, estimating the copper content in leaves and classifying the vegetation with different copper contamination using remote sensing could also be the basis of a good future study.

In terms of other advantages, this study would help the research on remote sensing of stressed vegetation. For example, similar changes in reflectance and vegetation indices may be resulted from copper stress as well as other stress [39–41], but this study provided some physical foundations to distinguish copper stress from other stress.

In this study, the soil before treatments was collected from vegetable garden without any contamination. In general, the copper content in this kind of soil is <2 mg/kg [42,43]. The increments of copper content in the soil after treatments are 25, 50, 100, . . . , 4800 mg/kg. Hence, the copper content in the soil at a minimum stress level exceeds that in the normal soil by 10 times. The initial copper content in soil almost has no impact. In fact, the use of the numbers, including 25, 50, 100, . . . , 4800, are simply labels to distinguish the different stress levels. In general, the vegetation can be regarded as copper-stressed vegetation if the growth and biochemical components are obviously influenced when excessive copper is present in the soil, water and the atmosphere [4,44,45]. Kabata-Pendias et al. [44] indicated that vegetation activities would be stressed when the copper content in soil amounted to two to 10 times that in normal soil. Hence, the vegetation could be called copper-stressed vegetation in this paper.

Compared with normal leaves, copper-stressed leaves have been verified that the leaf structure is disrupted and copper content significantly increases [4,45]. The data in this study also illustrates these facts to some degree. Moreover, the simulated reflectance generated by the measured contents of biochemical components (the measurements are regarded as correct) and the standard PROSPECT-5 model was away from the measured reflectance. Based on above facts, we set two hypothesis: (1) the disruption of leaf structure resulted in the increase of N; (2) the increase of copper content resulted in more absorption of energy. Thus, the modification strategy is estimating N in the wavelength range where copper absorption is near to zero, and estimating the specific absorption coefficient related to copper using correct N. This modification strategy is regarded as reasonable. In this sense, this approach also has some physical basis. In the future, others may propose more reasonable approaches to get similar output reflectance. However, the mentioned two hypothesis have not been absolutely verified since Figure 5c is not a good proof and it shows that the approach is not statistically robust. K_{cu} reflects the overall effects from copper stress, not only the absorption of copper ions. Moreover, the tolerance to copper stress varies with different vegetation. From this point of view, the approach in this study is empirical, and not a true physical modification approach. As previously mentioned, other modification approaches may provide the same total output reflectance with that in this study.

For example, if the contents of biochemical components were measured incorrectly, the obtained results in this study, in terms of the accuracy, may be obtained using standard PROSECT-5 model by changing the *N* value and the effective content of water in the leaf, and maybe of pigments as well.

There is a difference between this study and the studies on remote sensing of normal vegetation. The samples with copper-stressed vegetation are much more difficult to acquire due to experimental complexity. Therefore, the number of samples in many studies related to copper-stressed vegetation is similar to this study (e.g., [46,47]). Even so, we have also realized that more datasets should be used to improve the specific absorption coefficient related to copper (K_{cu}) to further advance the new model. Both the modified model and PROSPECT-5 are based on the leaf scale. LAI (Leaf Area Index) and view and illumination geometry are usually in canopy models, which were not considered in this study. The application of the modified model on remote sensing image needs the coupling with reflectance models at the canopy scale, for example, the SAIL model. The coupling with canopy reflectance model could provide an effective and rapid way for monitoring copper-stressed vegetation on a large scale using remote sensing image, which is also the basis for good future work.

6. Conclusions

This study developed a new model for copper-stress leaves based on PROSPECT-5, and validations were conducted. In this study, the reflectance at 400–510 nm was used to estimate leaf structure parameters (*N*) of copper-stress leaves. The absorption related to copper was considered, and the specific absorption coefficient related to copper (K_{cu}) was estimated. The new model includes six inputs: leaf structure parameters (*N*); chlorophyll content; carotenoid content; equivalent water thickness; dry matter content; and; copper content. Factors that influence the reflectance of copper-stressed leaves are considered as much as possible. The new model shows better performance than PROSPECT-5 on copper-stressed leaves. Hence, this study solved some problems left by previous studies and developed a better model for simulating the copper-stressed leaves, and provides theoretical support for the research on copper-stressed vegetation using remote sensing. Moreover, it has potential significance for prospecting copper deposit and monitoring environmental pollution caused by copper. However, due to the difficulties on the acquirement of datasets of copper-stressed vegetation, more samples should be acquired and used to improve the specific absorption coefficient related to copper (K_{cu}) and the accuracy of new model in the future.

Acknowledgments: This study was supported by the National Key Research and Development Program of China (2016YFD0300603-2), the National Natural Science Foundation of China and the Science and Technology Facilities Council of the United Kingdom (6151101278), the National Science and Technology Major Project of China (04-Y20A36-9001-15/17), and the China Scholarship Council (CSC). The authors would like to thank Feret at the IRSTEA, France, who proposed PROSPECT-4 and PROSPECT-5, for his valuable suggestions on this study. Thank Henan Normal University for the help on data collection. The authors also would like to thank the three anonymous reviewers for their valuable comments and suggestions.

Author Contributions: Chengye Zhang and Huazhong Ren proposed the idea and processed the data, and also participated in the writing of the paper; Yanzhen Liang analyzed the data and the results; Suhong Liu participated in the collection of the data and processed the data; Qiming Qin also analyzed the data and the results; Okan K. Ersoy contributed to the writing of the paper.

Conflicts of Interest: The authors declare no conflict of interest.

References

1. Lulla, K. Some observations on geobotanical remote sensing and mineral prospecting. *Can. J. Remote Sens.* **1985**, *11*, 17–38. [CrossRef]
2. Labovitz, M.L.; Masuoka, E.J. The influence of auto-correlation in signature extraction—An example from a geobotanical investigation of Cotter Basin, Montana. *Int. J. Remote Sens.* **1984**, *5*, 315–332. [CrossRef]
3. Hede, A.N.H.; Koike, K.; Kashiwaya, K.; Sakurai, S.; Yamada, R.; Singer, D.A. How can satellite imagery be used for mineral exploration in thick vegetation areas? *Geochem. Geophys. Geosyst.* **2017**, *18*, 584–596. [CrossRef]

4.   Wang, J.; Wang, T.; Shi, T.; Wu, G.; Skidmore, A.K. A Wavelet-based area parameter for indirectly estimating copper concentration in carex leaves from canopy reflectance. *Remote Sens.* **2015**, *7*, 15340–15360. [CrossRef]

5.   Kong, W.; Huang, W.; Zhou, X.; Song, X.; Casa, R. Estimation of carotenoid content at the canopy scale using the carotenoid triangle ratio index from in situ and simulated hyperspectral data. *J. Appl. Remote Sens.* **2016**, *10*. [CrossRef]

6.   Liu, S.; Liu, X.; Hou, J.; Chi, G.; Cui, B. Study on the spectral response of *Brassica Campestris* L. leaf to the copper pollution. *Sci. China Technol. Sci.* **2008**, *51*, 202–208. [CrossRef]

7.   Emengini, E.J.; Blackburn, G.A.; Theobald, J.C. Discrimination of plant stress caused by oil pollution and waterlogging using hyperspectral and thermal remote sensing. *J. Appl. Remote Sens.* **2013**, *7*. [CrossRef]

8.   Gitelson, A.A.; Gritz, Y.; Merzlyak, M.N. Relationships between leaf chlorophyll content and spectral reflectance and algorithms for non-destructive chlorophyll assessment in higher plant leaves. *J. Plant Physiol.* **2003**, *160*, 271–282. [CrossRef] [PubMed]

9.   Herrmann, I.; Berenstein, M.; Paz-Kagan, T.; Sade, A.; Karnieli, A. Spectral assessment of two-spotted spider mite damage levels in the leaves of greenhouse-grown pepper and bean. *Biosyst. Eng.* **2017**, *157*, 72–85. [CrossRef]

10. Sanchez, R.A.; Hall, A.J.; Trapani, N.; Dehunau, R.C. Effects of water-stress on the chlorophyll content, nitrogen level and photosynthesis of leaves of 2 maize genotypes. *Photosynth. Res.* **1983**, *4*, 35–47. [CrossRef] [PubMed]

11. Jacquemoud, S.; Baret, F. PROSPECT—A model of leaf optical-properties spectra. *Remote Sens. Environ.* **1990**, *34*, 75–91. [CrossRef]

12. Jacquemoud, S.; Verhoef, W.; Baret, F.; Bacour, C.; Zarco-Tejada, P.J.; Asner, G.P.; Francois, C.; Ustin, S.L. PROSPECT plus SAIL models: A review of use for vegetation characterization. *Remote Sens. Environ.* **2009**, *113*, S56–S66. [CrossRef]

13. Allen, W.A.; Gausman, H.W.; Richardson, A.J.; Thomas, J.R. Interaction of isotropic light with a compact plant leaf. *J. Opt. Soc. Am.* **1969**, *59*, 1376–1379. [CrossRef]

14. Allen, W.A.; Gausman, H.W.; Richardson, A.J. Mean effective optical constants of cotton leaves. *J. Opt. Soc. Am.* **1970**, *60*, 542–547. [CrossRef]

15. Fourty, T.; Baret, F.; Jacquemoud, S.; Schmuck, G.; Verdebout, J. Leaf optical properties with explicit description of its biochemical composition: Direct and inverse problems. *Remote Sens. Environ.* **1996**, *56*, 104–117. [CrossRef]

16. Jacquemoud, S.; Ustin, S.L.; Verdebout, J.; Schmuck, G.; Andreoli, G.; Hosgood, B. Estimating leaf biochemistry using the PROSPECT leaf optical properties model. *Remote Sens. Environ.* **1996**, *56*, 194–202. [CrossRef]

17. Baret, F.; Fourty, T. Estimation of leaf water content and specific leaf weight from reflectance and transmittance measurements. *Agronomie* **1997**, *17*, 455–464. [CrossRef]

18. Fourty, T.; Baret, F. On spectral estimates of fresh leaf biochemistry. *Int. J. Remote Sens.* **1998**, *19*, 1283–1297. [CrossRef]

19. Jacquemoud, S.; Bacour, C.; Poilve, H.; Frangi, J.P. Comparison of four radiative transfer models to simulate plant canopies reflectance: Direct and inverse mode. *Remote Sens. Environ.* **2000**, *74*, 471–481. [CrossRef]

20. Le Maire, G.; Francois, C.; Dufrene, E. Towards universal broad leaf chlorophyll indices using PROSPECT simulated database and hyperspectral reflectance measurements. *Remote Sens. Environ.* **2004**, *89*, 1–28. [CrossRef]

21. Feret, J.; Francois, C.; Asner, G.P.; Gitelson, A.A.; Martin, R.E.; Bidel, L.P.R.; Ustin, S.L.; le Maire, G.; Jacquemoud, S. PROSPECT-4 and 5: Advances in the leaf optical properties model separating photosynthetic pigments. *Remote Sens. Environ.* **2008**, *112*, 3030–3043. [CrossRef]

22. Bousquet, L.; Lacherade, S.; Jacquemoud, S.; Moya, I. Leaf BRDF measurements and model for specular and diffuse components differentiation. *Remote Sens. Environ.* **2005**, *98*, 201–211. [CrossRef]

23. Zhu, Y.; Qu, Y.; Liu, S.; Chen, S. A reflectance spectra model for copper-stressed leaves: Advances in the PROSPECT model through addition of the specific absorption coefficients of the copper ion. *Int. J. Remote Sens.* **2014**, *35*, 1356–1373. [CrossRef]

24. Jancso, G. Effect of D and O-18 isotope substitution on the absorption spectra of aqueous copper sulfate solutions. *Radiat. Phys. Chem.* **2005**, *74*, 168–171. [CrossRef]

25. Kubalova, I.; Ikeda, Y. Chlorophyll measurement as a quantitative method for the assessment of cytokinin-induced green foci formation in tissue culture. *J. Plant Growth Regul.* **2017**, *36*, 516–521. [CrossRef]

26. Arevalo-Gardini, E.; Arevalo-Hernandez, C.O.; Baligar, V.C.; He, Z.L. Heavy metal accumulation in leaves and beans of cacao (*Theobroma cacao* L.) in major cacao growing regions in Peru. *Sci. Total Environ.* **2017**, *605*, 792–800. [CrossRef] [PubMed]

27. Hosgood, B.; Jacquemoud, S.; Andreoli, G.; Verdebout, J.; Pedrini, G.; Schmuck, G. *Leaf Optical Properties Experiment 93 (LOPEX93)*; European Commission: Brussels, Belgium, 1994.
28. OPTICALEAF-Database. Available online: http://opticleaf.ipgp.fr/index.php?page=database (accessed on 10 February 2017).
29. Li, X.; Liu, X.; Liu, M.; Wang, C.; Xia, X. A hyperspectral index sensitive to subtle changes in the canopy chlorophyll content under arsenic stress. *Int. J. Appl. Earth Obs.* **2015**, *36*, 41–53. [CrossRef]
30. Zhou, C. Research on Retrieval Method of Heavy Metal Content of Vegetation Using Hyperspectral Remote Sensing. Ph.D. Thesis, Jilin University, Changchun, China, 2016.
31. Wu, W. *Plant Physiology*, 2nd ed.; Science Press: Beijing, China, 2008; ISBN 9787030224132.
32. Van der Meer, F.D.; van der Werff, H.M.A.; van Ruitenbeek, F.J.A.; Hecker, C.A.; Bakker, W.H.; Noomen, M.F.; van der Meijde, M.; Carranza, E.J.M.; de Smeth, J.B.; Woldai, T. Multi- and hyperspectral geologic remote sensing: A review. *Int. J. Appl. Earth Obs.* **2012**, *14*, 112–128. [CrossRef]
33. Zhang, C.; Qin, Q.; Chen, L.; Wang, N.; Zhao, S.; Hui, J. Rapid determination of coalbed methane exploration target region utilizing hyperspectral remote sensing. *Int. J. Coal Geol.* **2015**, *150*, 19–34. [CrossRef]
34. Bishop, J.L.; Parente, M.; Catling, D. Juventae Chasma as Potential MSL Landing Site. In Proceeding of the 2nd MSL Landing Site Workshop, Old Town Pasadena, CA, USA, 23–25 October 2007; Available online: https://marsoweb.nas.nasa.gov/landingsites/msl/workshops/2nd_workshop/talks/Bishop_Juventae.pdf (accessed on 13 November 2017).
35. Qu, Y.; Liu, S.; Li, X. A novel method for extracting leaf-level solar-induced fluorescence of typical crops under Cu stress. *Spectrosc. Spectr. Anal.* **2012**, *32*, 1282–1286.
36. Chou, S.; Chen, J.M.; Yu, H.; Chen, B.; Zhang, X.; Croft, H.; Khalid, S.; Li, M.; Shi, Q. Canopy-level photochemical reflectance index from hyperspectral remote sensing and leaf-level non-photochemical quenching as early indicators of water stress in maize. *Remote Sens.* **2017**, *9*, 794. [CrossRef]
37. Rascher, U.; Alonso, L.; Burkart, A.; Cilia, C.; Cogliati, S.; Colombo, R.; Damm, A.; Drusch, M.; Guanter, L.; Hanus, J.; et al. Sun-induced fluorescence—A new probe of photosynthesis: First maps from the imaging spectrometer HyPlant. *Glob. Chang. Biol.* **2015**, *21*, 4673–4684. [CrossRef] [PubMed]
38. Zarco-Tejada, P.J.; Miller, J.R.; Mohammed, G.H.; Noland, T.L. Chlorophyll fluorescence effects on vegetation apparent reflectance: I. Leaf-level measurements and model simulation. *Remote Sens. Environ.* **2000**, *74*, 582–595. [CrossRef]
39. Ghulam, A.; Fishman, J.; Maimaitiyiming, M.; Wilkins, J.L.; Maimaitijiang, M.; Welsh, J.; Bira, B.; Grzovic, M. Characterizing crop responses to background ozone in open-air agricultural field by using reflectance spectroscopy. *IEEE Geosci. Remote Sens.* **2015**, *12*, 1307–1311. [CrossRef]
40. Zhang, C.; Ren, H.; Qin, Q.; Ersoy, O.K. A new narrow band vegetation index for characterizing the degree of vegetation stress due to copper: The copper stress vegetation index (CSVI). *Remote Sens. Lett.* **2017**, *8*, 576–585. [CrossRef]
41. Yang, X.; Chen, L. Evaluation of automated urban surface water extraction from Sentinel-2A imagery using different water indices. *J. Appl. Remote Sens.* **2017**, *11*. [CrossRef]
42. Fan, Y.; Lu, Y.; Li, F.; Xue, J.; Li, X.; Zhong, B.; Peng, L. Study on distribution of available copper content in soil of navel orange orchards in southern Jiangxi Province. *J. Fruit Sci.* **2015**, *32*, 69–73. [CrossRef]
43. Li, Z.; Yi, W. Present status and improvement countermeasures of the soil fertility of vegetable fields in Guangzhou suburb. *Guangdong Agric. Sci.* **2008**, *2*, 43–46. [CrossRef]
44. Kabata-Pendias, A.; Pendias, H. *Trace Element in Soil and Plants*; CRC Press: Boca Raton, FL, USA, 1984.
45. Hall, J. Cellular mechanisms for heavy metal detoxification and tolerance. *J. Exp. Bot.* **2002**, *53*, 1–11. [CrossRef] [PubMed]
46. Asmaryan, S.; Warner, T.A.; Muradyan, V.; Nersisyan, G. Mapping tree stress associated with urban pollution using the WorldView-2 Red Edge band. *Remote Sens. Lett.* **2013**, *4*, 200–209. [CrossRef]
47. Rathod, P.H.; Brackhage, C.; van der Meer, F.D.; Mueller, I.; Noomen, M.F.; Rossiter, D.G.; Dudel, G.E. Spectral changes in the leaves of barley plant due to phytoremediation of metals—Results from a pot study. *Eur. J. Remote Sens.* **2005**, *48*, 283–302. [CrossRef]

remote sensing

MDPI

Article

Retrieval of Aerosol Optical Depth in the Arid or Semiarid Region of Northern Xinjiang, China

Xinpeng Tian [1], Sihai Liu [2,*], Lin Sun [3,*] and Qiang Liu [1,4]

[1] College of Global Change and Earth System Science, Beijing Normal University, Beijing 100875, China;
    tian_xp@mail.bnu.edu.cn (X.T.); toliuqiang@bnu.edu.cn (Q.L.)
[2] Satellite Environment Center, Ministry of Environmental Protection of China, Beijing 100094, China
[3] Geomatics College, Shandong University of Science and Technology, Qingdao 266590, China
[4] State Key Laboratory of Remote Sensing Science, Jointly Sponsored by Beijing Normal University and
    Institute of Remote Sensing and Digital Earth of Chinese Academy of Sciences, Beijing 100875, China
*   Correspondence: liusihan1200@163.com (S.L.); sunlin6@126.com (L.S.);
    Tel.: +86-010-5880-2190 (S.L.); +86-0532-8803-2922 (L.S.)

Received: 28 November 2017; Accepted: 26 January 2018; Published: 29 January 2018

Abstract: Satellite remote sensing has been widely used to retrieve aerosol optical depth (AOD), which is an indicator of air quality as well as radiative forcing. The dark target (DT) algorithm is applied to low reflectance areas, such as dense vegetation, and the deep blue (DB) algorithm is adopted for bright-reflecting regions. However, both DT and DB algorithms ignore the effect of surface bidirectional reflectance. This paper provides a method for AOD retrieval in arid or semiarid areas, in which the key points are the accurate estimation of surface reflectance and reasonable assumptions of the aerosol model. To reduce the uncertainty in surface reflectance, a minimum land surface reflectance database at the spatial resolution of 500 m for each month was constructed based on the moderate-resolution imaging spectroradiometer (MODIS) surface reflectance product. Furthermore, a bidirectional reflectance distribution function (BRDF) correction model was adopted to compensate for the effect of surface reflectance anisotropy. The aerosol parameters, including AOD, single scattering albedo, asymmetric factor, Ångström exponent and complex refractive index, are determined based on the observation of two sunphotometers installed in northern Xinjiang from July to August 2014. The AOD retrieved from the MODIS images was validated with ground-based measurements and the Terra-MODIS aerosol product (MOD04). The 500 m AOD retrieved from the MODIS showed high consistency with ground-based AOD measurements, with an average correlation coefficient of ~0.928, root mean square error (RMSE) of ~0.042, mean absolute error (MAE) of ~0.032, and the percentage falling within the expected error (EE) of the collocations is higher than that for the MOD04 DB product. The results demonstrate that the new AOD algorithm is more suitable to represent aerosol conditions over Xinjiang than the DB standard product.

Keywords: BRDF; aerosol; MODIS; sunphotometer; arid/semiarid

1. Introduction

Xinjiang province in northwest China is part of the Central Asian dust storm area, which is one of the main sources of dust aerosols [1]. In recent years, many Xinjiang cities, especially the capital, Urumqi, have suffered a severe deterioration in air quality with significant contributions from atmospheric particulates [2]. Aerosols can significantly influence the ecosystem, climate and hydrological cycle by affecting radiative forcing [3] and its relation with the air quality indicators proportionated for sustainable development [4]. High aerosol pollution events have a wide-ranging impact on visibility [5] and human health [6]. Scientific data about the spatial and temporal dynamic of dust aerosol in Xinjiang are needed by the local government to facilitate development of policies to protect the ecosystem and diminish dust storms. Due to the lack of ground stations in this vast area, satellite retrieval is the practical way to provide the spatial and temporal distribution of aerosol optical depth (AOD) [7–9].

Many satellite sensors have released AOD products, including the total ozone mapping spectrometer (TOMS) [10], geostationary operational environmental satellite (GOES) [11], ozone monitoring instrument (OMI) [12], medium resolution imaging spectroradiometer (MERIS) [13], advanced very high resolution radiometer (AVHRR) [14], multi-angle imaging spectroradiometer (MISR) [15], sea-viewing wide field-of-view sensor (SeaWiFS) [16], moderate-resolution imaging spectroradiometer (MODIS) [17], and visible infrared imaging radiometer suite (VIIRS) [18]. Yet, the quality of AOD products over arid/semiarid areas, such as Xinjiang, is relatively low due to a large bias in the surface reflectance estimation as well as the aerosol model used in the retrieval algorithms. The operational MODIS AOD product over land is based on two algorithms: the dark target (DT) and deep blue (DB) algorithms [19]. In the Xinjiang area, the MODIS AOD product is mostly retrieved with the DB algorithm which is applied over bright areas, where the surface reflectance is relatively high, and distinguishing atmospheric aerosol contributions from the satellite sensor energy is difficult.

In the DB algorithm, for arid and semiarid regions, the surface reflectance was determined based on a pre-calculated surface reflectance database, which was compiled based on the minimum reflectivity method at the resolution of $0.1° \times 0.1°$ for each season using MODIS images [20]. The derived surface reflectance database therefore depends on the scattering angle, normalized difference vegetation index (NDVI), and season. Most of the validation studies concluded that, in general, MODIS DB retrieved aerosol products were comparable to aerosol robotic network (AERONET) data, and an expected error (EE) envelope could be defined that contained approximately 50–70% of the matchups [21]. Bilal and Nichol reported up to 75–80% of the DB retrievals within the EE [22]. One of the major error sources for the DB algorithm is the difference between the surface reflectance corresponding to the images and that from the pre-calculated database, as a result of the anisotropic surface reflectance. Another crucial aspect in AOD retrieval is the aerosol model. In the DB algorithm, the microphysical and optical properties of aerosols are based on a cluster analysis of the global AERONET database through 2010 [17]. However, the aerosol characteristics vary locally, and they cannot be accurately described using a global aerosol model, which increases the uncertainty in AOD retrieval [23].

In this paper, a monthly minimum land surface reflectance (MLSR) database for Xinjiang area (band: blue; resolution: 500 m; time span: 2010–2014) was established using MODIS surface reflectance product (MOD09A1), based on the minimum synthesis technique (MST) method to reduce the impact of cloud contamination [20,24,25]. A bidirectional reflectance distribution function (BRDF) correction model for angle normalization was adopted to compensate the effect of surface reflectance anisotropy. In addition, the CE-318 sunphotometer data are used to determine the aerosol characteristics of the study area.

2. Field Measurements and Data Used

The CE-318 sunphotometer is an automatic ground-based radiometer measuring both direct solar irradiance and diffuse sky radiance for almucantar and principal solar planes with a 1.2° field of view. Although the channel wavelength configuration depends on the instrument version, filters at 440, 500, 675, 870, 940 and 1020 nm wavelengths are always present [26], with an uncertainty of 0.01–0.02 [27]. The Microtops II sunphotometer is a portable, manually-operated instrument which measures AOD through direct solar irradiance measurements in five wavebands (380, 500, 675, 870, and 1020 nm) with an uncertainty of ~0.015–0.02 [28,29].

In order to obtain the characteristic parameters of atmospheric aerosol in the study area, the CE-318 and Microtops II sunphotometers were used at two ground-observed sites (Dahuangshan and Wucaiwan site). The Dahuangshan site is located in a coal mining area with sparse vegetation coverage and the Microtops II sunphotometer observation was carried out from 10 July to 24 August

2014. The Wucaiwan site is located in a remote rural area near the Gurbantunggut Desert and the CE-318 sunphotometer observation was carried out from 15 July to 21 August 2014. A map of the study area is shown in Figure 1, and the detailed information of sites and instruments is presented in Table 1.

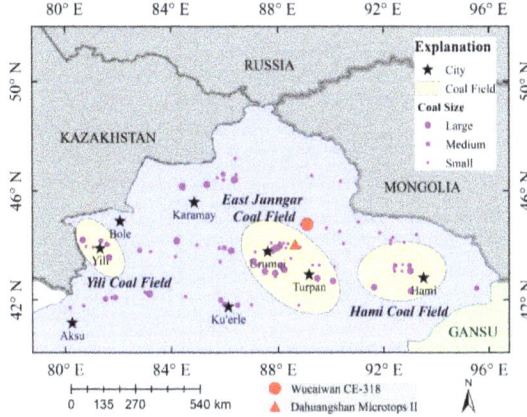

Figure 1. Map of the study area and the two ground-observed sites.

Table 1. The detailed information of sites and instruments.

Site	Instrument	Lon. (°E)	Lat. (°N)	Elevation (m)	Start Date	End Date
Dahuangshan	Microtops II	88.645	44.041	1018	2014/7/10	2014/8/24
Wucaiwan	CE-318	88.099	44.776	450	2014/7/15	2014/8/21

The MODIS (Terra) data products, namely calibrated radiance product (MOD02HKM), geolocation product (MOD03), surface reflectance product (MOD09A1), BRDF/albedo product (MCD43A1), and the Collection 6 (C6) aerosol product (MOD04), were obtained from the Level-1 and Atmosphere Archive & Distribution System (LAADS) Distributed Active Archive Center (DAAC) at the Goddard Space Flight Center (GSFC) (http://ladsweb.nascom.nasa.gov). The information of these satellite products and their applications is listed in Table 2.

Table 2. Moderate-resolution imaging spectroradiometer (MODIS) images used for aerosol optical depth (AOD) retrieval in this study.

Data Name	Date	Tile	N	Application
MOD09A1	2010–2014	H23-H25, V04-V05	240	Build surface reflectance and angle information database
MCD43A1	2014/07/10–2014/08/24	H23-H25, V04-V05	276	Build BRDF correction model
MOD02HKM	2014/07/10–2014/08/24		55	Calculate the top of the atmosphere reflectance
MOD03	2014/07/10–2014/08/24		55	Obtain geolocation data
MOD04	2014/07/10–2014/08/24		55	Validation

Note: N is the number of images.

3. Methodology

The top-of-atmosphere (TOA) reflectance (ρ_T), measured by a satellite sensor, is a function of surface and atmosphere optical parameters as well as solar/view zenith and azimuth angles; it can be estimated using Equation (1):

$$\rho_T(\tau_a, \theta_s, \theta_v, \varphi) = \rho_A(\tau_a, \theta_s, \theta_v, \varphi) + \rho_R(\theta_s, \theta_v, \varphi) + \frac{\rho_S}{1 - \rho_S \times S(\tau_a)} T(\tau_a, \theta_s) T(\tau_a, \theta_v) \tag{1}$$

where θ_s is the solar zenith angle, θ_v is the view zenith angle, φ is the relative azimuth angle, τ_a is the aerosol optical depth, ρ_A is the aerosol reflectance resulting from multiple scattering in the absence of molecules; ρ_R is the Rayleigh reflectance resulting from multiple scattering in the absence of aerosols; ρ_S is the surface reflectance; $T(\tau_a, \theta_s)$ and $T(\tau_a, \theta_v)$ are the transmissions of the atmosphere on the sun-surface path and the surface-sensor path, respectively; and $S(\tau_a)$ is the atmospheric backscattering ratio to account for multiple reflections between the surface and atmosphere.

Equation (1) indicates that the TOA reflectance comes from two parts: the pure atmospheric contribution (the first two terms in Equation (1)) and the combination of the atmosphere and land (the last term in Equation (1)). A key step of the satellite-retrieved AOD is to remove the surface contributions from the satellite image. An overview of our retrieval algorithm is represented as a dataflow diagram in Figure 2. This algorithm cannot retrieve cloud-contaminated pixels; therefore, before beginning the retrieval processing, we screened areas for the presence of clouds, and used the universal dynamic threshold cloud detection algorithm for cloud detection [30]. The surface reflectance for a given pixel was then determined from the pre-calculated MLSR and angle information database, which was built using MOD09A1 products. The aerosol model is an important parameter that affects the precision of AOD retrieval and must be considered carefully. In this study, it was established by the ground-based measurements data.

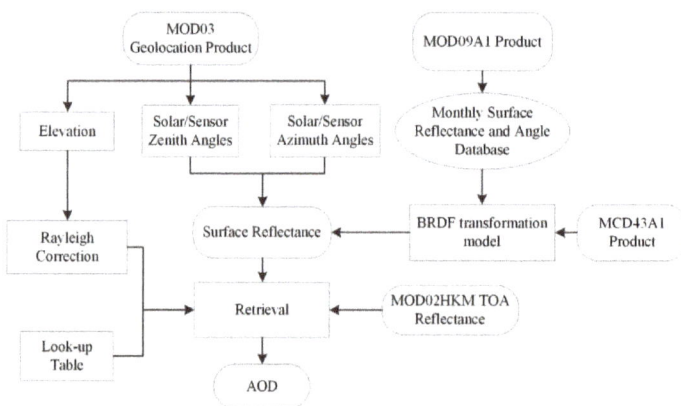

Figure 2. Flowchart of the aerosol retrieval algorithm in the study.

3.1. Construction of Surface Database

In order to remove the surface contributions from the satellite signal, a database of surface reflectance was built for the AOD retrieval. The MOD09A1 dataset provides surface reflectance at 500 m resolution in sinusoidal projection. Each MOD09A1 pixel has the best possible observation during an 8-day period as selected by high observation coverage, small view angle, absence of clouds or cloud shadow, and low aerosol loading. Validation has been carried out for the MOD09 which indicated that 50.52% of the observations in the blue band were within the MODIS theoretical uncertainty of $\pm(0.005 + 5\% \times \rho_S)$, where ρ_S is the surface reflectance [31]. In this study, MOD09A1 datasets for

5 years (2010–2014) and six tiles (H23-H25 and V04-V05) were used to build the surface reflectance database to support AOD retrieval.

Ideally, the MOD09A1 product should be the surface reflectance corrected for aerosol effect and cloud masked in the quality flag. However, detection of thin cloud or subpixel cloud are difficult, and there is also uncertainly in the aerosol effect correction [31]. Following the approach in the paper by Sun et al. [32], we adopted the criterion of minimum land surface reflectance (MLSR) for database synthesis because it is the most efficient way to avoid cloud contamination. In each 500 m pixel, the lowest surface reflectance value in the twenty images time series, i.e., four images per month for 5 successive years, was identified as the clearest observation. In order to correct the effect of surface bidirectional reflectance in AOD retrieval, the angle information in MOD09A1 was stored in the MLSR database along with the minimum reflectance. Figure 3 shows the distribution of surface reflectance in July, as well as the solar/view zenith and relative azimuth angles.

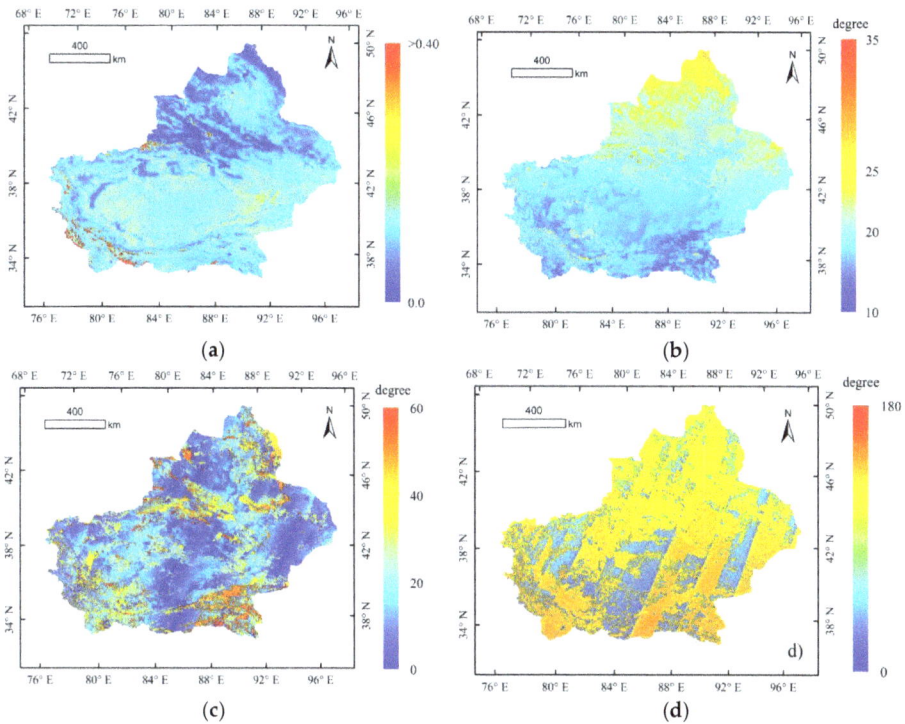

Figure 3. Example of a pre-calculated minimum land surface reflectance (MLSR) database using 5 years of MOD09A1 at 500 m resolution for July. (**a**): the surface reflectance at blue band; (**b**): the solar zenith angle; (**c**): the viewing zenith angle; (**d**): the relative azimuth angle.

3.2. BRDF Correction Surface Reflectance

In the aerosol retrieval algorithm, the accurate determination of the surface reflectance is one of the most crucial concerns. For the DB algorithm, the surface reflectance is prescribed by one of several methods, dependent on location, season, and land cover type, from a global surface reflectance database in visible bands. However, all methods do not consider the influence of imaging geometry. As the surface has anisotropic reflection characteristics and the geometric angles of the satellite sensors vary, the surface reflectance of the MLSR is different from that of images used to retrieve AOD. Thus, the pre-calculated MLSR database cannot be directly used for AOD retrieval. In this study, we adopted the kernel-driven BRDF model for the correction of the effects of anisotropic reflection of the surface. The correction model was proposed based on the kernel-driven BRDF model. The kernel-driven BRDF model is a semi-empirical models which is derived as a simplification of physically based BRDF models, with the merit of its linear form and small number of model parameters. It can be generally described by Equation (2) [33]:

$$R(\theta_s, \theta_v, \varphi, \lambda) = f_{iso}(\lambda) + f_{vol}(\lambda) \cdot K_{vol}(\theta_s, \theta_v, \varphi) + f_{geo}(\lambda) \cdot K_{geo}(\theta_s, \theta_v, \varphi) \tag{2}$$

where K_{vol} and K_{geo} are the volumetric and geometric kernels, respectively, which are functions of illumination and viewing geometry, describing volume and geometric scattering from surface elements f_{vol}, f_{geo} are the weights for the volumetric and geometric kernels, respectively, and f_{iso} is to the weight of the isotropic reflectance.

By simply normalizing the kernel weights with the isotropic weight, we define the anisotropy shape factors as $A_1 = f_{vol}/f_{iso}$ and $A_2 = f_{geo}/f_{iso}$. Then, the directional surface reflectance under geometric angle $\theta_{s1}, \theta_{v1}, \varphi_1$ can be predicted from the direction surface reflectance in the MLSR database under geometric angle $\theta_{s2}, \theta_{v2}, \varphi_2$, with Equation (3):

$$R_1(\theta_{s1}, \theta_{v1}, \varphi_1, \lambda) = \frac{1 + A_1 \cdot K_{vol}(\theta_{s1}, \theta_{v1}, \varphi_1) + A_2 \cdot K_{geo}(\theta_{s1}, \theta_{v1}, \varphi_1)}{1 + A_1 \cdot K_{vol}(\theta_{s2}, \theta_{v2}, \varphi_2) + A_2 \cdot K_{geo}(\theta_{s2}, \theta_{v2}, \varphi_2)} R_2(\theta_{s2}, \theta_{v2}, \varphi_2, \lambda) \tag{3}$$

Studies show that the combination of Ross–Thick and Li-SparseR (RTLSR) kernels works well with the observed data [34], and the f_{iso}, f_{vol}, and f_{geo} parameters of this combination are provided in the MCD43A1 products. However, as inversion of the BRDF is difficult, the BRDF parameters in MCD43A1 are noisy and cannot be directly used. Vermote et al. showed that the anisotropy of the surface is related to vegetation status [35]. So, in this study, we classify the study area into three categories, i.e., sparse vegetation, median vegetation and dense vegetation, according to the threshold of the Enhanced Vegetation Index (EVI). Statistical mean values of A_1 and A_2 were calculated for each category from pixels marked as best quality (QC = 0) in the MCD43A1 product. The mean values of A_1 and A_2 were then applied for all pixels in the category. The statistical mean values of A_1 and A_2 are shown in Table 3.

Table 3. The statistical mean values of A_1 and A_2 for three vegetation statuses.

Category	EVI	A_1	A_2
Sparse vegetation	EVI < 0.15	0.203	0.037
Median vegetation	0.15 < EVI < 0.60	0.438	0.173
Dense vegetation	EVI > 0.60	0.762	0.143

3.3. Aerosol Parameter Determination

To reduce the computation requirement, a look-up table (LUT) was constructed using the latest version (Version 2.1 Vector Code) of the 6S (second simulation of the satellite signal in the solar spectrum) (6SV) radiative transfer model (RTM) [36]. The 6SV code is a widely-used radiative transfer code that simulates the satellite signal accounting for elevated targets. One of the practical advantages of 6SV is that it provides standard atmosphere and aerosol models. The 6SV model was used to construct a LUT of atmosphere optical parameters according to the parameters values listed in Table 4.

Table 4. The parameters used in the look-up table construction.

Parameter	Number	Values
Band	1	Band 3 (Blue band)
AOD at 550 nm	15	0.0, 0.01, 0.05, 0.1, 0.2, 0.3, 0.4, 0.5, 0.6, 0.7, 0.8, 0.9, 1.0, 1.2, 1.5
Surface reflectance	18	0.0, 0.01, 0.02, 0.03, 0.04, 0.05, 0.06, 0.07, 0.08, 0.09, 0.10, 0.11, 0.12, 0.13, 0.14, 0.15, 0.18, 0.20
Solar zeniths (°)	14	0, 6, 12, 18, 24, 30, 36, 42, 48, 54, 60, 66, 72, 78
Satellite zeniths (°)	14	0, 6, 12, 18, 24, 30, 36, 42, 48, 54, 60, 66, 72, 78
Relative azimuths (°)	19	0, 10, 20, 30, 40, 50, 60, 70, 80, 90, 100, 110, 120, 130, 140, 150, 160, 170, 180

In this work, the CE-318 ground-based data were used to determine the aerosol microphysical and optical parameters, such as the Ångström exponent (AE), complex refractive index (RI), single scattering albedo (SSA), and asymmetric factor (g). The AE can be used to determine aerosol models; for example, the AE of desert aerosols is −1.0 to 0.5 [37]. For the desert model, the value of the real part of RI is 1.5–1.6 and the value of the imaginary part is approximately 0.01. Table 5 shows the values of the AE and RI in the study. The aerosol optical parameters, including the SSA and g at four wavelengths (i.e., 440, 675, 870, and 1020 nm; Figure 4), were retrieved from sky radiance almucantar measurements and direct sun measurements by a CE-318 sunphotometer. In this study, the first step is to calculate the average values of SSA and g at 440 and 675 nm, and then the inverse distance weighting interpolation [38] is adopted to obtain the values of SSA and g at 550 nm, which is 0.827 and 0.910, respectively.

Table 5. Calculation results of aerosol parameters of the Ångström exponent and complex refractive index.

Aerosol Optical Properties	Date/Value			
	15 July	3 August	4 August	21 August
Ångström exponent (440–870 nm)	0.467	0.409	0.441	0.443
Real part of RI	1.597	1.541	1.548	1.592
Imaginary part of RI	0.009	0.008	0.011	0.010

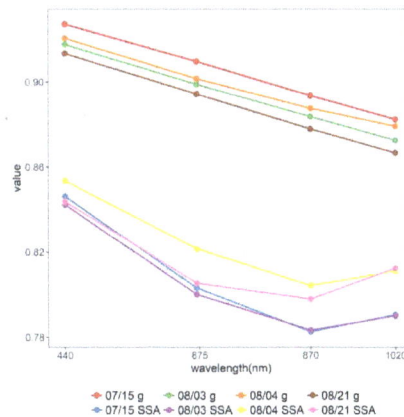

Figure 4. Time series of aerosol parameters of single scattering albedo (SSA) and g retrieved from sky radiance almucantar measurements and direct sun measurements. 15 July, 3 August, 4 August, and 21 August are the start date, the stable weather data, the maximum AOD date, and the end date, respectively.

3.4. Rayleigh Correction for Elevation Effect

Within the MODIS blue wavelength range, where Rayleigh scattering is relatively important compared to longer wavelengths, neglect of polarization in the radiative transfer code leads to significant errors in the calculated reflectances [39]. At sea level, the Rayleigh optical thickness (ROT) at visible channels can be estimated with an empirical function of wavelength λ [40]:

$$\tau_R(\lambda, Z = 0) = -0.00877 \times \lambda^{-4.05} \tag{4}$$

At a height above sea level, the ROT should be modulated by atmospheric pressure or elevation [41]:

$$\tau_R(\lambda, Z = z) = \tau_R(\lambda, Z = 0) \exp\left(\frac{-z}{8.5}\right) \tag{5}$$

where z is the ground height above sea level in kilometers, and 8.5 is the exponential scale height of the atmosphere. In this study, the MOD03 product was used for providing the altitude z for each pixel.

3.5. Error Indicators

The error statistics of the algorithm were verified by comparing the satellite-retrieved AOD with ground-based measurements from sunphotometers. We used the following four indicators to evaluate the error: the correlation coefficient (R), mean absolute error (MAE), root mean square error ($RMSE$), and EE. R is an indicator of relative agreement between satellite-retrieved AOD and ground-observed AOD; the MAE is the most natural measure of mean error magnitude; the $RMSE$ is used to measure the systematic and random differences between these two AOD observations; and the EE, representing the confidence envelopes of the retrieval algorithm, is used to evaluate the quality of a new algorithm relative to MODIS C6 AOD. Good matches of satellite-retrieved AOD are reported when the satellite-retrieved AOD falls within the envelope. The statistical indicators are defined as follows:

$$R = \frac{\sum\limits_{i=1}^{n} (x_i - \bar{x})(y_i - \bar{y})}{\sqrt{\sum\limits_{i=1}^{n} (x_i - \bar{x})^2 \sum\limits_{i=1}^{n} (y_i - \bar{y})^2}} \tag{6}$$

$$RMB = \bar{x}_i / \bar{y}_i \tag{7}$$

$$RMSE = \sqrt{\frac{1}{n}\sum\limits_{i=1}^{n} (x_i - y_i)^2} \tag{8}$$

$$MAE = \frac{1}{n}\sum\limits_{i=1}^{n} |x_i - y_i| \tag{9}$$

$$EE = \pm(0.05 + 0.15x_i) \tag{10}$$

where x_i is the ground-based sunphotometer measurement of AOD, y_i is the satellite-retrieved AOD; $\bar{x}$ and $\bar{y}$ are average values of x_i and y_i.

4. Results and Discussion

4.1. Spatial Distribution of AOD

In this study, 55 MOD02HKM images from July and August 2014 were selected to retrieve AOD. Figure 5 shows the MODIS standard false-color images (R, G, B: 2, 1, 4) in the northern Xinjiang area for 11 July, 15 July, 12 August and 21 August 2014. The landscapes of this area includes the Altai Mountains, Tianshan Mountains, and Gurbantunggut Desert, which are located in the Zhungeer basin and exhibit high surface reflectance in the blue band (Figure 3a). Figure 6 shows the spatial distribution of AOD retrieved from MODIS at a 500-m resolution, corresponding to the images in Figure 5. The MODIS C6 DB AOD products (10 km) are shown in Figure 7. The satellite-retrieved AOD from OLI data is more suitable than C6 DB AOD to represent the spatial pattern of aerosols over bright-reflecting source regions of northern Xinjiang. The revealed details in the aerosol distribution and variability are valuable in the study of transient aerosols. These results demonstrate that the new algorithm can achieve a continuous AOD distribution even in the bare land or desert areas, which have a high reflectance.

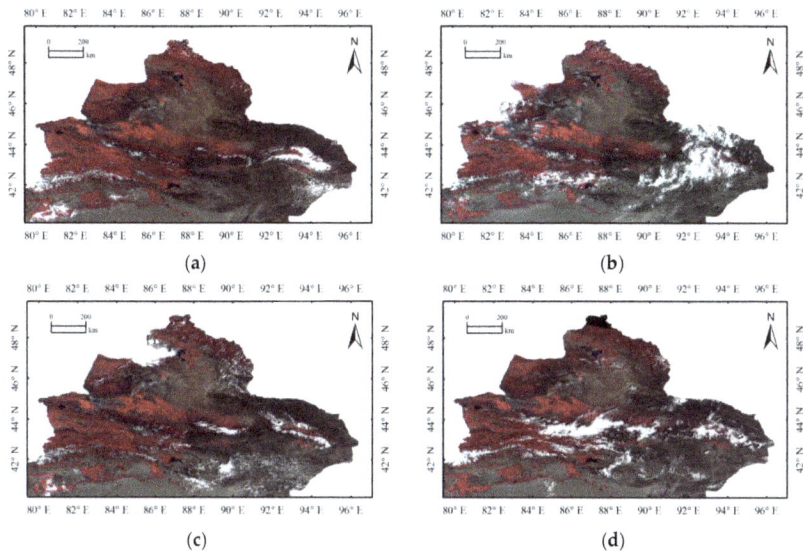

Figure 5. MODIS false-color images for the northern Xinjiang area (R, G, B = 2, 1, 4). (**a**): 11 July 2014; (**b**): 15 July 2014; (**c**): 12 August 2014; (**d**): 21 August 2014.

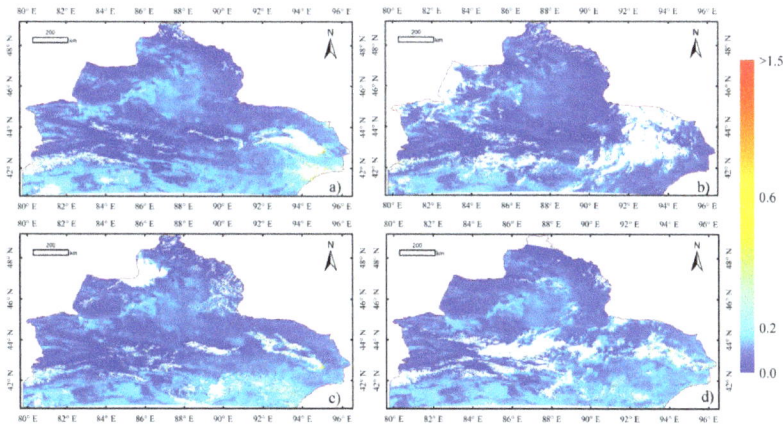

Figure 6. Retrieved AOD for the northern Xinjiang area. (**a**): 11 July 2014; (**b**): 15 July 2014; (**c**): 12 August 2014; (**d**): 21 August 2014.

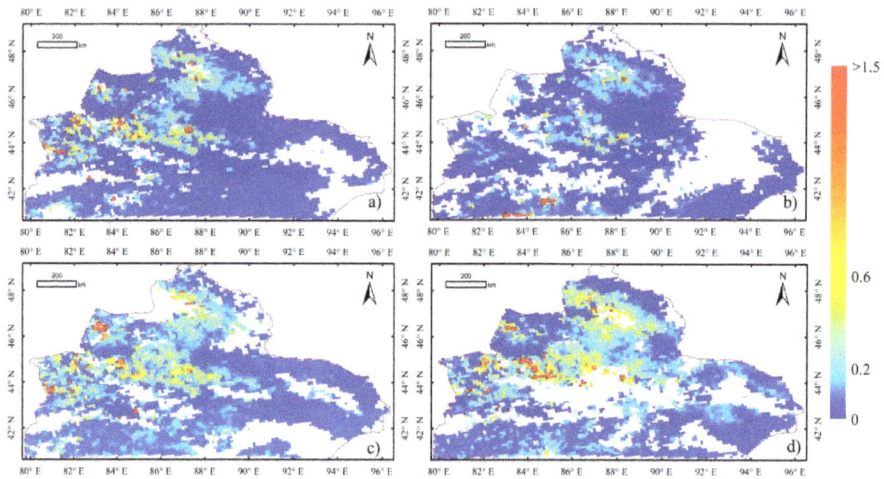

Figure 7. Distribution of MODIS 10-km deep blue (DB) AOD products for the northern Xinjiang area. (**a**): 11 July 2014; (**b**): 15 July 2014; (**c**): 12 August 2014; (**d**): 21 August 2014.

4.2. Validation

The satellite-retrieved AOD were validated at the two ground sites. As satellite-retrieved AOD retrievals are at 550 nm, the ground observations are not available at this wavelength, data are interpolated to 550 nm using the Ångström function [42], defined as

$$\tau_a(\lambda) = \beta \times \lambda^{-\alpha} \tag{11}$$

where $\tau_a(\lambda)$ is the AOD at wavelength λ, β is the turbidity factor, and α is the band index. β and α can be estimated from sunphotometer observations of AOD at two wavelengths λ_1 and λ_2, with the following expression:

$$\alpha = -\frac{\ln(\tau_a(\lambda_1)/\tau_a(\lambda_2))}{\ln(\lambda_1/\lambda_2)}, \ \beta = \frac{\tau_a(\lambda_1)}{\lambda_1^{-\alpha}} \tag{12}$$

Referring to the results of previous studies [43,44], the nearest available pair of wavelengths from CE-318 and Microtops II sunphotometer (normally 675 nm and either 440 or 500 nm) are used.

In this study, to match the instantaneous AOD value provided by satellites with the repeated measurements observed by sunphotometers, we followed the matchup methodology of Ichoku et al. [45]. The ground-observed data averaged within 30 min of the MODIS overpass are extracted and compared with MODIS AOD data averaged within 1.5 km (3 × 3 pixels) surrounding of the ground site. A total of 32 and 26 of the satellite-retrieved AOD observations coincided with Dahuangshan and Wucaiwan AOD measurements, as shown in Table 6. The AOD from the new algorithm achieved high correlation (~0.918–0.928) with low absolute error (~0.025–0.037), relative error (~13.9–16.7%) and the percentage falling within the EE of the collocations is ~96.9%, 96.2% at the Dahuangshan and Wucaiwan site, respectively. The total average absolute error and relative error were ~0.036 and ~16.6%, respectively, and with ~96.6% collocations falling within the EE envelope. These results indicate that the new algorithm could retrieve AOD with high accuracy and stability.

Table 6. Comparisons of retrieval accuracy between the new algorithm (New) and Terra-MODIS C6 DB product (DB).

Site	Count		R		Absolute Error		Relative Error (%)		r (%)	
	New	DB	New	DB	New	DB	New	DB	New	DB
Dahuangshan	32	29	0.928	0.774	0.037	0.047	13.9	17.4	96.9	90.6
Wucaiwan	26	24	0.918	0.931	0.025	0.023	16.7	15.7	96.2	96.2
Total	58	53	0.928	0.871	0.032	0.036	15.1	16.6	96.6	92.5

r is the percentage falling within the expected error (EE) of the collocations.

Figure 8 shows the scatter plots of the satellite-retrieved AOD against those obtained from the ground-based sunphotometer during the study period. Figure 8a is the results of the proposed algorithm. Figure 8b is the scatter plots of MODIS DB AOD at 10 km resolution against ground-based measurements. Figure 8c is the result of retrieved AOD without angle normalization, which used the same pre-calculated MLSR database as Figure 8a. It is evident that the new algorithm retrievals have the highest percentage within EE (~96.6%), highest correlations with AERONET AOD measurements (R = ~0.928) and smallest RMSE (~0.042) and MAE (~0.032). The AOD from MODIS DB was also highly correlated (R = 0.871), and the RMSE was 0.050 and MAE was 0.036. Nearly 92.5% of the collections fell within the EE envelope. Notably, the retrieved AOD values from new algorithm were close to the 1:1 line (slope = ~0.795, intercept = ~0.037), whereas the DB AODs were far from the line (slope = ~0.747, intercept = ~0.041). The retrieved AOD form without angle normalization has relatively low accuracy, with R = ~0.906, RMSE = ~0.055, and MAE = ~0.041. Only nearly 89.6% of the collocations fell within the EE envelope. This implies that angle normalization can improve the retrieval accuracy and is necessary for the AOD retrieval.

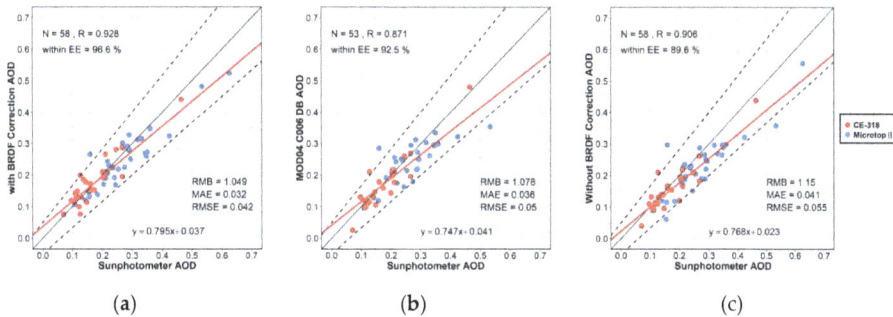

Figure 8. Validation of retrieved AOD from the (**a**) new algorithm (500 m), (**b**) MOD04 C6 DB algorithm (10 km), and (**c**) the new algorithm without angle normalization (500 m) against the ground-based sunphotometer AOD measurements. The black dashed lines are the EE lines, the black solid lines are the 1:1 line and red solid lines are the regression lines.

4.3. Uncertainty Analysis

Errors in AOD retrieval are attributed to several factors, including surface reflectance estimation and the aerosol model [46]. In this study, we assumed the change in surface reflectance within a few days to be negligible and thus determined the surface reflectance from the MLSR database and BRDF correction model. Changes in land cover doubtlessly lead to some errors in determining the true reflectance. Studies by Kaufman et al. show that errors of about 0.01 in assumed surface reflectance lead to errors on the order of 0.1 in AOD retrieval [47]. The land cover in Xinjiang has a strong seasonal cycle; most of the area is covered by sparse and dry vegetation between March and October, and snow is frequent in winter [48]. Figure 9a,b shows the variations in surface reflectance in the blue band around the two sites from MODIS data during the same month in different years. As shown in

Figure 9, the yearly variation in the same month was small and mainly in a range of 0.01, except for winter. The surface reflectance exhibited a stable variation during April to October, with an absolute difference value of ±0.01. However, the surface reflectance decreased by almost 0.35 from February to March; and increased by 0.40 from October to November. Such a variation is related to the snow melting and accumulation in March and November, respectively. In addition, due to the differences in snowfall in different years, the change in surface reflectance in winter is more obvious than in other seasons. Thus, it is not appropriate to determine surface reflectance through the MLSR database from November to February.

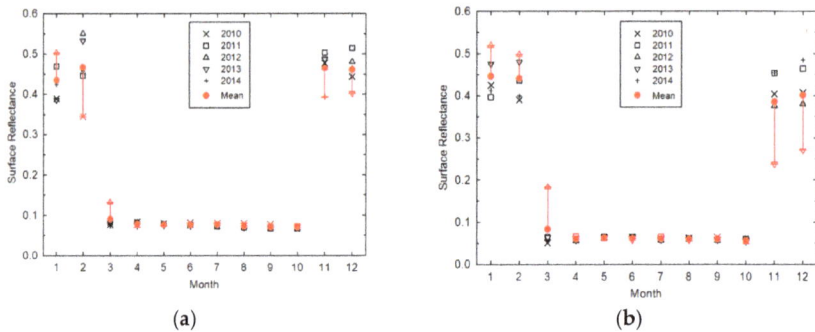

(a)   (b)

Figure 9. The variation of surface reflectance during the same month in different years near the ground sites. (**a**) Dahuangshan; (**b**) Wucaiwan. The error bars represent the maximum absolute error compared with the average value for 5 years.

5. Conclusions

In this study, a modified aerosol retrieval algorithm was proposed for retrieving AOD over the arid/semiarid region of northern Xinjiang from MODIS data at 500 m spatial resolution. The assumptions in this algorithm are as follows: the variation in surface reflectance is small over a month, and single scattering albedo (SSA) and asymmetry factors (g) are regionally constant for a few days. The MODIS surface reflectance products (MOD09A1) were used to determine the surface reflectance, and a LUT was constructed based on the 6SV RTM, which uses SSA and g values from ground-based measurements. The retrieved AOD values were validated by ground-based sunphotometer observations in two sites and compared with the MODIS DB AOD products. The results show that the new algorithm accurately retrieved aerosol AOD over the arid/semiarid region of northern Xinjiang, and the retrieved aerosol distribution contained more spatial details and variability than that of the DB AOD products. In this study, a BRDF correction model was applied to reduce the effect of surface reflectance anisotropy in AOD retrieval; thus, the retrieval ability is improved especially over the area with obvious surface bidirectional reflectance characteristics. However, according to the uncertainty analysis, the proposed algorithm has some limitations that should be improved: (1) Errors may occur by using constant values of SSA and g for the day of retrieval; (2) the MLSR database approach was sometimes found to be unsuccessful over snow surfaces, particularly when seasonal changes are significant, such as the snow melting in March and accumulation in November, respectively; (3) the BRDF correction only considered the mean values of anisotropy shape factors; thus, it ignored the variation of BRDF shape within the categories of land surface. These limitations will be explored in our future studies.

Acknowledgments: The authors acknowledge the NASA Goddard Space Flight Center for providing the MODIS data. This study was supported by the National Natural Science Foundation of China (41371356), the Fundamental Research Funds for the Central Universities (312231103) and the State's Key Project of Research and Development Plan of China (2016YFA0600103).

Author Contributions: Xinpeng Tian and Lin Sun performed the new AOD algorithm research and prepared the paper. Sihai Liu helped with the ground-based data and satellite remote sensing data collection and processing. Qiang Liu provided advice and suggestions.

Conflicts of Interest: The authors declare no conflict of interest.

References

1. Zhang, X.X.; Sharratt, B.; Chen, X.; Wang, Z.F.; Liu, L.Y.; Guo, Y.H.; Li, J.; Chen, H.S.; Yang, W.Y. Dust deposition and ambient PM 10 concentration in Northwest China: Spatial and temporal variability. *Atmos. Chem. Phys.* **2017**, *17*, 1699–1711. [CrossRef]
2. Mamtimin, B.; Meixner, F.X. Air pollution and meteorological processes in the growing dryland city of urumqi (Xinjiang, China). *Sci. Total Environ.* **2011**, *409*, 1277–1290. [CrossRef] [PubMed]
3. Ramanathan, V.; Crutzen, P.J.; Kiehl, J.T.; Rosenfeld, D. Atmosphere—Aerosols, climate, and the hydrological cycle. *Science* **2001**, *294*, 2119–2124. [CrossRef] [PubMed]
4. Klemm, R.J.; Mason, R.M. Aerosol research and inhalation epidemiological study (ARIES): Air quality and daily mortality statistical modeling—Interim results. *J. Air Waste Manag.* **2000**, *50*, 1433–1439. [CrossRef] [PubMed]
5. Cheung, H.C.; Wang, T.; Baumann, K.; Guo, H. Influence of regional pollution outflow on the concentrations of fine particulate matter and visibility in the coastal area of southern china. *Atmos. Environ.* **2005**, *39*, 6463–6474. [CrossRef]
6. Pope, C.A.; Ezzati, M.; Dockery, D.W. Fine particulate air pollution and life expectancies in the united states: The role of influential observations. *J. Air Waste Manag.* **2013**, *63*, 129–132. [CrossRef]
7. Mao, J.D.; Sheng, H.J.; Zhao, H.; Zhou, C.Y. Observation study on the size distribution of sand dust aerosol particles over Yinchuan, China. *Adv. Meteorol.* **2014**, *2014*, 157645. [CrossRef]
8. Xin, J.Y.; Du, W.P.; Wang, Y.S.; Gao, Q.X.; Li, Z.Q.; Wang, M.X. Aerosol optical properties affected by a strong dust storm over central and northern china. *Adv. Atmos. Sci.* **2010**, *27*, 562–574. [CrossRef]
9. Sun, L.; Li, R.B.; Tian, X.P.; Wei, J. Analysis of the temporal and spatial variation of aerosols in the Beijing-Tianjin-Hebei region with a 1 km aod product. *Aerosol. Air Qual. Res.* **2017**, *17*, 923–935. [CrossRef]
10. Torres, O.; Bhartia, P.K.; Herman, J.R.; Sinyuk, A.; Ginoux, P.; Holben, B. A long-term record of aerosol optical depth from toms observations and comparison to aeronet measurements. *J. Atmos. Sci.* **2002**, *59*, 398–413. [CrossRef]
11. Prados, A.I.; Kondragunta, S.; Ciren, P.; Knapp, K.R. Goes aerosol/smoke product (GASP) over north america: Comparisons to aeronet and modis observations. *J. Geophys. Res. Atmos.* **2007**, *112*. [CrossRef]
12. Torres, O.; Tanskanen, A.; Veihelmann, B.; Ahn, C.; Braak, R.; Bhartia, P.K.; Veefkind, P.; Levelt, P. Aerosols and surface UV products from ozone monitoring instrument observations: An overview. *J. Geophys. Res. Atmos.* **2007**, *112*. [CrossRef]
13. Vidot, J.; Santer, R.; Aznay, O. Evaluation of the meris aerosol product over land with aeronet. *Atmos. Chem. Phys.* **2008**, *8*, 7603–7617. [CrossRef]
14. Riffler, M.; Popp, C.; Hauser, A.; Fontana, F.; Wunderle, S. Validation of a modified AVHRR aerosol optical depth retrieval algorithm over central Europe. *Atmos. Meas. Tech.* **2010**, *3*, 1255–1270. [CrossRef]
15. Kahn, R.A.; Gaitley, B.J.; Garay, M.J.; Diner, D.J.; Eck, T.F.; Smirnov, A.; Holben, B.N. Multiangle imaging spectroradiometer global aerosol product assessment by comparison with the aerosol robotic network. *J. Geophys. Res. Atmos.* **2010**, *115*. [CrossRef]
16. Sayer, A.M.; Hsu, N.C.; Bettenhausen, C.; Jeong, M.J.; Holben, B.N.; Zhang, J. Global and regional evaluation of over-land spectral aerosol optical depth retrievals from seawifs. *Atmos. Meas. Tech.* **2012**, *5*, 1761–1778. [CrossRef]
17. Levy, R.C.; Mattoo, S.; Munchak, L.A.; Remer, L.A.; Sayer, A.M.; Patadia, F.; Hsu, N.C. The collection 6 modis aerosol products over land and ocean. *Atmos. Meas. Tech.* **2013**, *6*, 2989–3034. [CrossRef]
18. Jackson, J.M.; Liu, H.Q.; Laszlo, I.; Kondragunta, S.; Remer, L.A.; Huang, J.F.; Huang, H.C. Suomi-NPP VIIRS aerosol algorithms and data products. *J. Geophys. Res. Atmos.* **2013**, *118*, 12673–12689. [CrossRef]
19. Remer, L.A.; Mattoo, S.; Levy, R.C.; Munchak, L.A. Modis 3 km aerosol product: Algorithm and global perspective. *Atmos. Meas. Tech.* **2013**, *6*, 1829–1844. [CrossRef]

20. Hsu, N.C.; Tsay, S.C.; King, M.D.; Herman, J.R. Aerosol properties over bright-reflecting source regions. *IEEE Trans. Geosci. Remote Sens.* **2004**, *42*, 557–569. [CrossRef]

21. Wei, J.; Sun, L. Comparison and evaluation of different modis aerosol optical depth products over the Beijing-Tianjin-Hebei region in china. *IEEE J. Sel. Top. Appl. Earth Obs. Remote Sens.* **2017**, *10*, 835–844. [CrossRef]

22. Bilal, M.; Nichol, J.E. Evaluation of modis aerosol retrieval algorithms over the Beijing-Tianjin-Hebei region during low to very high pollution events. *J. Geophys. Res. Atmos.* **2015**, *120*, 7941–7957. [CrossRef]

23. Tao, M.H.; Wang, Z.F.; Tao, J.H.; Chen, L.; Wang, J.; Hou, C.; Wang, L.C.; Xu, X.G.; Zhu, H. How do aerosol properties affect the temporal variation of MODIS AOD bias in Eastern China? *Remote Sens. (Basel)* **2017**, *9*, 800. [CrossRef]

24. Koelemeijer, R.B.A.; de Haan, J.F.; Stammes, P. A database of spectral surface reflectivity in the range 335-772 nm derived from 5.5 years of GOME observations. *J. Geophys. Res. Atmos.* **2003**, *108*. [CrossRef]

25. Herman, J.R.; Celarier, E.A. Earth surface reflectivity climatology at 340–380 nm from toms data. *J. Geophys. Res. Atmos.* **1997**, *102*, 28003–28011. [CrossRef]

26. Holben, B.N.; Eck, T.F.; Slutsker, I.; Tanre, D.; Buis, J.P.; Setzer, A.; Vermote, E.; Reagan, J.A.; Kaufman, Y.J.; Nakajima, T.; et al. Aeronet—A federated instrument network and data archive for aerosol characterization. *Remote Sens. Environ.* **1998**, *66*, 1–16. [CrossRef]

27. Eck, T.F.; Holben, B.N.; Reid, J.S.; Dubovik, O.; Smirnov, A.; O'Neill, N.T.; Slutsker, I.; Kinne, S. Wavelength dependence of the optical depth of biomass burning, urban, and desert dust aerosols. *J. Geophys. Res. Atmos.* **1999**, *104*, 31333–31349. [CrossRef]

28. Knobelspiesse, K.D.; Pietras, C.; Fargion, G.S.; Wang, M.H.; Frouin, R.; Miller, M.A.; Subramaniam, A.; Balch, W.M. Maritime aerosol optical thickness measured by handheld sun photometers. *Remote Sens. Environ.* **2004**, *93*, 87–106. [CrossRef]

29. Morys, M.; Mims, F.M.; Hagerup, S.; Anderson, S.E.; Baker, A.; Kia, J.; Walkup, T. Design, calibration, and performance of microtops II handheld ozone monitor and sun photometer. *J. Geophys. Res. Atmos.* **2001**, *106*, 14573–14582. [CrossRef]

30. Sun, L.; Wei, J.; Wang, J.; Mi, X.T.; Guo, Y.M.; Lv, Y.; Yang, Y.K.; Gan, P.; Zhou, X.Y.; Jia, C.; et al. A universal dynamic threshold cloud detection algorithm (UDTCDA) supported by a prior surface reflectance database. *J. Geophys. Res. Atmos.* **2016**, *121*, 7172–7196. [CrossRef]

31. Vermote, E.; Kotchenova, S. Mod09 User's Guide (J/OL). Available online: http://modis-sr.ltdri.org (accessed on 20 November 2017).

32. Sun, L.; Wei, J.; Bilal, M.; Tian, X.P.; Jia, C.; Guo, Y.M.; Mi, X.T. Aerosol optical depth retrieval over bright areas using landsat 8 oli images. *Remote Sens. (Basel)* **2016**, *8*, 23. [CrossRef]

33. Roujean, J.L.; Leroy, M.; Deschamps, P.Y. A bidirectional reflectance model of the earths surface for the correction of remote-sensing data. *J. Geophys. Res. Atmos.* **1992**, *97*, 20455–20468. [CrossRef]

34. Lucht, W.; Schaaf, C.B.; Strahler, A.H. An algorithm for the retrieval of albedo from space using semiempirical brdf models. *IEEE Trans. Geosci. Remote Sens.* **2000**, *38*, 977–998. [CrossRef]

35. Vermote, E.; Justice, C.O.; Breon, F.M. Towards a generalized approach for correction of the BRDF effect in modis directional reflectances. *IEEE Trans. Geosci. Remote Sens.* **2009**, *47*, 898–908. [CrossRef]

36. Vermote, E.F.T.D.; Deuze, J.L.; Herman, M.; Morcrette, J.J. Second Simulation of a Satellite Signal in the Solar Spectrum-Vector (6SV). Available online: http://6s.ltdri.org/files/tutorial/6S_Manual_Part_1.pdf (accessed on 23 November 2017).

37. Tanre, D.; Kaufman, Y.J.; Holben, B.N.; Chatenet, B.; Karnieli, A.; Lavenu, F.; Blarel, L.; Dubovik, O.; Remer, L.A.; Smirnov, A. Climatology of dust aerosol size distribution and optical properties derived from remotely sensed data in the solar spectrum. *J. Geophys. Res. Atmos.* **2001**, *106*, 18205–18217. [CrossRef]

38. Watson, D.F.; Philip, G.M. A refinement of inverse distance weighted interpolation. *Geoprocessing* **1985**, *2*, 315–327.

39. Miishchenko, M.I.; Travis, L.D. Light-scattering by polydispersions of randomly oriented spheroids with sizes comparable to wavelengths of observation. *Appl. Opt.* **1994**, *33*, 7206–7225. [CrossRef] [PubMed]

40. Bodhaine, B.A.; Wood, N.B.; Dutton, E.G.; Slusser, J.R. On rayleigh optical depth calculations. *J. Atmos. Ocean. Technol.* **1999**, *16*, 1854–1861. [CrossRef]

41. Bucholtz, A. Rayleigh-scattering calculations for the terrestrial atmosphere. *Appl. Opt.* **1995**, *34*, 2765–2773. [CrossRef] [PubMed]

42. Angstrom, A. The parameters of atmospheric turbidity. *Tellus* **1964**, *16*, 64–75. [CrossRef]
43. Adames, A.F.; Reynolds, M.; Smirnov, A.; Covert, D.S.; Ackerman, T.P. Comparison of moderate resolution imaging spectroradiometer ocean aerosol retrievals with ship-based sun photometer measurements from the around the americas expedition. *J. Geophys. Res. Atmos.* **2011**, *116*. [CrossRef]
44. Tian, X.P.; Sun, L. Retrieval of aerosol optical depth over arid areas from modis data. *Atmosphere (Basel)* **2016**, *7*, 134. [CrossRef]
45. Ichoku, C.; Chu, D.A.; Mattoo, S.; Kaufman, Y.J.; Remer, L.A.; Tanre, D.; Slutsker, I.; Holben, B.N. A spatio-temporal approach for global validation and analysis of modis aerosol products. *Geophys. Res. Lett.* **2002**, *29*. [CrossRef]
46. Chu, D.A.; Kaufman, Y.J.; Zibordi, G.; Chern, J.D.; Mao, J.; Li, C.C.; Holben, B.N. Global monitoring of air pollution over land from the earth observing system-terra moderate resolution imaging spectroradiometer (MODIS). *J. Geophys. Res. Atmos.* **2003**, *108*, 4661. [CrossRef]
47. Kaufman, Y.J.; Wald, A.E.; Remer, L.A.; Gao, B.C.; Li, R.R.; Flynn, L. The modis 2.1-mu m channel—Correlation with visible reflectance for use in remote sensing of aerosol. *IEEE Trans. Geosci. Remote Sens.* **1997**, *35*, 1286–1298. [CrossRef]
48. Wang, X.W.; Xie, H.J.; Liang, T.G. Evaluation of modis snow cover and cloud mask and its application in northern Xinjiang, China. *Remote Sens. Environ.* **2008**, *112*, 1497–1513. [CrossRef]

remote sensing

MDPI

Article

Estimating Surface Downward Shortwave Radiation over China Based on the Gradient Boosting Decision Tree Method

Lu Yang [1], Xiaotong Zhang [1,*], Shunlin Liang [2], Yunjun Yao [1], Kun Jia [1] and Aolin Jia [1]

[1]   State Key Laboratory of Remote Sensing Science, Faculty of Geographical Science,
      Beijing Normal University, Beijing 100875, China; lyang201314@163.com (L.Y.); boyyunjun@163.com (Y.Y.);
      jiakun@bnu.edu.cn (K.J.); aolin@mail.bnu.edu.cn (A.J.)
[2]   Department of Geographical Sciences, University of Maryland, College Park, MD 20742, USA;
      sliang@umd.edu
*   Correspondence: xtngzhang@bnu.edu.cn;

Received: 4 November 2017; Accepted: 22 January 2018; Published: 26 January 2018

Abstract: Downward shortwave radiation (DSR) is an essential parameter in the terrestrial radiation budget and a necessary input for models of land-surface processes. Although several radiation products using satellite observations have been released, coarse spatial resolution and low accuracy limited their application. It is important to develop robust and accurate retrieval methods with higher spatial resolution. Machine learning methods may be powerful candidates for estimating the DSR from remotely sensed data because of their ability to perform adaptive, nonlinear data fitting. In this study, the gradient boosting regression tree (GBRT) was employed to retrieve DSR measurements with the ground observation data in China collected from the China Meteorological Administration (CMA) Meteorological Information Center and the satellite observations from the Advanced Very High Resolution Radiometer (AVHRR) at a spatial resolution of 5 km. The validation results of the DSR estimates based on the GBRT method in China at a daily time scale for clear sky conditions show an R^2 value of 0.82 and a root mean square error (RMSE) value of 27.71 $W \cdot m^{-2}$ (38.38%). These values are 0.64 and 42.97 $W \cdot m^{-2}$ (34.57%), respectively, for cloudy sky conditions. The monthly DSR estimates were also evaluated using ground measurements. The monthly DSR estimates have an overall R^2 value of 0.92 and an RMSE of 15.40 $W \cdot m^{-2}$ (12.93%). Comparison of the DSR estimates with the reanalyzed and retrieved DSR measurements from satellite observations showed that the estimated DSR is reasonably accurate but has a higher spatial resolution. Moreover, the proposed GBRT method has good scalability and is easy to apply to other parameter inversion problems by changing the parameters and training data.

Keywords: downward shortwave radiation; machine learning; gradient boosting regression tree; AVHRR; CMA

---

1. Introduction

Downward shortwave radiation (DSR) is a key parameter in the land-atmosphere interaction, which largely controls human life and ecosystems due to its important role in energy cycles [1,2], the hydrological cycle [3,4], the carbon cycles [5,6], and solar energy utilizations [7–13]. Therefore, knowledge of DSR is essential for improving our understanding of the Earth's climate and potential climatic changes [14]. A number of gridded global DSR products exist from remote sensing, reanalysis, and general circulation models (GCMs). Satellite remote sensing is one of the most practical ways to derive DSR measurements with relatively higher spatial resolution and accuracy.

Currently, DSR data can be obtained in three ways. The first is through collection from ground measurements. This method is characterized by high precision and uneven geographic distribution.

The second is estimation from reanalysis data and simulations from GCMs, which have relatively low spatial resolution and accuracy [15–17]. Examples include the ERA-Interim provided by European Center for Medium-Range Weather Forecast (ECMWF), the Japanese 55-year Reanalysis (JRA-55) provided by Japan Meteorological Agency, and the Modern-Era Retrospective analysis for Research and Applications (MERRA) reanalysis dataset provided by NASA. The third way is retrieval from remote sensing data [18–20], which can provide spatio-temporal continuous DSR estimates with relatively higher precision. Commonly used remote sensing datasets of surface solar radiance include the Global Energy and Water Cycle Experiment-Surface Radiation Budget (GEWEX-SRB), the International Satellite Cloud Climatology Project-Flux Data (ISCCP-FD), the University of Maryland-Shortwave Radiation Budget (UMD-SRB), and the Earth's Radiant Energy System (CERES). Each type of DSR data from a different source has advantages and limitations: the ground measurements provide accurate but sparse spatial coverage, whereas products from the two other methods may have larger uncertainties. The ground measurements are always used to evaluate the other two types of DSR estimates. GCMs are widely believed to have an advantage in simulating global scale climate changes [21]. A reanalysis product is a combination of a model and measurements. It uses observations to constrain the dynamic model to optimize complete coverage and accuracy [22]. DSR retrievals from remote sensing data always have a relatively higher accuracy than those from reanalysis data and simulations from GCMs. These DSR products have been widely evaluated using ground measurements [23–26]. For example, Zhang et al. [26] evaluated four current representative existing remote sensing products using 1151 sites from the Global Energy Balance Archive and the China Meteorological Administration (CMA). The results implied that DSR estimates from remotely sensed data were more accurate than those acquired from reanalysis and simulations from GCMs. The maximum spatial resolution of these four products is 0.5°, and the temporal resolution is thrice-hourly. Although the current global radiation products have finer temporal resolution, they have lower spatial resolutions, which limit their application [27]. Therefore, it is still necessary to generate higher spatial resolution DSR estimates using satellite observation.

Several algorithms have been developed for retrieving DSR measurements. The first way is to estimate DSR based on statistical models [28–33]. Perez et al. [31] developed a simple solar radiation forecast model using sky cover predictions. Yang et al. [32] used a hybrid model with CMA routine data to estimate DSR, and the validation results of this proposed model against ground measurements collected in Tibetan Plateau were better than satellite estimations from existing satellite products. Wang et al. [33] used a statistical model to establish the relationship between top of atmosphere (TOA) reflectance and net surface shortwave radiation using multiple regression and revised methods, and they then compared the precision of these methods using various parameters. Empirical statistical models usually construct a regression model directly using observed data and measured DSR values. These models are easy to apply but are disadvantaged by their lack of universality; the relationship established in a particular atmospheric condition or region may not be applicable in another area. The second method to retrieve DSR measurements is to estimate them based on parametric physical modeling methods [34–39]. Li et al. [37] proposed a parameterized model in which the normalized net surface shortwave radiation flux of the top incident irradiance of the atmosphere was used to establish a parametric relationship with the planetary albedo. Qin et al. [38] used satellite atmospheric and land products—including ozone thickness, precipitable water, aerosol loading, cloud water path, clouds effective particle radius, cloud fraction, and ground surface albedo—to establish a physically based parameterization model. They then used the model to estimate surface solar irradiance with a mean RMSE of approximately 100 W/m^2 and 35 W/m^2 on an instantaneous and daily mean basis, respectively. López et al. [39] proposed a new, simple parametric physical model to estimate global solar radiance under cloudy sky conditions. These methods often construct a physical model by simulating direct interaction between solar radiation and the atmosphere. This requires many parameters (e.g., aerosol optical depth, surface albedo, and moisture). It is obvious that model accuracy depends on these parameters.

Machine learning methods, which learn the relationship between inputs and outputs by fitting a flexible model directly from the data, are some of the most widely used methods to estimate DSR [40–45]. Wang [43] proposed a method try to derive DSR measurements using Moderate Resolution Imaging Spectroradiometer (MODIS) data (e.g., atmospheric profile product and surface reflectance) based on an artificial neural network (ANN) model. The validation results against ground measurements showed that the maximum root mean square error (RMSE) was less than 45 W·m^{-2}. Qin et al. [44] used an ANN-based method to establish the relationship between the measured monthly mean of daily global solar radiation levels and available remote sensing products with the aim of estimating global solar radiation. Zhou et al. [45] suggested that the Random Forest (RF) model was another feasible way to estimate DSR using satellite observations. These machine learning methods have their own advantages and disadvantages. For example, the attractiveness of an ANN is nonlinearity and high parallelism [46], and the RF cannot extrapolate beyond the training data and may not interpret well for conditions with few samples [47]. Although machine learning may provide powerful methods for estimating DSR from remote sensing data due to their ability to perform adaptive and nonlinear data fitting [48–50], the accuracy of the results is limited and many machine learning methods are prone to the phenomenon of overfitting. This can be avoid by using the gradient boosting regression tree (GBRT) method [51]. In addition, the GBRT can efficiently provide high accuracy. However, it has not been widely used for estimating DSR.

The objective of this study is to use a machine learning method, the GBRT, to obtain high accuracy DSR estimates from remote sensing and surface observed data under both clear and cloudy sky conditions in China. Moreover, this study aims to compare the DSR estimates from the GBRT with estimated values from classical ANN and existing remote sensing and reanalysis data.

The paper is organized as follows: Section 2 provides a brief introduction to the ground measurement and remote sensing data used. Section 2 also describes the methods used. Section 3 presents the results and an analysis. The conclusions are presented in Section 4.

2. Materials and Methods

2.1. Materials

2.1.1. Ground Measurements

The measurements of daily DSR used in this study were supplied by the CMA Meteorological Information Center. DSR was first measured in 1957, and its measurement was gradually collected at a total of 122 stations. However, the measurement at some stations have stopped sometime in the past. In 1994, there were 96 stations remaining to measure DSR. Quality control of the CMA DSR data was performed before the release; this included a spatial and temporal consistency check and manual inspection and correction [52]. Previous studies showed that the systematic errors in radiation measurements due to technical failure and operation-related problems are not rare [53,54]. Hence, a critical quality control procedure was performed to the ground measurements from the CMA before they were used in this study. The procedure is as described by Zhang [26]. Figure 1 shows the geographical distributions of the sites from the CMA. For more detailed information about the radiation data, it is possible to refer to the data description at the website http://data.cma.cn/.

This study used the daily DSR data collected from 96 radiation stations in China from 2001 to 2003. The daily DSR data from 2001 and 2002 were used to train the models, and the daily DSR data from 2003 was used to validate the model.

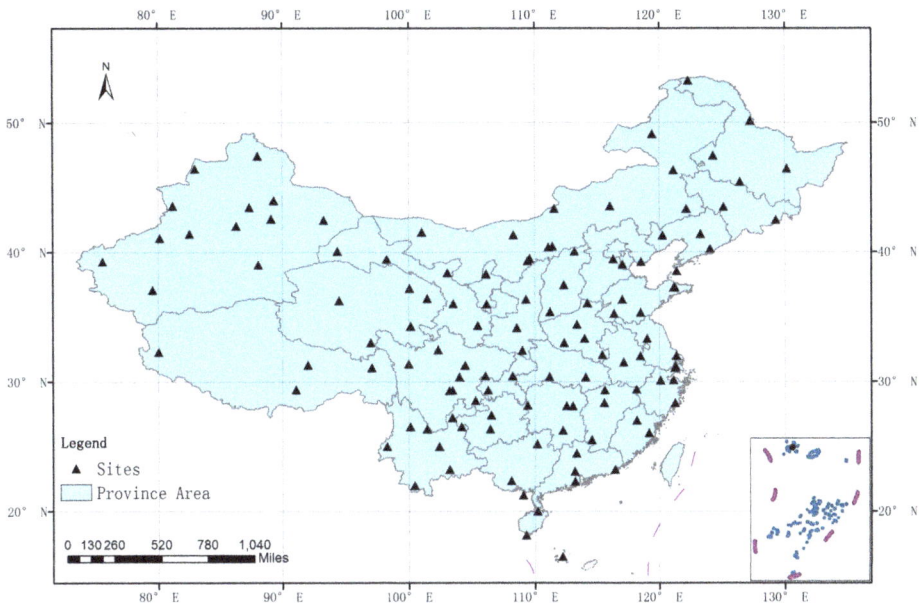

Figure 1. Spatial distribution of the radiation sites provided by the China Meteorological Administration (CMA) Meteorological Information Center.

2.1.2. Satellite Data

The National Oceanic and Atmospheric Administration (NOAA) Climate Data Records (CDR) of Visible and Near Infrared Reflectance from the Advanced Very High Resolution Radiometer (AVHRR) and the NASA Langley Research Center (LaRC) Cloud and Clear Sky Radiation Properties dataset were used in the paper. The two satellite datasets are from the Advanced Very High Resolution Radiometer (AVHRR) Global Area Coverage (GAC) Level 1B data, which has been quality controlled. These were taken from the NOAA-16 sun-synchronous orbit satellite observations provided by the NOAA CDR program. The NASA LaRC Cloud and Clear Sky Radiation Properties dataset is generated using the CERES Cloud Mask and Cloud Property Retrieval System (CCPRS) [55]. The NOAA CDR of Visible and Near Infrared Reflectance from AVHRR was calibrated by a multiple invariant Earth target calibration approach [56,57]. The NASA LaRC Cloud and Clear Sky Radiation Properties dataset was generated using algorithms initially designed for application to the Tropical Rainfall Measurement Mission (TRMM) and Moderate Resolution Imaging Spectroradiometer (MODIS) imagery within the NASA Clouds and the Earth's Radiant Energy System (CERES) program [58]. The spatial and temporal resolution of the dataset is about 4 km at the nadir and one day, respectively. Variables of the radiation properties dataset include cloud and clear sky pixel detection, cloud optical depth, cloud particle effective radius, land and sea surface temperature retrieval, shortwave broadband albedo, etc. [58]. Two variables including the calibrated 0.63 micron channel reflectance (channel 1) and the calibrated 0.86 micron channel reflectance (channel 2) were utilized for DSR estimation in this study [59]. Table 1 lists the corresponding information extracted from the AVHRR dataset used in this study.

Table 1. Input settings of the GBRT-based downward shortwave radiation (DSR) clear and cloudy sky models.

Inputs Data	Model	Unit	Range
Solar zenith angle	Clear and cloudy sky	Degrees	0–180
Viewing zenith angle	Clear and cloudy sky	Degrees	0–90
Relative azimuth angle	Clear and cloudy sky	Degrees	0–180
Top of atmosphere shortwave broadband albedo	Clear and cloudy sky	N/A	0–1.5
Reflectance of channel 1 and 2 of AVHRR	Clear and cloudy sky	Percent	0–12.5
Brightness temperature of channel 4 and 5 of AVHRR	Clear and cloudy sky	Degrees/Kelvins	160–340
Cloud optical depth	Cloudy sky	N/A	0–150
Cloud mask	Clear and cloudy sky	N/A	0–1

2.1.3. DSR Products

The two DSR products, the MERRA and the GEWEX-SRB DSR, were used in the paper. The MERRA product is a second reanalysis project from NASA for the satellite era (i.e., from 1979 to the present) using an updated new version of the Goddard Earth Observing System Data Assimilation System Version 5 (GEOS-5) [60]. The spatial resolution of the daily MERRA DSR estimate is 0.5° × 0.667°. The GEWEX-SRB radiation product from remotely sensed data used here was from the NASA/GEWEX-SRB shortwave version 3.0. The primary inputs to produce the data include shortwave and longwave radiances derived from International Satellite Cloud Climatology Project (ISCCP) pixel-level (DX) data, cloud and surface properties derived from the same source, temperature and moisture profiles, etc. [61]. The GEWEX-SRB DSR product was provided with a temporal resolution of 1 day and a spatial resolution of 1° from July 1983 to December 2008.

2.2. Methods

2.2.1. Gradient Boosting Regression Tree

The GBRT is a powerful, advanced statistical method widely used in classification and prediction. Because it does not require making assumptions on the data, it is extensively used in certain fields, such as in the optimization of recommendation systems [62,63], visual tracking algorithms [64], and traffic systems [65–68]. The attractiveness of GBRT comes from its ability to deal with the uneven distribution of data attributes, its lack of limitation for any hypothesis of input data, its better predictive capacity than a single decision tree, its power to deal with larger data size, and its transparency in terms of model development.

The GBRT produces competitive, highly robust, and interpretable procedures for both regression and classification. This was a method first proposed by Friedeman [51]. The core idea of this model is to generate a strong classifier by constructing an M amount of different weak classifiers through multiple iterations in order to reach the final combination. Each iteration is designed to improve the previous result by reducing the residuals of the previous model and establishing a new combination model in the gradient direction of the residual reduction. To describe the accuracy of the model, a loss function defined as $L(y, F)$ is introduced. The frequently employed loss functions include squared-error and absolute error [51]. Suppose that $\{x_i, y_i\}_{i=1}^{N}$ is the training sample. The x represents explanatory variables. The y represents the response variable. N is the number of the training sample. Let the M different individual decisions trees be represented by $\{h(x; \alpha_i)\}_{i=1}^{M}$, which is the parameterized function of the explanatory variables x and is characterized by $\alpha = \{\alpha_m\}_{m=1}^{M}$. β is the weight of each classifier, and α is the classifier parameter. Each tree divides the input space into the number of independent areas numbered J, as in $R_{1m}, \ldots, R_{jm}$. Each R_{jm} has a corresponding predicted value γ_{jm}. If the x-value is in the area R_{jm}, it means $x \in R_{jm}$ and the constant I equals 1. However, the constant

$I = 0$. Hence, the function ($F(x)$), which is an approximation function of the response variable. It can be written as follows:

$$\begin{cases} F(x) = \sum_{m=1}^{M} \beta_m h\ (x; \alpha_m) \\ h(x; \alpha_m) = \sum_{j=1}^{J} \gamma_{jm} I\ (x \in R_{jm}), where\ I = 1\ if\ x \in R_{jm}; I = 0,\ \text{otherwise} \end{cases} \quad (1)$$

The general process of GBRT shown in Figure 2 and more detail of GBRT can be find in Hastie et al. [69] and Ridgeway [70].

1. Initialize $F_0(x)$ to be a constant, $F_0(x)$=arg min$_\rho \sum_{i=1}^{N} L(y_i, \rho)$.

2. For m=1 to M do:
   For i =1 to N do :

3. Compute the negative gradient

$$\tilde{y}_{im} = -\left[\frac{\partial L(y_i, F(x_i))}{\partial F(x_i)} \right]_{F(x)=F_{m-1}(x-1)}$$

4. End;

5. Fit a regression tree $h(x; \alpha_m)$ to predict the targets $\tilde{y}_{im}$ from covariates x_i for all training data.

6. The α_m can be obtained as followed:

$$\alpha_m = \arg\min_{\alpha,\ \beta} \sum_{i=1}^{N} \left[\tilde{y}_{im} - \beta h(x_i; \alpha_m) \right]^2$$

7. Compute a gradient descent step size as:

$$\rho_m = \arg\min_\rho \sum_{i=1}^{N} L(y_i, F_{m-1}(x_i) + \rho h(x_i; \alpha_m))$$

8. Update the model as :

$$F_m(x) = F_{m-1}(x) + \rho_m h(x_i; \alpha_m)$$

End

Output the final model $F_M(x)$

Figure 2. The main procedures of the gradient boosting regression tree (GBRT) method.

The GBRT model can be constructed in three steps: (1) the preparation of the training database, (2) the architecture design and training phase, and (3) the application of the GBRT method. The next step is then to divide the data into clear sky and cloudy sky conditions according to the NOAA CDR of cloud mask data. If the pixel was marked as "cloud" by AVHRR data, it means it is under cloudy conditions. Otherwise, there is clear sky conditions. The GBRT-based DSR clear and cloudy sky model were trained using cloud mask data provided by the AVHRR data.

The performance of the DSR estimates was tested using the holdout method, which is a simple type of cross-validation. The dataset was randomly stratified into two groups, with 80% made part of the training dataset and 20% made part of the testing dataset. The main procedures are as follows.

(1) Extracting the TOA radiance from the NOAA CDR of Visible and Near Infrared Reflectance from AVHRR;

(2) Extracting the cloud properties from the NASA LaRC Cloud and Clear Sky Radiation Properties dataset;

(3) Training the clear and cloudy sky models. The inputs of the clear sky model include the solar zenith angle, viewing zenith angle, relative Azimuth angle, TOA shortwave broadband albedo, reflectance (from channel 1 and 2) of AVHRR, and the brightness temperature (from channel 4 and 5) of AVHRR. The input of the cloudy sky model used the same input variables as the clear sky model and cloud optical depth;

(4) Configuring the model coefficients. The optimal parameterization scheme was determined by looping in each parameter threshold. Table 2 shows the parameter setting details to determine the optimal parameterization for both the clear sky and cloudy sky conditions through the evaluation results (highest R^2 value and lowest bias and RMSE values) of the testing dataset for each loop;

(5) Evaluating against the ground measurements.

Figure 3 shows the flowchart of the proposed GBRT model used in this study.

Table 2. Parameters setting to determine the optimal parameters for the GBRT model.

Parameters	Threshold	Intervals
The number of iterations	50–300	50
Shrinkage	0.1–1	0.3
The depth of the tree	6–9	1
Sampling rate	0.2–1	0.2

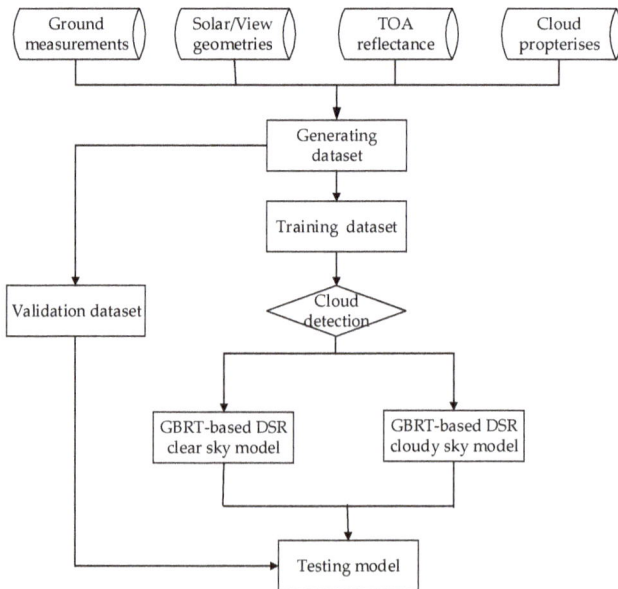

Figure 3. Flowchart of the GBRT method.

2.2.2. Artificial Neural Networks

ANNs are used as an empirical statistical method in a variety of applications such as classification, pattern recognition, forecasting, optimization, etc. [71–73]. An ANN model can be any model in which the output variables are computed from the input variables using compositions or connections of basic functions. In this research, a feedforward backpropagation neural network consisting of several layers of neurons was used. A neuron is a simplified mathematical model of a biological neuron, and a connection is a unique information transport link from a sending to a receiving neuron. Figure 4 shows a structural diagram of the ANN used in this study. The ANN model used here consists of three layers of neurons: input layers, hidden layers, and an output layer. Input $\{x_j\}_{j=1}^m$ is transmitted through a connection that multiplies its strength by a weight represented by $\{w_{ij}\}_{i=1}^k$. This gives the value $x_i w_{ij}$, which is an argument to a transfer function f that yields an output y_i.

$$y_i = f(\sum_{j=1}^m w_{ij}x_j) \tag{2}$$

where i is the index of neuron in the hidden layer and j is the index of inputs to the neural network. A typical feedforward network trained with a resilient backpropagation algorithm [74,75] is employed to estimate DSR in this paper.

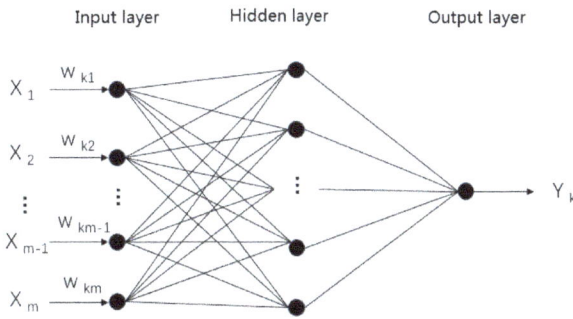

Figure 4. Artificial neural network (ANN) structure used in this study.

2.3. Constructing the Model

According to the characteristic variables in Table 1, corresponding data was extracted to establish the training dataset. Daily observed data from the CMA Meteorological Information Center from 2001 and 2002 were used as the response values (true values) of the training dataset. Information from the AVHRR cloudy and clear sky pixel detection was used to divide the training dataset into a cloudy sky training dataset and a clear sky training dataset. In addition, the missing values were removed both from the training and validation dataset.

2.3.1. Constructing the GBRT-Based DSR Model

The key step in building an efficient GBRT model is finding the optimal architecture. Building the GBRT model in a stage-wise fashion and regenerating the model minimizes the expected value of a certain loss function. After adding many trees to the model, the fitted model should have a small training error. However, it is important to remember that the generalization ability does not improve in direct proportion with the size of the fitted model; if the model is overfitted and possesses an extremely small error with the training dataset, its generalization ability will be poor. The performance of the GBRT model is influence by these four parameters as follows: the number of iterations, shrinkage, the depth of the tree, and the sampling rate [63]. As the number of iterations increases, model complexity will also increase, leading to poor prediction performance on the test dataset. Determining the appropriate number of iterations is essential to minimize future risks in prediction. Overfitting can be avoided by limiting the number of iterations and reducing the contribution of each tree. This is also known as shrinkage (or learning rate). There is a tradeoff between the number of iterations and the learning rate. A lower learning rate value means that the model is more robust but has a slower computing speed. The size of each tree is called the depth of the tree. The depth of the tree refers to the number of nodes in a tree. This parameter depends on the number of data points and the characteristic variables of the data. In theory, if the value of this parameter is too large, the model will run at a slower rate. The sampling rate is the ratio of the subsample to the total number of training instances. When set to 0.5, it means that the model randomly collected half the data instances to grow trees. This will prevent overfitting. This procedure should be used with adjusting the learning rate and the number of iterations.

In the present case, successive performance testing showed that an architecture of 250 trees with a tree depth of 6, a sampling rate of 0.6, and a learning rate of 0.1 was optimal to estimate the DSR under clear sky conditions. These values are 250, 6, 0.8, and 0.1, respectively, under cloudy sky conditions.

Considering that cloud optical depth is related to DSR under cloudy sky conditions, the cloud optical depth was chosen as the input data for the cloudy sky model. This was different from the input data used for the clear sky model. Table 1 shows the input data of the GBRT-based DSR model under clear sky and cloudy sky conditions. The debugging procedure for key parameters such as the number of trees, the size of each tree, the learning rate, and the subsample ratio was described earlier.

2.3.2. Constructing the ANN-Based DSR Model

The ANN training databases in this study were the same as those used in the GBRT model. The architecture is mainly defined by the number of layers, the number of neurons in each layer, and the transition function associated with each neuron. As for other parameters (e.g., initial weighting), details of these will not be shown in this paper. In the present case, successive performance testing has shown that an architecture with one hidden layer is sufficient to estimate DSR. The number of nodes in the input layer was set to nine nodes, and the number of nodes in the output layer was set to one. After testing, the number of the nodes in the hidden layer was 12 under clear sky conditions and 14 under cloudy sky conditions. The transfer function of the hidden layer was a tan-sigmoid transfer function, and those of the other two layers were linear functions under both clear sky and cloudy sky conditions. Theoretically, various sets of functions such as step, linear, and no linear functions could be used as the transfer function of different layers. However, the tan-sigmoid (for the hidden layer) and linear (for the input and output layers) types were most commonly used in the literature [71].

3. Results and Analysis

The estimated daily and monthly mean DSR based on the GBRT method were not only evaluated against ground measurements but also compared with the evaluation results from those estimated from the ANN-based DSR model. Additionally, the estimated DSR values were also compared with current existing DSR products from the GEWEX-SRB and the MERRA. The validation results were shown in terms of bias, RMSE, and correlation coefficient (R^2).

3.1. Validation with Ground Measurements

3.1.1. Validation at a Daily Time Scale

The ground measurements at the selected 96 stations collected from CMA in 2003 were compared to the grid points of the estimated DSR based on the GBRT method. The performance of the GBRT-based DSR clear sky model using the training dataset and the validation dataset is shown in Figure 5. As shown in Figure 5, the daily estimated DSR correlates well with ground measurements under clear sky conditions. The daily DSR estimates under the clear sky conditions for the training dataset have an overall RMSE value of 19.05 W·m^{-2} (19.06%), a bias value of 0.00 W·m^{-2} (2.41%), and an R^2 value of 0.92. These values were 27.71 W·m^{-2} (38.38%), -2.53 W·m^{-2} (1.37%), and 0.82, respectively, for the validation dataset. The validation results at a daily time scale demonstrate that the GBRT is a practically applicable and effective method for estimating DSR under clear sky conditions using satellite observations from AVHRR data.

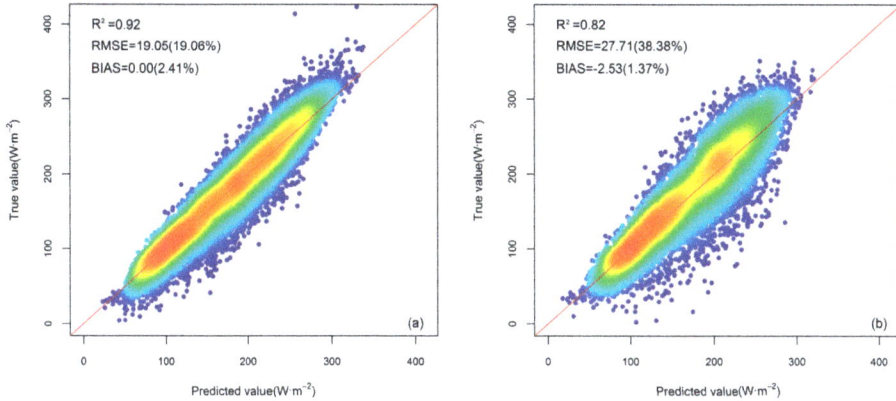

Figure 5. (**a**) Evaluation results of the training set's daily estimated DSR based on the GBRT-based clear sky model against ground measurements in 2001 and 2002. (**b**) Evaluation results of the validation set's daily estimated DSR based on the GBRT-based clear sky model against ground measurements in 2003. The number in the parentheses is the percent bias or root mean square error (RMSE) value.

Figure 6 presents the evaluation results of the GBRT-based DSR cloudy sky model using the training dataset and the validation dataset. The daily DSR estimates for the training dataset under the cloudy sky conditions have an overall RMSE value of 33.37 $W \cdot m^{-2}$ (30.21%), an R^2 value of 0.79, and a bias value of 0.01 $W \cdot m^{-2}$ (4.74%). These values for the validation dataset were 42.97 $W \cdot m^{-2}$ (34.57%), 0.64, and -2.83 $W \cdot m^{-2}$ (1.45%), respectively. The accuracy was slightly lower than that of the clear sky model, which may be related to the influence of clouds [76].

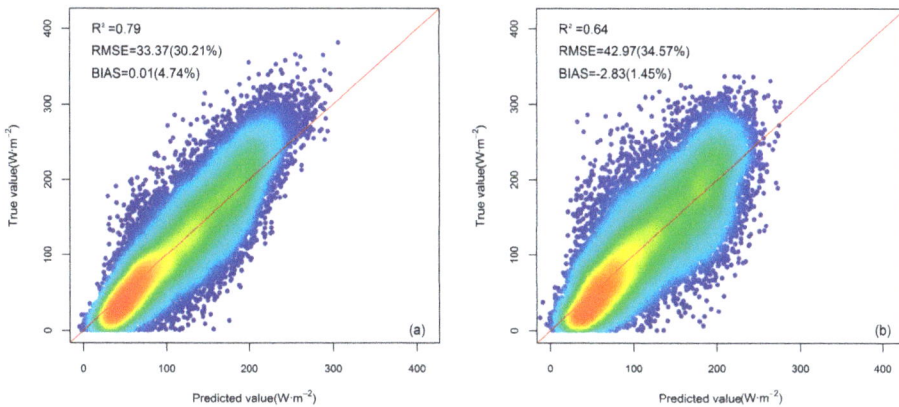

Figure 6. (**a**) Evaluation results of the training set's daily estimated DSR based on the GBRT-based cloudy sky model against ground measurements in 2001 and 2002. (**b**) Evaluation results of the validation set's daily estimated DSR based on the GBRT-based cloudy sky model against ground measurements in 2003. The number in the parentheses is the percent bias or RMSE value.

When building the models for DSR estimation, we found that channel 4 and 5 influence the model accuracy. Figures 7 and 8 show a comparison of the evaluation results without considering

AVHRR channels 4 and 5 under clear and cloudy sky conditions, respectively. As shown in Figure 7, the daily DSR estimates without considering these two channels under clear sky conditions of the training dataset have an overall RMSE value of 26.52 W·m^{-2} (23.93%), a bias value of -0.26 W·m^{-2} (3.25%), and an R^2 value of 0.85. It can be concluded that the clear sky model yields higher accuracy if AVHRR channels 4 and 5 are considered. Similar results were also found under cloudy sky conditions. The daily DSR estimates without considering these two channels under cloudy sky conditions of the training dataset have an overall RMSE value of 37.52 W·m^{-2} (31.86%), a bias value of 0.16 W·m^{-2} (4.66%), and an R^2 value of 0.73. A potential reason for this may be the total atmospheric water vapor effect on DSR estimation, which may be to cause large uncertainties. Previous studies showed that AVHRR channels 4 and 5 have been widely used to retrieve the total atmospheric water vapor [77,78].

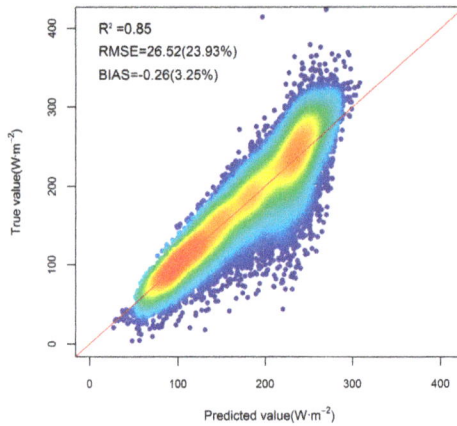

Figure 7. Validation results of the estimated daily DSR based on the GBRT model under clear sky conditions without considering Advanced Very High Resolution Radiometer (AVHRR) channels 4 and 5 as the input variables. The number in the parentheses is the percent bias or RMSE value.

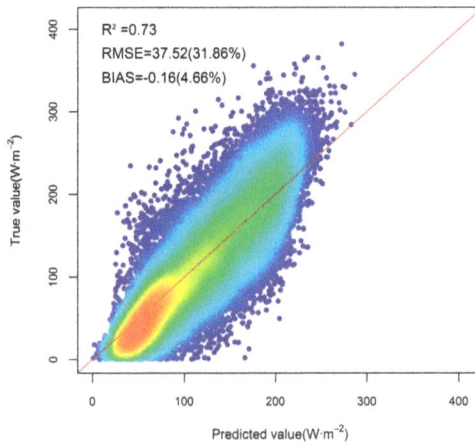

Figure 8. Validation results of the estimated daily DSR based on the GBRT model without considering AVHRR channels 4 and 5 as the input variables under cloudy sky conditions. The number in the parentheses is the percent bias or RMSE value.

3.1.2. Validation at a Monthly Time Scale

To further show the relative accuracy of the GBRT method, we also validated the estimated DSR at a monthly time scale. To perform the comparison, monthly DSR estimates were obtained by averaging the daily DSR data of each month. Figure 9 shows the evaluation results of the training dataset and validation dataset based on the GBRT model of 2003 at a monthly time scale. The monthly estimated DSR of the training dataset has an overall RMSE value of 14.22 $W \cdot m^{-2}$ (12.50%), a bias value of -0.30 $W \cdot m^{-2}$ (2.04%), and an R^2 value of 0.94. These values were 15.40 $W \cdot m^{-2}$ (12.93%), -2.25 $W \cdot m^{-2}$ (1.01%), and 0.92, respectively, for the validation dataset. Like the validation results at a daily time scale, the validation results at a monthly time scale showed that the GBRT model is reasonably accurate.

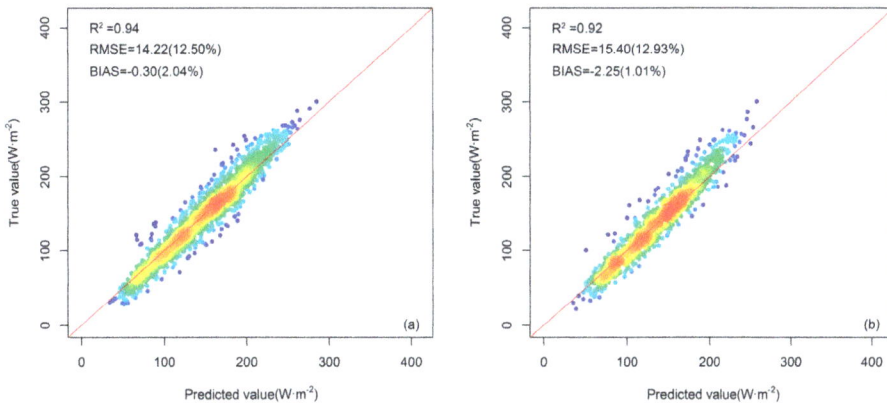

Figure 9. (**a**) Evaluation results of the training set's estimated monthly mean DSR based on the GBRT-based DSR model against ground measurements in 2001 and 2002. (**b**) Evaluation results of the validation set's estimated monthly mean DSR based on the GBRT-based DSR model against ground measurements in 2003. The number in the parentheses is the percent bias or RMSE value.

3.2. Comparison with the ANN-Based Method

3.2.1. Validation at a Daily Time Scale

Figure 10a,b shows the evaluation results of the estimated daily DSR of the training and the validation dataset based on the ANN-based DSR model under clear sky and cloudy sky conditions. The daily DSR estimates based on the ANN-based clear sky model of the training dataset have an overall RMSE value of 26.53 $W \cdot m^{-2}$ (41.84%) and a bias value of -0.09 $W \cdot m^{-2}$ (0%). These values were 27.15 $W \cdot m^{-2}$ (46.07%) and -3.67 $W \cdot m^{-2}$ (1.60%), respectively, for the validation dataset. Although the RMSE of the estimated daily DSR of the validation dataset was slightly lower than that of the GBRT model, the mean absolute bias of the ANN-based model was 3.67 $W \cdot m^{-2}$ (1.60%), which is larger than that of the GBRT model (2.53 $W \cdot m^{-2}$ (1.37%)) (Table 3). The evaluation results of the ANN-based cloudy sky model are shown in Figure 11. The daily DSR estimates based on the training dataset's ANN cloudy sky DSR model have an overall RMSE value of 42.07 $W \cdot m^{-2}$ (33.99%) and a bias value of 0.17 $W \cdot m^{-2}$ (3.13%). These values were 42.39 $W \cdot m^{-2}$ (34.50%) and -4.35 $W \cdot m^{-2}$ (0.17%), respectively, for the validation dataset. According to the comparison results shown in Figures 10 and 11 and Table 3, it was clear that the predictive abilities of the GBRT model are better than the ANN model at a daily time scale.

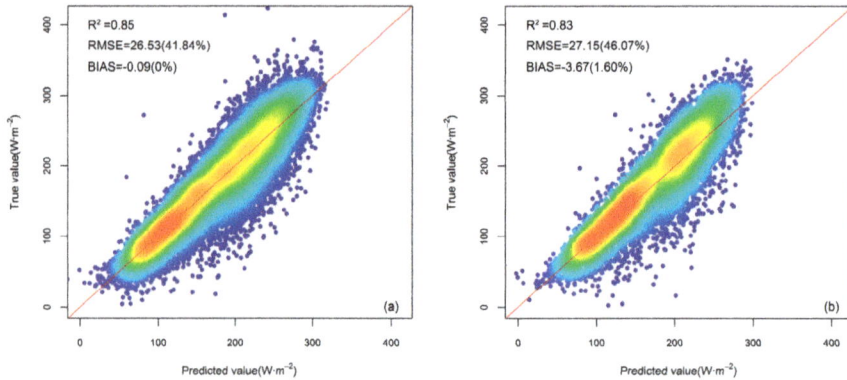

Figure 10. (**a**) Evaluation results of the training dataset's daily estimated DSR based on the ANN-based clear sky model against ground measurements in 2001 and 2002. (**b**) Evaluation results of the validation dataset's daily estimated DSR based on the ANN-based clear sky model against ground measurements in 2003. The number in the parentheses is the RMSE value.

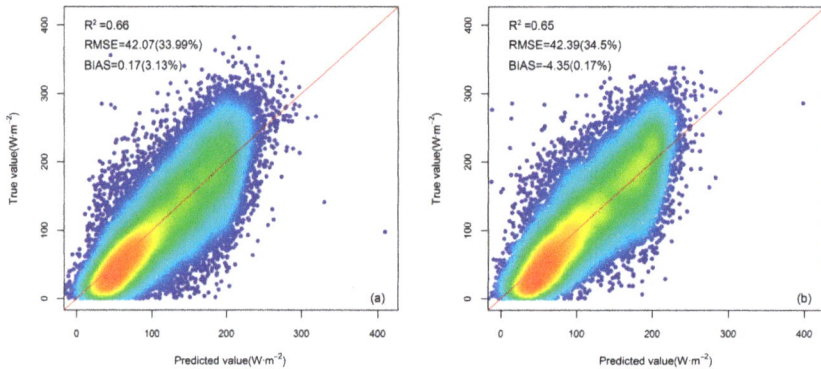

Figure 11. (**a**) Evaluation results of the training dataset's daily estimated DSR based on the ANN-based cloudy sky model against ground measurements in 2001 and 2002. (**b**) Evaluation results of the validation dataset's daily estimated DSR based on the ANN-based cloudy sky model against ground measurements in 2003. The number in the parentheses is the percent bias or RMSE value.

Table 3. Comparison of the results of the ANN and GBRT models at a daily time scale (using measurements from 2001 and 2002 as the training dataset and measurements from 2003 as the validation dataset). The number in the parentheses is the percent bias or RMSE value.

Sky Condition	Dataset	Method	R^2	RMSE ($W \cdot m^{-2}$)	Bias ($W \cdot m^{-2}$)
Clear sky	Training set	GBRT	0.92	19.05 (19.06%)	0 (2.41%)
		ANN	0.85	26.53 (41.84%)	−0.09 (0%)
	Validation set	GBRT	0.82	27.71 (38.38%)	−2.53 (1.37%)
		ANN	0.83	27.15 (46.07%)	−3.67 (1.60%)
Cloudy sky	Training set	GBRT	0.79	33.37 (30.21%)	0.01 (4.74%)
		ANN	0.66	42.07 (33.99%)	0.17 (3.13%)
	Validation set	GBRT	0.64	42.97 (34.57%)	−2.83 (1.45%)
		ANN	0.65	42.39 (34.50%)	−4.35 (0.17%)

3.2.2. Validation at a Monthly Time Scale

Similar to what we did with the GBRT model, we also validated the estimated DSR at a monthly time scale to further show the accuracy of the ANN method. To perform the comparison, monthly DSR estimates were calculated by averaging the daily DSR of each month. Figure 12a shows the evaluation results of the training dataset based on the ANN-based DSR model in 2003 at a monthly time scale. The R^2 was 0.88, which was lower than that of GBRT model. The RMSE was 18.95 W·m^{-2} (15.81%) larger than that of GBRT model. The evaluation results of the validation dataset's monthly estimated DSR based on the ANN-based DSR model is shown in Figure 12b. The R^2 was 0.87, and the RMSE was 20.05 W·m^{-2} (16.20%). As in the evaluation results at a daily time scale, it is obvious that the GBRT model performs better than the ANN model at a monthly time scale.

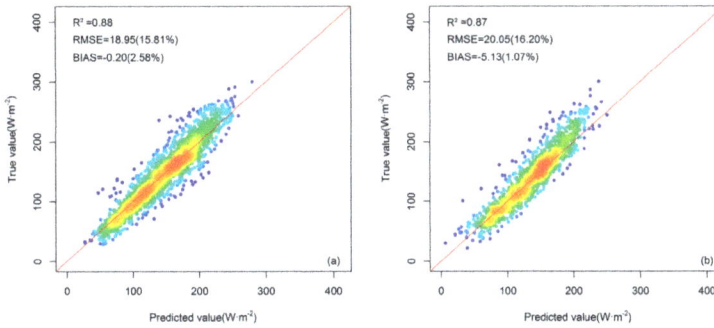

Figure 12. (a) Evaluation results of the training set's estimated monthly mean DSR based on the ANN-based DSR model against ground measurements in 2001 and 2002. (b) Evaluation results of the validation set's estimated monthly mean DSR based on the ANN-based DSR model against ground measurements in 2003. The number in the parentheses is the percent bias or RMSE value.

Although the DSR estimates based on the GBRT model at both daily and monthly time scales were relatively higher accuracy than those from the ANN-based model, the machine learning methods including GBRT and ANN are sensitive to the choice of parameters. Therefore, the parameters chosen for these two machine learning methods may influence the accuracy of the DSR estimates. In this study, the optimal parameterization scheme was determined by looping in each parameter threshold. Advanced methods for deriving the optimal parameters for both GBRT and ANN should be tested in the future.

3.3. Comparison with Existing DSR Products

3.3.1. Mapping DSR over China

To demonstrate the applicability of the GBRT-based DSR model for regional mapping, the surface monthly mean DSR was estimated based on the GBRT method in the mainland of China in March 2003. Figure 13a shows the estimated results for monthly DSR in March 2003. The GEWEX-SRB and MERRA monthly DSR for the same month are also shown in Figure 13b,c for comparison. According to these three figures, it can be concluded that the spatial distribution of estimated DSR based on the GBRT method is similar to that from the GEWEX-SRB. However, large discrepancies occurred in the comparison with the MERRA. Moreover, the DSR estimates from the GBRT model provide more details compared to the other two existing DSR products.

Figure 13d,e shows the differences between the monthly mean DSR estimates from the GBRT model and those from the GEWEX-SRB and the MERRA, respectively. Before comparison, the DSR estimates from the GBRT model and the MERRA were projected onto a 1° spatial resolution using bilinear interpolation to match the resolution of the GEWEX-SRB data. As shown in the Figure 13, the differences between the GBRT-based DSR estimates and the GEWEX-SRB DSR product were smaller than that between the GBRT-based DSR estimates and the MERRA DSR product. The maximum differences between the GBRT-based DSR estimates and the GEWEX-SRB DSR product were found in the Tibetan Plateau. The maximum differences between the GBRT-based DSR estimates and the MERRA DSR product were found in southeast China, which were greater than 100 W·m^{-2} at some areas. The large discrepancies in the Tibetan Plateau may be related to the high elevation of the area. Yang et al. [32] pointed out that the discrepancies among the satellite products were always larger in highly variable terrain and smaller for non-variable terrain. The large differences in southeast China were probably due to inappropriate representation of aerosols and clouds, as well as their interactions with the algorithms used for this region [79,80]. In this area, heavy pollution is occurring due to rapid economic development and high population density. However, the DSR comparison of the GBRT model and current existing products were only performed for one month. This may also cause large uncertainties. Therefore, further investigations should be conducted for DSR estimation in the future if long-term DSR estimates are generated based on the GBRT method.

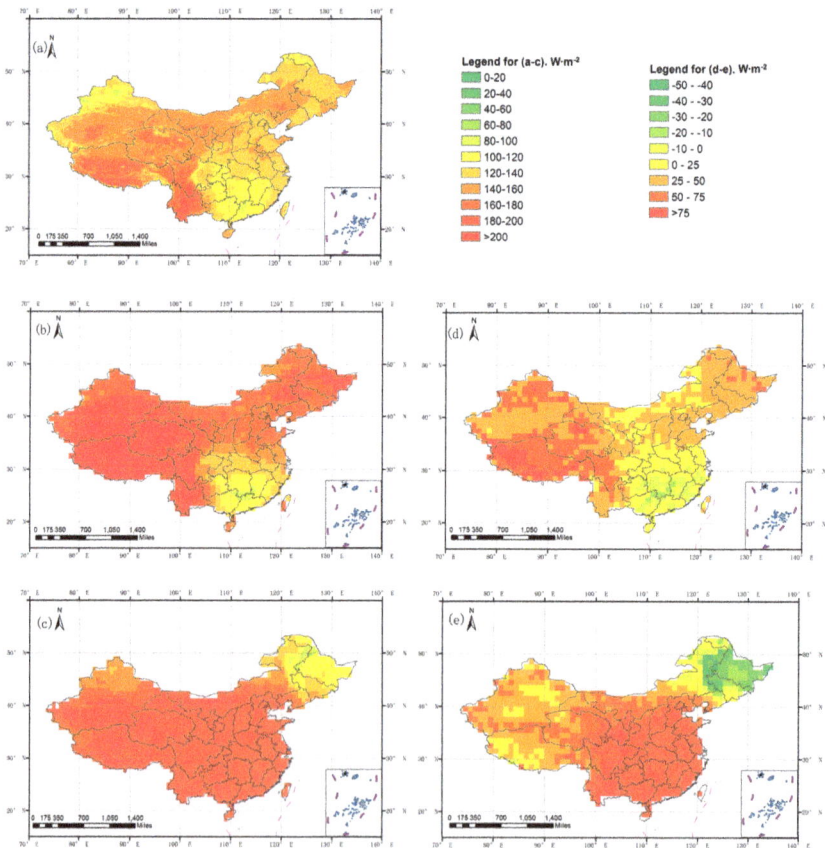

Figure 13. The spatial distribution of the DSR estimates from (**a**) the GBRT model, (**b**) the GEWEX-SRB, and (**c**) the MERRA in March 2003. (**d**) The differences between monthly mean DSR estimates of the GEWEX-SRB and the GBRT model (i.e., the GEWEX-SRB estimates minus the GBRT-based estimates) in March 2003. (**e**) The differences between monthly mean DSR estimates of the MERRA and the GBRT model (i.e., the MERRA estimates minus the GBRT-based estimates) in March 2003.

3.3.2. Validation with Ground Measurements

To further show the accuracy of the DSR estimates based on the GBRT method, we also compared the evaluation results of the GBRT-based daily estimated DSR against ground measurements from CMA in 2003 with those of current existing DSR products from the GEWEX-SRB and the MERRA. As shown in Figure 14, the daily estimated DSR based on the GBRT method correlates very well with the ground measurements, with an RMSE value of 31.65 W·m^{-2} (21.34%) and a bias value of 0.86 W·m^{-2} (1.50%). These values were 40.82 W·m^{-2} (30.93%) and 27.39 W·m^{-2} (17.86%), respectively, for the GEWEX-SRB-based estimates, and 74.2 W·m^{-2} (39.40%) and 57.27 W·m^{-2} (30.06%), respectively, for the MERRA-based estimates. It was obvious that the evaluation results of the GBRT-based DSR model were better than those of the other two products. However, the spatial representativeness of ground measurements is a potential error source for DSR evaluation.

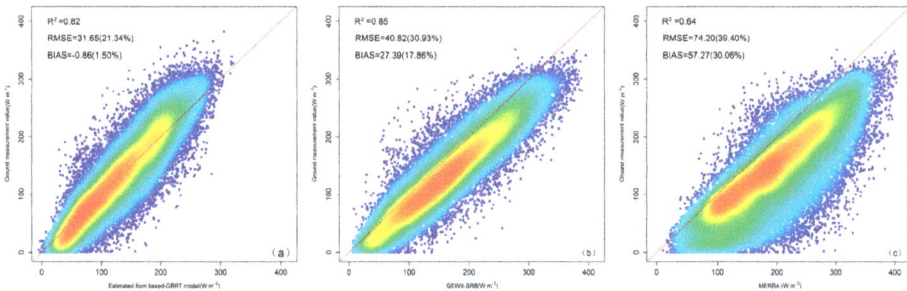

Figure 14. Scatter plots comparing the results from (**a**) the GBRT-based DSR model, as well as the DSR products (**b**) the Global Energy and Water Cycle Experiment-Surface Radiation Budget (GEWEX-SRB) and (**c**) Modern-Era Retrospective analysis for Research and Applications (MERRA) against ground measurements in 2003. The number in the parentheses is the percent bias or RMSE value.

As pointed out by Hakuba et al. [81], the monthly and annual mean representation error at the surface sites with respect to their 1° surroundings are, on average, 3.7% (4 W·m^{-2}) and 2% (3 W·m^{-2}), respectively. The DSR estimates from the GBRT model and current existing radiation products have different spatial resolutions. Therefore, the regional dependence of errors of coarse-resolution satellite products for complex terrain may cause large discrepancies.

4. Conclusions

DSR is an essential parameter in the terrestrial radiation budget and a necessary input for land-surface process models. Although several radiation products using satellite observations have been released, their coarse spatial resolution and low accuracy limit their application. Therefore, high spatio-temporal resolution and high accuracy DSR is still required for many applications. To achieve this goal, a fast, accurate, and robust GBRT method that has the ability to handle different types of input variables and model complex relations was developed to estimate DSR using satellite observations from AVHRR.

The estimated DSR was evaluated using the ground measurements from CMA and was compared with one remote sensing DSR product (the GEWEX-SRB) and one reanalysis DSR product (the MERRA). The daily estimated DSR had an overall R^2 value of 0.82, an RMSE of 27.71 W·m^{-2} (38.38%), and a bias of −2.53 W·m^{-2} (1.37%) under clear sky conditions, and an R^2 of 0.64, an RMSE of 42.97 W·m^{-2} (34.57%), and a bias of −2.83 W·m^{-2} (1.45%) under cloudy sky conditions. Comparison of the DSR estimates with the reanalyzed and the retrieved DSR values from satellite observation showed that the estimated DSR values are reasonably accurate but with higher spatial resolution. However, the DSR comparison of the GBRT model and current existing products was only performed for one month, which may cause large uncertainties. Beside this, measurement errors (e.g., instrument sensitivity, drift, and urbanization effects) and spatial representativeness of surface measurements are potential sources of error in DSR estimation [81]. Therefore, further investigations should be conducted for DSR estimation in the future if long-term DSR estimates are generated based on the GBRT method.

The strengths of GBRT are accuracy, speed, and robustness [51]. To show the advantages of GBRT, an ANN model was built. The results were compared between the GBRT-based DSR model and the ANN-based DSR model under clear and cloudy sky conditions, as shown in Section 3.2. The daily validation analysis showed that the maximum RMSE for GBRT-based and ANN-based clear sky model was less than 28 W·m^{-2}, but the bias of the GBRT-based clear sky model (-2.53 W·m^{-2}) was less than that of the ANN-based clear sky model (-3.67 W·m^{-2}). Similar results were also found for the cloudy sky model. The ANN has two known disadvantages: it needs a relatively long processing time to train a model with many input variables, and it behaves unpredictably when overestimation occurs during the training stage [82]. In contrast, the GBRT was evaluated as a promising machine learning approach in terms of processing speed and accuracy. All experiments were conducted on a Windows 7 Intel(R) Core(TM) i7-6700 CPU, 3.4 GHz, 20.00 GB RAM processor. The means for the elapsed time of completion of the GBRT clear sky model and the GBRT cloudy sky model were within 10 seconds. Therefore, we conclude that the GBRT method performs better than the ANN method for DSR estimation in this study. As it is well known, the mechanisms of machine learning methods are often considered to be black boxes, and the training procedure is sensitive to the choice of parameters. These limitations may influence the accuracy of the DSR estimates.

The contributions of this study demonstrate that the GBRT is efficient and practical for estimating DSR using remote sensing and ground observation data. Simultaneously, this method has a very good development procedure for defining training data and generating parameters. The method also has more extensive applicability than other current methods. The proposed GBRT-based method can also be used for the retrieval of other land surface variables.

Acknowledgments: This work was supported in part by the National Key Research and Development Program of China (No. 2016YFA0600102 and No. 2017YFA0603002) and in part by the National Natural Science Foundation of China under grant 41571340. The surface observation data of surface incident solar radiation was downloaded from CMA (http://cdc.nmic.cn/home.do). The AVHRR data was downloaded from NOAA National Climatic Data Center (NCDC) (website: https://www.ncdc.noaa.gov/cdr).

Author Contributions: Xiaotong Zhang and Shunlin Liang designed the experiment. Xiaotong Zhang and AoLin Jia collected the required data. YunJun Yao and Kun Jia preprocessed the data. Lu Yang and Xiaotong Zhang performed the experiment. Lu Yang, Xiaotong Zhang, and Shunlin Liang conducted the analysis.

Conflicts of Interest: The authors declare no conflict of interest.

References

1. Lu, N.; Liu, R.; Liu, J.; Liang, S. An algorithm for estimating downward shortwave radiation from GMS 5 visible imagery and its evaluation over China. *J. Geophys. Res. Atmos.* **2010**, *115*. [CrossRef]
2. Gupta, S.K.; Ritchey, N.A.; Wilber, A.C.; Whitlock, C.H.; Gibson, G.G.; Stackhouse, P.W., Jr. A Climatology of Surface Radiation Budget Derived from Satellite Data. *J. Clim.* **1999**, *12*, 2691–2710. [CrossRef]
3. Gautam, R.; Hsu, N.C.; Lau, K.M.; Kafatos, M. Aerosol and rainfall variability over the Indian monsoon region: Distributions, trends and coupling. *Ann. Geophys.* **2009**, *27*, 3691–3703. [CrossRef]
4. Wild, M.; Ohmura, A.; Gilgen, H.; Roeckner, E. Validation of General Circulation Model Radiative Fluxes Using Surface Observations. *J. Clim.* **1995**, *8*, 1309–1324. [CrossRef]
5. Running, S.W.; Thornton, P.E.; Nemani, R.; Glassy, J.M. *Global Terrestrial Gross and Net Primary Productivity from the Earth Observing System*; Springer: New York, NY, USA, 2000; pp. 44–57.
6. Running, S. W.; Nemani, R.; Glassy, J.M.; Thornton, P.E. *MODIS Daily Photosynthesis (PSN) and Annual Net Primary Production (NPP) Product (MOD17). Algorithm Theoretical Basis Document*; NASA Goddard Space Flight Center: Greenbelt, MD, USA, 1999.
7. Mondol, J.D.; Yohanis, Y.G.; Norton, B. Solar radiation modelling for the simulation of photovoltaic systems. *Renew. Energy* **2008**, *33*, 1109–1120. [CrossRef]

8. Perez, R.; Seals, R.; Zelenka, A. Comparing satellite remote sensing and ground network measurements for the production of site/time specific irradiance data. *Sol. Energy* **1997**, *60*, 89–96. [CrossRef]

9. Blanc, P.; Gschwind, B.T.; Lefèvre, M.; Wald, L. The HelioClim Project: Surface Solar Irradiance Data for Climate Applications. *Remote Sens.* **2011**, *3*, 343–361. [CrossRef]

10. Kaplanis, S.; Kaplani, E. A model to predict expected mean and stochastic hourly global solar radiation *I(h;nj)* values. *Renew. Energy* **2007**, *32*, 1414–1425. [CrossRef]

11. Wong, L.T.; Chow, W.K. Solar radiation model. *Appl. Energy* **2001**, *69*, 191–224. [CrossRef]

12. Salcedo-Sanz, S.; Casanova-Mateo, C.; Pastor-Sánchez, A.; Sánchez-Girón, M. Daily global solar radiation prediction based on a hybrid Coral Reefs Optimization—Extreme Learning Machine approach. *Sol. Energy* **2014**, *105*, 91–98. [CrossRef]

13. Mellit, A.; Benghanem, M.; Arab, A.H.; Guessoum, A. A simplified model for generating sequences of global solar radiation data for isolated sites: Using artificial neural network and a library of Markov transition matrices approach. *Sol. Energy* **2005**, *79*, 469–482. [CrossRef]

14. Houghton, J.T. Climate change 2001: The scientific basis. *Neth. J. Geosci.* **2001**, *87*, 197–199.

15. Zhang, X.; Liang, S.; Song, Z.; Niu, H.; Wang, G.; Tang, W.; Chen, Z.; Jiang, B. Local Adaptive Calibration of the Satellite-Derived Surface Incident Shortwave Radiation Product Using Smoothing Spline. *IEEE Trans. Geosci. Remote Sens.* **2016**, *54*, 1156–1169. [CrossRef]

16. Zhang, X.; Liang, S.; Wang, G.; Yao, Y.; Jiang, B.; Cheng, J. Evaluation of the Reanalysis Surface Incident Shortwave Radiation Products from NCEP, ECMWF, GSFC, and JMA Using Satellite and Surface Observations. *Remote Sens.* **2016**, *8*, 225. [CrossRef]

17. Zhang, X.; Liang, S.; Zhou, G.; Wu, H.; Zhao, X. Generating Global LAnd Surface Satellite incident shortwave radiation and photosynthetically active radiation products from multiple satellite data. *Remote Sens. Environ.* **2014**, *152*, 318–332. [CrossRef]

18. Lu, X. Estimation of the Instantaneous Downward Surface Shortwave Radiation Using MODIS Data in Lhasa for All-Sky Conditions. Master's Thesis, Clark University, Worcester, MA, USA, 2016.

19. Barzin, R.; Shirvani, A.; Lotfi, H. Estimation of daily average downward shortwave radiation from MODIS data using principal components regression method: Fars province case study. *Int. Agrophys.* **2017**, *31*, 23–34. [CrossRef]

20. Liu, H.; Pinker, R.T. Radiative fluxes from satellites: Focus on aerosols. *J. Geophys. Res.* **2008**, *113*. [CrossRef]

21. Stone, P.H.; Risbey, J.S. On the limitations of general circulation climate models. *Geophys. Res. Lett.* **2013**, *17*, 2173–2176. [CrossRef]

22. Betts, A.K.; Zhao, M.; Dirmeyer, P.A.; Beljaars, A.C.M. Comparison of ERA40 and NCEP/DOE near-surface data sets with other ISLSCP-II data sets. *J. Geophys. Res. Atmos.* **2006**, *111*. [CrossRef]

23. Rossow, W.B.; Zhang, Y.C. Calculation of surface and top of atmosphere radiative fluxes from physical quantities based on ISCCP data sets: 2. Validation and first results. *J. Geophys. Res. Atmos.* **1995**, *100*, 1167–1197. [CrossRef]

24. Gui, S.; Liang, S.; Wang, K.; Li, L.; Zhang, X. Assessment of Three Satellite-Estimated Land Surface Downwelling Shortwave Irradiance Data Sets. *IEEE Geosci. Remote Sens. Lett.* **2010**, *7*, 776–780. [CrossRef]

25. Jia, B.; Xie, Z.; Dai, A.D.; Shi, C.; Chen, F. Evaluation of satellite and reanalysis products of downward surface solar radiation over East Asia: Spatial and seasonal variations. *J. Geophys. Res. Atmos.* **2013**, *118*, 3431–3446. [CrossRef]

26. Zhang, X.; Liang, S.; Wild, M.; Jiang, B. Analysis of surface incident shortwave radiation from four satellite products. *Remote Sens. Environ.* **2015**, *165*, 186–202. [CrossRef]

27. Kim, H.-Y.; Liang, S. Development of a hybrid method for estimating land surface shortwave net radiation from MODIS data. *Remote Sens. Environ.* **2010**, *114*, 2393–2402. [CrossRef]

28. Yang, K.; Koike, T.; Ye, B. Improving estimation of hourly, daily, and monthly solar radiation by importing global data sets. *Agric. For. Meteorol.* **2006**, *137*, 43–55. [CrossRef]

29. Cano, D.; Monget, J.M.; Albuisson, M.; Guillard, H.; Regas, N.; Wald, L. A method for the determination of the global solar radiation from meteorological satellite data. *Sol. Energy* **2010**, *37*, 31–39. [CrossRef]

30. Tang, W.J.; Yang, K.; Qin, J.; Min, M. Development of a 50-year daily surface solar radiation dataset over China. *Sci. China Earth Sci.* **2013**, *56*, 1555–1565. [CrossRef]

31. Perez, R.; Moore, K.; Wilcox, S.; Renné, D.; Zelenka, A. Forecasting solar radiation—Preliminary evaluation of an approach based upon the national forecast database. *Sol. Energy* **2007**, *81*, 809–812. [CrossRef]

32. Yang, K.; He, J.; Tang, W.; Qin, J.; Cheng, C.C.K. On downward shortwave and longwave radiations over high altitude regions: Observation and modeling in the Tibetan Plateau. *Agric. For. Meteorol.* **2010**, *150*, 38–46. [CrossRef]

33. Dongdong, W.; Shunlin, L.; Tao, H.; Qinqing, S. Estimation of Daily Surface Shortwave Net Radiation from the Combined MODIS Data. *IEEE Trans. Geosci. Remote Sens.* **2015**, *53*, 5519–5529. [CrossRef]

34. Rigollier, C.; Lefèvre, M.; Wald, L. The method Heliosat-2 for deriving shortwave solar radiation from satellite images. *Sol. Energy* **2004**, *77*, 159–169. [CrossRef]

35. Muneer, T.; Younes, S.; Munawwar, S. Discourses on solar radiation modeling. *Renew. Sustain. Energy Rev.* **2007**, *11*, 551–602. [CrossRef]

36. Mueller, R.W.; Matsoukas, C.; Gratzki, A.; Behr, H.D.; Hollmann, R. The CM-SAF operational scheme for the satellite based retrieval of solar surface irradiance—A LUT based eigenvector hybrid approach. *Remote Sens. Environ.* **2009**, *113*, 1012–1024. [CrossRef]

37. Li, Z.; Leighton, H.; Cess, R.D. Surface net solar radiation estimated from satellite measurements: Comparisons with tower observations. *J. Clim.* **1993**, *6*, 1764–1772. [CrossRef]

38. Qin, J.; Tang, W.; Yang, K.; Lu, N.; Niu, X.; Liang, S. An efficient physically based parameterization to derive surface solar irradiance based on satellite atmospheric products. *J. Geophys. Res. Atmos.* **2015**, *120*, 4975–4988. [CrossRef]

39. López, G.; Batlles, F.J. Estimating Solar Radiation from MODIS Data. *Energy Procedia* **2014**, *49*, 2362–2369. [CrossRef]

40. Mellit, A.; Eleuch, H.; Benghanem, M.; Elaoun, C.; Pavan, A.M. An adaptive model for predicting of global, direct and diffuse hourly solar irradiance. *Energy Convers. Manag.* **2010**, *51*, 771–782. [CrossRef]

41. Jiang, Y. Prediction of monthly mean daily diffuse solar radiation using artificial neural networks and comparison with other empirical models. *Energy Policy* **2008**, *36*, 3833–3837. [CrossRef]

42. Voyant, C.; Muselli, M.; Paoli, C.; Nivet, M.L. Optimization of an artificial neural network dedicated to the multivariate forecasting of daily global radiation. *Energy* **2011**, *36*, 348–359. [CrossRef]

43. Wang, T.; Yan, G.; Chen, L. Consistent retrieval methods to estimate land surface shortwave and longwave radiative flux components under clear-sky conditions. *Remote Sens. Environ.* **2012**, *124*, 61–71. [CrossRef]

44. Qin, J.; Chen, Z.; Yang, K.; Liang, S.; Tang, W. Estimation of monthly-mean daily global solar radiation based on MODIS and TRMM products. *Appl. Energy* **2011**, *88*, 2480–2489. [CrossRef]

45. Zhou, Q.; Flores, A.; Glenn, N.F.; Walters, R.; Han, B. A machine learning approach to estimation of downward solar radiation from satellite-derived data products: An application over a semi-arid ecosystem in the U.S. *PLoS ONE* **2017**, *12*. [CrossRef] [PubMed]

46. Jain, A.K.; Mao, J.; Mohiuddin, K.M. Artificial Neural Networks: A Tutorial. *Computer* **1996**, *29*, 31–44. [CrossRef]

47. McInerney, D.O.; Nieuwenhuis, M. A comparative analysis of kNN and decision tree methods for the Irish National Forest Inventory. *Int. J. Remote Sens.* **2009**, *30*, 4937–4955. [CrossRef]

48. Mubiru, J.; Banda, E.J.K.B. Estimation of monthly average daily global solar irradiation using artificial neural networks. *Sol. Energy* **2008**, *82*, 181–187. [CrossRef]

49. Lam, J.C.; Wan, K.K.W.; Yang, L. Solar radiation modelling using ANNs for different climates in China. *Energy Convers. Manag.* **2008**, *49*, 1080–1090. [CrossRef]

50. Kanamitsu, M.; Ebisuzaki, W.; Woollen, J.; Yang, S.K.; Hnilo, J.J.; Fiorino, M.; Potter, G.L. NCEP–DOE AMIP-II Reanalysis (R-2). *Bull. Am. Meteorol. Soc.* **2002**, *83*, 1631–1643. [CrossRef]

51. Friedman, J.H. Greedy function approximation: A gradient boosting machine. *Ann. Stat.* **2001**, *29*, 1189–1232. [CrossRef]

52. Ma, Y.Z.; Liu, X.N.; Xu, S. The description of Chinese radiation data and their quality control procedures. *Meteorol. Sci.* **1998**, *2*, 53–56. (In Chinese)

53. Moradi, I. Quality control of global solar radiation using sunshine duration hours. *Energy* **2009**, *34*, 1–6. [CrossRef]

54. Tang, W.; Yang, K.; He, J.; Qin, J. Quality control and estimation of global solar radiation in China. *Sol. Energy* **2010**, *84*, 466–475. [CrossRef]

55. Minnis, P.; Bedka, K.; Yost, C.R.; Bedka, S.T.; Scarino, B.A.; Khlopenkov, K.; Khaiyer, M.M. *A Consistent Long-Term Cloud and Clear-Sky Radiation Property Dataset from the Advanced Very High Resolution Radiometer (AVHRR), Climate Algorithm Theoretical Basis Document*; NOAA Climate Data Record Program

CDRP-ATBD-0826 Rev.1; CDR Program Library: South Carolina, SC, USA, 2016; Available online: http://www.ncdc.noaa.gov/cdr/operationalcdrs.html (accessed on 20 August 2017).

56. Bhatt, R.; Doelling, D.R.; Scarino, B.R.; Gopalan, A.; Haney, C.O.; Minnis, P.; Bedka, K.M. A Consistent AVHRR Visible Calibration Record Based on Multiple Methods Applicable for the NOAA Degrading Orbits. Part I: Methodology. *J. Atmos. Ocean. Technol.* **2016**, *33*, 2499–2515. [CrossRef]

57. Doelling, D.R.; Bhatt, R.; Scarino, B.R.; Gopalan, A.; Haney, C.O.; Minnis, P.; Bedka, K.M. A Consistent AVHRR Visible Calibration Record Based on Multiple Methods Applicable for the NOAA Degrading Orbits. Part II: Validation. *J. Atmos. Ocean. Technol.* **2016**, *33*, 2517–2534. [CrossRef]

58. Minnis, P.B.; Kristopher; The NOAA CDR Program. *NOAA Climate Data Record (CDR) of Cloud and Clear-Sky Radiation Properties*; Version 1.0; NOAA National Centers for Environmental Information, 2015. https://data.nodc.noaa.gov/cgi-bin/iso?id=gov.noaa.ncdc:C00876 (accessed on 20 August 2017).

59. Doelling, D.M.; Patrick; The NOAA CDR Program. *NOAA Climate Data Record (CDR) of Visible and Near Infrared Reflectance from GOES and AVHRR*; Version 1.0[C00860]; NOAA National Centers for Environmental Information., 2015. Available online: https://data.nodc.noaa.gov/cgi-bin/iso?id=gov.noaa.ncdc:C00860 (accessed on 20 August 2017).

60. Rienecker, M.M.; Suarez, M.J.; Gelaro, R.; Todling, R.; Bacmeister, J.; Liu, E.; Bosilovich, M.G.; Schubert, S.D.; Takacs, L.; Kim, G.K. MERRA: NASA's Modern-Era Retrospective Analysis for Research and Applications. *J. Clim.* **2011**, *24*, 3624–3648. [CrossRef]

61. Zhang, T.; Stackhouse, P.W.; Gupta, S.K.; Cox, S.J.; Mikovitz, J.C.; Srb, N.G. *The Effect of Cloud Fraction on the Radiative Energy Budget: The Satellite-Based GEWEX-SRB Data vs. the Ground-Based BSRN Measurements*; American Geophysical Union: Washington, DC, USA, 2011.

62. Wang, Y.; Feng, D.; Li, D.; Chen, X.; Zhao, Y.; Niu, X. A mobile recommendation system based on logistic regression and Gradient Boosting Decision Trees. In Proceedings of the 2016 International Joint Conference on Neural Networks (IJCNN), Vancouver, BC, Canada, 24–29 July 2016; pp. 1896–1902.

63. Dror, G.; Koren, Y.; Maarek, Y.; Szpektor, I. I Want to answer; who has a question?: Yahoo! answers recommender system. In Proceedings of the 17th ACM SIGKDD International Conference on Knowledge Discovery and Data Mining, San Diego, CA, USA, 21–24 August 2011; pp. 1109–1117.

64. Son, J.; Jung, I.; Park, K.; Han, B. Tracking-by-Segmentation with Online Gradient Boosting Decision Tree. In Proceedings of the IEEE International Conference on Computer Vision, Santiago, Chile, 7–13 December 2015; pp. 3056–3064.

65. Zhang, Y.; Haghani, A. A gradient boosting method to improve travel time prediction. *Transp. Res. Part C Emerg. Technol.* **2015**, *58*, 308–324. [CrossRef]

66. Chung, Y.S. Factor complexity of crash occurrence: An empirical demonstration using boosted regression trees. *Accid. Anal. Prev.* **2013**, *61*, 107–118. [CrossRef] [PubMed]

67. Xia, Y.; Jungangb, C.H.E.N. Traffic Flow Forecasting Method Based on Gradient Boosting Decision Tree. In Proceedings of the 5th International Conference on Frontiers of Manufacturing Science and Measuring Technology, Taiyuan, China, 24–25 June 2017.

68. Ding, C.; Wang, D.; Ma, X.; Li, H. Predicting Short-Term Subway Ridership and Prioritizing Its Influential Factors Using Gradient Boosting Decision Trees. *Sustainability* **2016**, *8*, 1100. [CrossRef]

69. Hastie, T.; Tibshirani, R.; Friedman, J. The elements of statistical learning. 2001. *Technometrics* **2001**, *45*, 267–268.

70. Ridgeway, G. Generalized Boosted Models: A Guide to the GBM Package. *Update* **2005**, *1*, 1–12.

71. Suzuki, K. *Artificial Neural Networks—Methodological Advances and Biomedical Applications*; Intech: Rijeka, Croatia, 2011; Available online: https://www.intechopen.com/books/citations/artificial-neural-networks-methodological-advances-and-biomedical-applications (accessed on 10 September 2017).

72. Yadav, A.K.; Chandel, S.S. Solar energy potential assessment of western Himalayan Indian state of Himachal Pradesh using J48 algorithm of WEKA in ANN based prediction model. *Renew. Energy* **2015**, *75*, 675–693. [CrossRef]

73. Şahin, M. Comparison of modelling ANN and ELM to estimate solar radiation over Turkey using NOAA satellite data. *Int. J. Remote Sens.* **2013**, *34*, 7508–7533. [CrossRef]

74. Riedmiller, M.; Braun, H. A direct adaptive method for faster backpropagation learning: The RPROP algorithm. In Proceedings of the IEEE International Conference on Neural Networks, San Francisco, CA, USA, 28 March–1 April 1993; Volume 581, pp. 586–591.

Remote Sens. **2018**, *10*, 185

75. Qian, Y.; Jia, Z.; Jiong, Y.U.; Yang, F. Application of BP-ANN to classification of hyperspectral grassland in desert. *Comput. Eng. Appl.* **2011**, *47*, 225–228.

76. Hatzianastassiou, N.; Matsoukas, C.; Fotiadi, A.; Pavlakis, K.G. Global distribution of Earth's surface shortwave radiation budget. *Atmos. Chem. Phys. Discuss.* **2005**, *5*, 2847–2867. [CrossRef]

77. Sobrino, J.A.; Raissouni, N.; Simarro, J.; Nerry, F. Atmospheric water vapor content over land surfaces derived from the AVHRR data: Application to the Iberian Peninsula. *IEEE Trans. Geosci. Remote Sens.* **1999**, *37*, 1425–1434. [CrossRef]

78. Sobrino, J.A.; Jimenez, J.C.; Raissouni, N.; Soria, G. A simplified method for estimating the total water vapor content over sea surfaces using NOAA-AVHRR channels 4 and 5. *IEEE Trans. Geosci. Remote Sens.* **2002**, *40*, 357–361. [CrossRef]

79. Xia, X.A.; Wang, P.C.; Chen, H.B.; Liang, F. Analysis of downwelling surface solar radiation in China from National Centers for Environmental Prediction reanalysis, satellite estimates, and surface observations. *J. Geophys. Res. Atmos.* **2006**, *111*. [CrossRef]

80. Wu, F.; Fu, C. Assessment of GEWEX/SRB version 3.0 monthly global radiation dataset over China. *Meteorol. Atmos. Phys.* **2011**, *112*, 155. [CrossRef]

81. Hakuba, M.Z.; Folini, D.; Sanchez-Lorenzo, A.; Wild, M. Spatial representativeness of ground-based solar radiation measurements. *J. Geophys. Res. Atmos.* **2013**, *118*, 8585–8597. [CrossRef]

82. Verrelst, J.; Muñoz, J.; Alonso, L.; Delegido, J.; Rivera, J.P.; Camps-Valls, G.; Moreno, J. Machine learning regression algorithms for biophysical parameter retrieval: Opportunities for Sentinel-2 and -3. *Remote Sens. Environ.* **2012**, *118*, 127–139. [CrossRef]

remote sensing

MDPI

Article

A Lookup-Table-Based Approach to Estimating Surface Solar Irradiance from Geostationary and Polar-Orbiting Satellite Data

Hailong Zhang [1], Chong Huang [2], Shanshan Yu [1], Li Li [1], Xiaozhou Xin [1,*] and Qinhuo Liu [1,*]

[1] State Key Laboratory of Remote Sensing Science, Jointly Sponsored by the Institute of Remote Sensing and Digital Earth, Chinese Academy of Sciences, and Beijing Normal University, Beijing 100101, China; zhlnjnu@163.com (H.Z.); yushan0427@163.com (S.Y.); lili3982@radi.ac.cn (L.L.)

[2] State Key Laboratory of Resources and Environmental Information System, Institute of Geographical Sciences and Natural Resources Research, Chinese Academy of Sciences, Beijing 100101, China; huangch@lreis.ac.cn

* Correspondence: xin_xzh@163.com (X.X.); Liuqh@radi.ac.cn (Q.L.);
   Tel.: +86-10-6487-9382 (X.X.); +86-10-6484-9840 (Q.L.)

Received: 9 December 2017; Accepted: 16 February 2018; Published: 7 March 2018

Abstract: Incoming surface solar irradiance (SSI) is essential for calculating Earth's surface radiation budget and is a key parameter for terrestrial ecological modeling and climate change research. Remote sensing images from geostationary and polar-orbiting satellites provide an opportunity for SSI estimation through directly retrieving atmospheric and land-surface parameters. This paper presents a new scheme for estimating SSI from the visible and infrared channels of geostationary meteorological and polar-orbiting satellite data. Aerosol optical thickness and cloud microphysical parameters were retrieved from Geostationary Operational Environmental Satellite (GOES) system images by interpolating lookup tables of clear and cloudy skies, respectively. SSI was estimated using pre-calculated offline lookup tables with different atmospheric input data of clear and cloudy skies. The lookup tables were created via the comprehensive radiative transfer model, Santa Barbara Discrete Ordinate Radiative Transfer (SBDART), to balance computational efficiency and accuracy. The atmospheric attenuation effects considered in our approach were water vapor absorption and aerosol extinction for clear skies, while cloud parameters were the only atmospheric input for cloudy-sky SSI estimation. The approach was validated using one-year pyranometer measurements from seven stations in the SURFRAD (SURFace RADiation budget network). The results of the comparison for 2012 showed that the estimated SSI agreed with ground measurements with correlation coefficients of 0.94, 0.69, and 0.89 with a bias of 26.4 W/m^2, −5.9 W/m^2, and 14.9 W/m^2 for clear-sky, cloudy-sky, and all-sky conditions, respectively. The overall root mean square error (RMSE) of instantaneous SSI was 80.0 W/m^2 (16.8%), 127.6 W/m^2 (55.1%), and 99.5 W/m^2 (25.5%) for clear-sky, cloudy-sky (overcast sky and partly cloudy sky), and all-sky (clear-sky and cloudy-sky) conditions, respectively. A comparison with other state-of-the-art studies suggests that our proposed method can successfully estimate SSI with a maximum improvement of an RMSE of 24 W/m^2. The clear-sky SSI retrieval was sensitive to aerosol optical thickness, which was largely dependent on the diurnal surface reflectance accuracy. Uncertainty in the pre-defined horizontal visibility for 'clearest sky' will eventually lead to considerable SSI retrieval error. Compared to cloud effective radius, the retrieval error of cloud optical thickness was a primary factor that determined the SSI estimation accuracy for cloudy skies. Our proposed method can be used to estimate SSI for clear and one-layer cloud sky, but is not suitable for multi-layer clouds overlap conditions as a lower-level cloud cannot be detected by the optical sensor when a higher-level cloud has a higher optical thickness.

Keywords: surface solar irradiance; geostationary satellite; polar orbiting satellite; LUT method; SURFRAD

1. Introduction

Surface Solar Irradiance (SSI) is commonly referred to as the amount of downward solar energy incident to a horizontal surface, and is a major component of the surface energy balance that governs the exchange processes of energy between Earth's land surface and atmosphere [1,2]. SSI is required by land-surface models, hydrological models, and ecological models to simulate land–atmosphere interactions [3,4]. Accurate observation and estimation of global energy spatial-temporal distribution is essential for climate change monitoring and forecasting [5].

A non-uniform spatial and temporal distribution of SSI has large effects on regional and global climates. However, sparse networks of ground SSI measurements are insufficient for modeling land-surface processes and Earth radiation budget research. Furthermore, fewer surface stations are located in mountainous areas, yet SSI is highly dependent on topography and features larger temporal and spatial variations than horizontal surfaces [6]. Numerous attempts have been made at estimating SSI from satellite data on local or regional scales with multi-scale temporal resolutions in order to overcome the limitations of in situ records [7–15]. Perez et al. (1997) demonstrated that satellite-derived irradiation is more accurate compared to interpolation techniques obtained from station measurements if the distance from the station exceeds 34 km for hourly irradiation and 50 km for daily irradiances [16].

Global SSI datasets have been available since the 1990s at different spatiotemporal resolutions based on multi-source remotely sensed data. These include the International Satellite Cloud Climatology Project (ISCCP) [17], the Earth Radiation Budget Experiment (ERBE) [18], the National Centers for Environmental Prediction and National Center for Atmospheric Research Reanalysis Project [19], the Global Energy and Water Cycle Experiment Surface Radiation Budget (GEWEX-SRB), the Clouds and the Earth's Radiant Energy System (CERES) [20], Satellite Application Facility on Climate Monitoring Solar Surface Radiation Heliosat (CM SAF SARAH) [21], and Global Land Surface Satellite (GLASS) products [22]. The above satellites and their parameter-based meteorological products provide long-term, multiple time-scale global SSI data, but are generally associated with coarser spatial resolutions (e.g., >1°), excluding GLASS and CM SAF SARAH, which have a 5-km resolution and bias estimation. The majority of these products cannot meet the requirements for studying land-surface processes and fail to describe the spatial changes with sufficient accuracy due to their coarse spatial resolutions [14]. Zhang et al. (2015) evaluated four products using 1151 ground sites and found that SSI was generally overestimated by approximately 10 W/m^2, while the averaged global annual mean SSI from the ground-measured-calibrated value was 180.6 W/m^2 [23]. Differences range from 10 to 30 W/m^2, with maximum discrepancies in areas of high cloud cover in the tropics between the ISCCP and ERBE datasets [24]. These differences may be partly due to spatial resolution; in fact, Pinker and Laszlo found an average difference of about 8–9% in daily surface irradiance when adjusting the resolution from 8 to 50 km [25]. Another possible explanation for the bias found in these products can be attributed to the cloud fractional cover and aerosol optical depth [6].

Satellite-based SSI products are useful for historical global SSI analysis, while the general circulation model (GCM) and the numerical weather prediction (NWP) model can be used to estimate SSI at timespans ranging from days to decades using projection scenarios of emissions and land use. The errors obtained from NWP models are generally less than 50 W/m^2 or exceed 200 W/m^2 for clear-sky and cloudy-sky conditions, respectively [26]. Lara-Fanego et al. (2012) found the forecast errors produced by Weather Research and Forecasting (WRF) to be 2% under clear-sky conditions and 18% for cloudy skies [27]. Due to the coarse resolution of most WRF models and the GCM, detailed cloud properties and Earth's energy budget have not been clearly demonstrated. Accurately estimated "kilometer-level" SSI datasets are necessary to overcome the limitations of cloud representations in the climate model.

Besides "single point" ground observations and "kilometer-level" SSI datasets, SSI can be directly retrieved using the relationship between SSI and the top-of-atmosphere (TOA) radiance measured by satellite sensors [4,11,28] or indirectly retrieved through rigorous radiative transfer models (RTMs). However, RTMs are disadvantageous since they are generally time-consuming and require a substantial amount of unavailable detailed atmospheric profile data, and are thus not convenient for applications to large areas with a fine-resolution grid. Semi-empirical models have been developed that contain meteorological variable inputs and feature a parameterized hybrid model with simplified atmospheric transmittance. Some of these models include the pre-computed lookup tables (LUT) method based on RTMs, which has a reduced computational time at the expense of accuracy. These studies considered the extinction and absorption of solar radiation caused by aerosols, water vapor (PW), ozone, and clouds [4,7,14,29].

SSI estimation under cloudy skies is much more complex compared to clear-sky models. The performance of physical SSI models under cloudy skies is largely dominated by cloud macrophysical and microphysical properties, such as cloud fractional cover (CFC), cloud optical thickness (COT), and cloud effective particle radius (ER) with high variability in space and time [30,31]. The Moderate Resolution Imaging Spectroradiometer (MODIS) onboard the Terra and Aqua satellites provides detailed and consistent atmospheric, terrestrial, and oceanic products, and studies have been developed for SSI mapping from pairs of MODIS products [2,32]. Barzin et al. (2017) proposed a combination of principal components analysis (PCA) and regression models for estimating daily average downward solar radiation using MODIS data for ten synoptic stations in Fars Province, Iran, with a root mean square error (RMSE) of 0.9–2.04 MJ/(m^2·d) [33]. The largest uncertainties resulting from SSI retrieval arise from inadequate information on cloud properties. Many studies have taken advantage of the fine spatial resolution and higher temporal resolution (5–30 min) of geostationary satellites to derive inhomogeneous and rapidly changing atmospheric parameters. The HELIOSAT algorithm uses a simplified parameterization for cloud transmission, also denoted the cloud index, derived from geostationary satellite measurements and a clear-sky model to calculate sky SSI [34]. Newly improved HELIOSAT-based models have been proposed [35]. An Artificial Neural Network (ANN) can be used to predict SSI simulation from meteorological parameters and satellite images using training data [12,36]. Few studies have focused on SSI estimation for different cloud phases despite the thermodynamic effects of cloud processes being significantly different [14].

The bispectral solar reflectance method has been widely used for retrieving COT and ER from passive satellite multispectral imagers [37]. It has been employed in cloud property retrieval for MODIS [38], the Advanced Very High Resolution Radiometer (AVHRR) [39], and the Spinning Enhanced Visible and Infrared Imager (SEVIRI) [40]. However, the passive optical remote-sensing-based pixel-level COT retrieval uncertainty will be larger than 10% for clouds having COT >70 (MODIS cloud optical properties product Algorithm Theoretical Basis Document for collection 6, MOD06/MYD06-ATBD. See details from https://modis-atmos.gsfc.nasa.gov/sites/default/files/ModAtmo/C6MOD06OPUser-Guide.pdf). The millimeter-wavelength cloud-profiling radar (CPR) on CloudSat and the cloud-aerosol lidar with orthogonal polarization (CALIOP) on Cloud-Aerosol Lidar and Infrared Pathfinder Satellite Observation (CALIPSO) provides an opportunity for detailing the structures of clouds, but the temporal resolution of them is too low to monitor the rapid changes of clouds.

This paper presents a lookup-table-based method for all-sky SSI retrieval from combined polar-orbiting and geostationary satellites with rapid retrieval of changing cloud micro-physical properties. The cloud properties retrieval was based on an assumption of a homogeneous one-layer cloud model, and the method is valid for pixels with a solar zenith angle less than 81.4 to better match the "daylight" region as referenced by MOD06/MYD06-ATBD. The SSI estimation approach we proposed should be applicable to clear sky and cloudy sky having COT <70 with less COT retrieval uncertainty as indicated by MOD06/MYD06-ATBD. This paper is organized as follows: Section 2 describes our method and the datasets used in our study to estimate SSI. Validation and a comparison of the results are provided in Sections 3 and 4. Conclusions are provided in Section 5.

The variations in cloud vertical structures and morphology affect the atmospheric circulation, radiation budget, and satellite-retrieved cloud properties. Most of the sensor-received radiance came from the top of cloud for conditions in which upper optical thick cloud overlaps the lower optical thin cloud. The maximum difference of sensor-received radiance and surface-received radiance is about 4 W/m^2 (7%) and 75 W/m^2 (60%) for upper cloud having COT of 70 when lower cloud COT changes from 1 to 100 (Figures 1 and 2. Sensor and surface-received radiance were estimated using the following parameters: solar zenith angle is 30°, surface albedo is 0.2, cloud phase is water cloud, upper cloud top height is 8 km, upper cloud base height is 6 km, lower cloud top height is 3 km, lower cloud base height is 1 km, cloud particle effective radius is 6 μm). For upper cloud having higher COT, lower-level cloud has a minor impact on satellite-retrieved cloud properties but a larger impact on SSI. Our proposed method for SSI estimation is suitable for one-layer cloud sky, and larger errors may be introduced for multi-layer clouds overlap conditions.

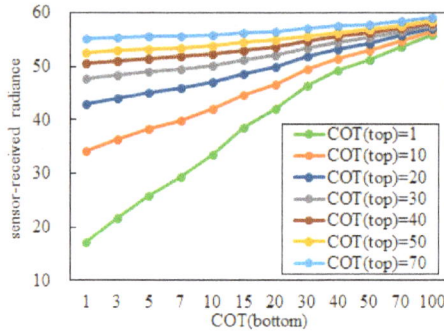

Figure 1. Estimated sensor-received radiance for multi-layer clouds overlap conditions.

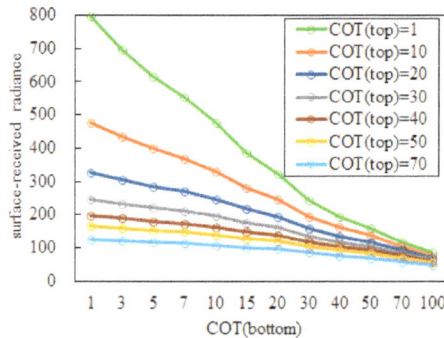

Figure 2. Estimated surface-received radiance for multi-layer clouds overlap conditions.

2. Materials and Methods

2.1. Materials

2.1.1. Geostationary Images

The data used in this study was acquired from the third-generation GOES-13 (Geostationary Operational Environmental Satellite System) satellite operated by the national environmental satellite, data, and information service of the National Oceanic and Atmospheric Administration (NOAA). GOES-13 was used for weather forecasting, severe storm tracking, and meteorology research. GOES-13 was launched on 24 May 2006, and is positioned at 75°W, 35,786 km over the Equator. The imager on-board GOES-13 scans Earth's surface every 30 min and provides five spectral channels. The nadir spatial resolution is 1 km for the visible channel (0.65 μm), 4 km for three thermal infrared channels

(3.9 μm, 6.48 μm, and 10.7 μm), and 8 km for channel 6 (13.3 μm). Details can be seen from http://www.ssec.wisc.edu/datacenter/standard_GOES8-15.html, and data can be downloaded freely from http://www.class.ncdc.noaa.gov/saa/products/welcome.

2.1.2. Ancillary Input Data

The MCD43D (V006) surface albedo product derived from the combined Terra and Aqua satellites was used in this study. The bidirectional reflectance distribution function (BRDF) was estimated from all cloud-free observations during a 16-day period. The MCD43D product incorporates the Climate Modeling Grid (CMG) structure and the pixel resolution is 1000 m. The broadband (0.2–4.0 μm) surface albedo for clear skies was calculated as the interpolation between the white-sky and black-sky albedo values dependent on the aerosol optical depth and solar zenith. Only the white-sky albedo product was used for cloudy skies due to minor differences discovered when introducing black-sky albedo for direct beam reflection [41].

The NCEP Climate Forecast System Reanalysis (CFSR) data created using the National Centers for Environmental Prediction (NCEP) Climate Forecast System version 2 (CFSv2) (https://rda.ucar.edu/datasets/ds094.1/#!description) was used in our study. The files in this dataset were grouped by month. The grid spacing was 0.205~0.204° from 0°E to 359.795°E, and 89.843°N to 89.843°S (1760 × 880 Longitude/Gaussian Latitude) [42]. Ground surface temperature was selected to be an ancillary input for the cloud effective radius retrieval method for the GEOS-13 infrared channel. The precipitable water of the entire atmosphere was selected to drive the SSI retrieval algorithm, since the 12.0-μm channel was replaced by a 13.3-μm channel. Furthermore, the GOES-13 satellite and the retrieval of precipitable water from the "split window" method (using channels 11.0 μm and 12.0 μm) was inapplicable.

The global 1-km Shuttle Radar Topography Mission with 30 arc-second resolution data (SRTM30) was used to represent the surface elevation (http://vterrain.org/Elevation/SRTM/) required for the retrieval of SSI.

2.1.3. Pyranometer Data for Validation

The Surface Radiation Budget Network (SURFRAD) was established in 1993 with the support of NOAA's Office of Global Programs. Its primary mission was to support climate research using accurate, continuous, and long-term measurements of the surface radiation budget over the United States. Seven SURFRAD stations are currently operating in climatologically diverse regions in the United States, including Fort Peck, Montana (FPK), Table Mountain, Colorado (TBL), Bondville, Illinois (BON), Goodwin Creek, Mississippi (GWN), Penn State, Pennsylvania (PSU), Desert Rock, Nevada (DRA), and Sioux Falls, South Dakota (SXF). The downwelling global solar irradiance (0.28–3 μm) is measured by a pyranometer (model SpectroSun SR-75) with reported uncertainties of ±2% to ±5% [43]. SURFRAD data are provided daily with a sample rate of 1 min (https://www.esrl.noaa.gov/gmd/grad/surfrad/). The maintenance and quality control of these measurements follow World Meteorological Organization (WMO) standards.

2.2. Methods

SSI is retrievable by assuming a homogeneous and plane-parallel atmospheric layer without considering the three-dimensional effects. The discrimination of clear and cloudy conditions was implemented by a cloud detection procedure, and the cloud thermodynamic phase was retrieved using IR channels. The cloud parameters (cloud optical thickness and effective particle radius) were inversed from the visible channel and IR channels based on the previous work of Nakajima [37,39]. SSI was estimated using a LUT-based method with the atmospheric and land-surface parameters derived above. The proposed SSI retrieval scheme is given in Figure 3. Clear and cloudy skies were first labeled using the cloud detection procedure. Aerosol optical depth (AOD) and precipitable water were retrieved for clear skies and cloud microphysical parameters were derived for cloudy skies using the pre-calculated LUT. SSI was calculated using the LUT for both clear and cloudy skies. Cloud detection is briefly described in Section 2.2.1. The details for retrieving AOD and cloud microphysical parameters are described in Sections 2.2.2 and 2.2.3.

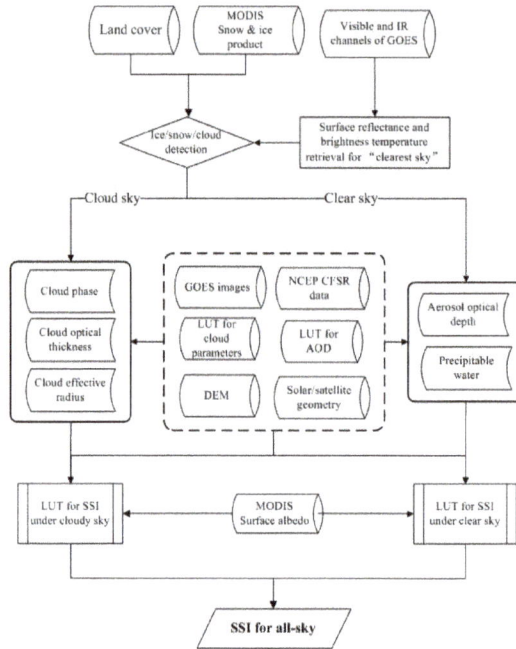

Figure 3. Flow chart of surface solar irradiance (SSI) retrieval from geostationary and polar-orbiting satellite data. MODIS: Moderate Resolution Imaging Spectroradiometer; GOES: Geostationary Operational Environmental Satellite; NCEP CFSR: National Centers for Environmental Prediction Climate Forecast System Reanalysis; LUT: lookup table; AOD: aerosol optical density; DEM: digital elevation model.

2.2.1. Pre-Processing the Images

Surface reflectance can be estimated from the visible band's at-sensor spectral radiance under clear-sky conditions through atmospheric radiative transfer models, such as the Santa Barbara DISORT Atmospheric Radiative Transfer (SBDART) [44]. The top-of-atmosphere reflectance is converted into surface reflectance given the solar-sensor geostationary viewing geometry, Rayleigh scattering, well-mixed gaseous absorption, ozone and water vapor absorption, and aerosol extinction through atmospheric correction. An aerosol visibility of 100 km and a rural model is used to represent clear atmospheric conditions. The water vapor and other trace gases are initialized with the default values of the SBDART model. Surface reflectance is determined by the minimum reflectance retrieved from the visible band taken at the same local time per daylight hour over a temporal period of one month for cloud-free detection due to the difficulty of discriminating the "clearest" atmospheric conditions. Details of the proposal have been discussed by Liang et al. (2006) [28] and Zhang et al. (2014) [4]. A 30° threshold on the glint cone angle was applied to avoid sun-glint affecting water surfaces, and a lower reflectance threshold of 0.005 was applied for the land surface to exclude cloud shadow pixels [45].

Cloud detection was performed pixel-by-pixel using the coupled Cloud Depiction and Forecast System model using the reflectance of visible bands and the brightness temperature of infrared bands [46]. This procedure incorporated temporal differencing, dynamic thresholding, and spectral discrimination to detect clouds with the appropriate optical thickness.

2.2.2. Aerosol Optical Depth Estimation

Aerosol plays a key role in Earth's radiation budget by scattering and absorbing solar and terrestrial radiation. The single broadband visible channel of most geostationary satellites is not sufficient to retrieve the aerosol size and single scattering albedo, although they are important for radiation extinction. The AOD was retrieved using the visible band of GOES with a pre-calculated LUT. The dimensions of the LUT are summarized in Table 1. The rural type was defined as incorporated in the SBDART radiative transfer code with a single scattering albedo at 0.55 μm of 0.9558 and an asymmetry factor of 0.6891. The standard atmospheric profile of the midlatitude summer model was used as the default.

Table 1. LUT dimensions for AOD retrieval.

Input Variable	Value Range	Increment
Solar zenith angle	0–89°	5°
Viewing zenith angle	0–89°	5°
Relative azimuth angle	0–180°	30°
Aerosol horizontal visibility	5, 10, 20, 30, 40, 50, 70, 100 km	-
Aerosol type	Rural	-
Water vapor	0.01–5.0 g/cm^2	0.5
Surface altitude	0–6 km	1 km
Surface reflectance	0–1.0	0.1

The LUT was pre-generated using the SBDART model for a range of discrete atmospheric and land-surface values to improve the calculation efficiency without reducing accuracy. SBDART was numerically integrated with Discrete Ordinate Radiative Transfer (DISORT), which assumes a plane-parallel radiative transfer in a vertically inhomogeneous atmosphere. The number of streams for radiance computations was 20 for the zenith angle and azimuth angle. The surface reflectance and cloud mask was determined for each pixel as described in Section 2.2.1. Aerosol horizontal visibility (VIS) was computed for every cloud-free pixel using the rural aerosol model and the given solar position, satellite position, amount of water vapor, and surface altitude. A linear interpolation of the lookup table entries to the actual aerosol visibility was used in this study. Once the VIS was known, the AOD (at 550 nm) was estimated using the following equation [44] (http://www.ncgia.ucsb.edu/projects/metadata/standard/uses/sbdart.htm):

$$AOD_{(0.55\,\mu m)} = 3.912 \times \frac{1.05 \times W + 1.51 \times (1 - W)}{VIS} \tag{1}$$

where W is a weighting factor, which is a piecewise function depending on the value of VIS and is given by the following equation:

$$\begin{cases} W = \left(\frac{1/VIS - 1/23}{1/5 - 1/23} \right), & 5 < VIS < 23 \\ W = 1, & VIS < 5 \\ W = 0, & VIS > 23 \end{cases} \tag{2}$$

2.2.3. Retrieving Cloud Microphysical Properties

The cloud thermodynamic phase, cloud optical thickness (COT), and effective particle radius were used to describe the radiative properties of clouds in the solar spectral region. The thermodynamic phases of the cloud were classified as: water clouds, ice clouds, mixed clouds, and undetected clouds following the cloud phase determination proposed by Choi et al. (2007) [47]. The retrieval method was based on the theory that cloud reflectance at non-absorbing wavelengths of the visible band is strongly related to COT, while the reflection at the absorbing wavelengths of the near infrared bands is primarily a function of ER [37]. In this study, the visible channel was used to derive COT and the IR3.9

channel was chosen to obtain ER. The radiance received by the sensor at 3.9 μm ($L_{3.9}^{obs}$) was composed of solar reflection, cloud thermal radiance, and ground thermal radiance for thin clouds. The radiance for 0.65 μm and 3.9 μm is given simply as follows:

$$L_{0.65}^{obs} = L_{0.65}^{cloud} + L_{0.65}^{sr} \tag{3}$$

$$L_{3.9}^{obs} = L_{3.9}^{cloud} + L_{3.9}^{sr} + L_{3.9}^{th(cloud)} + L_{3.9}^{th(sr)} \tag{4}$$

where $L_{0.65}^{cloud}$ and $L_{3.9}^{cloud}$ are the cloud-reflected radiance at the VIS and IR3.9 channels, respectively, $L_{0.65}^{sr}$ and $L_{3.9}^{sr}$ are the ground-reflected radiance at the VIS and IR3.9 channels, respectively. $L_{3.9}^{th(cloud)}$ and $L_{3.9}^{th(sr)}$ are the cloud and ground thermal radiance, respectively. $L_{3.9}^{sr}$ and $L_{3.9}^{th(sr)}$ were assumed to be 0 for thick clouds (COT >16). $L_{3.9}^{sr}$ and $L_{3.9}^{th(sr)}$ were simulated based on the Planck function of ground temperature (T_g) and cloud-top temperature (T_c) to remove the thermal effects of ER retrieval for thin clouds (COT <16). T_g data were derived from the NCEP-CFSv2 dataset, and T_c was approximated by the cloud-top brightness temperature given by the IR channel at 10.7 μm.

COT and ER were retrieved using LUTs generated from the SBDART one-dimensional radiative transfer code. The LUTs were calculated for different values of COT, ER, solar zenith angle, satellite viewing angle, relative azimuth angle, surface albedo, surface temperature, and cloud-top temperature using the different spectral response functions of the visible and infrared bands (Table 2). These calculations were carried out under the following assumptions: (1) there is only one single layer of clouds for every pixel, (2) the clouds are homogeneous, plane-parallel, and cover the whole pixel, and (3) ice clouds are composed of spherical particles. COT and ER were assigned to be the average value of water and ice clouds for mixed-phase clouds.

Table 2. Characteristics of LUT for the retrieval of cloud microphysical parameters.

Input Variable	Value Range	Increment
Solar zenith angle	0–89°	5°
Viewing zenith angle	0–89°	5°
Relative azimuth angle	0–180°	30°
COT	0.5, 1, 2, 5, 8, 11, 15, 20, 30, 50, 70, 100	-
ER (μm)	Water cloud: 2, 4, 8, 16, 32 Ice cloud: 2, 4, 8, 16, 32, 64	-
Surface albedo	0–1.0	0.1
Surface temperature (K)	280–320	2
Cloud-top temperature	195–300	5
Cloud phase	Water, ice	-

2.2.4. All-Sky SSI Estimation

SSI was estimated separately for clear and cloudy skies with different input data using AOD data and cloud physical parameters efficiently derived from geostationary images (as discussed in Sections 2.2.2 and 2.2.3). CO_2 and ozone were set to default values in SSI estimation since they had a negligible impact. PW and aerosol had a considerable influence on SSI in cloud-free conditions. Clouds played a dominant role in SSI during cloudy-sky conditions, and PW was set at 2.9 g/cm² as defined in the standard atmospheric profile of the midlatitude summer model. Aerosol horizontal visibility was set to 100 km for the SSI estimation of cloudy skies since AOD was insignificant compared to clouds and difficult to derive under cloudy conditions.

The all-sky SSI estimation was derived using LUTs generated for clear and cloudy skies. The common variables used for the LUTs were the solar zenith angle, surface altitude, and surface albedo. The LUT atmospheric variables for clear skies were PW and aerosol visibility, while the LUT for cloudy skies contained cloud phase, COT, and ER. The SSI for "mixed-phase clouds" was assigned to be the averaged SSI estimation for water and ice clouds. The SSI for "undetected cloud phase" pixels was calculated using the LUT of water clouds. The range of values and the increments of the above variables were the same as in Tables 1 and 2. The instantaneous SSI was estimated by linear interpolation from the lookup table once the above input data were known.

3. Results

In this section, the algorithm discussed above is evaluated using the data from seven SURFRAD stations during the entire year of 2012 and a comparison is performed with other SSI estimates. The performance of the SSI estimate is evaluated using three metrics: the mean bias error (MBE, in W/m^2), RMSE (in W/m^2), and correlation coefficient (R^2).

Huang et al. (2016) [48] demonstrated that the observed SSI averaged over 30 min was optimal for a comparison with kilometer-level satellite-based SSI estimation. Therefore, we adopted half-hour averaged SSI observations centered at the acquired time of the satellite images to evaluate the satellite-derived instantaneous SSI estimation. The validation results gathered from seven SURFRAD stations in 2012 under clear- and cloudy-sky conditions are displayed in Figure 4 and the statistics are compared in Table 3. The overall root mean square error (RMSE) values were 99.5 W/m^2 (25.5%), 80.0 W/m^2 (16.8%), and 127.6 W/m^2 (55.1%) for all-sky, clear-sky, and cloudy-sky conditions, respectively. The validation revealed a positive bias of 26.4 W/m^2 (5.5%) and a negative bias of -5.9 W/m^2 (-2.6%) for clear and cloudy skies. The RMSE values for all-sky ranged from 83.3 W/m^2 (21.7%) to 132.1 W/m^2 (32.5%), the RMSE values for clear skies ranged from 61.4 W/m^2 (11.8%) to 118.6 W/m^2 (24.7%), and the RMSE values for cloudy skies ranged from 98.5 W/m^2 (45.2%) to 141.5 W/m^2 (65.2%).

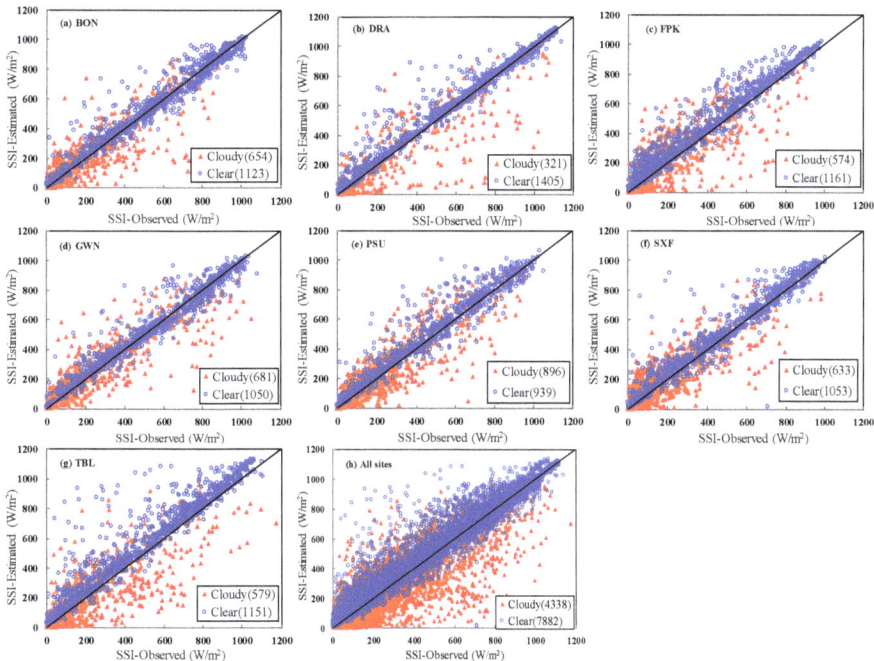

Figure 4. Validation results for the instantaneous surface solar irradiance estimated at seven Surface Radiation Budget Network (SURFRAD) stations by the scheme proposed in this study. BON: Bondville, Illinois; DRA: Desert Rock, Nevada; FPK: Fort Peck, Montana; GWN: Goodwin Creek, Mississippi; PSU: Penn State, Pennsylvania; SXF: Sioux Falls, South Dakota; TBL: Table Mountain, Colorado.

Table 3. Overview of statistics comparing estimated SSI and SURFRAD measurements for the year of 2012.

Site	Latitude	Longitude	Clear Sky			Cloudy Sky			All Sky		
			R^2	BIAS W/m² (%)	RMSE W/m² (%)	R^2	BIAS W/m² (%)	RMSE W/m² (%)	R^2	BIAS W/m² (%)	RMSE W/m² (%)
BON	40.06°N	88.37°W	0.96	6.4 (1.3)	61.0 (12.6)	0.72	4.97 (2.4)	111.4 (53.6)	0.92	4.5 (1.2)	83.3 (21.7)
DRA	36.63°N	116.02°W	0.97	19.0 (3.6)	61.4 (11.8)	0.63	−54.5 (−18.6)	178.2 (60.9)	0.92	5.3 (1.1)	94.8 (19.8)
FPK	48.31°N	105.10°W	0.94	43.8 (10.9)	79.9 (19.9)	0.60	−0.1 (−0.04)	141.5 (65.2)	0.87	29.3 (8.6)	104.4 (30.7)
GWN	34.25°N	89.87°W	0.96	2.1 (0.4)	62.2 (12.1)	0.76	−2.2 (−0.9)	122.3 (47.9)	0.91	0.4 (0.1)	90.7 (22.1)
PSU	40.72°N	77.93°W	0.93	25.7 (5.7)	81.5 (18.2)	0.80	6.0 (2.8)	98.5 (45.2)	0.90	16.1 (4.8)	90.2 (26.9)
SXF	43.73°N	96.62°W	0.92	20.6 (4.3)	81.5 (17.1)	0.69	5.7 (2.7)	111.5 (53.8)	0.90	15.0 (4.0)	93.9 (25.1)
TBL	48.31°N	105.24°W	0.90	65.2 (13.6)	118.6 (24.7)	0.60	−28.7 (−11.1)	155.6 (60.0)	0.84	33.8 (8.3)	132.1 (32.5)
All			0.94	26.4 (5.5)	80.0 (16.8)	0.69	−5.9 (−2.6)	127.6 (55.1)	0.89	14.9 (3.8)	99.5 (25.5)

These statistics indicate that the quality of the retrieval was better for clear skies, which had a correlation coefficient (R^2) ranging from 0.9 to 0.96, in comparison to cloudy skies for all stations, which featured a correlation coefficient ranging from 0.60 to 0.80; this was true for both systematic bias and scatter (Figure 4). The largest RMSE values for clear- and all-sky conditions both occurred at Table Mountain, while the smallest RMSE values for clear- and all-sky conditions occurred at Desert Rock. All stations exhibited a positive bias for clear- and all-sky conditions.

Further investigation was carried out in our study due to a larger positive bias being discovered in Table Mountain (TBL) compared to other stations with clear skies. The surface of the TBL station in Colorado was mixed by rocks, sparse grasses, desert shrubs, and small cactus, and the surface altitude was 1689 m. The positive bias was partially due to the errors of cloud detection for a mixed surface with a higher altitude. Some pixels covered by thin clouds or haze were classified as "clear sky", and thus resulted in an overestimation of SSI. Nevertheless, as is well-known, the aerosol "dark target" approach is only valid for a dense dark vegetation (DDV) surface, and it is inappropriate for the TBL station with a lower vegetation fractional cover, and thus will generally lead to substantial errors in the retrieved AOD.

On the "station observation" scale, clouds generally deviate much more under the horizontal/vertical homogeneity assumption of the SSI estimation approach than other atmospheric variables, such as aerosol and total water vapor. The inhomogeneous properties of clouds may cause substantial errors in retrieving cloud optical thickness from visible channels with a 1-km resolution and an effective particle radius from an infrared channel with approximately 4 km from satellite data. The larger discrepancies for cloudy-sky SSI estimation may be attributed to the horizontal/vertical inhomogeneity of clouds and the spatial observing scale mismatches in sensor footprints between ground-observed and satellite-retrieved data. The negative effects of the mismatches will be enlarged for a lower solar zenith and viewing zenith, resulting in poorer SSI estimation and evaluation accuracy, especially for partially covered clouds or broken clouds.

4. Discussion

4.1. Comparison with Other SSI Estimates

SSI estimation with in situ observations at SURFRAD sites were collected in order to compare the accuracy of our proposed algorithm with previous studies that estimate SSI from geostationary and polar-orbiting satellite data. The results are listed in Tables 4 and 5.

Zhang et al. (2014) used a LUT-based method from geostationary satellite images to estimate incident shortwave radiation at 5-km resolution, which was validated with observation data at seven SURFRAD sites of 2008 (Table 4) [4]. The results revealed that the RMSE values produced by our proposed method were less than the values provided by Zhang's estimation, apart from the validation at GWN, which had RMSE values of 90.7 W/m^2 and 86 W/m^2, respectively. Our proposed model exhibited an overall positive bias at all seven sites, while Zhang's model provided a negative bias at DRA and TBL. The largest bias in our model was 33.8 W/m^2 at TBL, compared with -55 W/m^2 produced by Zhang's method at DRA.

Table 4. Overview of error statistics for all-sky SSI for the year of 2008 (Zhang et al., 2014). RMSE: root mean square error.

Site	R^2	BIAS (W/m^2)	RMSE (W/m^2)
BON	0.86	20	100
DRA	0.88	-55	119
FPK	0.82	5.5	111
GWN	0.92	1.7	86
PSU	0.87	12	100
SXF	0.86	14	102
TBL	0.77	-8.7	140

Qin et al. (2015) developed a physical parameterization to estimate SSI from MODIS atmospheric and land products and the retrievals were validated against in situ measurements at SURFRAD for three years (2006–2008) (Table 5) [29]. Different validation results can be examined between our model and Qin's method. Qin's method yielded a better performance for clear sky at all sites. The unfavorable comparison results may attribute to the inaccurate input data of our model with total precipitable water at approximately 20-km resolution and the uncertainty of retrieved AOD which will be discussed later. Our model provided an improved performance with a lesser RMSE of 1–12 W/m^2 compared to Qin's method for all-sky conditions at BON, FPK, PSU, SXF, and GWN, which used input data from Aqua, while Qin's method yielded values in agreement at DRA, TBL, and GWN with input data from Terra. All three methods yielded a poorer validation at TBL; this might have been caused by pyranometer calibration accuracy, climatic conditions, and mixed ground cover. Furthermore, our proposed model indicated a lesser RMSE of about 2–4 W/m^2 compared to the method provided by Tang et al. (2016) [14], which combined an artificial neural network and parameterization model for SSI estimation from multifunctional transport satellite (MTSAT) geostationary satellite images and MODIS atmospheric and land products. The overall accuracy of our model with an RMSE of 99.5 W/m^2 (25.5%) is comparable to the MODIS-products-driven Breathing Earth System Simulator (BESS) shortwave products with an RMSE of 111.1 W/m^2 (22.6%) and 137.1 W/m^2 (31.7%) for temperate and continental climate zones, respectively [49].

As indicated by Yeom, a reduced RMSE of about 10 W/m^2 can be found with the spatial resolution changed from 1 km to 5 km [13]. Considering the estimated SSI of our model with 1-km resolution and the referenced studies with 5-km resolution, we can draw a conclusion that the performance of our proposed scheme was comparable with or even more accurate than state-of-the-art satellite-based SSI retrieval models.

Table 5. Overview of error statistics for all-sky SSI derived from MODIS products for the years 2006–2008 (Qin et al., 2015).

Site	Clear Sky (W/m^2)				All Sky (W/m^2)			
	Terra		Aqua		Terra		Aqua	
	BIAS	RMSE	BIAS	RMSE	BIAS	RMSE	BIAS	RMSE
BON	11.5	41.1	15.3	54.4	4.0	86.3	7.6	95.0
DRA	−11.3	41.9	8.4	34.4	−11.0	55.0	8.1	69.7
FPK	20.2	43.8	29.9	49.0	−4.3	105.6	7.8	95.1
GWN	21.7	47.2	24.7	56.7	17.1	72.4	22.0	92.8
PSU	27.0	57.8	25.1	59.7	21.8	101.7	15.1	99.0
SXF	17.7	43.3	19.1	47.0	−5.3	101.0	−2.2	98.8
TBL	2.1	37.6	7.1	42.9	−17.7	113.6	−1.9	123.0

4.2. Error Analysis in SSI Retrieval

Aerosol and clouds are the primary atmospheric parameters (besides the solar zenith and surface altitude) that affect SSI for clear and cloudy skies. The retrieval uncertainty of these two parameters will be discussed in this section.

The diurnal change of the underlying surface reflectance is a key parameter for AOD retrieval, and it is gathered by searching for the minimum value of surface reflectance within a 30-day period. The surface reflectance was inversed using a lookup-table-based method and a horizontal visibility set to 100 km (which was approximated to be 0.06 of the AOD value using the relationship between the VIS and AOD as indicated by Equations (1) and (2)). However, the assumption will inevitably introduce some uncertainty since a great spatial and temporal variation of aerosol has been discovered. The AOD data from SURFRAD sites generated from visible Multi-Filter Rotating Shadow band Radiometers (MFRSR) were collected in our study to further investigate the changes in AOD. The statistical results of observed AOD gathered from six SURFRAD sites are displayed in Figure 5 (except for PSU since there was no observed AOD data in 2012).

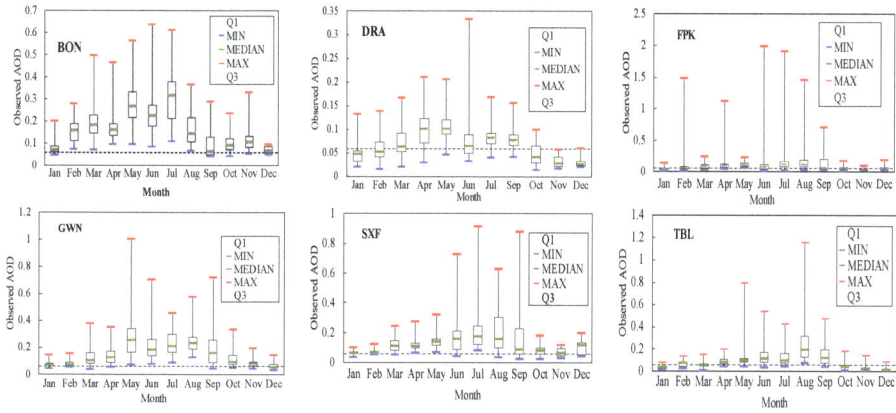

Figure 5. Statistical results of observed AOD at six SURFRAD stations. The dashed lines on the vertical axis are AOD values of 0.06 for pre-defined clearest-sky conditions of diurnal surface reflectance retrieval.

A high yearly temporal variation can be found at all six sites. A maximum AOD value was found in summer and a minimum value in winter, and a maximum variability of AOD occurred during the summer months. The pre-defined AOD of 0.06 for clearest-sky conditions had: an overall underestimation at BON, SXF, and GWN for the entire year of 2012; an overall underestimation in summer and overestimation in winter at TBL and DRA; and an overall underestimation in summer at FPK. A lower assumed AOD may have been responsible for the SSI overestimation in our proposed method. The surface reflectance was overestimated due to the lower predefined AOD value for clearest-sky conditions, which in turn resulted in underestimations of the retrieved AOD and the overall overestimation of SSI for clear skies. The surface reflectance was invariable for every month, and larger uncertainty had arisen for snow seasons with higher reflectance and vegetation-growing seasons with a lower reflectance. Furthermore, an overestimation of surface reflectance caused an underestimation of cloud optical thickness and an underestimation of cloud transmittance for transparency or semi-transparency. Besides the uncertainty of surface reflectance, the aerosol attenuation effects were influenced by a large solar zenith angle and bright surface.

Cloud microphysical parameters, such as cloud optical thickness, thermal phase, and effective particle radius, are important variables in estimating SSI. Large negative effects on the performance of the SSI retrievals were possibly caused by cloud parameter retrieval errors due to the inhomogeneity and spatiotemporal variation of clouds. The sensitivity of SSI to cloud optical thickness and the effective radius of water clouds is presented in Figure 6. SSI was estimated using the rigorous radiative transfer model (SBDART) with the input variables set as: a midlatitude summer atmospheric profile with total precipitable water of 2.92 g/cm^2, a total ozone column of 320 DU, a solar zenith angle of $30°$, a surface elevation of 0 km, a surface albedo of 0.2, and a horizontal visibility of 100 km. A positive relationship between the effective particle radius and SSI can be seen, while a negative relationship between the SSI and cloud optical thickness can be observed. SSI will change about 88 W/m^2 and 90 W/m^2 with the effective particle radius range of 2 μm to 30 μm for water and ice clouds, respectively. SSI is not dependent on the effective radius when the cloud particle radius is greater than 20 μm since the variation of SSI does not exceed 5 W/m^2. The SSI will change by about 548 and 600 W/m^2 for a cloud optical thickness range of 2 to 30 for water and ice clouds, respectively. It can be concluded that the error in SSI estimation for cloudy skies is primarily affected by the uncertainty of the cloud optical thickness retrieval. There may be more than one solution for optical thickness and effective radius retrieval for optically thin clouds. The retrieved cloud optical thickness only represents 20–40% of the total optical thickness of the total cloud layer for clouds having COT ≥ 8, as indicated by Nakajima [37]. This situation can partially explain the positive bias of SSI estimation for cloudy skies.

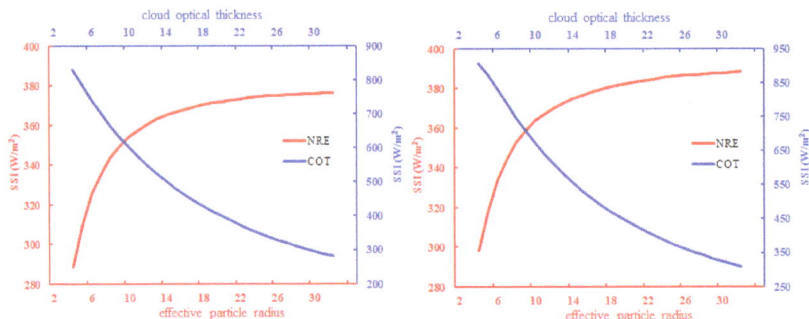

Figure 6. The sensitivity of SSI to cloud optical thickness (given an effective particle radius of 20 μm) and effective particle radius (given a cloud optical thickness of 20) for water clouds (**left**) and ice clouds (**right**).

An overall underestimation of SSI with a maximum bias of 62.5 W/m^2 and a minimum of 1 W/m^2 for water clouds is indicated in Figure 7 and Table 6. All scatterplots for water and ice clouds are uniformly distributed between the line of 1:1. The RMSE has a maximum value of 177.6 W/m^2 at DRA, thereby exceeding the values for clear skies (61.4 W/m^2) by a factor of three. The minimum RMSE value is 80.2 W/m^2 at SXF, which is nearly the same as clear skies. Compared with water clouds, ice clouds tend to have a larger RMSE and a lower correlation coefficient (Figure 8, Table 6). The validation results for ice cloud cases reveal a negative bias at DRA and TBL, which have an elevation greater than 1000 m. The largest RMSE for ice clouds was 211.3 W/m^2, which is about 75.3% of the systematic error. The relative accuracy of the modeled SSI for ice cloud cases is lower than 40%. This might be caused by the ice crystal density, particle size, shape, or direction, which are difficult to derive and describe for accurate scattering computation.

Besides the uncertainty of the input variables derived from satellite data, large uncertainty may arise from the assumption of plane-parallel, homogeneous atmospheric conditions. Furthermore, a one-dimensional atmospheric transfer model cannot deal with the geometrical effects of scattering from a higher solar zenith. The model is less reliable when the sub-pixel is partially cloudy, or when a rapid change in the atmospheric profile occurs during satellite observations.

Table 6. Validation results at seven SURFRAD stations for water and ice clouds.

Site	Water Clouds				Ice Clouds				Mixed Clouds	Undetected Clouds
	R^2	Bias W/m^2 (%)	RMSE W/m^2 (%)	NO.	R^2	Bias W/m^2 (%)	RMSE W/m^2 (%)	NO.	NO.	NO.
BON	0.74	−8 (−3.5)	106 (49.5)	280	0.66	27 (19.8)	93 (68.7)	121	222	31
DRA	0.70	−63 (−21.0)	178 (59.6)	149	0.38	−44 (−15.7)	211 (75.3)	78	74	20
FPK	0.62	−25 (−10.0)	157 (63.7)	187	0.54	15 (6.6%)	150 (67.2)	133	220	34
GWN	0.78	−31 (−10.6)	136 (46.3)	326	0.75	45 (32.0)	101 (71.2)	130	181	44
PSU	0.81	−1 (−0.42)	105 (43.7)	576	0.81	44 (35.8)	83 (67.7)	83	217	20
SXF	0.80	−5 (−2.6)	80 (43.0)	233	0.49	29 (16.5)	135 (78.1)	109	243	47
TBL	0.64	−38 (−12.9)	169 (57.5)	197	0.61	−0.2 (−0.1)	118 (61.1)	186	173	20

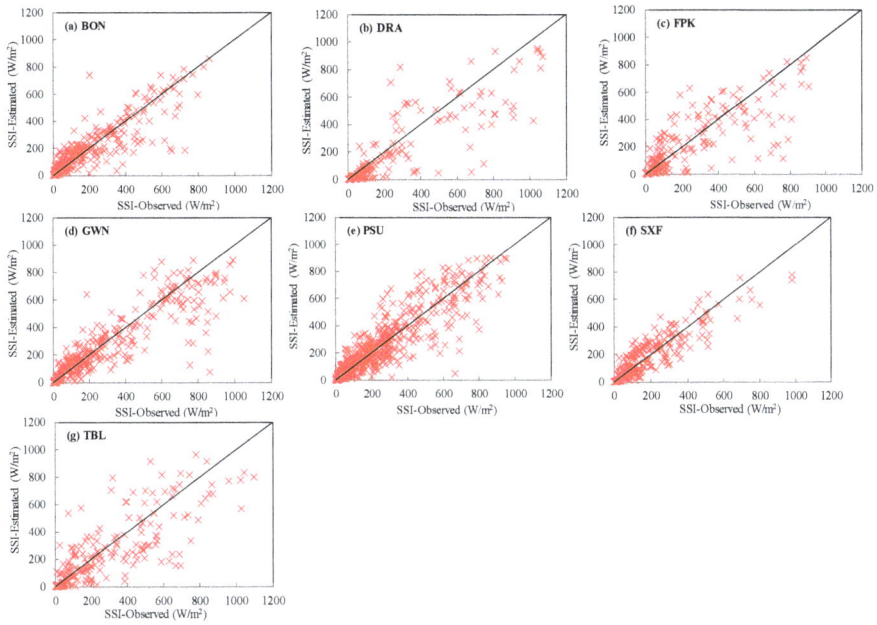

Figure 7. Validation of SSI for water cloud cases at SURFRAD sites (mixed and undetected clouds were not included in the comparison).

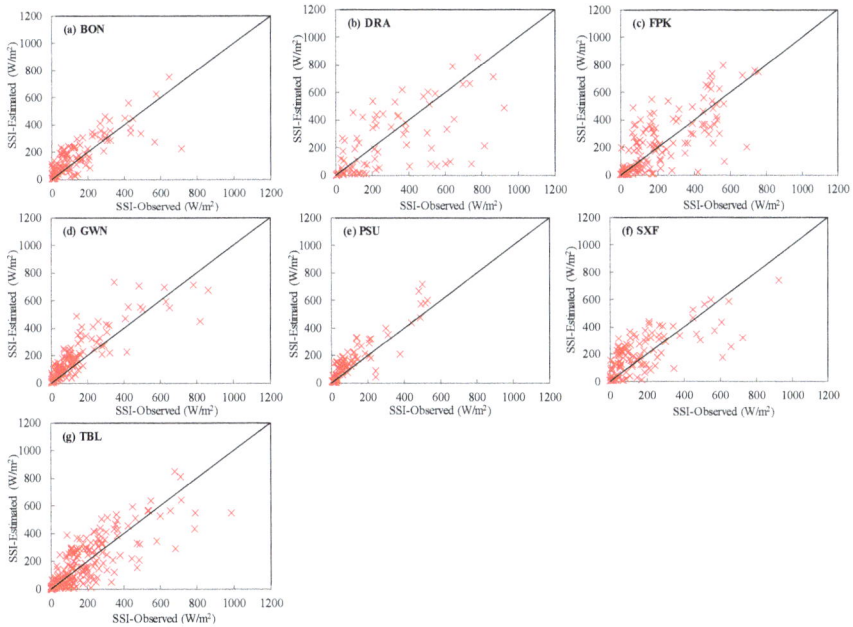

Figure 8. Validation of SSI for ice cloud cases at SURFRAD sites (mixed and undetected clouds were not included in the comparison).

Remote Sens. **2018**, *10*, 411

5. Conclusions

The paper describes a novel approach to estimating surface solar irradiance (SSI) from a combined geostationary satellite image, MODIS land-surface albedo, and NCEP CFSR data. Aerosol optical thickness and cloud parameters (cloud phase, effective particle radius, and cloud optical thickness) were directly retrieved from the visible channel of geostationary satellite images for clear and cloudy skies. Total precipitable water was derived from the NCEP data, and other atmospheric variables, such as ozone, carbon dioxide, and trace gases, were not considered in our SSI estimation. The SSI was obtained by searching for and linearly interpolating a pre-calculated lookup table, which was created using the one-dimensional radiative transfer model for computational efficiency at the cost of calculation accuracy.

The validation was performed via station observations at SURFRAD and other developed algorithms were used with input from satellite data on an instantaneous basis to evaluate the performance of the estimates. The results demonstrated that our method could effectively retrieve instantaneous SSI with correlation coefficients of 0.94, 0.69, and 0.89, and an overall RMSE of 80.0 W/m^2 (16.8%), 127.6 W/m^2 (55.1%), and 99.5 W/m^2 (25.5%) for clear-sky, cloudy-sky, and all-sky conditions, respectively. Our algorithm generally overestimated the SSI for clear- and all-sky conditions. Uncertainty analysis revealed that the accuracy of AOD retrieval was largely dependent upon diurnal surface reflectance. An overestimation of surface reflectance resulted in an underestimation of AOD and led to an overestimation of SSI. Large uncertainty may arise for optically thin clouds due to the ambiguous solutions for cloud optical thickness and effective radius. The RMSE for ice clouds is generally larger than water clouds since the radiative transfer process for ice clouds is mainly affected by ice crystal shape and particle size, which are difficult to directly retrieve with acceptable accuracy. In summary, our proposed method holds great promise for accurately estimating regional or global SSI and conducting research on Earth's energy budget using products from geostationary satellites, such as FY2, Himawari-8, MTG, and MODIS.

Acknowledgments: This study was financially supported by the Strategic Priority Research Program of Chinese Academy of Sciences (Grant No. XDA15012002, XDA15007502) and the National Natural Science Foundation of China (Grant No. 41771394, 41471335, 41201352). NCEP data are available via https://rda.ucar.edu/datasets/ds094.1/#!description. The in situ SSI data collected at the SURFRAD station in the United States are available at https://www.esrl.noaa.gov/gmd/grad/surfrad/. GOES satellite imaging radiometer data are freely downloadable from https://www.class.ncdc.noaa.gov/saa/products/welcome.

Author Contributions: Hailong Zhang and Qinhuo Liu designed the research. Xiaozhou Xin processed the satellite and observation data. Chong Huang evaluated and analyzed the model's performance. Shanshan Yu and Li Li designed the cloud detection method.

Conflicts of Interest: The authors declare no conflict of interest.

References

1. Wild, M. Global dimming and brightening: A review. *J. Geophys. Res. Atmos.* **2009**, *114*. [CrossRef]
2. Houborg, R.; Soegaard, H.; Emmerich, W.; Moran, S. Inferences of all-sky solar irradiance using Terra and Aqua MODIS satellite data. *Int. J. Remote Sens.* **2007**, *28*, 4509–4535. [CrossRef]
3. Huang, J.; Yu, H.; Guan, X.; Wang, G.; Guo, R. Accelerated dryland expansion under climate change. *Nat. Clim. Chang.* **2016**, *6*, 166–171. [CrossRef]
4. Zhang, X.; Liang, S.; Zhou, G.; Wu, H.; Zhao, X. Generating Global LAnd Surface Satellite incident shortwave radiation and photosynthetically active radiation products from multiple satellite data. *Remote Sens. Environ.* **2014**, *152*, 318–332. [CrossRef]
5. Trenberth, K.E. An imperative for climate change planning: Tracking Earth's global energy. *Curr. Opin. Environ. Sustain.* **2009**, *1*, 19–27. [CrossRef]
6. Alexandri, G.; Georgoulias, A.K.; Meleti, C.; Balis, D.; Kourtidis, K.A.; Sanchez-Lorenzo, A.; Trentmann, J.; Zanis, P. A high resolution satellite view of surface solar radiation over the climatically sensitive region of Eastern Mediterranean. *Atmos. Res.* **2017**, *188*, 107–121. [CrossRef]

7. Yang, K.; Koike, T.; Ye, B. Improving estimation of hourly, daily, and monthly solar radiation by importing global data sets. *Agric. For. Meteorol.* **2006**, *137*, 43–55. [CrossRef]
8. Wang, L.; Kisi, O.; Zounemat-kermani, M.; Ariel, G.; Zhu, Z.; Gong, W. Solar radiation prediction using different techniques: Model evaluation and comparison. *Renew. Sustain. Energy Rev.* **2016**, *61*, 384–397. [CrossRef]
9. Shamim, M.A.; Remesan, R.; Bray, M.; Han, D. An improved technique for global solar radiation estimation using numerical weather prediction. *J. Atmos. Sol. Terr. Phys.* **2015**, *129*, 13–22. [CrossRef]
10. Zou, L.; Wang, L.; Lin, A.; Zhu, H.; Peng, Y.; Zhao, Z. Estimation of global solar radiation using an arti fi cial neural network based on an interpolation technique in southeast China. *J. Atmos. Sol. Terr. Phys.* **2016**, *146*, 110–122. [CrossRef]
11. Xie, Y.; Sengupta, M.; Dudhia, J. A Fast All-sky Radiation Model for Solar applications (FARMS): Algorithm and performance evaluation. *Sol. Energy* **2016**, *135*, 435–445. [CrossRef]
12. Zou, L.; Wang, L.; Xia, L.; Lin, A.; Hu, B.; Zhu, H. Prediction and comparison of solar radiation using improved empirical models and Adaptive Neuro-Fuzzy Inference Systems. *Renew. Energy* **2017**, *106*, 343–353. [CrossRef]
13. Yeom, J.; Seo, Y.; Kim, D.; Han, K. Solar Radiation Received by Slopes Using COMS Imagery, a Physically Based Radiation Model, and GLOBE. *J. Sens.* **2016**, *2016*, 4834579. [CrossRef]
14. Tang, W.; Qin, J.; Yang, K.; Liu, S.; Lu, N.; Niu, X. Retrieving high-resolution surface solar radiation with cloud parameters derived by combining MODIS and MTSAT data. *Atmos. Chem. Phys.* **2016**, *16*, 2543–2557. [CrossRef]
15. Arbizu-Barrena, C.; Ruiz-Arias, J.A.; Rodríguez-Benítez, F.J.; Pozo-Vázquez, D.; Tovar-Pescador, J. Short-term solar radiation forecasting by advecting and diffusing MSG cloud index. *Sol. Energy* **2017**, *155*, 1092–1103. [CrossRef]
16. Perez, R.; Seals, R.; Zelenka, A. Comparing satellite remote sensing and ground network measurements for the production of site/time specific irradiance data. *Sol. Energy* **1997**, *60*, 89–96. [CrossRef]
17. Bishop, J.K.B.; Rossow, W.B. Spatial and Temporal Variability of Global Surface Solar Irradiance. *J. Geophys. Res.* **1991**, *96*, 16839–16858. [CrossRef]
18. Li, Z.; Leighton, H. Global climatologies of solar radiation budgets at the surface and in the atmosphere from 5 years of ERBE data. *J. Geophys. Res. Atmos.* **1993**, *98*, 4919–4930. [CrossRef]
19. Kalnay, E.; Kanamitsu, M.; Kistler, R.; Collins, W.; Deaven, D.; Gandin, L.; Iredell, M.; Saha, S.; White, G.; Woollen, J.; et al. The NCEP/NCAR 40-Year Reanalysis Project. *Bull. Am. Meteorol. Soc.* **1996**, *77*, 437–471. [CrossRef]
20. Kato, S.; Loeb, N.G.; Rose, F.G.; Doelling, D.R.; Rutan, D.A.; Caldwell, T.E.; Yu, L.; Weller, R.A. Surface Irradiances Consistent with CERES-Derived Top-of-Atmosphere Shortwave and Longwave Irradiances. *J. Clim.* **2012**, *26*, 2719–2740. [CrossRef]
21. Müller, R.; Pfeifroth, U.; Träger-Chatterjee, C.; Trentmann, J.; Cremer, R. Digging the METEOSAT treasure-3 decades of solar surface radiation. *Remote Sens.* **2015**, *7*, 8067–8101. [CrossRef]
22. Liang, S.; Zhao, X.; Liu, S.; Yuan, W.; Cheng, X.; Xiao, Z.; Zhang, X.; Liu, Q.; Cheng, J.; Tang, H.; et al. A long-term Global LAnd Surface Satellite (GLASS) data-set for environmental studies. *Int. J. Digit. Earth* **2013**, *6*, 5–33. [CrossRef]
23. Zhang, X.; Liang, S.; Wild, M.; Jiang, B. Analysis of surface incident shortwave radiation from four satellite products. *Remote Sens. Environ.* **2015**, *165*, 186–202. [CrossRef]
24. Seager, R.; Blumenthal, M.B. Modeling Tropical Pacific Sea Surface Temperature with Satellite-Derived Solar Radiative Forcing. *J. Clim.* **1994**, *7*, 1943–1957. [CrossRef]
25. Pinker, R.T.; Laszlo, I. Effects of Spatial Sampling of Satellite Data on Derived Surface Solar Irradiance. *J. Atmos. Ocean. Technol.* **1991**, *8*, 96–107. [CrossRef]
26. Mathiesen, P.; Kleissl, J. Evaluation of numerical weather prediction for intra-day solar forecasting in the continental United States. *Sol. Energy* **2011**, *85*, 967–977. [CrossRef]
27. Lara-Fanego, V.; Ruiz-Arias, J.A.; Pozo-Vázquez, D.; Santos-Alamillos, F.J.; Tovar-Pescador, J. Evaluation of the WRF model solar irradiance forecasts in Andalusia (southern Spain). *Sol. Energy* **2012**, *86*, 2200–2217. [CrossRef]

28. Liang, S.; Zheng, T.; Liu, R.; Fang, H.; Tsay, S.-C.; Running, S. Estimation of incident photosynthetically active radiation from Moderate Resolution Imaging Spectrometer data. *J. Geophys. Res.* **2006**, *111*, D15208. [CrossRef]

29. Qin, J.; Tang, W.; Yang, K.; Lu, N.; Niu, X.; Liang, S. An efficient physically based parameterization to derive surface solar irradiance based on satellite atmospheric products. *J. Geophys. Res. Atmos.* **2015**, *120*, 4975–4988. [CrossRef]

30. Sun, Z.; Liu, A. Fast scheme for estimation of instantaneous direct solar irradiance at the earth's surface. *Sol. Energy* **2013**, *98*, 125–137. [CrossRef]

31. Alexandri, G.; Georgoulias, A.K.; Zanis, P.; Katragkou, E.; Tsikerdekis, A.; Kourtidis, K.; Meleti, C. On the ability of RegCM4 regional climate model to simulate surface solar radiation patterns over Europe: An assessment using satellite-based observations. *Atmos. Chem. Phys.* **2015**, *15*, 13195–13216. [CrossRef]

32. Chen, M.; Zhuang, Q.; He, Y. An efficient method of estimating downward solar radiation based on the MODIS observations for the use of land surface modeling. *Remote Sens.* **2014**, *6*, 7136–7157. [CrossRef]

33. Barzin, R.; Shirvani, A.; Lotfi, H. Estimation of daily average downward shortwave radiation from MODIS data using principal components regression method: Fars province case study. *Int. Agrophys.* **2017**, *31*, 23–34. [CrossRef]

34. Cano, D.; Monget, J.; Albuisson, M.; Guillard, H.; Regas, N.; Wald, L. A method for the determination of the global solar radiation from meteorological satellite data. *Sol. Energy* **1986**, *37*, 31–39. [CrossRef]

35. Castelli, M.; Stöckli, R.; Zardi, D.; Tetzlaff, A.; Wagner, J.E.; Belluardo, G.; Zebisch, M.; Petitta, M. The HelioMont method for assessing solar irradiance over complex terrain: Validation and improvements. *Remote Sens. Environ.* **2014**, *152*, 603–613. [CrossRef]

36. Linares-Rodriguez, A.; Ruiz-Arias, J.A.; Pozo-Vazquez, D.; Tovar-Pescador, J. An artificial neural network ensemble model for estimating global solar radiation from Meteosat satellite images. *Energy* **2013**, *61*, 636–645. [CrossRef]

37. Nakajima, T.; King, M.D.; Nakajima, T.; King, M.D. Determination of the Optical Thickness and Effective Particle Radius of Clouds from Reflected Solar Radiation Measurements. Part I: Theory. *J. Atmos. Sci.* **1990**, *47*, 1878–1893. [CrossRef]

38. Platnick, S.; King, M.D.; Ackerman, S.A.; Menzel, W.P.; Baum, B.A.; Riédi, J.C.; Frey, R.A. The MODIS cloud products: Algorithms and examples from terra. *IEEE Trans. Geosci. Remote Sens.* **2003**, *41*, 459–472. [CrossRef]

39. Nakajima, T.Y.; Nakjma, T. Wide-Area Determination of Cloud Microphysical Properties from NOAA AVHRR Measurements for FIRE and ASTEX Regions. *J. Atmos. Sci.* **1995**, *52*, 4043–4059. [CrossRef]

40. Roebeling, R.A.; Feijt, A.J.; Stammes, P. Cloud property retrievals for climate monitoring: Implications of differences between Spinning Enhanced Visible and Infrared Imager (SEVIRI) on METEOSAT-8 and Advanced Very High Resolution Radiometer (AVHRR) on NOAA-17. *J. Geophys. Res. Atmos.* **2006**, *111*, 1–51. [CrossRef]

41. Deneke, H.M.; Feijt, A.J.; Roebeling, R.A. Estimating surface solar irradiance from METEOSAT SEVIRI-derived cloud properties. *Remote Sens. Environ.* **2008**, *112*, 3131–3141. [CrossRef]

42. Saha, S.; Moorthi, S.; Wu, X.; Wang, J.; Nadiga, S.; Tripp, P.; Behringer, D.; Hou, Y.T.; Chuang, H.Y.; Iredell, M.; et al. The NCEP climate forecast system version 2. *J. Clim.* **2014**, *27*, 2185–2208. [CrossRef]

43. Augustine, J.A.; DeLuisi, J.J.; Long, C.N. SURFRAD—A National Surface Radiation Budget Network for Atmospheric Research. *Bull. Am. Meteorol. Soc.* **2000**, *81*, 2341–2357. [CrossRef]

44. Ricchiazzi, P.; Yang, S.; Gautier, C.; Sowle, D. SBDART: A Research and Teaching Software Tool for Plane-Parallel Radiative Transfer in the Earth's Atmosphere. *Bull. Am. Meteorol. Soc.* **1998**, *79*, 2101–2114. [CrossRef]

45. Popp, C.; Hauser, A.; Foppa, N.; Wunderle, S. Remote sensing of aerosol optical depth over central Europe from MSG-SEVIRI data and accuracy assessment with ground-based AERONET measurements. *J. Geophys. Res. Atmos.* **2007**, *112*, 1–16. [CrossRef]

46. D'Entremont, R.P.; Gustafson, G.B. Analysis of Geostationary Satellite Imagery Using a Temporal-Differencing Technique. *Earth Interact.* **2003**, *7*, 1–25. [CrossRef]

47. Choi, Y.S.; Ho, C.H.; Ahn, M.H.; Kim, Y.M. An exploratory study of cloud remote sensing capabilities of the Communication, Ocean and Meteorological Satellite (COMS) imagery. *Int. J. Remote Sens.* **2007**, *28*, 4715–4732. [CrossRef]

48. Huang, G.; Li, X.; Huang, C.; Liu, S.; Ma, Y.; Chen, H. Representativeness errors of point-scale ground-based solar radiation measurements in the validation of remote sensing products. *Remote Sens. Environ.* **2016**, *181*, 198–206. [CrossRef]

49. Ryu, Y.; Jiang, C.; Kobayashi, H.; Detto, M. MODIS-derived global land products of shortwave radiation and diffuse and total photosynthetically active radiation at 5 km resolution from 2000. *Remote Sens. Environ.* **2018**, *204*, 812–825. [CrossRef]

remote sensing

MDPI

Article

Estimation of Daily Average Downward Shortwave Radiation over Antarctica

Yingji Zhou [1,2], Guangjian Yan [1,2,*], Jing Zhao [3], Qing Chu [1,2], Yanan Liu [1,2], Kai Yan [1,2], Yiyi Tong [1,2], Xihan Mu [1,2], Donghui Xie [1,2] and Wuming Zhang [1,2]

[1] State Key Laboratory of Remote Sensing Science, Jointly Sponsored by Beijing Normal University and Institute of Remote Sensing and Digital Earth of Chinese Academy of Sciences, Beijing 100875, China; zhouyj_41@163.com (Y.Z.); striving321@126.com (Q.C.); 201731170031@mail.bnu.edu.cn (Y.L.); 201121170050@mail.bnu.edu.cn (K.Y.); tongyiyi0311@163.com (Y.T.); muxihan@bnu.edu.cn (X.M.); xiedonghui@bnu.edu.cn (D.X.); wumingz@bnu.edu.cn (W.Z.)

[2] Beijing Engineering Research Center for Global Land Remote Sensing Products, Institute of Remote Sensing Science and Engineering, Faculty of Geographical Science, Beijing Normal University, Beijing 100875, China;

[3] Harbin Institute of Technology Shenzhen Graduate School, Shenzhen 518000, China; 201131170021@mail.bnu.edu.cn

* Correspondence: gjyan@bnu.edu.cn; Tel.: +86-10-5880-2085

Received: 11 January 2018; Accepted: 19 February 2018; Published: 9 March 2018

Abstract: Surface shortwave (SW) irradiation is the primary driving force of energy exchange in the atmosphere and land interface. The global climate is profoundly influenced by irradiation changes due to the special climatic condition in Antarctica. Remote-sensing retrieval can offer only the instantaneous values in an area, whilst daily cycle and average values are necessary for further studies and applications, including climate change, ecology, and land surface process. When considering the large values of and small diurnal changes of solar zenith angle and cloud coverage, we develop two methods for the temporal extension of remotely sensed downward SW irradiance over Antarctica. The first one is an improved sinusoidal method, and the second one is an interpolation method based on cloud fraction change. The instantaneous irradiance data and cloud products are used in both methods to extend the diurnal cycle, and obtain the daily average value. Data from South Pole and Georg von Neumayer stations are used to validate the estimated value. The coefficient of determination (R^2) between the estimated daily averages and the measured values based on the first method is 0.93, and the root mean square error (RMSE) is 32.21 W/m^2 (8.52%). As for the traditional sinusoidal method, the R^2 and RMSE are 0.68 and 70.32 W/m^2 (18.59%), respectively The R^2 and RMSE of the second method are 0.96 and 25.27 W/m^2 (6.98%), respectively. These values are better than those of the traditional linear interpolation (0.79 and 57.40 W/m^2 (15.87%)).

Keywords: downward shortwave radiation; daily average value; Antarctica; sinusoidal method; cloud fraction; interpolation

1. Introduction

Solar shortwave (SW) radiation reaching the surface of the Earth is the primary energy source, which plays a significant role in surface energy balance, temperature variations, hydrological cycle, and terrestrial net primary productivity [1–3]. Although the annual change is small, the impact on the global climate is difficult to ignore for an 'amplification effect' [4–6].

Antarctica is the coldest, highest, driest, and windiest continent in the Earth [7]. The surface changes that are caused by irradiation in the Antarctic area affect the entire planet by the ice albedo feedback mechanism [4,8,9]. Therefore, studying the changes in the SW irradiation in Antarctica is significant [10,11].

Irradiation flux data are required in many regional climate system models and applications as input parameters [12–14]. However, the irradiation that was measured by ground observation stations has been proven to be spatially inadequate [15]. Satellite remote sensing technique is a suitable way to obtain solar irradiance data at continent scale [16]. The estimated values based on remote-sensing images are instantaneous. Climate, ecology, and land surface process models require daily average or diurnal cycle data. Directly integrating daily solar irradiation values on the basis of few instantaneous irradiance values is inaccurate. Therefore, numbers of methods, including empirical method [17], sinusoidal method [18–20], meteorological parameter interpolation method [21,22], lookup table (LUT) method [23], quadratic polynomial regression method [24], and polar orbit and static satellite data fusion method, have been developed to obtain daily values [25].

The empirical method uses a large number of ground entity sample data to establish the empirical relationship between instantaneous irradiance and daily average irradiation to calculate daily values [17]. On the one hand, this method has high accuracy. On the other hand, this method relies on a large amount of surface data, which means that it is limited when ground stations are inadequate.

The sinusoidal model method assumes that surface irradiation follows the sinusoidal curve on the time scale, which means that daily average irradiation can be calculated on the basis of instantaneous irradiance and satellite overpass time [19]. This method also assumes that the change in solar zenith angle (SZA) is the main factor for the daily variation in irradiance. However, in polar regions, this assumption is not true due to the large SZA and perpetual day. Furthermore, this method ignores the cloud influence, and it is only applicable under clear sky conditions.

The direct meteorological parameter interpolation method extends meteorological parameters by linear interpolation to estimate irradiance, using temporal scaling-up meteorological data [21]. This method is easy to operate, although the number of instantaneous meteorological values is necessary to increase to achieve good results. However, in polar regions, long-term meteorological data are difficult to obtain.

The LUT method uses surface reflectivity (clear sky condition) to establish a LUT of various atmospheric conditions. This method utilises the table to find the atmospheric visibility of other observation time points, produces the linear interpolation every 30 min and obtains the instantaneous value of irradiance at that moment [23]. This method requires long-term data accumulation to obtain the surface reflectivity of the entire area, which means that it is inapplicable in Antarctica.

The quadratic polynomial regression method has a similar curve shape to the sine function method; it assumes that solar irradiance is zero at sunrise and sunset [24]. However, this assumption is unreliable because of the perpetual day in Antarctica.

The polar-orbit satellite and geostationary satellite data fusion method considers cloud coverage and uses the geostationary satellite to map the cloud [25]. However, no geostationary satellite data are available over Antarctica.

As mentioned above, no suitable temporal extension method exists for the study of surface SW radiation over Antarctica. This study intends to improve two of the methods above. One is called improved sinusoidal method, which considers the small diurnal change in SZA. The other is called cloud fraction (CF) parameter interpolation, which can handle the rapid cloud coverage change in a day. The data and two improved methods are described in Section 2. The interpolation results and validation are introduced in Section 3. The discussion is presented in Section 4, and the conclusion is in Section 5.

2. Materials and Methods

2.1. Data

2.1.1. Cloud Data and Instantaneous Irradiance Data

The cloud data in this study are mainly based on Suomi National Polar-Orbiting Partnership (S-NPP) satellite cloud product, which uses inversion of Visible Infrared Imaging Radiometer Suite (VIIRS) data. The S-NPP satellite has the same orbital plane as the Terra and Aqua satellites. Its orbital height is approximately 824 km, and its corresponding orbital period is approximately 101 min. The VIIRS sensor has a field of view of 112.56° and a scan width of approximately 3040 km [26,27]. This wide scan area can offer considerable data at high latitudes and allow numerous cloud products in a single day. The satellite passes the study area more than five times a day (Figure 1). Consequently, much data are collected to show the change in cloud coverage and reduce the uncertainty when integrating. In this study, we select geographic positioning products, cloud optical thickness (COT) and cloud base height (CBH) of VIIRS, and calculate hemispherical effective CF (HECF) and regional CF (RCF) (shown in Appendix A). We use the data from December 2013 to February 2014, from December 2014 to February 2015, from December 2015 to February 2016, and from December 2016 to January 2017. The total number of cloud data is 13,986. The number of cloud data near the Georg von Neumayer (GVN) station is 2665 and that near the South Pole (SPO) station is 3331.

We select high spatial resolution (1 km) instantaneous estimated flux data, which are mainly calculated by Santa Barbara DISORT Atmospheric Radiative Transfer (SBDART)-CF model, on the basis of our previous research [28]. SBDART-CF is based on the traditional one-dimensional radiative transfer model SBDART and classifies the actual sun/cloud-viewing geometric conditions into nine subtypes. The main input parameters to the model are listed in Appendix A. The data cover the whole Antarctica area and can consider clear and cloudy sky conditions. The time scale and total number of instantaneous irradiance data are the same as cloud data.

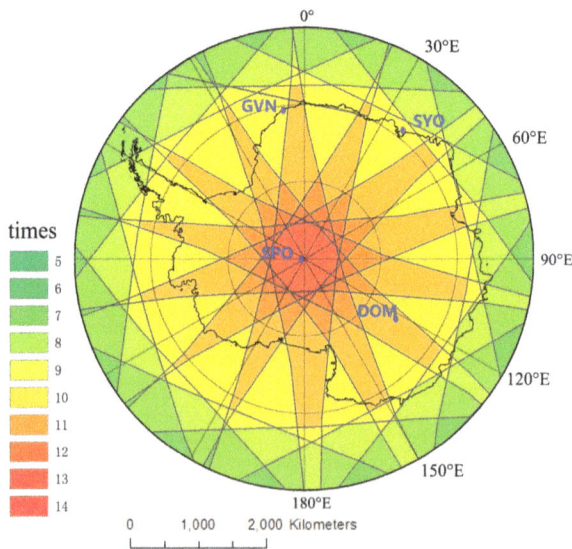

Figure 1. Overpass times of the Suomi National Polar-Orbiting Partnership satellite in one day and distribution of Baseline Surface Radiation Network stations in Antarctica.

2.1.2. Ground Station Data

We use ground measured data of solar radiation from the Baseline Surface Radiation Network (BSRN) observing stations to validate our estimated results. Four stations (shown in Figure 1) of BSRN exist in Antarctica [29]. The BSRN offers the link to download the data (ftp://ftp.bsrn.awi.de). We select the GVN and SPO stations to validate our estimated value (shown in Table 1). The GVN station was established by Germany in 1981, and it has provided radiation observation data from 1992 [30]. SPO station founded by the United States (US) in November 1956, is near the South Pole, and it has offered radiation data from 1992 [31]. We use the Shortwave downward (GLOBAL) radiation included in the 'LR 0100 + LR 0300' data set, which means basic radiation and other radiation measurements. The time resolution of the global radiation is 1 min. 333 days in the two stations are used to validate our result.

Table 1. Information of surface stations in the Antarctic Continent.

Station Name	Abbreviation	Latitude	Longitude	Elevation (m)	Surface Condition
Georg von Neumayer	GVN	70.65°S	8.25°W	42	Ice sheet
South Pole	SPO	89.98°S	24.80°W	2800	Glaciers and deposits

2.2. Temporal Scaling-Up Method

2.2.1. Calculation of the Diurnal Variation Range of SZA

In most areas, the diurnal variation in SZA is larger than 90°; it is the most important factor that affects downward shortwave irradiance [28]. The range of SZA diurnal variation is determined by latitude and date. We demonstrate the changes in DSR and SAZ on 17 October 2013 and 22 December 2013 at the GVN and SPO stations, respectively, to explain the diurnal variation in shortwave irradiance on the Antarctic surface in summer at different latitudes (shown in Figure 2).

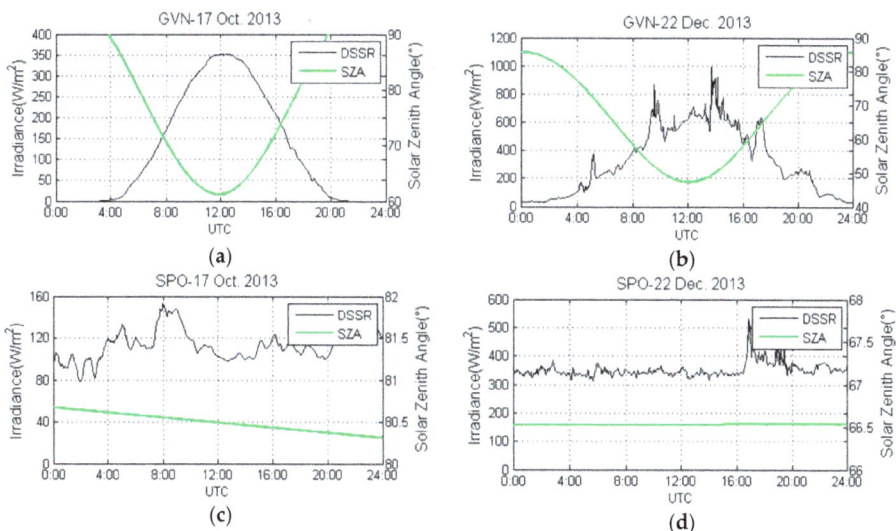

Figure 2. (a) Diurnal variation range of solar zenith angle (SZA) and downward surface shortwave irradiance (DSSR) at the Georg von Neumayer (GVN) station on 17 October 2013; (b) Diurnal variation range of SZA and DSSR at the GVN station on 22 December 2013; (c) Diurnal variation range of SZA and DSSR at the South Pole (SPO) station on 17 October 2013; and, (d) Diurnal variation range of SZA and DSSR at SPO station on 22 December 2013.

The first step in our method is to calculate the range of diurnal SZA variation, sunrise time point and sunset time point [32]. Our sensitivity analysis indicates that if the range is larger than 10°, then SZA remains an important parameter when calculating radiation value. Consequently, we improve the sinusoidal method. If the change range is less than 10°, then cloud coverage becomes the most important influencing factor, which makes cloud coverage fraction interpolation a better choice.

2.2.2. Improved Sinusoidal Method

Traditional sinusoidal method firstly obtains satellite passing time point and the instantaneous irradiance value at this time point. Sunrise and sunset time points are then captured. The instantaneous values are determined with Formula (1) to calculate the maximum radiation in one day.

$$R_{n\_max} = \frac{R_{overpass}}{\sin\left(\frac{t_{overpass} - t_{rise}}{t_{set} - t_{rise}} \pi\right)} \tag{1}$$

The time point of satellite passing is marked as $t_{overpass}$, and the instantaneous value is marked as $R_{overpass}$. This formula leads to overestimated data of radiation because the SZA has small range of change in Antarctica. We improve this formula as Formula (2).

$$R_{n\_max} = \frac{R_{overpass}}{a * \sin\left[b * \left(\frac{t_{overpass} - t_{rise}}{t_{set} - t_{rise}}\right) \pi + c\right] + d} \tag{2}$$

A larger SZA in a high latitude area is considered, and parameter 'a' is inserted to compress the sine function. Downward irradiance is always positive during polar days; therefore, parameter 'd' is introduced to match the case. Parameters 'b' and 'c' are inserted to balance the different periods of time when SZA changes when compared with low-latitude areas. The curve can be fitted by least square method, after which the values of parameters 'a,' 'b,' 'c' and 'd' of each pixel in the area can be determined. The values of 'a,' 'b,' 'c', and 'd' are input into Formula (3), and we can obtain the irradiance value at any time in one day.

$$R_t = R_{overpass} * \frac{a * \sin\left[b * \left(\frac{t - t_{rise}}{t_{set} - t_{rise}}\right) \pi + c\right] + d}{a * \sin\left[b * \left(\frac{t_{overpass} - t_{rise}}{t_{set} - t_{rise}}\right) \pi + c\right] + d} \tag{3}$$

2.2.3. Cloud Coverage Fraction Interpolated Method

The improved sine curve model can match the case in most Antarctic areas, except near the South Pole. In this area, the daily change in SZA is insignificant. Instead, the main element that impacts downward shortwave irradiance is the cloud change. In our previous study, the most sensitive cloud parameters for irradiance estimation are HECF, COT, and RCF [28,33]. Consequently, we use satellite data to calculate instantaneous parameters to generate accurate data in this area. Other values at different times can be calculated with Formula (4).

$$P(t) = \frac{t_{i+1} - t}{t_{i+1} - t_i} * P(t_i) + \frac{t - t_i}{t_{i+1} - t_i} * P(t_{i+1}) \tag{4}$$

't' is the time point between two overpass times, when the latest polar-orbiting satellite overpass time is 't_i' and the next overpass time is 't_{i+1}', 'P(t)' is the cloud parameter value at time 't', and 'P(t_i)', 'P(t_{i+1})' are the values at overpass time. We can let 't' to represent 24 h in one day.

We can then calculate the hourly downward shortwave irradiance flux by SBDART-CF model. In this way, the daily change in downward irradiance in this area can be revealed accurately.

2.2.4. Modelling Daily Solar Radiation

The process of calculating daily average downward shortwave irradiation values is described in Figure 3. The interpolation based on the improved sinusoidal method can integrate the values of irradiance from sunrise to sunset (00:00 to 24:00 during polar day) with Formula (5).

$$R_{daily} = \int_{t_{rise}}^{t_{set}} R_t(t)dt \tag{5}$$

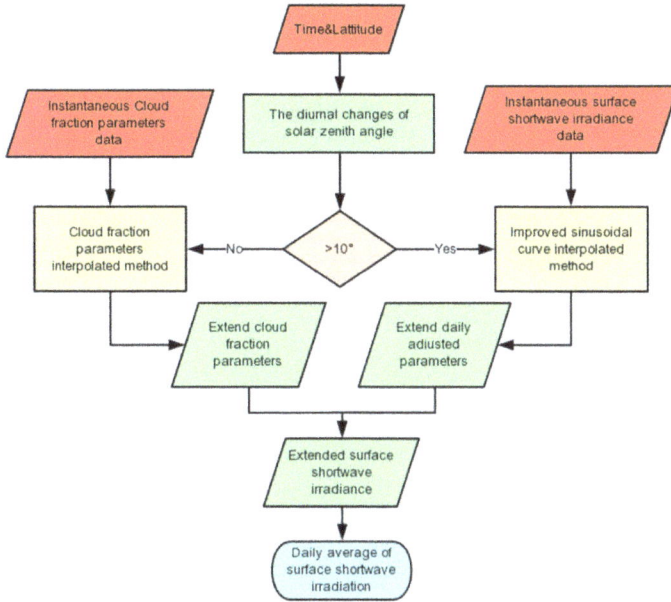

Figure 3. Process of calculating daily average downward shortwave irradiation values.

The cloud coverage fraction interpolated method can calculate the total radiation using Formula (6).

$$R_{daily} = \int_{t_{rise}}^{t_1} R_{t1}(t)dt$$
$$+ \sum_{i=1}^{n-1} \left\{ \int_{t_i}^{t_{i+1}} \left[\frac{t_{i+1}-t}{t_{i+1}-t_i} * R_{t_i}(t) + \frac{t-t_i}{t_{i+1}-t_i} * R_{t_i}(t) \right] dt \right\} \tag{6}$$
$$+ \int_{t_n}^{t_{set}} R_{t_n}(t)dt$$

We obtain hourly instantaneous values, and the other values between two instantaneous values are calculated with weighted average method. The daily average value can be estimated with Formula (7).

$$R_{daily\_avg} = R_{daily} / (t_{set} - t_{rise}) \tag{7}$$

3. Results

3.1. Diurnal Variation in SZA

We can calculate the diurnal variation in SZA at different dates and latitudes. During one year, the polar days and nights alternately show up in the southern hemisphere high-latitude area. We can calculate the range of SZA in one day on the basis of dates and latitudes. We can then decide which method should be used in specific area and date. Figure 4 shows the SZA change through the year at latitudes of 60°S, 70°S, 80°S, and 90°S.

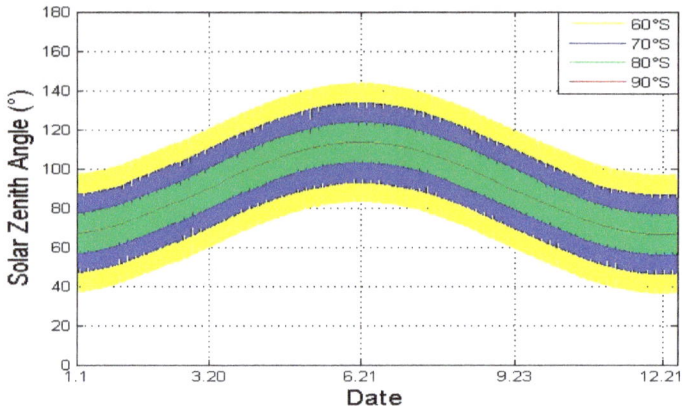

Figure 4. Change range of solar zenith angle at latitudes of 60°S, 70°S, 80°S, and 90°S in all year.

3.2. Diurnal Cycle of Interpolated Irradiance at Different Stations

The areas around the GVN and SPO stations are taken as examples. The diurnal variation in instantaneous downward shortwave irradiance estimated by two interpolation methods is shown below.

The interpolation results of the area around the GVN station are shown in Figure 5. Midnight sun occurs at the GVN station on that day; thus, the shortwave downward irradiance has a positive value over the entire 24-h period. Eight times of S-NPP transiting and corresponding products in the day at 0:47, 02:33, 14:32, 16:09, 17:15, 19:28, 21:10, and 22:47 are available. Although only 24 values are shown, the sinusoidal method can generate the irradiance value at any time in a day.

The shortwave irradiance around the GVN station shows a significant sinusoidal variation in one day. The maximum value appears near 14:00, and the values in the morning and evening are obviously lower. The lowest value is approximately 30 W/m² , which matches the value that was measured by the GVN ground station.

Figure 6 shows the diurnal cycle results of surface shortwave irradiance where the SPO station is located. The cloud coverage parameter interpolation on 2 January 2015, indicates that the SPO station has 10 cloud parameter products on this day, and the satellite transit times are 0:03, 01:45, 03:22, 05:04, 06:47, 08:29, 11:54, 13:31, 18:38, and 22:03. The shortwave irradiance is stable around the SPO station, which is between 400 and 480 W/m². In fact, the more times the satellite can transit, the closer to actual situations the simulation cloud movement condition can be. On the contrary, when the satellite transits less, the insufficient consideration of temporal and spatial changes in clouds will lead to error.

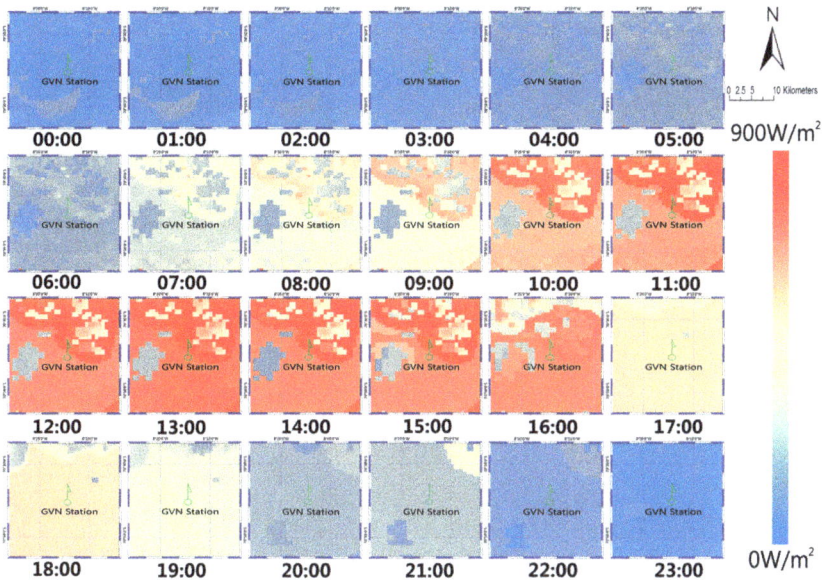

Figure 5. Results of hourly interpolation in the Georg von Neumayer station area on 9 December 2014.

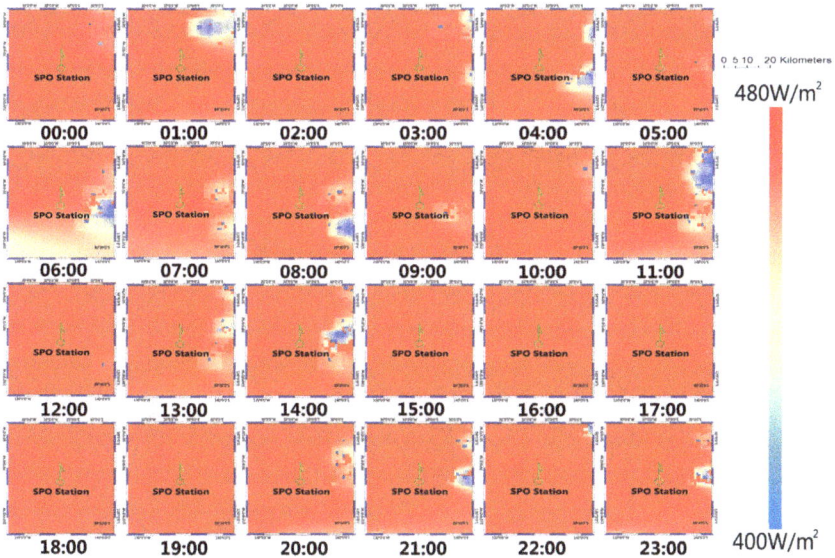

Figure 6. Results of hourly interpolation in the South Pole station area on 2 January 2015.

3.3. Average Daily Irradiation at Different Stations

We can calculate the total irradiation value of a day around the ground station with Formula (5) or (6) (shown in Figure 7). In polar days, we consider the length of day as 24 h; in other days, we can obtain the time points of sunrise and sunset [32]. Theoretically, the higher the temporal resolution of the instantaneous irradiance is, the more accurate the daily total and daily average values can be.

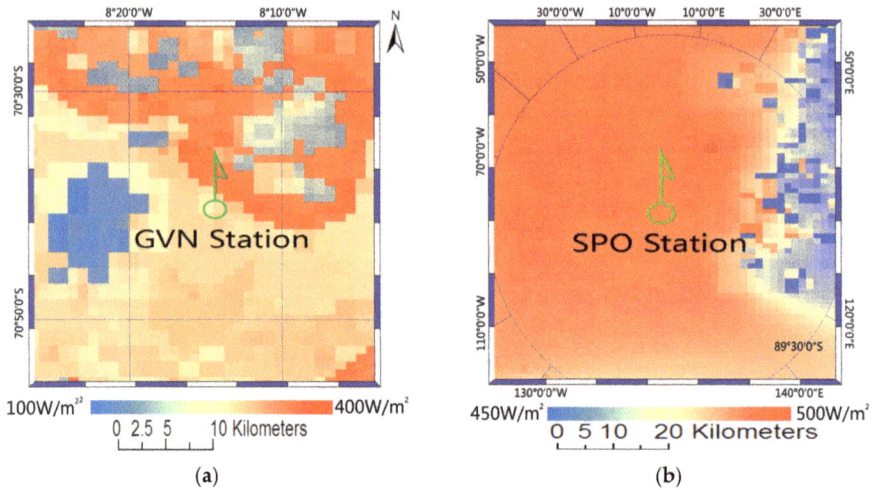

Figure 7. (**a**) Daily average value of solar global irradiation over horizontal surface in the Georg von Neumayer station area on 9 December 2014; and, (**b**) Daily average value of solar global irradiation over horizontal surface in the South Pole station area on 2 January 2015.

We use the data from December 2013 to February 2014, from December 2014 to February 2015, from December 2015 to February 2016, and from December 2016 to January 2017 to validate our estimated daily average values. The data quality control process is performed before validation to ensure that invalid data are deleted. We filter the data for SZA less than 90° and delete values that are less than zero. The measured value for some time periods is 0, which may be due to the snow cover being caused on the instrument surface. The irradiance values for these periods are also removed. We calculated the interpolation values in the neighbouring pixel (10 km × 10 km) and the daily average values measured by the ground station in summer are compared and analyzed. After data control, the number of daily average value in the SPO station is 326, and that in the GVN station is 332. In the formulas below, the ground-observed daily average irradiation data will be noted as R_O, and the estimated daily average irradiation data will be noted as R_e. The mean values of the two distributions are noted as R_{Om} and R_{em}. The total number of data is noted as N. Three statistical metrics are used to evaluate the estimates: the coefficient of determination (R^2) calculated by Formula (8); the root mean square error (RMSE), as calculated by Formulas (9) and (10) and Mean bias error (MBE) calculated by Formulas (11) and (12) [34–37]. The RMSE and MBE are expressed here both in percent and absolute unit, as shown in Figure 8. The red line demonstrates the equation which includes Re and Rg, while the blue line is the demonstration when 'Re = Rg' for comparison.

$$R^2 = \left[\frac{\sum_1^N (R_e - R_{em}) * (R_g - R_{gm})}{\sum_1^N (R_e - R_{em})^2 * (R_g - R_{gm})^2} \right]^2 \tag{8}$$

$$RMSE = \sqrt[2]{\frac{\sum_1^N (R_e - R_g)^2}{N}} \tag{9}$$

$$RMSE(\%) = \frac{100}{R_{gm}} \sqrt[2]{\frac{\sum_1^N (R_e - R_g)^2}{N}} \tag{10}$$

$$MBE = \frac{\sum_1^N (R_e - R_g)}{N} \tag{11}$$

$$\mathrm{MBE}(\%)\frac{100}{R_{gm}}*\frac{\sum_1^N\left(R_e-R_g\right)}{N} \tag{12}$$

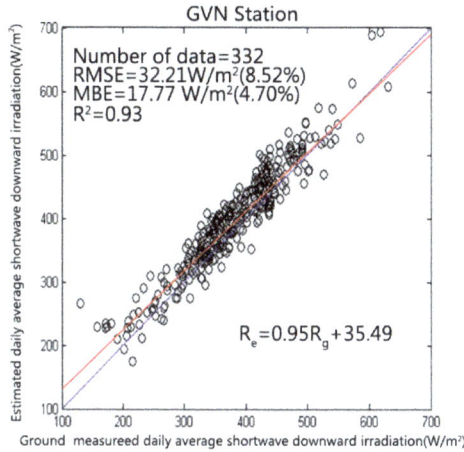

(a)

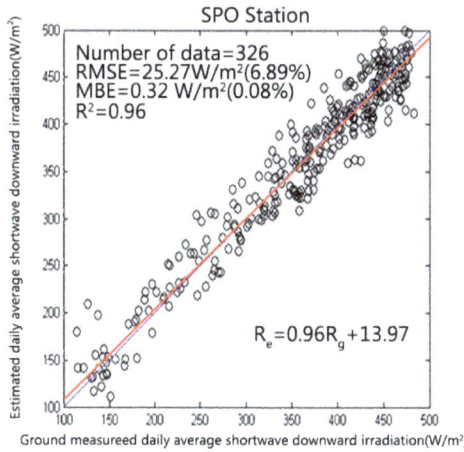

(b)

Figure 8. (**a**) Comparison of the estimated downward shortwave irradiation from the improved sinusoidal method and the ground-measured downward shortwave irradiation in the Georg von Neumayer station; and, (**b**) Comparison of the estimated downward shortwave irradiation from the cloud coverage fraction interpolated method and the ground-measured downward shortwave irradiation in the South Pole station area.

4. Discussion

4.1. Comparison of the Algorithm of the National Aeronautics and Space Administration (NASA)'s Surface Solar Radiation Budget Data Set

The U.S. NASA's Surface Solar Radiation budget data set (https://gewex-srb.larc.nasa.gov/) produced for the Global Energy and Water Exchanges Programme provides daily average shortwave (SW) downward solar irradiation flux starting no later than July 1983 and extending to December 2007.

SW surface radiation budget data sets are derived on a $1° \times 1°$ global grid with two sets of algorithms, known as primary SW algorithm and Langley parameterized SW algorithms (LPSA) [38,39].

The primary SW algorithm gains irradiances from logarithmically averaged three-hourly International Satellite Cloud Climatology Project data. The satellite configuration consists of five geostationary satellites and at least one polar-orbiting satellite. The measurements taken at 00:00, 03:00, 06:00, 09:00, 12:00, 15:00, 18:00, and 21:00 UTC are included. The daily average irradiation is averaged by them. The LPSA using the values of daily average irradiation is computed by averaging measured daytime instantaneous (3 h) reflectance, weighted by the instantaneous value of cos SZA.

4.2. Comparison with Traditional Interpolation Methods

We intend to use two different interpolation methods to obtain the daily average values with high accuracy of DSSR in the study area. In 1979, Tarpley indicated that the sinusoidal formula of the SZA can be used to approximate the daily variation IN solar radiation [19]. The diurnal variation in surface shortwave irradiance is not a simple sinusoidal model due to geographical location, atmospheric conditions, and other factors, which means that the model needs to be corrected.

Lagouarde and Brunet advanced a sinusoidal model that can describe the diurnal variation in surface temperature and considers the time span and amplitude of curve [20]. Bisht proposed a diurnal sinusoidal model of surface net radiation, which is suitable for interpolation with MODIS data under clear sky conditions [18]. Wang proposed a modified sinusoidal method, supposing the values of irradiance at sunrise and sunset are zero [23]. However, the curve amplitude is smaller due to the smaller diurnal variation in SZA in Antarctica. In addition, no sunrise or sunset exists due to the polar day in summer, which makes the curve time span longer, even one cycle longer than the traditional curve. We set parameters '*a*' and '*d*' to adjust the curve amplitude in Formula (2), and parameters '*b*' and '*c*' to control time span and cycle.

The transit time is more than four in one day; therefore, we can set four parameters to adjust the sinusoidal curve. Table 2 shows the estimated daily average downward shortwave irradiation based on the traditional sinusoidal curve (the R^2 is near 0.68, and the RMSE is 70.32 W/m^2). The parameters of the linear fits $R_e = C_1 + C_2 \times R_g$ for the two methods in the GVN station are also shown in Table 2.

Table 2. Comparison between the daily average values estimated by four different interpolation methods and ground measurements from two stations.

Station	Method	R^2	C_1 (W/m^2)	C_2	RMSE (W/m^2)	RMSE (%)	MBE (W/m^2)	MBE (%)
GVN station	Improved sinusoidal curve	0.93	35.49	0.95	32.21	8.52	17.77	4.70
	Traditional sinusoidal curve	0.68	55.50	1.09	70.32	18.59	36.39	9.62
SPO station	Cloud coverage fraction Interpolation	0.96	13.97	0.96	25.27	6.98	0.32	0.08
	Linear interpolation	0.79	28.51	1.04	57.40	15.87	30.18	8.34

In Antarctica, the sensitivity analysis method is used to prove that the cloud coverage fraction is the main parameter when diurnal changes in SZA are small. Unlike in the linear interpolation method, whether interpolation or calculation should be performed firstly does not matter in the cloud coverage fraction interpolation. In our calculations, the relationship between cloud coverage and downward shortwave irradiance values in the SBDART-CF model is nonlinear, which leads to different results. Table 2 shows the results from the linear interpolation method. If cloud coverage does not change over time, the results of two methods should be similar. Therefore, the cloud coverage interpolation method is more suitable when the cloud cover changes. The parameters of the linear fits $R_e = C_1 + C_2 \times R_g$ for the two methods in SPO station are shown in Table 2.

4.3. Limitation and Further Study

In Figure 5, the DSR is not continuous in space, because when we fit the four parameters, the influence of the cloud is ignored in some areas and enlarged in other areas. An apparent disadvantage of the sinusoidal method is that the cloud coverage in every transmit time will affect the four parameters in Formula (2).

In Figure 8a, the results of daily average irradiation values are slightly overestimated. In the experiments, the changes of irradiance near the station are not only caused by the change of SZA, but also by cloud conditions. Consequently, this method is more suitable under clear sky conditions. We need more cloud data from polar satellite to capture the change of cloud. S-NPP is the only satellite on which this paper relies. Obtaining data from more satellites can reduce the error between the interpolation results and the real-time measured values.

The cloud coverage interpolation method is computationally intensive and inefficient. With the accumulation of future cloud data, a look-up table between cloud coverage changes and the DSR in this area can be established to speed up the operation.

The methods of this study are not suitable for low and middle latitude areas because of inadequate satellite transits. However, the fusion of multi-source orbit satellite cloud data can cover the shortage of inadequate transits to some extent. Thus, the high spatial resolution geostationary satellite will be helpful in making our methods suitable for lower latitude areas [40,41].

More profound problems must be solved in the future. Firstly, the terrain effect in the research area is not considered in this paper. The influence of different topography on downward shortwave irradiance in Antarctica should be considered in further studies when combined with our previous studies [42]. Secondly, the influence of surface weather conditions is ignored owing to the lack of the surface weather data. Thirdly, the data for interpolation are available only after 2013 because of the limitation of the satellite launch time. In further studies, AVHRR and MODIS satellite sensors can be used [43], and the surface shortwave radiation estimate data set with long-term data series based on the method that is provided in this paper can be established.

5. Conclusions

In this paper, two interpolation methods to estimate daily average values of downward shortwave irradiation are mainly discussed. On the basis of the sinusoidal method provided by previous research, this research has improved the traditional sinusoidal interpolation method. Meanwhile, we present new cloud fraction parameter interpolation method to consider the cloud condition change in Antarctica. Four parameters are introduced ('*a*', '*b*', '*c*' and '*d*') and are fitted in the improved sinusoidal method, with the data support of the S-NPP satellite that passes through polar areas several times a day. In this manner, the traditional sine curve model is improved to prevent overestimation. For cloud coverage fraction parameter interpolation method, the S-NPP satellite cloud product provides cloud fraction data of the hemisphere space, which are the input parameters to obtain the interpolation results.

According to the validation by the data from two BSRN surface stations, the R^2 of the first method is 0.93; the RMSE is 32.21 W/m^2 (8.52%) and the MBE is 17.77 W/m^2 (4.70%) in GVN station area. The R^2, RMSE and MBE of the second method are 0.96, 25.27 W/m^2 (6.98%) and 0.32 W/m^2 (0.08%), respectively, in the SPO station area. When compared with existing methods, our methods are more accurate than the traditional sinusoidal method (R^2 = 0.68; RMSE = 70.32 W/m^2 (18.59%) and MBE = 36.39 W/m^2 (9.62%)) in GVN station area and the direct linear interpolation method (R^2 = 0.79; RMSE = 57.40 W/m^2 (15.87%); MBE = 38.18 W/m^2 (8.34%)) in SPO station area.

Acknowledgments: The work is funded partially by the Key Program of Natural Science Foundation of China (Grant No. 41331171and No. 41301357) and Natural Science Foundation of China (Grant No. 61227806).

Author Contributions: Yingji Zhou, Guangjian Yan, Jing Zhao, designed the experiments; Yingji Zhou, Yanan Liu, Qing Chu, performed the experiments and analyzed the data; Kai Yan and Yiyi Tong contributed to Sections 3.2 and 4.2. All authors contributed in writing the paper.

Conflicts of Interest: The authors declare no conflict of interest. The founding sponsors had no role in the design of the study, in the collection, analyses, or interpretation of data; in the writing of the manuscript, and decision to publish the results.

Appendix A

The SBDART-CF model was improved by Santa Barbara DISORT Atmospheric Radiative Transfer (SBDART), which has been established to simulate radiative transfer. To consider the effect of cloud radiative forcing, we should analyze whether the directions of the sun and sensor are obscured by clouds. We can calculate HECF and RCF by Formulas (A1) and (A2) [33].

$$\text{HECF} = \sum_{i=0}^{n} \frac{s * \cos^3 \theta_i}{2\pi(1 - \cos\alpha)h_i^2} \tag{A1}$$

$$\text{RCF} = \frac{n * s}{A} \tag{A2}$$

In the above two formulas, s is the area of target pixel, and A is the area of a 40 km × 40 km slide window; n is the number of Nadir-view cloud pixels; θ_i is the angle between cloud pixel and target pixel; the $2\pi(1 - \cos\alpha)$ and h_i is the attitude between cloud pixel and target pixel (shown in Figure A1) [33].

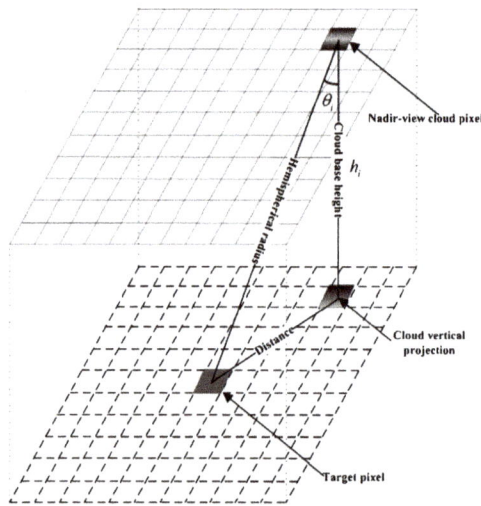

Figure A1. Cloud pixel and target pixel in the slide window.

We can calculate the instantaneous irradiance values by using SBDART-CF model, with the main input parameters listed in Table A1. If the surface is Lambert, the surface irradiance flux can be calculated as Formulas (A3)–(A7).

$$F(\mu_i) = F_0(\mu_i) + F_m(\mu_i) \tag{A3}$$

$$F_m(\mu_i) = \frac{r_s\rho}{1 - r_s\rho}\mu_i E_0 \gamma(\mu_i) \tag{A4}$$

In Formulas (A3) and (A4), μ_i is the cosine of the SZA; r_s is the surface reflectance; $F_0(\mu_i)$ represents the downward surface irradiance flux when r_s is 0. $F_m(\mu_i)$ is the irradiance that is scattered multiple times between ground and atmosphere; ρ is the spherical albedo of the atmosphere; E_0 is the solar irradiance at the top of the atmosphere and $\gamma(\mu_i)$ is the total atmospheric transmittance of the solar direction (including both direct and diffuse transmittance).

$$F_{0\_clr} = Dir_{clr} + Dif_{cld} * HECF + Dif_{clr} * (1 - HECF) \tag{A5}$$

$$F_{0\_cld} = Dir_{cld} + Dif_{cld} * HECF + Dif_{clr} * (1 - HECF) \tag{A6}$$

In Formulas (A5) and (A6), F_{0_clr} and F_{0_cld} represent the downward surface irradiance flux without ground contribution in clear sky and cloudy sky, respectively. The Dir_{clr} and Dir_{cld} are direct irradiance in clear sky and cloudy sky conditions, respectively. The Dif_{cld} includes scattering from clouds, and the Dif_{clr} is scattering from other atmospheric molecules in hemispherical space.

$$F_m = \left[F_{0_{clr}} * (1 - RCF) + F_{0_{cld}} * RCF \right] * \frac{r_s * \left[(1 - HECF) * \rho_a + HECF * \rho_c \right]}{1 - r_s * \left[(1 - RCF) * \rho_a + RCF * \rho_c \right]} \tag{A7}$$

In Formula (A7), we assume that the surface is Lambert with stable reflectance r_s to calculate the multiple scattering value F_m between the surface and the atmosphere. Meanwhile, the SBDART-CF model considers the hemispheric partly cloudy condition. We classify the actual sun/cloud-viewing geometric conditions into nine subtypes (shown in Figure A2).

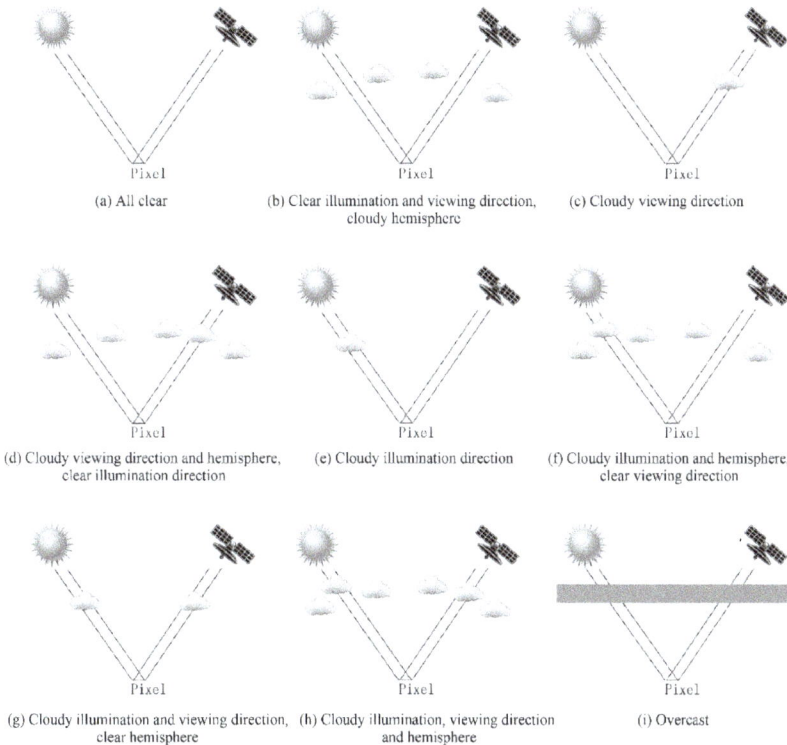

(a) All clear

(b) Clear illumination and viewing direction, cloudy hemisphere

(c) Cloudy viewing direction

(d) Cloudy viewing direction and hemisphere, clear illumination direction

(e) Cloudy illumination direction

(f) Cloudy illumination and hemisphere, clear viewing direction

(g) Cloudy illumination and viewing direction, clear hemisphere

(h) Cloudy illumination, viewing direction and hemisphere

(i) Overcast

Figure A2. The Classification for sun-sensor-hemisphere cloud cover conditions.

We list the main input parameters of the SBDART-CF model. In our method the cloud cover parameters (CBH, COT, HECF and RCF) are included. The output data are instantaneous downward shortwave irradiance values.

Table A1. Main Input Parameters of the SBDART-CF Model.

Main Input Parameter	Description	Unit
SZA	solar zenith angle	°
Albedo	surface albedo	-
VIS	visibility	km
COT	cloud optical thickness	-
CBH	cloud base height	km
Alt	altitude	km
HECF	hemispheric cloud fraction	-
RCF	regional cloud fraction	-

References

1. Platt, T. Primary production of the ocean water column as a function of surface light intensity: Algorithms for remote sensing. *Deep Sea Res. Part A Oceanogr. Res. Pap.* **1986**, *33*, 149–163. [CrossRef]
2. Shook, K.; Pomeroy, J. Synthesis of incoming shortwave radiation for hydrological simulation. *Hydrol. Res.* **2011**, *42*, 433–446. [CrossRef]
3. Wild, M.; Ohmura, A.; Schär, C.; Müller, G.; Folini, D.; Schwarz, M.; Hakuba, M.Z.; Sanchez-Lorenzo, A. The global energy balance archive (GEBA) version 2017: A database for worldwide measured surface energy fluxes. *Earth Syst. Sci. Data* **2017**, *9*, 601. [CrossRef]
4. Budyko, M.I. The effect of solar radiation variations on the climate of the earth. *Tellus* **1969**, *21*, 611–619. [CrossRef]
5. Kiehl, J.T.; Trenberth, K.E. Earth's annual global mean energy budget. *Bull. Am. Meteorol. Soc.* **1997**, *78*, 197–208. [CrossRef]
6. Foukal, P.; Fröhlich, C.; Spruit, H.; Wigley, T. Variations in solar luminosity and their effect on the earth's climate. *Nature* **2006**, *443*, 161–166. [CrossRef] [PubMed]
7. Hansom, J.D.; Gordon, J. *Antarctic Environments and Resources: A Geographical Perspective*; Routledge: New York, NY, USA; Oxfordshire, UK, 2014.
8. Hall, A. The role of surface albedo feedback in climate. *J. Clim.* **2004**, *17*, 1550–1568. [CrossRef]
9. Holland, M.M.; Bitz, C.M. Polar amplification of climate change in coupled models. *Clim. Dyn.* **2003**, *21*, 221–232. [CrossRef]
10. Stanhill, G.; Cohen, S. Recent changes in solar irradiance in Antarctica. *J. Clim.* **1997**, *10*, 2078–2086. [CrossRef]
11. King, J.; Connolley, W. Validation of the surface energy balance over the antarctic ice sheets in the UK meteorological office unified climate model. *J. Clim.* **1997**, *10*, 1273–1287. [CrossRef]
12. Stocker, T. *Climate Change 2013: The Physical Science Basis: Working Group I Contribution to the Fifth Assessment Report of the Intergovernmental Panel on Climate Change*; Cambridge University Press: Cambridge, UK, 2014.
13. Wild, M.; Folini, D.; Henschel, F.; Fischer, N.; Müller, B. Projections of long-term changes in solar radiation based on cmip5 climate models and their influence on energy yields of photovoltaic systems. *Sol. Energy* **2015**, *116*, 12–24. [CrossRef]
14. Trnka, M.; Žalud, Z.; Eitzinger, J.; Dubrovský, M. Global solar radiation in central European lowlands estimated by various empirical formulae. *Agric. For. Meteorol.* **2005**, *131*, 54–76. [CrossRef]
15. Journée, M.; Bertrand, C. Improving the spatio-temporal distribution of surface solar radiation data by merging ground and satellite measurements. *Remote Sens. Environ.* **2010**, *114*, 2692–2704. [CrossRef]
16. Pinker, R.; Frouin, R.; Li, Z. A review of satellite methods to derive surface shortwave irradiance. *Remote Sens. Environ.* **1995**, *51*, 108–124. [CrossRef]
17. Liang, S.; Zheng, T.; Liu, R.; Fang, H.; Tsay, S.C.; Running, S. Estimation of incident photosynthetically active radiation from moderate resolution imaging spectrometer data. *J. Geophys. Res. Atmos.* **2006**, *111*. [CrossRef]
18. Bisht, G.; Venturini, V.; Islam, S.; Jiang, L. Estimation of the net radiation using modis (moderate resolution imaging spectroradiometer) data for clear sky days. *Remote Sens. Environ.* **2005**, *97*, 52–67. [CrossRef]
19. Tarpley, J. Estimating incident solar radiation at the surface from geostationary satellite data. *J. Appl. Meteorol.* **1979**, *18*, 1172–1181. [CrossRef]
20. Lagouarde, J.; Brunet, Y. A simple model for estimating the daily upward longwave surface radiation flux from noaa-avhrr data. *Int. J. Remote Sens.* **1993**, *14*, 907–925. [CrossRef]

21. Van Laake, P.E.; Sanchez-Azofeifa, G.A. Mapping par using modis atmosphere products. *Remote Sens. Environ.* **2005**, *94*, 554–563. [CrossRef]

22. Niu, X. *Radiative Fluxes and Albedo Feedback in Polar Regions*; University of Maryland: College Park, MD, USA, 2011.

23. Wang, D.; Liang, S.; Liu, R.; Zheng, T. Estimation of daily-integrated par from sparse satellite observations: Comparison of temporal scaling methods. *Int. J. Remote Sens.* **2010**, *31*, 1661–1677. [CrossRef]

24. Xu, X.; Du, H.; Zhou, G.; Mao, F.; Li, P.; Fan, W.; Zhu, D. A method for daily global solar radiation estimation from two instantaneous values using modis atmospheric products. *Energy* **2016**, *111*, 117–125. [CrossRef]

25. Chen, L.; Yan, G.; Ren, H.; Wang, T. A simple fusion algorithm of polar-orbiting and geostationary satellite data for the estimation of surface shortwave fluxes. In Proceedings of the 2016 IEEE International Geoscience and Remote Sensing Symposium (IGARSS), Beijing, China, 10–15 July 2016; IEEE: Piscataway, NJ, USA, 2016; pp. 2657–2660.

26. Cao, C.; De Luccia, F.J.; Xiong, X.; Wolfe, R.; Weng, F. Early on-orbit performance of the visible infrared imaging radiometer suite onboard the suomi national polar-orbiting partnership (s-npp) satellite. *IEEE Trans. Geosci. Remote Sens.* **2014**, *52*, 1142–1156. [CrossRef]

27. Yan, K.; Park, T.; Chen, C.; Xu, B.; Song, W.; Yang, B.; Zeng, Y.; Liu, Z.; Yan, G.; Knyazikhin, Y.; et al. Generating global products of lai and fpar from snpp-viirs data: Theoretical background and implementation. *IEEE Trans. Geosci. Remote Sens.* **2018**, *PP*, 1–19. [CrossRef]

28. Zhao, J.; Yan, G.; Jiao, Z.; Chen, L.; Chu, Q. Enhanced shortwave radiative transfer model based on sbdart. *J. Remote Sens.* **2017**, *6*, 853–863.

29. König-Langlo, G.; Sieger, R.; Schmithüsen, H.; Bücker, A.; Richter, F.; Dutton, E. *The Baseline Surface Radiation Network and Its World Radiation Monitoring Centre at the Alfred Wegener Institute*; World Meteorological Organization: Geneva, Switzerland, 2013.

30. König-Langlo, G.; Loose, B. The meteorological observatory at neumayer stations (gvn and nm-ii) Antarctica. *Polarforschung2006* **2007**, *76*, 25–38.

31. Dutton, E.G. Basic and other measurements of radiation at station south pole (1998-06). *Earth Environ. Sci.* **2007**. [CrossRef]

32. Grena, R. An algorithm for the computation of the solar position. *Sol. Energy* **2008**, *82*, 462–470. [CrossRef]

33. Chen, L.; Yan, G.; Wang, T.; Ren, H.; Calbó, J.; Zhao, J.; McKenzie, R. Estimation of surface shortwave radiation components under all sky conditions: Modeling and sensitivity analysis. *Remote Sens. Environ.* **2012**, *123*, 457–469. [CrossRef]

34. Gueymard, C.A. A review of validation methodologies and statistical performance indicators for modeled solar radiation data: Towards a better bankability of solar projects. *Renew. Sustain. Energy Rev.* **2014**, *39*, 1024–1034. [CrossRef]

35. Román, R.; Antón, M.; Valenzuela, A.; Gil, J.; Lyamani, H.; De Miguel, A.; Olmo, F.; Bilbao, J.; Alados-Arboledas, L. Evaluation of the desert dust effects on global, direct and diffuse spectral ultraviolet irradiance. *Tellus B Chem. Phys. Meteorol.* **2013**, *65*, 19578. [CrossRef]

36. Mateos, D.; Bilbao, J.; Kudish, A.; Parisi, A.; Carbajal, G.; Di Sarra, A.; Román, R.; De Miguel, A. Validation of omi satellite erythemal daily dose retrievals using ground-based measurements from fourteen stations. *Remote Sens. Environ.* **2013**, *128*, 1–10. [CrossRef]

37. De Miguel-Bilbao, S.; Ramos, V.; Blas, J. Assessment of polarization dependence of body shadow effect on dosimetry measurements in 2.4 ghz band. *Bioelectromagnetics* **2017**, *38*, 315–321. [CrossRef] [PubMed]

38. Pinker, R.; Laszlo, I. Modeling surface solar irradiance for satellite applications on a global scale. *J. Appl. Meteorol.* **1992**, *31*, 194–211. [CrossRef]

39. Gupta, S.K.; Kratz, D.P.; Stackhouse, P.W., Jr.; Wilber, A.C. *The Langley Parameterized Shortwave Algorithm (LPSA) for Surface Radiation Budget Studies*; version 1.0; Langley Research Center: Hampton, VA, USA, 2001.

40. Yang, J.; Zhang, Z.; Wei, C.; Lu, F.; Guo, Q. Introducing the new generation of Chinese geostationary weather satellites–fengyun 4 (fy-4). *Bull. Am. Meteorol. Soc.* 2016. [CrossRef]

41. Bessho, K.; Date, K.; Hayashi, M.; Ikeda, A.; Imai, T.; Inoue, H.; Kumagai, Y.; Miyakawa, T.; Murata, H.; Ohno, T. An introduction to himawari-8/9—Japan's new-generation geostationary meteorological satellites. *J. Meteorol. Soc. Jpn. Ser. II* **2016**, *94*, 151–183. [CrossRef]

42. Yan, G.; Wang, T.; Jiao, Z.; Mu, X.; Zhao, J.; Chen, L. Topographic radiation modeling and spatial scaling of clear-sky land surface longwave radiation over rugged terrain. *Remote Sens. Environ.* **2016**, *172*, 15–27. [CrossRef]

43. Jiao, Z.; Yan, G.; Zhao, J.; Wang, T.; Chen, L. Estimation of surface upward longwave radiation from modis and viirs clear-sky data in the Tibetan plateau. *Remote Sens. Environ.* **2015**, *162*, 221–237. [CrossRef]

remote sensing

MDPI

Article

A Multi-Scale Validation Strategy for Albedo Products over Rugged Terrain and Preliminary Application in Heihe River Basin, China

Xingwen Lin [1,3], Jianguang Wen [1,2,*], Qinhuo Liu [1,2,*], Qing Xiao [1], Dongqin You [1,2], Shengbiao Wu [1,3], Dalei Hao [1,3] and Xiaodan Wu [4]

[1] State Key Laboratory of Remote Sensing Science, Institute of Remote Sensing and Digital Earth, Chinese Academy of Sciences, Beijing 100101, China; linxw@radi.ac.cn (X.L.); xiaoqing@radi.ac.cn (Q.X.); youdq@radi.ac.cn (D.Y.); wusb@radi.ac.cn (S.W.); haodl@radi.ac.cn (D.H.)
[2] Joint Center for Global Change Studies (JCGCS), Beijing 100875, China
[3] University of Chinese Academy of Sciences, Beijing 100049, China
[4] College of Earth Environmental Sciences, Lanzhou University, Gansu 730000, China; upcwuxiaodan@163.com
* Correspondence: wenjg@radi.ac.cn (J.W.); liuqh@radi.ac.cn (Q.L.);
  Tel.: +86-1346-638-3594 (J.W.); +86-1391-029-3801 (Q.L.)

Received: 1 December 2017; Accepted: 19 January 2018; Published: 24 January 2018

Abstract: The issue for the validation of land surface remote sensing albedo products over rugged terrain is the scale effects between the reference albedo measurements and coarse scale albedo products, which is caused by the complex topography. This paper illustrates a multi-scale validation strategy specified for coarse scale albedo validation over rugged terrain. A Mountain-Radiation-Transfer-based (MRT-based) albedo upscaling model was proposed in the process of multi-scale validation strategy for aggregating fine scale albedo to coarse scale. The simulated data of both the reference coarse scale albedo and fine scale albedo were used to assess the performance and uncertainties of the MRT-based albedo upscaling model. The results showed that the MRT-based model could reflect the albedo scale effects over rugged terrain and provided a robust solution for albedo upscaling from fine scale to coarse scale with different mean slopes and different solar zenith angles. The upscaled coarse scale albedos had the great agreements with the simulated coarse scale albedo with a Root-Mean-Square-Error (RMSE) of 0.0029 and 0.0017 for black sky albedo (BSA) and white sky albedo (WSA), respectively. Then the MRT-based model was preliminarily applied for the assessment of daily MODerate Resolution Imaging Spectroradiometer (MODIS) Albedo Collection V006 products (MCD43A3 C6) over rugged terrain. Results showed that the MRT-based model was effective and suitable for conducting the validation of MODIS albedo products over rugged terrain. In this research area, it was shown that the MCD43A3 C6 products with full inversion algorithm, were generally in agreement with the aggregated coarse scale reference albedos over rugged terrain in the Heihe River Basin, with the BSA RMSE of 0.0305 and WSA RMSE of 0.0321, respectively, which were slightly higher than those over flat terrain.

Keywords: land surface albedo; multi-scale validation; rugged terrain; MRT-based model; MCD43A3 C6

---

1. Introduction

Land surface shortwave albedo is defined as the fraction of incident solar irradiance reflected by Earth's surface over the shortwave band (0.3–3 μm) in the whole solar spectrum [1]. It is a key climate-regulating parameter that determines the amount of solar radiation absorbed by the land surface at regional and global scales [2,3]. Remote sensing satellites provide a practical method to estimate land surface albedos because of their large spatial scale coverage and a high revisit frequency [4]. However, the retrieved albedos suffer from large uncertainties due to the inherent complexity of the physical processes and their parameterization of retrieval algorithms [5]. Thus, it is critical to evaluate the performance of the retrieved albedos prior to their wide application.

Many scientists have focused on assessing the accuracy of albedo products in recent decades over flat and homogeneous land surfaces. Taking the MODIS albedo products validation as an example, the validation results showed that the MODIS albedo products displayed high accuracy with an uncertainty (Root-Mean-Square-Error, RMSE) below 0.03 at the snow-free land covers and 0.07 at the snow-covered land surface, respectively, when the validation activities occurred in the homogeneous land surface or in sites with high spatial representativeness [6–19]. Generally, the albedo product can be assessed by direct comparison with in situ albedos at the sites where the land surface is sufficiently homogeneous [6,9,10,20]. However, in the heterogeneous land surface, the in situ albedo cannot be directly compared with albedo products because of the scale mismatching between the in situ albedo and the albedo products, unless that in situ albedo can be considered with high spatial representativeness over the sampled area [10,16,20,21]. The scale mismatching will result in about 15% disagreement between the MODIS albedo and in situ albedo [6,9,22]. As the sample sites with limited spatial representativeness, multi-points albedo observing is generally adapted to capture the spatial distribution characteristics of the albedo over the sampled area. The simplest and most efficient method is to average these albedos within the area and as the reference truth to compare with the albedo products [23]. Alternatively, the multi-scale validation strategy provides a solution to deal with the scale mismatching over a heterogeneous land surface by introducing fine scale albedo products (e.g., the Enhanced Thematic Mapper plus (ETM+) or China HJCCD (HJ) albedo) as an upscaling bridge between in situ albedo and coarse scale satellite albedo products (e.g., MODIS and Global LAnd Surface Satellite (GLASS) albedo) [8,17,24,25]. In situ albedos are used to calibrate the fine scale albedo. Then, the calibrated fine scale albedos are aggregated to the coarse scale and for albedo validation. Previous studies have indicated that the upscaling of albedo from fine scale to coarse scale is highly linear over flat terrains [8,16,17,24–26]. Therefore, in the case of flat terrain, the linear weighted average model was considered as a good performance model for upscaling the fine scale albedo to a coarse scale.

As a special heterogeneous land surface, the topography has vast effects on the land surface albedo [24,26]. Topographic slope, aspect, shadow, and solar location influence albedo values and their spatial distribution when compared with that over flat terrain [13,27–30]. The coarse scale albedo decreased with the increase of the slope facing away from the sun and increased when facing toward the sun [30,31], and generally showed larger values over the slope facing toward the sun than that facing away from the sun, especially, in the shadowing case [27,30,32]. The complex topography leads to the intensive scale effects on albedo products among different spatial resolutions over the rugged terrain [17,33]. However, neglecting the scale effect caused by the complex topography in albedo products results in unreliable validation results [33–35]. Peng et al., (2014) assessed the MODIS products by using the multi-scale validation strategy with the HJ albedo as the bridge to aggregate the in situ albedo linearly to the MODIS pixel scale. The uncertainty distribution analysis showed that the largest scaling uncertainty was at the pixels over rugged terrain and its uncertainty of MODIS was as high as 0.07, when neglecting the scale effects at the upscaling progress over rugged terrain [17]. Therefore, neither the direct comparison nor the linear upscaling model in the multi-scale validation strategy were suitable for the coarse scale albedo products validation over rugged terrain [33]. The albedo spatial scale issue caused by topography should be emphasized in the multi-scale validation strategy over rugged terrain.

The objective of this paper was to develop an upscaling method in the procedure of the multi-scale validation strategy for albedo products validation over rugged terrain. Simulated data with different mean slopes and solar zenith angle over nine Digital Elevation Models (DEM) were used to validate the albedo upscaling method. Based on the proposed upscaling method, the aggregated HJ albedo, which had been validated by the in situ albedo over the Heihe River Basin, was used as the reference truth for the MODIS albedo products preliminary validation. The paper is organized as follows: Section 2 describes the multi-scale validation strategy, including the upscaling method and the fine scale albedo products retrieval algorithm; Section 3 describes the experimental area and validation dataset; Section 4 shows the performance of the albedo upscaling model and the preliminary validation results for MODIS albedo products. The discussions are summarized in Section 5. Finally, a brief conclusion is drawn in Section 6.

2. Multi-Scale Validation Methodology

2.1. Multi-Scale Validation Procedure over Rugged Terrain

The general multi-scale validation strategy includes three key procedures over rugged terrain. The first one is the retrieval and accurate evaluation of fine scale albedo. The second is the calibration of fine scale albedo products. Finally, the third process is to scale up the fine scale albedo products to the coarse scale [17,24]. The multi-scale validation strategy over rugged terrain has similar procedures than those over flat terrain, which is shown in Figure 1. The DEM was used here to calculate the topographic factors (e.g., slope, aspect), and to couple with fine scale reflectance to retrieve fine scale albedo. The fine scale albedo was assessed by direct comparison with in situ albedos and was calibrated for reducing its uncertainties. An albedo upscaling model, which was based on the mountain radiation transfer theory (MRT-based albedo upscaling model), was proposed to aggregate the fine scale albedo to a coarse spatial scale. Consequently, the aggregated albedos could be directly compared with the coarse scale albedo products. To implement a successful multi-scale validation strategy over rugged terrain, two key issues should be solved including the albedo upscaling method and the fine scale albedo-generated algorithm on sloping surfaces.

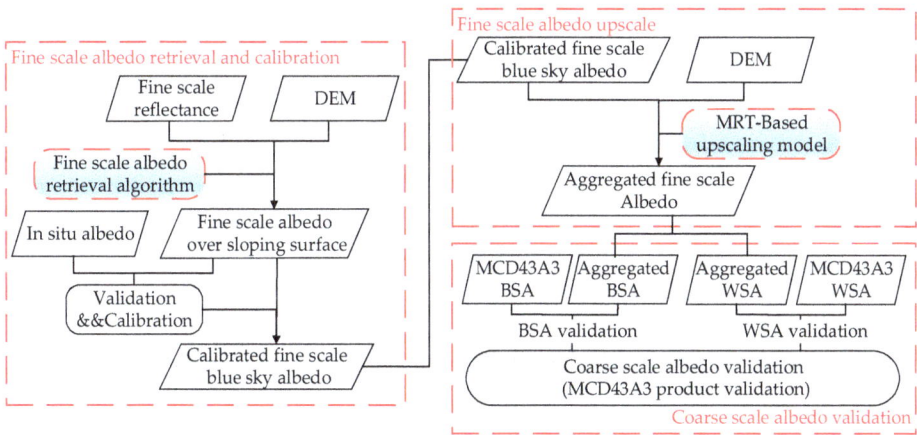

Figure 1. The coarse albedo validation procedure over rugged terrain.

2.2. MRT-Based Albedo Upscaling Model over Rugged Terrain

Coarse scale albedo can be defined as the ratio of the reflected solar radiant flux and incident solar radiant flux. Assuming that the coarse scale pixel is horizontal overall and the topographic effects between the coarse scale pixels can be ignored, the incident solar radiant flux of a coarse scale pixel can be expressed as the sum of direct and diffuse radiant flux at the micro-slope. Therefore, the coarse scale reflected radiant flux was calculated by summing up the micro-slope reflected radiant flux. If the micro-slope albedo is known, the coarse scale albedo can be expressed as:

$$A_c = \frac{\phi_c^\uparrow}{\phi_c^\downarrow} = \frac{\sum\limits_{k=1}^{N} \partial_k d\phi_k^{T\downarrow}}{\phi_{sc}^\downarrow + \phi_{dc}^\downarrow} \tag{1}$$

where A_c is the coarse scale albedo; $\phi_c^\uparrow$, $\phi_c^\downarrow$ are the coarse scale pixel's reflected and incident solar radiant flux, respective; N is the amount of micro-slope within the corresponding area of coarse scale pixel; ∂_k and $d\phi_k^{T\downarrow}$ are the albedo and incident radiant flux of the kth micro-slope, respectively. $\phi_{sc}^\downarrow$, $\phi_{dc}^\downarrow$ are the coarse scale direct solar radiant flux and diffuse radiant flux, respectively. They can be expressed as:

$$\phi_{sc}^\downarrow = E_s Area_C \tag{2}$$

$$\phi_{dc}^\downarrow = E_d Area_C \tag{3}$$

$$Area_C = N d A_{th} \tag{4}$$

where E_s, E_d are the incident direct solar irradiance and diffuse irradiance on the horizontal plane; $Area_C$ is the area of the coarse scale pixel, which can be expressed as the sum of the projected area (dA_{th}) of the micro-slope on horizontal plane.

For each micro-slope, $d\phi_k^{T\downarrow}$ can be expressed as the sum of direct radiant flux, diffuse radiant flux, and terrain radiant flux reflected by adjacent terrain (Figure 2) [32,36]. The incident direct and diffuse solar radiant flux of a coarse scale pixel can be expressed as the sum of the micro-slope's solar direct and diffuse radiant flux on the projected horizontal plane, respectively. Hence, Equation (1) can be displayed as:

$$A_C = \frac{\sum\limits_{k=1}^{N} [d\phi_{sk}^{T\uparrow} + d\phi_{dk}^{T\uparrow} + d\phi_{ak}^{T\uparrow}]}{N(E_s + E_d)dA_{th}} \tag{5}$$

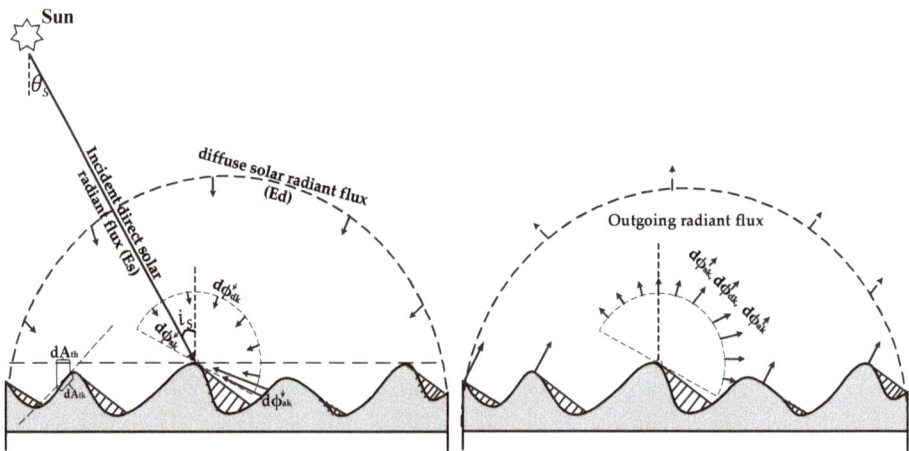

Figure 2. The contributions of incident and outgoing radiant flux over a coarse scale pixel; direct, diffuse and terrain irradiances.

The superscript 'T' means that the pixel is on a sloping surface; the subscript 'k' means that it is the kth fine scale micro-slope; $d\phi_{sk}^{T\uparrow}, d\phi_{dk}^{T\uparrow}, d\phi_{ak}^{T\uparrow}$ are the reflected direct solar radiant flux, diffuse sky flux, and terrain radiant flux of the kth micro-slope, respectively. They can be written as:

$$d\phi_{sk}^{T\uparrow} = E_{sk}^h \left(\frac{\cos i_{sk}}{\cos \theta_{sk}}\right) \Theta_k dA_{tk} \alpha_{bk}^T \tag{6}$$

$$d\phi_{dk}^{T\uparrow} = E_{dk}^h dA_{th} V_d \alpha_{wk}^T \tag{7}$$

$$d\phi_{ak}^{T\uparrow} = [E_{sk}^h \left(\frac{\cos i_{sk}}{\cos \theta_{sk}}\right) \Theta_k dA_{tk} + E_{dk}^h dA_{th} V_d] V_t \alpha_{wk}^T \tag{8}$$

$$dA_{th} = dA_{tk} \cos \alpha_k \tag{9}$$

where $\cos i_{sk}$, $\cos \theta_{sk}$ are the cosine of the relative incident angle on sloping surface and the incident zenith angle on horizontal surface; Θ_k is the binary coefficient in the direction of the sun which is used to show whether the pixel is shadowed by the terrain; E_{sk}^h, E_{dk}^h are the incident solar direct and diffuse irradiance on horizontal surface; dA_{tk} is the area of the kth micro-slope surface; V_d is the sky-view factor, which is the sky portion seen from a specific surface [37]; and V_t is the portion of adjacent terrain seen from a surface (the methods for calculating these parameters are listed in Appendix A). α_k is the slope of kth micro-slope surface; a_{bk}^T, a_{wk}^T are the micro-slope's directional-hemisphere reflectance (also as black-sky albedo, BSA) and bi-hemispherical reflectance (white-sky albedo, WSA) on a sloping surface, respectively. Inserting Equations (6)–(9) into (5), the coarse scale albedo can be rewritten as:

$$A_C = \frac{\sum\limits_{k=1}^{N} (E_{sk}^h \frac{\cos i_{sk}}{\cos \theta_{sk} \cos \alpha_k} \Theta_k a_{bk}^T + E_{dk}^h V_d a_{wk}^T + [E_{sk}^h \frac{\cos i_{sk}}{\cos \theta_{sk} \cos \alpha_k} \Theta_k + E_{dk}^h V_d] V_{tk} a_{wk}^T)}{N(E_{sk}^h + E_{dk}^h)} \tag{10}$$

We define a parameter S as the sky diffuse ratio factor, which is the proportion of diffuse solar irradiance to the total radiation. It can be calculated by the ratio of the diffuse radiance on a horizontal surface and the downward global solar radiance and exposes values between 0 and 1.

$$S = \frac{E_{dk}^h}{E_{sk}^h + E_{dk}^h} \tag{11}$$

Inserting Equation (11) into (10), the coarse spatial scale albedo can be simplified as:

$$A_C = \frac{1}{N} \sum\limits_{k=1}^{N} [(1-S)\frac{\cos i_{sk}}{\cos \theta_{sk} \cos \alpha_k} \Theta_k a_{bk}^T] + \\ \sum\limits_{k=1}^{N} [(1-S)\frac{\cos i_{sk}}{\cos \theta_{sk} \cos \alpha_k} \Theta_k V_{tk} a_{wk}^T + V_{dk} a_{wk}^T S + V_{dk} V_{tk} a_{wk}^T S] \tag{12}$$

Similar to the descriptions of the blue sky albedo of MODIS [38,39], the aggregated coarse scale albedo can also be approximated through a linear combination of BSA and WSA, weighted by the sky diffuse ratio factor. Thus, when S equals zero, it means that it has no diffuse skylight and the value of A_C can be considered as the coarse scale BSA (BSA_C). However, when S is 1, the A_C is the coarse scale WSA (WSA_C):

$$BSA_C = \frac{1}{N} \sum\limits_{k=1}^{N} [\frac{\cos i_{sk}}{\cos \theta_{sk} \cos \alpha_k} \Theta_k (a_{bk}^T + V_{tk} a_{wk}^T)] \tag{13}$$

$$WSA_C = \frac{1}{N} \sum\limits_{k=1}^{N} [\frac{\cos i_{sk}}{\cos \theta_{sk} \cos \alpha_k} \Theta_k V_{tk} a_{wk}^T + V_{dk} a_{wk}^T + V_{dk} V_{tk} a_{wk}^T] \tag{14}$$

It is obvious that the BSA and WSA have intense topographic effects as they depend on topographic factors and the solar incident angle [33]. Thus, they have scale effects in the validation of albedo over rugged terrain. Furthermore, Equations (13) and (14) provide the albedo upscaling method and function as the reference truth in albedo validation over rugged terrain.

2.3. Fine Scale Albedo Retrieval Algorithm over Rugged Terrain

Currently, the Bidirectional Reflectance Distribution Function-based (BRDF-based) albedo retrieval algorithms cannot be applied directly to fine scale albedo retrieval because of the lacking of enough fine scale multi-angle observations [40,41]. A feasible method is the direct retrieval algorithm such as the Angular Bin (AB) algorithm, which estimates the surface broadband albedo based on a single-date/angular observation [40–43]. In the AB algorithm, the incident and observing hemisphere were divided into several angular bins. The POLarization and Directionality of Earth Reflectance-BRDF (POLDER-BRDF) data were used as prior knowledge in this algorithm for extracting anisotropy reflectance information to build a Look-Up Table (LUT) of the land surface albedo at each angular bin [41]. Then, the multi-variant linear regression models were established at each angular bin, which linked the narrowband surface directional reflectance with broadband albedo, specifically, the shortwave WSA and BSA corresponding to the solar angle at local noon [40,41]. The multi-variant linear regression relationship can be expressed as:

$$\alpha_{AB} = C_0(\theta_s, \varphi_s, \theta_v, \varphi_v) + \sum_{i=1}^{n} C_i(\theta_s, \varphi_s, \theta_v, \varphi_v)\rho_i(\theta_s, \varphi_s, \theta_v, \varphi_v) \tag{15}$$

where α_{AB} represents the land surface fine scale albedo including BSA and WSA; $C_i(\theta_s, \varphi_s, \theta_v, \varphi_v)$ is the regression coefficient; and $\rho_i(\theta_s, \varphi_s, \theta_v, \varphi_v)$ is the surface directional reflectance at band i, which are the functions of solar/view angles ($\theta_s, \varphi_s, \theta_v, \varphi_v$) are the solar zenith angle, solar azimuth angle, view zenith angle, and view azimuth angle). The AB algorithm has the advantages of simple computation, fewer input parameters, and consideration of surface bidirectional and spectral characteristics [41]. Preliminary validation of the AB algorithm showed good applicability for albedo retrieval with an absolute error of 0.009 for vegetation, 0.012 for soil, and 0.030 for snow/ice [41–43].

However, the current AB algorithm had to be improved before being applied over rugged terrain as the regression models and the LUT were built on horizontal land surface. Therefore, the coefficient and the LUT were not suitable for the albedo retrieval over rugged terrain. One accessible approach to improve the AB algorithm over rugged terrain was to rebuild the regression models and LUT on the sloping surface. Under the assumptions that the BRDF shape was the same over the land cover, the BRDF shape for the slope was the BRDF for the rotated angles. Sloping surface incident/observation hemispheres were re-divided and rotated to the sloping surface at the sloping coordinate system. Thus, it can be re-written on a sloping surface as:

$$\alpha_{ABT} = C_0(i_s, \phi_s, i_v, \phi_v) + \sum_{i=1}^{n} C_i(i_s, \phi_s, i_v, \phi_v)\rho_i(i_s, \phi_s, i_v, \phi_v) \tag{16}$$

where α_{ABT} is the fine scale sloping surface albedo, i_s, ϕ_s, i_v, ϕ_v are the relative solar zenith angle, solar azimuth angle, view zenith angle and azimuth angle on the sloping coordinate system, respectively. $\rho_i(i_s, \phi_s, i_v, \phi_v)$ is the land surface reflectance on the sloping surface. $C_i(i_s, \phi_s, i_v, \phi_v)$ is the regression coefficient on a sloping coordinate system. The sloping surface reflectance can be retrieved by the Coupled-BRDF mountain radiation transfer model, which was developed by Wen in 2015 [44].

3. The Experimental Area and Dataset

3.1. Simulated Coarse Scale and Fine Scale Albedos

The simulated coarse scale albedos were used as reference data for comparison with the aggregated coarse scale albedos, which were upscaled from a simulated fine scale albedo by using the MRT-based upscaling model. The BSA can be simulated by integrating directional reflectance over the exitance hemisphere, and the WSA can be obtained by integrating the directional reflectance over all viewing and irradiance reflectance directions. The coarse scale directional reflectance can be simulated according to the radiosity theory [33,34,45], which coupled with the DEM and micro-slope reflectance.

In the coarse scale reflectance simulation, the micro-slope or the fine scale directional reflectance can be directly simulated by the PROSAIL model [46], which couples the PROSPECT leaf optical properties model [47] with the SAIL canopy reflectance model [48] and has been widely validated and applied to reflectance modeling studies [49]. For directional reflectance simulation over rugged terrain, the leaf inclination distribution function (LIDF) was assumed as a spherical type, and the incident/observation geometries were corrected to the sloping coordinate system to assess the photon path length alteration. Table 1 illustrates the parameters of the inputted PROSAIL model for land surface reflectance simulation over the sloping surface.

Table 1. The parameters and the values of the PROSAIL model.

Parameter	Value
Solar Zenith Angle, SZA (degree)	0–60 (Interval 10)
Solar Azimuth Angle, SAA (degree)	0–360 (Interval 30)
View Zenith Angle, VZA (degree)	0–90 (Interval 10)
View Azimuth Angle, VAA (degree)	0–360 (Interval 30)
Leaf Chlorophyll a + b Concentration, $Cab(\mu g/cm^2)$	40
carotenoid content, Car $(\mu g/cm^2)$	8
equivalent water thickness, Cw (g/cm^2)	0.01
dry matter content, Cm (g/cm^2)	0.009
scene leaf area index, LAI (m^2/m^2)	3
average leaf angle, ALA (degree)	30
structure coefficient (N)	1.5
hot-spot size parameter(m/m)	0.01
Soil brightness parameter	1

Nine DEMs (Table 2) were generated with different Gaussian height distributions to provide the various slope and aspect of micro-slope (shown in Figure 3). Specifically, the nine DEMs had the same reference template, which was simulated by a Gaussian random distribution model with one unit mean elevation and 0.25 unit standard error within the area of 100 units × 100 units. By linking with the real distance of 30 m, the simulated micro-slope could be considered with the 30 m resolution [33]. Additionally, the vertical elevation was exaggerated in 1, 10, and 20, respectively. Through appropriate exaggeration in the elevation and different Gaussian smoothing filtering parameters, the micro-slopes were generated with different slope and aspect. However, only the central part 510 m × 510 m (17 × 17 grid cells) of the above DEM was considered as a coarse scale pixel coverage to avoid the ambiguous calculation errors at the edge of the DEM. The mean slopes listed in Table 2 were defined as the average slope of the grids within the 17 × 17 grid cells. Since the BSA depends on the solar geometry (including SZA and SAA), the SZA varied from 0° to 60° with a 10° interval, and SAA varied from 0° to 360° with a 30° interval. Therefore, 819 BSAs and nine WSAs from nine DEM files were simulated.

Table 2. Mean slopes of the DEMs.

Filter	Exaggeration = 1	Exaggeration = 10	Exaggeration = 20
	Mean Slope	Mean Slope	Mean Slope
1 × 1	3.38	29.69	47.18
3 × 1	2.7	24.68	41.25
5 × 1	1.98	18.76	33.23

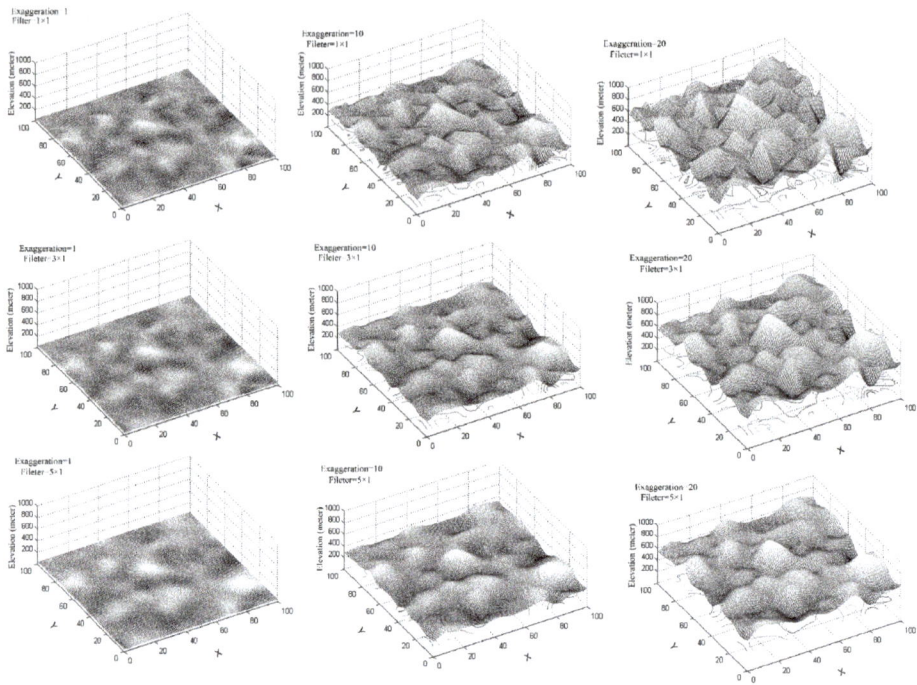

Figure 3. Simulated DEM with Gaussian height distributions.

3.2. In Situ Albedo Measurements

The upper stream of the Heihe River Basin (HRB) was selected as the study area for coarse scale albedo validation over rugged terrain. HRB is a typical inland river in the arid region of northwest China (97.1°E–102.0°E, 37.7°N–42.7°N). This region was selected as the core experimental watershed of Hi-WATER, because of the abundant accumulations of long temporal scale records since 1995 [50,51]. There are mountains in this region where the altitude ranges from 1025 m to 5076 m. The dominated land covers are cropland, Gobi Desert, grassland, forest, and high-latitude meadows.

A prototype watershed observation system has been established since 2009 [52]. Fifteen Automatic Weather Stations (AWS) were mounted in this observation system (Figure 4), which recorded the essential parameters every ten minutes including upwelling solar shortwave radiance (USR), and downwelling solar shortwave radiance (DSR). In situ albedo can be calculated as the ratio of USR and DSR at the local solar noon (11:30–12:30). Figure 4 illustrates the study area and the location of the selected AWSs.

The 15 AWSs in HRB are temporary sites where the measurements were only recorded at the periods of the Hi-WATER experiments. Two CNR pyranometers were mounted back to back between 2 m and 2.5 m above the top of the canopy at every AWS. The measurement footprints were circular and 80% of the signal came from a region with a diameter between 34.8 m and 43.5 m, which could be easily calculated as per Sailor's work [53]. The measurements recorded at 15 AWSs were considered with the high representativeness of mean albedo within the 30 m HJ pixel. Table 3 summarizes the details of the albedo measured sites used for validation.

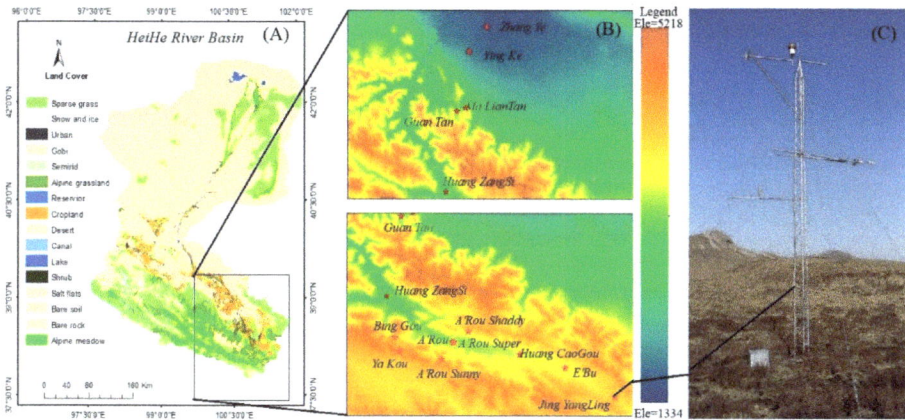

Figure 4. Study area: (**A**) the overview of the study area; (**B**) the location of the Automatic Weather Stations (AWSs) in the DEM imageries; and (**C**) one example of the AWSs.

Table 3. The description of the selected AWSs.

Site Name	Lat/Lon/Ele (Deg/Deg/M)	Land Covers	Slope (Deg)	Mean Slope (Deg)	Time Periods
A'rou Super	38.047/100.464/3017	Grassland	1.148	2.5	2013, 2014
A'rou Sunny	38.09/100.520/3579	Grassland	5.746	11.64	2013, 2014
A'rou Shady	37.984/100.411/3585	Grassland	9.871	15.703	2013, 2014
E'bu	37.949/100.915/3355	Grassland	2.316	3.376	2013, 2014
Huang ZangSi	38.225/100.192/2651	Cropland	4.764	6.503	2013, 2014
Huang CaoGou	38.003/100.731/3196	Grassland	6.968	3.917	2013, 2014
Jing YangLing	37.838/101.116/3793	Grassland	5.206	10.85	2013, 2014
Zhang Ye	38.975/100.446/1456	Cropland	4.847	2.882	2009, 2013, 2014
Hua ZhaiZi	38.765/100.319/1740	Desert	5.206	3.775	2009, 2013, 2014
Guan Tan	38.534/100.250/2839	Forest	14.943	11.418	2009
A'rou	38.051/100.457/2993	Grassland	3.554	3.671	2009
Bing Gou	38.067/100.222/3438	Grassland	4.274	10.573	2009
Ya Kou	38.014/100.242/4137	Grassland	8.432	9.504	2009
Ying Ke	38.858/100.41/1517	Cropland	1.148	2.296	2009
Ma LianTan	38.548/100.296/2827	Grassland	15.42	18.561	2009

Ten AWSs were mounted in the grassland, three AWSs were located in the cropland, one site in the desert, and one site in the forest. The slope of the AWSs and the mean slope of coarse scale pixels (500 m in this paper) were obtained from the DEM were listed in Table 3. According to the mean slope, if the mean slope was less than 5°, the coarse scale pixel was thought of as being of a gentle slope. If the mean slope was greater than 5° and less than 10°, the coarse scale pixel was a relatively rugged terrain. When the mean slope was greater than 10°, the coarse scale pixel was considered as steep terrain. The DEM were collected from the Shuttle Radar Topography Mission (SRTM) DEM data [54,55], and had a 30 m spatial resolution and UTM projection [56]. Slope, aspect, shaded factor, and other terrain parameters were derived from the DEM referenced to the algorithm in Dozier's work [37].

3.3. Satellite Imagery

The 500 m Collection V006 MODIS albedo product (MCD43A3 C6) was selected as the coarse scale albedo products for validation. The semi-empirical, kernel-driven BRDF model was the primary algorithm to retrieve surface BRDF and albedo at 16-day time periods [57]. A back-up algorithm was employed for deriving the albedo in a situation of insufficient angular sampling because of cloud cover or orbital constraints [28]. The MCD43A3 products provided shortwave broadband black-sky albedo (BSA) and white-sky albedo (WSA) at a 500 m resolution [38].

The fine scale remote sensing data were collected from the China HJ-1/CCD sensor which provides an opportunity to record earth land surface reflectance with high spatial resolution and a broad coverage in China [44]. The HJ-1/CCD can record 4-band images with a 30 m spatial scale and a revisiting circle of less than two days [17]. The HJ images were geo-rectified for matching with the DEM file.

4. Results

4.1. Assessment of the MRT-Based Upscaling Model

A high-quality albedo upscaling model is of great importance for the application of multi-scale validation strategy. The accuracy of the MRT-based albedo upscaling model was assessed through a comparison of the aggregated fine scale albedo with the reference coarse scale albedo. Moreover, to show the performance of the MRT-based upscaling method, the line weighted average model was also used to scale up the fine scale albedo to a coarse spatial scale, which is the most commonly-used model in the multi-scale validation method over flat terrain [8,9,24,58]. The bias, root-mean-square error (RMSE), mean absolute percent error (MAPE), and coefficient of determination (R^2) were used to show the accuracy of the aggregated coarse scale albedos, which are expressed as follows:

$$\text{Bias} = \sum_{i=1}^{n}(p_i - o_i)/n \tag{17}$$

$$\text{RMSE} = \sqrt{\sum_{i=1}^{n}(p_i - o_i)^2/n} \tag{18}$$

$$\text{MAPE} = \frac{100\%}{n}\sum_{i=1}^{n}|(p_i - o_i)|/\overline{o_i} \tag{19}$$

$$R^2 = \sum_{i=1}^{n}(p_i - \overline{p_i})(o_i - \overline{o_i})/\left[\sum_{i=1}^{n}(p_i - \overline{p_i})^2\sum_{i=1}^{n}(o_i - \overline{o_i})^2\right] \tag{20}$$

where $p_i, \overline{p_i}$ are the albedo observation and average albedo observation; $o_i, \overline{o_i}$ are the reference albedo measurement and average of reference albedo measurement; and n represents the amount of simulated albedo measurements.

4.1.1. Accuracy of Using MRT-based Albedo Upscaling Model

Figure 5 shows the scatter plots between the reference coarse scale albedo and aggregated coarse scale albedo as well as the histogram of bias distributions. The aggregated coarse scale BSAs showed significant agreements with the reference simulated coarse scale albedo with an RMSE of 0.0029, Bias of −0.0001, MAPE of 0.94%, and R^2 of 0.9962 (Figure 5A). The histogram showed that the bias reflected a normal distribution with the maximum bias less than 0.015 (Figure 5B). Figure 5C shows that the aggregated coarse scale WSAs had a similar accuracy with the BSAs. The bias was 0.0007, the MAPE was 0.79%, the RMSE was less than 0.0017, and the R^2 is as high as 0.9984. The bias in Figure 5D showed that the maximum bias among the nine simulated WSAs was less than 0.003 following the change of the mean slope. Overall, these results showed that the aggregated coarse scale albedos had fewer discrepancies with the reference coarse scale albedos.

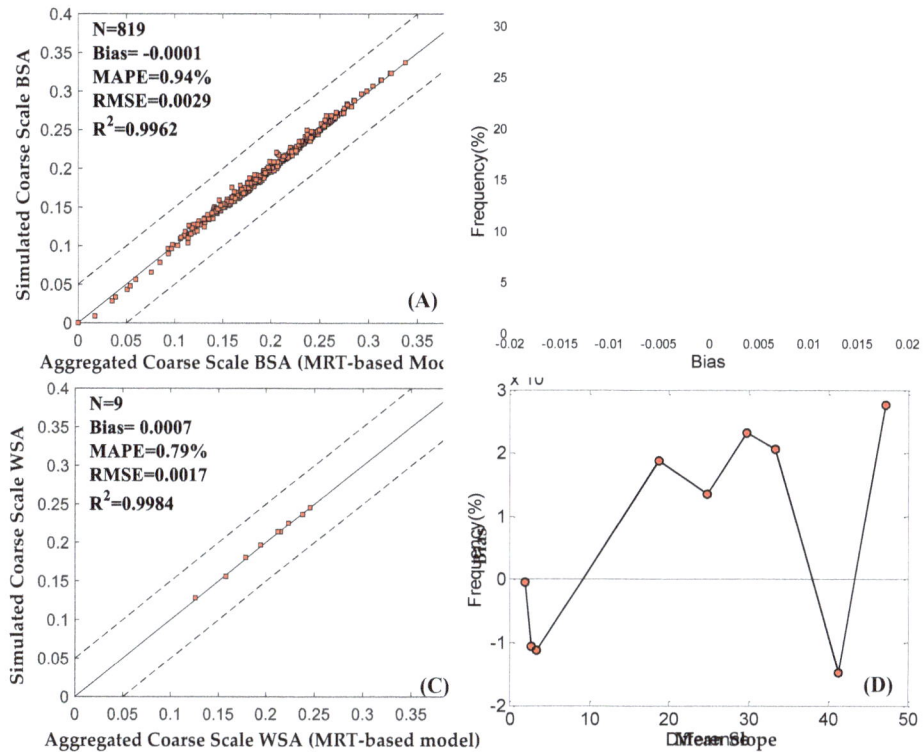

Figure 5. Scatter plots and the histogram evaluate the aggregated coarse scale albedo (using a Mountain-Radiation-Transfer (MRT)-based model) against the reference simulated coarse scale albedo for (**A**) the aggregated coarse scale BSA vs. the simulated coarse scale BSA; (**B**) the bias histogram of the aggregated coarse scale BSA minus the reference coarse scale BSA; (**C**) the aggregated coarse scale WSA vs. the simulated coarse scale WSA; and (**D**) the bias between the nine aggregated coarse scale WSAs and the reference simulated WSAs following the increase of the mean slope.

The BSA depends on the incident solar geometry and is intensely affected by complex terrain, therefore, the accuracy of aggregated coarse scale albedo varied with different SZAs and terrain slopes. Figure 6 shows the comparison between the aggregated coarse scale albedo and simulated coarse scale albedo at the solar zenith angles from 0° to 60° with an interval of 20°. It was obvious that the aggregated coarse scale BSAs had the significant agreement with the reference coarse scale BSAs at the four selected SZAs. With the increase of SZA, the RMSE values showed a slight increase from 0.0024 at SZA of 0° to 0.0025 at SZA of 40°. The Bias, MAPE, and R^2 had a little change when the solar zenith angle varied from 0° to 40°. Even though the SZA increased to 60°, the aggregated coarse scale BSAs still had significant agreement with the reference simulated coarse scale albedo with a bias of 0.0021, MAPE of 1.51%, RMSE of 0.0049, and R^2 of 0.9971. The bias distribution histogram also demonstrated that the maximum bias of the two types of BSAs was less than 0.015 at the four selected SZAs.

Figure 7 shows the comparison between the simulated coarse scale BSAs and reference coarse scale BSAs over four selected DEM types with different mean slopes. The mean slope of the DEM in Figure 7A was lower than 10°, which was considered as the flat terrain. The results showed that the aggregated coarse scale BSAs had great agreement with the reference simulated coarse scale albedo with an RMSE of 0.0001. When the slope is varied from greater than 10° and less than 20°, where the terrain can be considered as a gentle slope, the RMSE was increased to 0.0019 (Figure 7B). When the mean slope was greater than 20° and less than 30° (where the land surface can be considered some rugged), the RMSE increased to 0.0034 (Figure 7C). However, when the slope was larger than 30°, the RMSE still reminded as 0.0041 (Figure 7D). Though, the regression coefficient and R² were much closer to 1 at these four situations, the MAPE had an obvious increase from 0.05% to 2.04%. From the histograms of the mean slopes less than 30° (Figure 7A–C), we could see that more than 70% of the biases were distributed at the intervals less than 0.01. The maximum bias was smaller than 0.02 at the four mean slopes of the DEM. These four figures showed that the aggregated coarse scale BSAs had the great agreement with the simulated coarse scale BSAs at the four selected mean slopes.

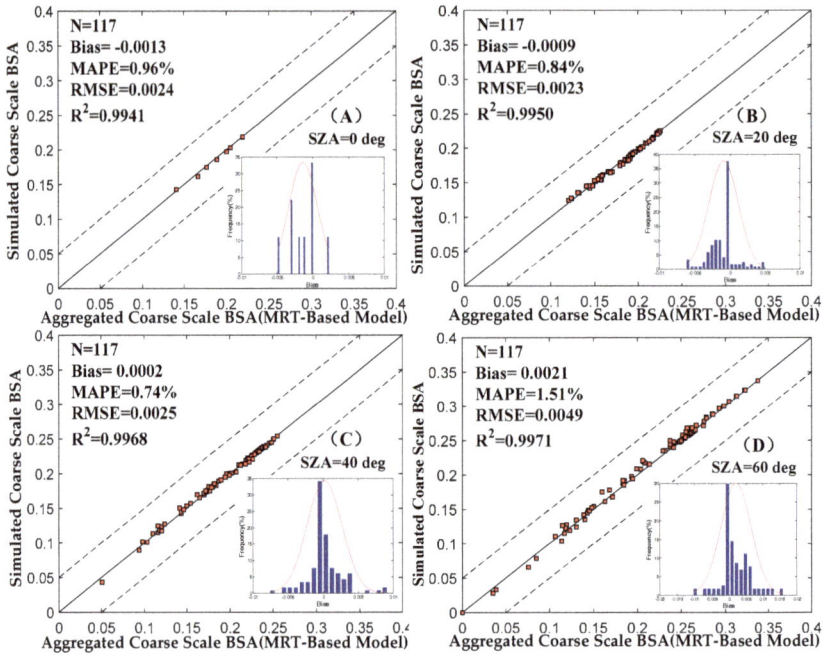

Figure 6. Validation results at different solar zenith angle: (**A**) the solar zenith angle is 0°; (**B**) the Solar Zenith Angle (SZA) is 20°; (**C**) the SZA is 40°; and (**D**) the SZA is 60°.

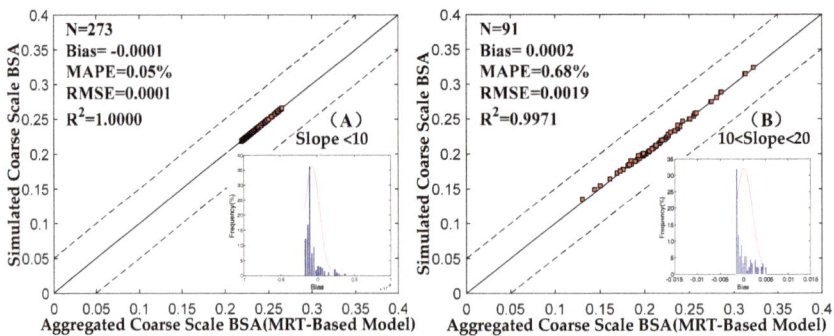

Figure 7. *Cont.*

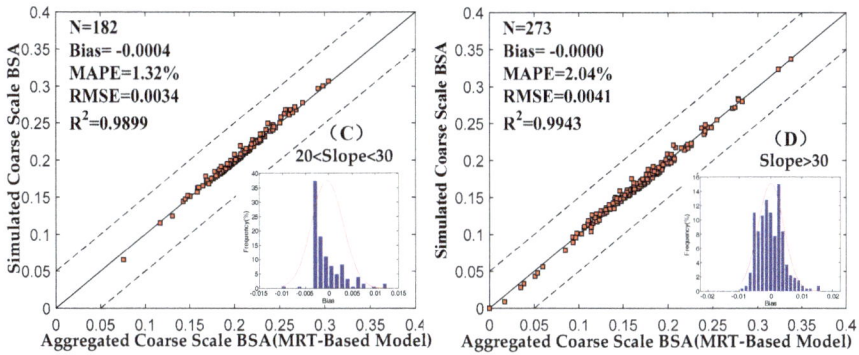

Figure 7. Validation of the MRT-based upscaling model at different slopes (**A**) the slope is smaller than 10 degree; (**B**) the slope is greater than 10 degree and less than 20 degree; (**C**) the slope is between 20 and 30 degrees; and (**D**) the slope is larger than 30 degrees.

4.1.2. Accuracy Analysis by using the Linear Weighted Average Upscaling Model

Figure 8 shows the comparison between the reference coarse scale albedos and the aggregated coarse scale albedos, which averaged the fine scale albedo within a coarse scale pixel. The linear averaged albedos showed a big discrepancy with the simulated reference coarse scale albedo with an RMSE of 0.0744, and a bias of −0.0481 overall, for the different slopes and incident solar geometries. The R^2 was reduced as low as 0.0223. Though 80% of the bias was distributed symmetrically in the interval of −0.10 to 0.15, the maximum bias increased to nearly 0.25. Additionally, the comparison with WSA showed similar results as the BSA with a bias of −0.0383, and RMSE of 0.0532, respectively. Figure 8D shows the changes of biases between the aggregated coarse scale WSAs and the reference simulated coarse scale WSAs following the increase of mean slope. Furthermore, it shows the obviously increase of the bias following the change of the mean slope with the maximum absolute bias greater than 0.12 at the place that the mean slope is increased to 47.18. The results showed that the linear weighted average model was not very suitable to deal with the scaling mismatching for coarse scale albedo validation over rugged terrain.

The statistics metrics, which are summarized in Table 4, record the comparison results between the reference coarse scale albedo and the aggregated coarse scale albedo which was aggregated by the MRT-based upscaling model and linear weighted average model, respectively. Overall, the MRT-based upscaled coarse scale BSAs had a great agreement with the simulated reference coarse scale BSAs. Following the increase of the solar zenith angle, the uncertainty of the MRT-based upscaled coarse scale BSAs had a slight increase as the RMSE increased from 0.0022 to 0.0049, and the MAPE increased from 0.71% to 1.50%. Meanwhile, the RMSE and MAPE also showed an obviously increase trend following the increase of mean slope of the DEM where the RMSE varied from 0.0001 to 0.0041, and the MAPE varied from 0.05% to 2.04%. However, the aggregated coarse scale BSAs upscaled by the linear average model had large uncertainty when compared with the reference simulated coarse scale BSAs. Following the increasing of SZA and mean slope of DEM, the bias, MAPE, and RMSE had a similar increasing trend as that of the aggregated coarse scale BSA upscaled by the MRT-based model. When the mean slope was smaller than 10°, the coarse scale BSAs aggregated by the linear average model had great consistency with the reference simulated coarse scale BSAs with an RMSE of 0.0043, bias of 0.0002, MAPE of 1.26%, and R^2 of 0.9028. When the slope is increased to larger than 30°, the absolute of the bias, RMSE, MAPE, and R^2 also increased as larger as 0.0860, 0.0809, 37.37%, and 0.1288, respectively.

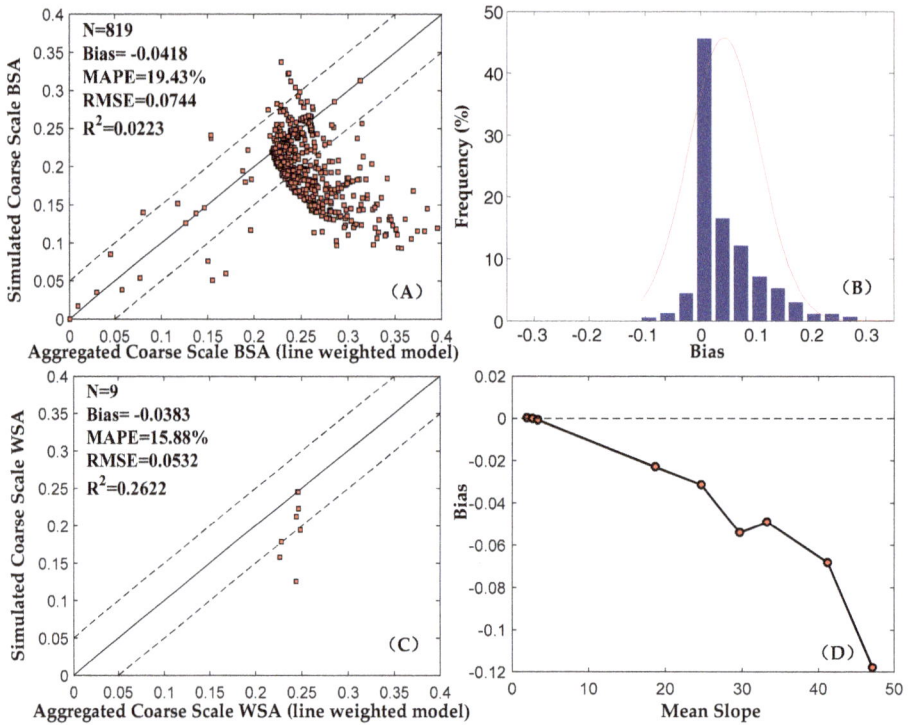

Figure 8. Scatter plots and the histogram evaluate the fine scale albedo against simulated coarse scale albedo for: (**A**) the aggregated fine scale BSA vs. the simulated coarse scale BSA; (**B**) the histogram of the aggregated fine scale BSA minus the coarse scale BSA; (**C**) the aggregated fine scale WSA vs. the simulated coarse scale WSA; and (**D**) the bias distribution of the aggregated fine scale WSA and the coarse scale WSA following the increase of the mean slopes.

Table 4. Statistical metrics from the validation between coarse scale BSA and aggregated fine scale BSA aggregated by different upscaling methods (α^* means the slope of the DEM).

Parameters	Aggregated Fine Scale BSA (MRT-Based Upscaling Model) vs. Simulated Coarse Scale BSA					Aggregated Fine Scale BSA (Linear Weighted Average Upscaling Model) vs. Simulated Coarse Scale Albedo BSA				
	N	Bias	MAPE	RMSE	R^2	N	Bias	MAPE	RMSE	R^2
SZA = 0°	117	−0.0013	0.96%	0.0024	0.9941	117	−0.0389	16.79%	0.0557	0.9634
SZA = 10°	117	0.0013	0.93%	0.0024	0.9942	117	−0.0393	16.93%	0.0573	0.9532
SZA = 20°	117	−0.0009	0.84%	0.0023	0.9950	117	−0.0416	17.68%	0.0634	0.9340
SZA = 30°	117	0.0004	0.71%	0.0022	0.9965	117	−0.0448	18.95%	0.0743	0.6190
SZA = 40°	117	−0.0002	0.74%	0.0025	0.9968	117	−0.0493	21.19%	0.0863	0.4593
SZA = 50°	117	−0.0010	0.89%	0.0030	0.9980	117	−0.0411	20.82%	0.0825	0.1448
SZA = 60°	117	−0.0021	1.50%	0.0049	0.9971	117	−0.0374	23.23%	0.0923	0.2497
$\alpha^* < 10°$	273	−0.0001	0.05%	0.0001	1	273	0.0002	1.26%	0.0043	0.9028
$10° < \alpha^* < 20°$	91	0.0002	0.68%	0.0019	0.9971	91	−0.0250	14.19%	0.0607	0.272
$20° < \alpha^* < 30°$	182	−0.0004	1.32%	0.0034	0.9899	182	−0.0462	20.14%	0.0677	0.0302
$\alpha^* > 30°$	273	−0.0000	2.04%	0.0041	0.9943	273	−0.0860	37.37%	0.0809	0.1288
Overall	819	−0.0001	0.94%	0.0029	0.9962	819	−0.0481	19.43%	0.0744	0.0233

The discrepancies between the MRT-based upscaled albedo and the reference albedo raised slightly following the increasing of the incident solar zenith angle and the mean slope, especially when the SZA was larger to 60° and the mean slope was larger than 30°. However, even though the SZA is increased to 60° and the mean slope increased to larger than 30°, the MRT-based albedo upscaling model showed the great performance over rugged terrain with a maximum bias and RMSE smaller than 0.0032 and 0.0049, respectively.

4.2. Application of MRT-Based Upscaling Model for MODIS Albedo Validation

4.2.1. Accuracy Assessment of HJ Albedo

The accuracy and uncertainty of fine scale HJ blue-sky albedos have an essential influence on coarse scale albedo validation in the multi-scale validation strategy [24]. Therefore, the HJ blue-sky albedos were assessed firstly with in situ albedos. The HJ blue sky albedo can be approximated through a linear combination of BSA and WSA, weighted by the fraction of actual direct to diffuse skylight. Figure 9A shows the comparison between the in situ albedo and HJ blue-sky albedo. It was obvious that the two albedos had a less discrepancy with a bias of 0.0028, RMSE of 0.0272, MAPE of 10.0%, and R^2 of 0.7470. Figure 9B shows the error histogram of bias. It was easily found that the biases were normally distributed significantly. Though 80% of the bias was distributed within the interval of −0.05 to 0.05 and the maximum bias was less than 0.07. The validation results indicated that within the 95% confidence interval (shown by the *p*-value, which is less than 0.005), the regression model could be used to describe the relationship between the fine scale HJ albedos and in situ albedos. Additionally, it can be used as the calibration model to calibrate the HJ albedos.

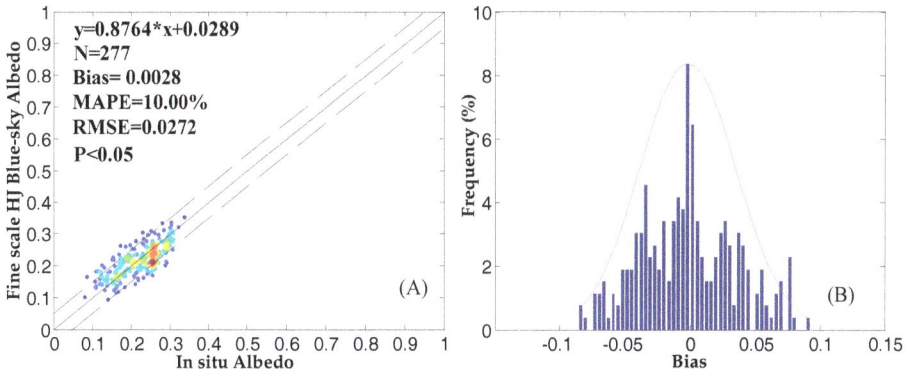

Figure 9. Scatter plots and bias histogram between fine scale albedos and the in situ albedos: (**A**) in situ albedo vs. the fine scale albedo; (**B**) Bias histogram of fine scale blue-sky albedo minus the in situ albedo The colors refer to the density of points (from highest (red) to lowest (blue)).

4.2.2. Comparison of the Aggregated Coarse Scale Albedo with MODIS Albedo Products

The MRT-based albedo upscaling model was used for aggregating the calibrated HJ albedo to a coarse spatial scale. Figure 10 displays the validation results by comparing the coarse scale MCD43A3 C6 albedos with the coarse scale aggregated HJ albedos. The results showed that MCD43A3 C6 BSAs with full inversion had agreement with the aggregated HJ BSAs, with an RMSE of 0.0305, bias of 0.008, MAPE of 13.50%, and R^2 of 0.6517. While, the MCD43A3 C6 BSAs with magnitude inversion showed slight differences with the aggregated HJ BSAs with an RMSE of 0.0511 and bias of 0.0097, MAPE of 19.17%, and R^2 of 0.3817. The MCD43A3 WSAs showed the similar accuracy and uncertainty as the BSAs, with an RMSE of 0.0321 for full inversion and an RMSE of 0.0531 for magnitude inversion. The results showed that the uncertainty of MCD43A3 C6 over rugged terrain was slightly larger than 0.02, which has been shown in many studies [15,16,18,20,22,59–61]. The two bias histograms showed that the biases were distributed symmetrically at the interval zero and the maximum bias were smaller than 0.1.

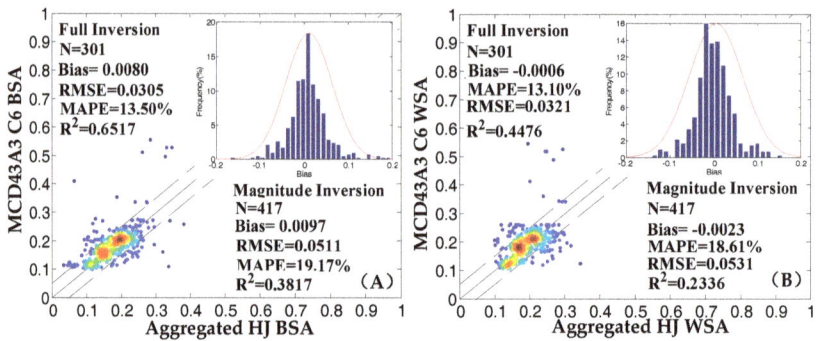

Figure 10. Coarse scale MCD43A3 C6 albedos validation by comparison with the aggregated coarse scale albedos: (**A**) the aggregated coarse scale BSAs vs. the MCD43A3 C6 BSAs; (**B**) the aggregated coarse scale WSAs vs. the MCD43A3 C6 WSAs. The colors refer to the density of points (from highest (red) to lowest (blue).

Figure 11 illustrates the MCD43A3 C6 products validation results at different mean slopes around the AWSs. This showed that the MCD43A3 C6 products had good agreements with the aggregated HJ albedos at the surface where the slope was lower than 5°. The bias, MAPE, RMSE and R^2 for the MCD43A3 full inversion BSAs were 0.0108, 10.88%, 0.0244, and 0.6136, respectively (Figure 11A). When the slope increased to greater than 5° and less than 10°, the RMSE increased to 0.0315 for BSA and 0.0308 for the WSAs of full inversion (Figure 11C,D). However, when the slope was larger than 10°, there were large discrepancies between the MCD43A3 C6 full inversion BSAs and the aggregated HJ BSAs with an RMSE of 0.0365 and MAPE of 19.71% (Figure 11E). The WSAs of full inversion showed a similar increase trend, with a bias of 0.0126, MAPE of 11.39% and RMSE of 0.0293 (Figure 11B), when the mean slope was lower than 5°. When the mean slope was raised to larger than 10°, the bias, MAPE, and RMSE increased to −0.0133, 15.45% and 0.0313, respectively (Figure 11F).

Generally, the MCD43A3 C6 albedos of magnitude inversion had lower quality than the albedos of full inversion. Following the increase of mean slope, the RMSE and MAPE displayed an outstanding rise for the two inversions MCD43A3 C6 BSAs, as well as the magnitude inversion WSAs, especially where the mean slope was greater than 10°. Over rugged terrain, the land surface albedos were seriously influenced by the complex interactions among the topography and the incident radiation from the sun location, the shadow, the atmosphere, and surface scattering irradiance from the adjacent terrain. The topographic effects led to more uncertainty in albedo remote sensing retrieval. Neglecting these effects, the albedo algorithm may lead to lower quality albedos.

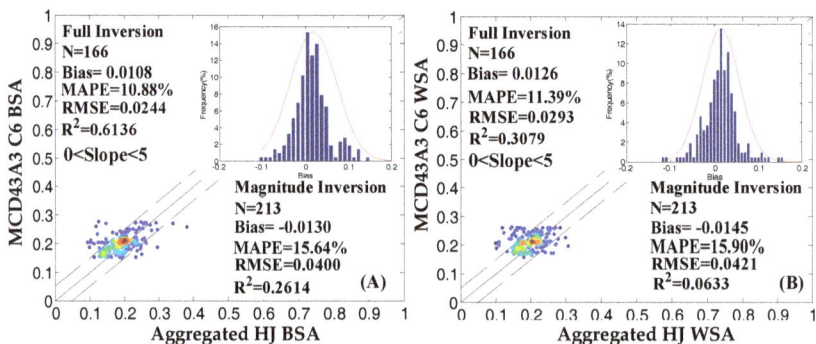

Figure 11. *Cont.*

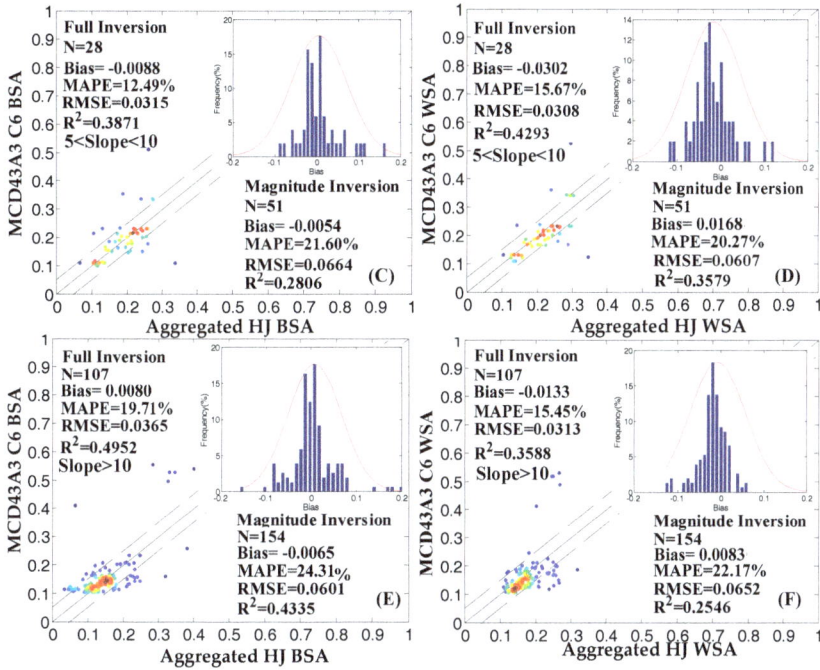

Figure 11. Coarse scale MCD43A3 C6 products validation by comparing with the aggregated fine scale albedo at different mean slopes. The colors refer to the density of points (from highest (red) to lowest (blue).

5. Discussion

The topography affects the land surface albedo. Compared with the validation over flat terrain, the albedo validation over rugged terrain should consider the scale effects. This paper proposed an MRT-based albedo upscaling model for dealing with the scale effects in the multi-scale validation strategy over rugged terrain. The simulated coarse scale albedos were used as the reference truth to assess the MRT-based albedo upscaling model.

The assessments showed that the coarse scale albedo upscaled by the MRT-based model had the great agreement with the reference coarse scale albedo in different solar zenith angle or terrain with different mean slopes (Table 4). The uncertainty of the aggregated coarse scale albedos had the slight variation following the change of SZA and mean slope. In particular, the coarse scale albedos aggregated by the MRT-based model still showed the great consistency with the reference coarse scale albedo when the mean slope was larger than 30° or the solar zenith angle was up to 60° (Figures 6 and 7). Adversely, the coarse scale albedo aggregated by the linear average model showed a larger discrepancy with the reference coarse scale albedo (Figure 8). Following the increase of solar zenith angle and the mean slope of the terrain, the discrepancy had an obviously increasing trend (Table 4). The MAPE and RMSE were raised to 37.37% and 0.0809, following the increase of the mean slope, respectively. Comparisons between the validation results of the two upscaled coarse scale albedos showed that the MAPEs increased obviously following the increase of mean slope (Table 4). When the mean slope was larger than 30°, the difference of MAPE increased larger than 35%. However, when the mean slope was lower than 10°, the MAPE did not show an obvious difference. Similarly, the difference between the two upscaled models also increased following the increase of the solar zenith angle. When the solar zenith angle was 60°, the difference between the two MAPEs was larger than 20%. Notably, the two MAPEs also showed a large difference when the solar zenith angle was 0°.

The simulated results showed that the SZA and mean slope had a significant influence on the scale effects of albedo over rugged terrain. When the land surface was flat or the mean slope was lower than 10°, the scale effects could be thought as very gentle, and either the MRT-based albedo upscaling model or the linear average upscaling model could be used to upscale fine scale albedo to a coarse scale. When the land surface was steep or the mean slope larger than 10°, the difference caused by scale effects was larger than 15% of MAPE. The solar zenith angle coupled with complex terrain can cause the mutual shadowing and re-distribution of download solar radiation. These characteristics resulted in the land surface being more heterogeneous and larger albedo scale effects.

The MRT-based upscaling model was applied to MCD43A3 C6 albedo products validation in HRB. Fine scale albedos were retrieved from the HJCCD sensor by the improved AB algorithm on a sloping surface and used as a bridge to deal with the scale mismatching between the in situ albedos and the MCD43A3 C6 albedos. The HJ albedos were evaluated and calibrated by in situ albedos for reducing the systemic errors and the accidental errors. Furthermore, they were upscaled to the coarse scale and were functioned as the reference truth over albedo validation. The validation results showed the MCD43A3 C6 products of full inversion had high agreements at the sites where the mean slope was lower than 5°. Following the increase of mean slope, the MCD43A3 C6 BSAs of full or magnitude inversion showed the outstanding increase of RMSE, with an RMSE varied from 0.0244 to 0.0365 for the full inversion BSA and 0.040 to 0.0601 for the magnitude inversion, respectively. The MCD43A3 of the full WSAs products also showed a similar result with the RMSE varying from 0.0293 to 0.0313 following the mean slope increasing from 0° to greater than 10°.

However, the multi-scale validation strategy used for albedo validation over rugged terrain also still faces challenges. The geometric misalignment between DEM and fine scale albedo had an intense influence on the application of the MRT-based albedo upscaling method. In this paper, the HJ reflectance data were geometrically rectified by the geo-referenced Landsat images and rechecked manually. Second, the MRT-based albedo upscaling model was built under the assumption that the coarse scale pixel was overall horizontal. To use the MRT-based method where the coarse scale pixel was not overall horizontal, the simple way was to change the angular effects in model development. Therefore, the MRT-based upscaling model can also be suitable for using in coarse scale albedo validation. The quality of fine scale albedo and the multiple land covers were also affected the validation results. Thus, in the future, we plan to improve the method's applicability and assess the accuracy of MCD43A3 C6 products comprehensively under multiple land covers and long temporal periods over rugged terrain at the global scale.

6. Conclusions

This paper proposed a multi-scale validation strategy for coarse scale albedo products validation over rugged terrain, where the fine scale albedos on sloping land surfaces were used as a bridge to upscale in situ albedo to coarse scale. In this paper, the MRT-based albedo upscaling method was developed and preliminarily applied for aggregating the fine scale albedos to coarse scale. The applicability and performance of the MRT-based upscaling model were assessed by comparing the upscaled coarse scale albedo with the simulated reference coarse scale albedo. The multi-scale validation strategy with an MRT-based albedo upscaling model was applied to assess the accuracy of MCD43A3 products over rugged terrain in the Heihe River Basin, China. The HJ satellite albedo was selected as the fine scale albedo and was upscaled to the coarse scale by the MRT-based albedo upscaling model.

The validation results showed that the coarse scale albedo validation are suffered the scale effects over rugged terrain. The scale effects increased following the increase of mean slope and the solar zenith angle over rugged terrain. Furthermore, it cannot be overlooked when the mean slope was larger than 10°. The MRT-based albedo upscaling model had the great performance in dealing with the issue of scale mismatching for albedo validation over rugged terrain.

The linear average model showed a similar performance with the albedo validation over flat terrain. However, over rugged terrain, the coarse scale albedos upscaled by the linear average model showed a bigger discrepancy than the simulated coarse scale albedos, which indicated that the linear average model had worse applicability for albedo validation over rugged terrain, especially, when the slope was larger than 10°.

The MRT-based albedo upscaling model was used to assess the MCD43A3 products. It showed great performance in aggregating the fine scale HJ albedo to coarse scale. The validation results showed that in the Heihe River Basin, the MCD43A3 BSA products with full inversion algorithm had an increased uncertainty following the increase of the mean slope, with an RMSE varying from 0.0244 to 0.0365, and a MAPE that varied from 10.88% to 19.71. Meanwhile, the MCD43A3 WSA products with the full inversion algorithm had an RMSE varying from 0.0293 to 0.0313 and MAPE from 11.39% to 15.49%.

Acknowledgments: This work was jointly supported by the Chinese Natural Science Foundation Project (NO.: 41671363, 41601380).

Author Contributions: Xingwen Lin and Jianguang Wen conceived, designed the method and relative experiments; Xingwen Lin, Jianguang Wen, Qinhuo Liu, Shengbiao Wu, Qing Xiao performed the experiments and were involved in the result analysis and discussion; Shengbiao Wu, Dongqin You and Dalei Hao, and Xiaodan Wu gave much helpful advice on the revising work; Xingwen Lin wrote the paper.

Conflicts of Interest: The authors declare no conflict of interest.

Appendix A

The total incident solar irradiance (E_T) on a micro-slope surface consists of three components: direct (E_s^T), diffuse solar irradiance, and the component from adjacent topography (E_a^T). It can be expressed as:

$$E_T = E_s^T + E_d^T + E_a^T \tag{A1}$$

In flat terrain, the direct solar irradiance can be described as the function of the exo-atmospheric solar irradiance and the solar zenith angle. Unlike the received solar beam at flat terrain, the direct solar irradiance varies with the angle formed by the solar beam and the normal to the surface [62]. In addition, the adjacent terrain may cast a shadow on the pixels, which results in receiving no direct radiance at all [63]. Considering these effects, the direct solar irradiance E_s^T can be described as the function of the relative solar illumination angle (i_s), the direct solar irradiance on flat terrain, and the shadow factors, which can be expressed as follows [32]:

$$E_s^T = E_s^h \frac{\cos i_s}{\cos \theta_s} \Theta \tag{A2}$$

where E_s^h is the incident solar irradiance on a horizontal plane; i_s is defined as the relative solar illumination angle over the incline surface, which can be calculated from the DEM slope and aspect angle (α, β) and the solar incident geometry (θ_s, φ_s) on a horizontal surface and can be shown as [64]

$$\cos i_s = \cos \theta_s \cos \alpha + \sin \theta_s \sin \alpha \cos(\beta - \varphi_s) \tag{A3}$$

where Θ is a binary coefficient, and is set to zero whenever a surface is shadowed by surrounding ridges (cast shadow) and set to one otherwise [65]. Compared with the incident diffuse solar irradiance on the horizontal surface, it may be reduced by the topographic effect due to the hemispherical sky dome being partially obstructed by the surrounding terrain and the tilt of the surface itself [66].

Generally, the diffuse irradiance was estimated from the incident solar diffuse irradiance obtained with a horizontal surface by a simple relation as

$$E_d^T = E_d^h V_d \tag{A4}$$

by Dozier's work [37]. The V_d was defined as the sky-view factor, which was the ratio of the diffuse sky irradiance at a point to that on an unobstructed horizontal surface. Several researches were targeted on the accuracy calculation of sky-view factors [67]. In this paper, this term can be expressed as

$$V_d = \left(\frac{1}{2\pi} \int_0^{2\pi} \left[\cos\alpha \, \sin^2 H_\varphi + \sin\alpha \cos(\varphi - \beta)(H_\varphi - \sin H_\varphi \cos H_\varphi)\right]d\varphi\right) \\ (1 + \cos^3\theta_s \cos^2 i_s)[1 + \sin^3(\frac{\alpha}{2})] \tag{A5}$$

by Wen's work [26]. Where H_φ is the horizontal angle from the zenith downward to the local horizon for direction φ. Adjacent terrain irradiance can be computed by a simple model by Hansen [68]. The model had been applied widely and taken into account of the average land surface reflectance of nearby pixels, the slope of the surface.

$$E_a^T = [E_s^T + E_d^T]V_t\rho_{mean} \tag{A6}$$

where ρ_{mean} is the average reflectance of nearby terrain; parameter V_t is the terrain view factor (range 0–1) calculated from the local slope or a horizon analysis. A simple relation between V_d and V_t can be simple calculated as the Equation (A7).

$$V_t = 1 - V_d \tag{A7}$$

References

1. Dickinson, R.E. Land surface processes and climate—Surface albedos and energy balance. *Adv. Geophys.* **1983**, *25*, 305–353.
2. Lucht, W.; Hyman, A.H.; Strahler, A.H.; Barnsley, M.J.; Hobson, P.; Muller, J.P. A comparison of satellite-derived spectral albedos to ground-based broadband albedo measurements modeled to satellite spatial scale for a semidesert landscape. *Remote Sens. Environ.* **2000**, *74*, 85–98. [CrossRef]
3. Lucht, W. Expected retrieval accuracies of bidirectional reflectance and albedo from EoS-MODIS and MISR angular sampling. *J. Geophys. Res. Atmos.* **1998**, *103*, 8763–8778. [CrossRef]
4. You, D.; Wen, J.; Xiao, Q.; Liu, Q.; Liu, Q.; Tang, Y.; Dou, B.; Peng, J. Development of a high resolution BRDF/albedo product by fusing airborne CASI reflectance with MODIS daily reflectance in the Oasis area of the Heihe river basin, China. *Remote Sens.* **2015**, *7*, 6784–6807. [CrossRef]
5. Li, X. Characterization, controlling, and reduction of uncertainties in the modeling and observation of land-surface systems. *Sci. China Earth Sci.* **2014**, *57*, 80–87. [CrossRef]
6. Jin, Y. Consistency of MODIS surface bidirectional reflectance distribution function and albedo retrievals: 2. Validation. *J. Geophys. Res.* **2003**, *108*. [CrossRef]
7. Wang, K.; Liu, J.; Zhou, X.; Sparrow, M.; Ma, M.; Sun, Z.; Jiang, W. Validation of the MODIS global land surface albedo product using ground measurements in a semidesert region on the Tibetan plateau. *J. Geophys. Res. Atmos.* **2004**, *109*, D05107. [CrossRef]
8. Salomon, J.G.; Schaaf, C.B.; Strahler, A.H.; Gao, F. Validation of the MODIS bidirectional reflectance distribution function and albedo retrievals using combined observations from the aqua and terra platforms. *IEEE Trans. Geosci. Remote Sens.* **2006**, *44*, 1555–1565. [CrossRef]
9. Susaki, J.; Yasuoka, Y.; Kajiwara, K.; Honda, Y.; Hara, K. Validation of MODIS albedo products of paddy fields in Japan. *IEEE Trans. Geosci. Remote Sens.* **2007**, *45*, 206–217. [CrossRef]
10. Román, M.O.; Schaaf, C.B.; Woodcock, C.E.; Strahler, A.H.; Yang, X.; Braswell, R.H.; Curtis, P.S.; Davis, K.J.; Dragoni, D.; Goulden, M.L. The MODIS (Collection V005) BRDF/albedo product: Assessment of spatial representativeness over forested landscapes. *Remote Sens. Environ.* **2009**, *113*, 2476–2498. [CrossRef]

11. Wang, K.; Liang, S.; Schaaf, C.L.; Strahler, A.H. Evaluation of moderate resolution imaging spectroradiometer land surface visible and shortwave albedo products at fluxnet sites. *J. Geophys. Res. Atmos.* **2010**, *115*, 1383–1392. [CrossRef]

12. Wang, X.; Zender, C.S. MODIS snow albedo bias at high solar zenith angles relative to theory and to in situ observations in greenland. *Remote Sens. Environ.* **2010**, *114*, 563–575. [CrossRef]

13. Wang, Z.; Schaaf, C.B.; Chopping, M.J.; Strahler, A.H.; Wang, J.; Román, M.O.; Rocha, A.V.; Woodcock, C.E.; Shuai, Y. Evaluation of moderate resolution imaging spectroradiometer (MODIS) snow albedo product (MCD43A) over tundra. *Remote Sens. Environ.* **2012**, *117*, 264–280. [CrossRef]

14. Stroeve, J.; Box, J.E.; Wang, Z.; Schaaf, C.; Barrett, A. Re-evaluation of MODIS MCD43 greenland albedo accuracy and trends. *Remote Sens. Environ.* **2013**, *138*, 199–214. [CrossRef]

15. Román, M.O.; Gatebe, C.K.; Shuai, Y.; Wang, Z.; Gao, F.; Masek, J.G.; He, T.; Liang, S.; Schaaf, C.B. Use of in situ and airborne multiangle data to assess MODIS- and Landsat-based estimates of directional reflectance and albedo. *IEEE Trans. Geosci. Remote Sens.* **2013**, *51*, 1393–1404. [CrossRef]

16. Wang, Z.S.; Schaaf, C.B.; Strahler, A.H.; Chopping, M.J.; Roman, M.O.; Shuai, Y.M.; Woodcock, C.E.; Hollinger, D.Y.; Fitzjarrald, D.R. Evaluation of MODIS albedo product (MCD43A) over grassland, agriculture and forest surface types during dormant and snow-covered periods. *Remote Sens. Environ.* **2014**, *140*, 60–77. [CrossRef]

17. Peng, J.; Liu, Q.; Wen, J.; Liu, Q.; Tang, Y.; Wang, L.; Dou, B.; You, D.; Sun, C.; Zhao, X.; et al. Multi-scale validation strategy for satellite albedo products and its uncertainty analysis. *Sci. China Earth Sci.* **2014**, *58*, 573–588. [CrossRef]

18. Campagnolo, M.L.; Sun, Q.; Liu, Y.; Schaaf, C.; Wang, Z.; Román, M.O. Estimating the effective spatial resolution of the operational BRDF, albedo, and NADIR reflectance products from MODIS and VIIRS. *Remote Sens. Environ.* **2016**, *175*, 52–64. [CrossRef]

19. Liu, Y.; Wang, Z.; Sun, Q.; Erb, A.M.; Li, Z.; Schaaf, C.B.; Zhang, X.; Román, M.O.; Scott, R.L.; Zhang, Q. Evaluation of the VIIRS BRDF, albedo and NBAR products suite and an assessment of continuity with the long term MODIS record. *Remote Sens. Environ.* **2017**, *201*, 256–274. [CrossRef]

20. Cescatti, A.; Marcolla, B.; Vannan, S.K.S.; Pan, J.Y.; Román, M.O.; Yang, X.; Ciais, P.; Cook, R.B.; Law, B.E.; Matteucci, G. Intercomparison of MODIS albedo retrievals and in situ measurements across the global fluxnet network. *Remote Sens. Environ.* **2012**, *121*, 323–334. [CrossRef]

21. Lin, X.; Wen, J.; Tang, Y.; Ma, M.; You, D.; Dou, B.; Wu, X.; Zhu, X.; Xiao, Q.; Liu, Q. A web-based land surface remote sensing products validation system (LAPVAS): Application to albedo product. *Int. J. Digit. Earth* **2017**, *11*, 1–21. [CrossRef]

22. Mira, M.; Weiss, M.; Baret, F.; Courault, D.; Hagolle, O.; Gallego-Elvira, B.; Olioso, A. The MODIS (Collection V006) BRDF/albedo product MCD43D: Temporal course evaluated over agricultural landscape. *Remote Sens. Environ.* **2015**, *170*, 216–228. [CrossRef]

23. Wu, X.; Wen, J.; Xiao, Q.; Liu, Q.; Peng, J.; Dou, B.; Li, X.; You, D.; Tang, Y.; Liu, Q. Coarse scale in situ albedo observations over heterogeneous snow-free land surfaces and validation strategy: A case of modis albedo products preliminary validation over Northern China. *Remote Sens. Environ.* **2016**, *184*, 25–39. [CrossRef]

24. Liang, S.L.; Fang, H.L.; Chen, M.Z.; Shuey, C.J.; Walthall, C.; Daughtry, C.; Morisette, J.; Schaaf, C.; Strahler, A. Validating MODIS land surface reflectance and albedo products: Methods and preliminary results. *Remote Sens. Environ.* **2002**, *83*, 149–162. [CrossRef]

25. Disney, M.; Lewis, P.; Thackrah, G.; Quaife, T.; Barnsley, M. Comparison of MODIS broadband albedo over an agricultural site with ground measurements and values derived from earth observation data at a range of spatial scales. *Int. J. Remote Sens.* **2004**, *25*, 5297–5317. [CrossRef]

26. Wen, J.; Liu, Q.; Liu, Q.; Xiao, Q.; Li, X. Parametrized BRDF for atmospheric and topographic correction and albedo estimation in Jiangxi rugged terrain, China. *Int. J. Remote Sens.* **2009**, *30*, 2875–2896. [CrossRef]

27. Schaaf, C.; Li, X.; Strahler, A. Topographic effects on bidirectional and hemispherical reflectances calculated with a geometric-optical canopy model. *IEEE Trans. Geosci. Remote Sens.* **1994**, *32*, 1186–1193. [CrossRef]

28. Strahler, A.H.; Muller, J.; Lucht, W.; Schaaf, C.; Tsang, T.; Gao, F.; Li, X.; Lewis, P.; Barnsley, M.J. MODIS BRDF/albedo product: Algorithm theoretical basis document version 5.0. *MODIS Doc.* **1999**, *23*, 42–47.

29. Liu, J.; Schaaf, C.; Strahler, A.; Jiao, Z.; Shuai, Y.; Zhang, Q.; Roman, M.; Augustine, J.A.; Dutton, E.G. Validation of moderate resolution imaging spectroradiometer (MODIS) albedo retrieval algorithm: Dependence of albedo on solar zenith angle. *J. Geophys. Res.* **2009**, *114*. [CrossRef]

30. Cherubini, F.; Vezhapparambu, S.; Bogren, W.; Astrup, R.; Strømman, A.H. Spatial, seasonal, and topographical patterns of surface albedo in norwegian forests and cropland. *Int. J. Remote Sens.* **2017**, *38*, 4565–4586. [CrossRef]

31. Wen, J.; Zhao, X.; Liu, Q.; Tang, Y.; Dou, B. An improved land-surface albedo algorithm with DEM in rugged terrain. *IEEE Geosci. Remote Sens. Lett.* **2014**, *11*, 883–887.

32. Proy, C.; Tanre, D.; Deschamps, P.Y. Evaluation of topographic effects in remotely sensed data. *Remote Sens. Environ.* **1989**, *30*, 21–32. [CrossRef]

33. Wen, J.G.; Liu, Q.H. Scale effect and scale correction of land-surface albedo in rugged terrain. *Int. J. Remote Sens.* **2009**, *30*, 5397–5420. [CrossRef]

34. Liang, S.; Lewis, P.; Dubayah, R.; Qin, W.; Shirey, D. Topographic effects on surface bidirectional reflectance scaling. *J. Remote Sens.* **1997**, *1*, 82–93.

35. Jin, Y.; Schaaf, C.B.; Gao, F.; Li, X.; Strahler, A.H.; Lucht, W.; Liang, S. Consistency of MODIS surface bidirectional reflectance distribution function and albedo retrievals: 1. Algorithm performance. *J. Geophys. Res. Atmos.* **2003**, *108*, 347–362. [CrossRef]

36. Allen, R.G.; Trezza, R.; Tasumi, M. Analytical integrated functions for daily solar radiation on slopes. *Agric. For. Meteorol.* **2006**, *139*, 55–73. [CrossRef]

37. Dozier, J.; Frew, J. Rapid calculation of terrain parameters for radiation modeling from digital elevation data. *IEEE Trans. Geosci. Remote Sens.* **1990**, *28*, 963–969. [CrossRef]

38. Lewis, P.; Barnsley, M.J. Influence of the sky radiance distribution on various formulations of the earth surface albedo. In Proceedings of the International Symposium on Physical Measurements and Signatures in Remote Sensing, Paris, France, 17–21 January 1994.

39. MODIS User Guide V006. Available online: https://www.umb.edu/spectralmass/terra_aqua_modis/modis_brdf_albedo_product_mcd43 (accessed on 19 November 2017).

40. Qu, Y.; Liang, S.L.; Liu, Q.; He, T.; Liu, S.H.; Li, X.W. Mapping surface broadband albedo from satellite observations: A review of literatures on algorithms and products. *Remote Sens.* **2015**, *7*, 990–1020. [CrossRef]

41. Qu, Y.; Liu, Q.; Liang, S.; Wang, L. Direct-estimation algorithm for mapping daily land-surface broadband albedo from MODIS data. *IEEE Trans. Geosci. Remote Sens.* **2014**, *52*, 907–919. [CrossRef]

42. Liu, Q.; Wang, L.Z.; Qu, Y.; Liu, N.F.; Liu, S.H.; Tang, H.R.; Liang, S.L. Preliminary evaluation of the long-term glass albedo product. *Int. J. Digit. Earth* **2013**, *6*, 69–95. [CrossRef]

43. Liang, S.; Zhao, X.; Liu, S.; Yuan, W.; Cheng, X.; Xiao, Z.; Zhang, X.; Liu, Q.; Cheng, J.; Tang, H.; et al. A long-term global land surface satellite (glass) data-set for environmental studies. *Int. J. Digit. Earth* **2013**, *6*, 5–33. [CrossRef]

44. Wen, J.; Liu, Q.; Tang, Y.; Dou, B.; You, D.; Xiao, Q.; Liu, Q.; Li, X. Modeling land surface reflectance coupled BRDF for HJ-1/CCD data of rugged terrain in Heihe river basin, china. *IEEE J. Sel. Top. Appl. Earth Obs. Remote Sens.* **2015**, *8*, 1–13. [CrossRef]

45. Borel, C.C.; Gerstl, S.A.W.; Powers, B.J. The radiosity method in optical remote-sensing of structured 3-D surfaces. *Remote Sens. Environ.* **1991**, *36*, 13–44. [CrossRef]

46. Duan, S.-B.; Li, Z.-L.; Wu, H.; Tang, B.-H.; Ma, L.; Zhao, E.; Li, C. Inversion of the prosail model to estimate leaf area index of maize, potato, and sunflower fields from unmanned aerial vehicle hyperspectral data. *Int. J. Appl. Earth Obs.* **2014**, *26*, 12–20. [CrossRef]

47. Jacquemoud, S.; Baret, F. Prospect: A model of leaf optical properties spectra. *Remote Sens. Environ.* **1990**, *34*, 75–91. [CrossRef]

48. Verhoef, W. Light scattering by leaf layers with application to canopy reflectance modeling: The sail model. *Remote Sens. Environ.* **1984**, *16*, 125–141. [CrossRef]

49. Darvishzadeh, R.; Matkan, A.A.; Ahangar, A.D. Inversion of a radiative transfer model for estimation of rice canopy chlorophyll content using a lookup-table approach. *IEEE J. Sel. Top. Appl. Earth Obs. Remote Sens.* **2012**, *5*, 1222–1230. [CrossRef]

50. Li, X.; Cheng, G.; Liu, S.; Xiao, Q.; Ma, M.; Jin, R.; Che, T.; Liu, Q.; Wang, W.; Qi, Y.; et al. Heihe watershed allied telemetry experimental research (hiwater): Scientific objectives and experimental design. *Bull. Am. Meteorol. Soc.* **2013**, *94*, 1145–1160. [CrossRef]

51. Li, X.; Li, X.; Li, Z.; Ma, M.; Wang, J.; Xiao, Q.; Liu, Q.; Che, T.; Chen, E.; Yan, G. Watershed allied telemetry experimental research. *J. Geophys. Res. Atmos.* **2009**, *114*, 2191–2196. [CrossRef]

52. Ma, M.; Che, T.; Li, X.; Xiao, Q.; Zhao, K.; Xin, X. A prototype network for remote sensing validation in China. *Remote Sens.* **2015**, *7*, 5187–5202. [CrossRef]

53. Sailor, D.J.; Resh, K.; Segura, D. Field measurement of albedo for limited extent test surfaces. *Sol. Energy* **2006**, *80*, 589–599. [CrossRef]

54. Farr, T.G.; Werner, M.; Kobrick, M. *The Shuttle Radar Topography Mission: Introduction to Special Session*; EGS-AGU-EUG Joint Assembly: Nice, France, 2003; pp. 37–55.

55. Global Land Cover Facility. Available online: http://glcf.umd.edu/data/srtm/ (accessed on 19 November 2017).

56. Sun, G.; Ranson, K.J.; Kharuk, V.I.; Kovacs, K. Validation of surface height from shuttle radar topography mission using shuttle laser altimeter. *Remote Sens. Environ.* **2003**, *88*, 401–411. [CrossRef]

57. Schaaf, C.B.; Gao, F.; Strahler, A.H.; Lucht, W.; Li, X.W.; Tsang, T.; Strugnell, N.C.; Zhang, X.Y.; Jin, Y.F.; Muller, J.P.; et al. First operational BRDF, albedo NADIR reflectance products from MODIS. *Remote Sens. Environ.* **2002**, *83*, 135–148. [CrossRef]

58. Jiao, Z.; Wang, J.; Xie, L.; Zhang, H.; Yan, G.; He, L.; Li, X. Initial validation of MODIS albedo product by using field measurements and airborne multiangular remote sensing observations. *J. Remote Sens.* **2005**, *9*, 64–72.

59. Schaaf, C.B.; Wang, Z.; Strahler, A.H. Commentary on Wang and Zender—MODIS snow albedo bias at high solar zenith angles relative to theory and to in situ observations in greenland. *Remote Sens. Environ.* **2011**, *115*, 1296–1300. [CrossRef]

60. Goldberg, M.D.; Xue, H.; Bloom, H.J.; Wang, J.; Jin, H.; Bi, J. Validation of the MODIS albedo product and improving the snow albedo retrieval with additional AMSR-E data in Qinghai-Tibet plateau. *SPIE Opt. Eng. Appl.* **2010**, *7811*, 781108. [CrossRef]

61. Román, M.O.; Gatebe, C.K.; Schaaf, C.B.; Poudyal, R.; Wang, Z.; King, M.D. Variability in surface brdf at different spatial scales (30 m–500 m) over a mixed agricultural landscape as retrieved from airborne and satellite spectral measurements. *Remote Sens. Environ.* **2011**, *115*, 2184–2203. [CrossRef]

62. Dozier, J.; Outcalt, S.I. An approach toward energy balance simulation over rugged terrain. *Geogr. Anal.* **1979**, *11*, 65–85. [CrossRef]

63. Duguay, C.R.; Ledrew, E.F. Estimating surface reflectance and albedo from Landsat-5 thematic mapper over rugged terrain. *Photogramm. Eng. Remote Sens.* **1992**, *58*, 551–558.

64. Smith, J.A.; Lin, T.L.; Ranson, K.J. The lambertian assumption and Landsat data. *Photogramm. Eng. Remote Sens.* **1980**, *46*, 1183–1189.

65. Richter, R. Correction of atmospheric and topographic effects for high spatial resolution satellite imagery. *Int. J. Remote Sens.* **1997**, *18*, 1099–1111. [CrossRef]

66. Sandmeier, S.; Itten, K.I. A physically-based model to correct atmospheric and illumination effects in optical satellite data of rugged terrain. *IEEE Trans. Geosci. Remote Sens.* **1997**, *35*, 708–717. [CrossRef]

67. Kimes, D.S.; Kirchner, J.A. Modeling the effects of various radiant transfers in mountainous terrain on sensor response. *IEEE Trans. Geosci. Remote Sens.* **1981**, *2*, 100–108. [CrossRef]

68. Hansen, L.B.; Kamstrup, N.; Hansen, B.U. Estimation of net short-wave radiation by the use of remote sensing and a digital elevation model—A case study of a high arctic mountainous area. *Int. J. Remote Sens.* **2002**, *23*, 4699–4718. [CrossRef]

remote sensing

MDPI

Article

Simulation and Analysis of the Topographic Effects on Snow-Free Albedo over Rugged Terrain

Dalei Hao [1,2], Jianguang Wen [1,2,3,*], Qing Xiao [1,2], Shengbiao Wu [1,2], Xingwen Lin [1,2], Baocheng Dou [1], Dongqin You [1] and Yong Tang [1]

[1] State Key Laboratory of Remote Sensing Science, Institute of Remote Sensing and Digital Earth, Chinese Academy of Sciences, Beijing 100101, China; haodl@radi.ac.cn (D.H.); xiaoqing@radi.ac.cn (Q.X.); wushengbiao90@163.com (S.W.); linxw@radi.ac.cn (X.L.); dou3516@163.com (B.D.); youdq@radi.ac.cn (D.Y.); tangyong@radi.ac.cn (Y.T.)
[2] University of Chinese Academy of Sciences, Beijing 100049, China
[3] Joint Center for Global Change Studies (JCGCS), Beijing 100875, China
* Correspondence: wenjg@radi.ac.cn; Tel.: +86-10-6484-2510

Received: 30 November 2017; Accepted: 8 February 2018; Published: 11 February 2018

Abstract: Topography complicates the modeling and retrieval of land surface albedo due to shadow effects and the redistribution of incident radiation. Neglecting topographic effects may lead to a significant bias when estimating land surface albedo over a single slope. However, for rugged terrain, a comprehensive and systematic investigation of topographic effects on land surface albedo is currently ongoing. Accurately estimating topographic effects on land surface albedo over a rugged terrain presents a challenge in remote sensing modeling and applications. In this paper, we focused on the development of a simplified estimation method for snow-free albedo over a rugged terrain at a 1-km scale based on a 30-m fine-scale digital elevation model (DEM). The proposed method was compared with the radiosity approach based on simulated and real DEMs. The results of the comparison showed that the proposed method provided adequate computational efficiency and satisfactory accuracy simultaneously. Then, the topographic effects on snow-free albedo were quantitatively investigated and interpreted by considering the mean slope, subpixel aspect distribution, solar zenith angle, and solar azimuth angle. The results showed that the more rugged the terrain and the larger the solar illumination angle, the more intense the topographic effects were on black-sky albedo (BSA). The maximum absolute deviation (MAD) and the maximum relative deviation (MRD) of the BSA over a rugged terrain reached 0.28 and 85%, respectively, when the SZA was 60° for different terrains. Topographic effects varied with the mean slope, subpixel aspect distribution, SZA and SAA, which should not be neglected when modeling albedo.

Keywords: land surface albedo; snow-free albedo; rugged terrain; topographic effects; black-sky albedo (BSA)

1. Introduction

Land surface albedo, defined as the fraction of incident solar radiation (0.3–3 μm) reflected by land surfaces [1,2], is one of the most significant geophysical variables affecting the Earth's climate and controlling the surface radiation budget. It plays a crucial role in a variety of models, including general circulation models, land surface climate models, energy balance models, hydrology models, and biosphere models. Land surface albedo under ambient light conditions, also known as blue-sky albedo, is a combination of directional hemispheric reflectance, known as black-sky albedo (BSA), and bihemispherical reflectance, known as white-sky albedo (WSA); blue sky albedo takes into account the proportion of diffused skylight illumination and the solar zenith angle [3–6]. Over recent decades, land surface albedo remote sensing estimation algorithms have been developed, demonstrating that a bihemispherical integration method using the bidirectional reflectance distribution function (BRDF) [7–9] has a robust performance and is widely used in albedo estimation. This method for estimating albedo generally assumes that the land surface terrain is flat and homogeneous [10,11], and albedo products have been created with this method using different satellite datasets, such as the Moderate Resolution Imaging Spectroradiometer (MODIS) [12], the polarization and directionality of Earth reflectances (POLDER) [13], the multi-angle imaging spectroradiometer (MISR) [14], and the Clouds and the Earth's Radiant Energy System (CERES) [15].

Spatial heterogeneities, which are comprised of complex topography and heterogeneous land cover, complicate the estimation of land surface albedo [16]. Directly applying albedo estimation methods that are suitable for flat terrain to a rugged terrain leads to large errors [17,18]. Actually, topography plays different roles in albedo estimation at different spatial scales [19]. For a single slope, physical-based BRDF models, such as the improved four-scale geometric-optical model for sloping terrain (GOST) [20], the improved soil-leaf-canopy radiative transfer model for sloping terrain (SLCT) [21], the improved Li-Strahler geometric-optical canopy model for sloping terrain (GOMST) [8], the vertical vegetation model (VVM) [7], the path length correction (PLC) model [22], have been developed based on the radiative transfer principle and the geometric optical theory, which depend on the orientation of the plant stand and the particular configuration of the sun direction and the terrain slope [7,8]. By integrating the BRDF over the exitance hemisphere for a single irradiance direction, the BSA for a single slope can be obtained; by integrating the BRDF over all viewing and irradiance directions, the WSA for a single slope can be calculated. Investigations have shown that the albedo for a single slope is related to the slope and aspect of the single slope [7,8,23,24]. With an increase in slope, the albedo becomes sensitive to the aspect of the slope, and the slopes facing away from the sun may display larger albedos than those of the sunward facing slopes due to increased mutual shadowing [8]. In addition, the terrain shadowing and diffused radiation from the adjacent slopes significantly influences the single slope albedo [8]. For a rugged terrain, the topographic effects on albedo generally focus on the integrated effects caused by subpixel slopes within one remote sensing pixel [18,19]. Neglecting the subpixel topography variability in albedo estimation over a rugged terrain leads to significant deviations [17,19,25], which can reach a relative error of 33% for a mean slope of 40° [19]. It has been shown that MODIS albedo retrievals are also highly sensitive to subpixel topography, and the MODIS albedo over a rugged terrain can change up to 100% for spruce vegetation in winter [26].

Albedo depends on both land surface characteristics and the atmosphere. Topography alters land surface characteristics and solar illumination geometry. Thus, the topographic effects on albedo over a rugged terrain are related to the spatial distribution characteristics of the subpixel topography, the solar zenith angle (SZA), and the solar azimuth angle (SAA) [17,19,25]. Compared to a single slope, it is difficult to investigate the variation in albedo over a rugged terrain by integrating the effects of subpixel slopes. The lack of a rigorous and effective physical BRDF/albedo model for rugged terrains has contributed to this challenge. Wen et al. [19] developed a land surface albedo estimation and scale correction method over a rugged terrain by the hemispheric integration of surface reflectance over a rugged terrain, where the rugged terrain reflectance was estimated based on the subpixel reflectance. Considering that the analysis of topographic effects on land surface albedo required huge amounts of typical albedo data over the rugged terrain to ensure the reliability of the analysis results, this method was inconvenient for the large-scale simulation of surface albedo because the rugged terrain reflectances under the entire hemispheric view space were required during each albedo calculation under different SZAs and SAAs. Therefore, we dedicated ourselves to developing a simplified method to estimate the albedo over a rugged terrain directly by the subpixel albedo in this paper based on the same idea in Wen et al. [19].

In this paper, we focused on quantitatively investigating topographic effects on snow-free albedo over a rugged terrain based on the developed method. The BSA was related to both the topography and the solar illumination geometry. The shadows induced by adjacent topographies immensely affected the BSA. However, the WSA was independent of solar illumination geometry and shadow effects. Therefore, to emphasize the topographic effects on albedo, BSA was selected instead of WSA in this paper. This paper was organized as follows. First, a BSA estimation method for rugged terrain was proposed in Section 2. Section 3 described the DEM datasets and the generation method of the reference BSA dataset. The proposed BSA estimation method was validated in Section 4.1; Section 4.2 quantitatively investigated topographic effects on snow-free BSA given variations in mean slope, subpixel aspect distribution, SZA, and SAA. Finally, a conclusion was provided in Section 5.

2. Methods

2.1. BSA over a Rugged Terrain

The rugged terrain, which is comprised of a number of subpixel slopes with different slopes and aspects, is shown in Figure 1a, where it is assumed that the entire terrain is horizontal. The SZA and SAA are denoted by θ_s and ϕ_s, respectively, with respect to the horizontal plane.

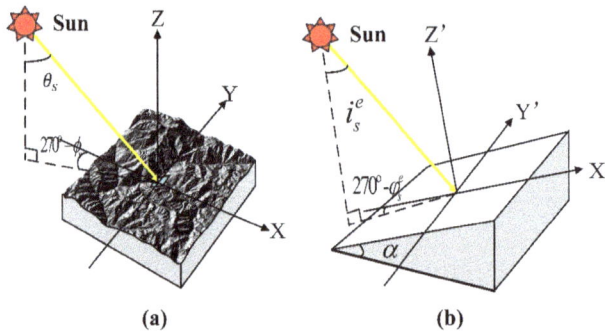

Figure 1. (a) Rugged terrain with a large number of subpixel slopes; (b) Virtually-smoothed single slope.

According to the definition of albedo [1], the BSA $\rho_{BSA\_low}(\theta_s, \phi_s)$ over a rugged terrain under a clear sky is calculated by:

$$\rho_{BSA\_low}(\theta_s, \phi_s) = \frac{\Phi_e^b}{\Phi_i^b} \tag{1}$$

where Φ_e^b and Φ_i^b represent the reflected and incident radiation flux over the whole pixel under a clear sky, respectively. The incident radiation flux over a rugged terrain is equal to the incident radiation flux received by the corresponding projected horizontal pixel. Therefore, Φ_i^b is equal to

$$\Phi_i^b = E_s \cos\theta_s \int \cos a_j dA_{tj} \tag{2}$$

where E_s represents the direct solar irradiance independent of topography, $\int \cos a_j dA_{tj}$ represents the area of the projected horizontal pixel, a_j denotes the slope of the subpixel slope, dA_{tj} denotes the incremental surface area of the subpixel slope and the subscript tj is the jth subpixel slope.

2.2. BSA Estimation Method Derivation

The subpixel slope BSA upscaling approach, based on the radiosity theory [27], is a feasible scheme to estimate the BSA over a rugged terrain. However, with the upscaling approach, many parameters are required, including reflectance characteristics, vegetation structural parameters, illumination conditions, and the radiation of each subpixel slope, which results in more uncertainties and computational complexities. The parameterized idea adopted by Wen [19] is an alternative method to estimate BSA due to its simplicity and efficiency.

This idea assumes that a virtual slope with slope α and aspect β exists, where incoming and outgoing radiation are the same as the sum of those over a rugged terrain, as shown in Figure 1. Thus, Φ_e^b can be expressed as:

$$\Phi_e^b = E_s \cos i_s^e \rho_{BSA\_eq}(i_s^e, \varphi_s^e) A_e(\theta_s, \phi_s) \tag{3}$$

where $\rho_{BSA\_eq}(i_s^e, \varphi_s^e)$ represents the BSA of the virtual slope; i_s^e and φ_s^e represent the relative SZA and SAA corresponding to the virtual slope, respectively; and A_e denotes the area of the virtual slope surface, which depends on both the subpixel topography distribution and the solar illumination geometry.

According to the principle from the mountain radiative transfer theory [28], under a specific solar illumination geometry and by neglecting the multi-scattering effects from an adjacent terrain, the incremental reflected radiation flux of the jth subpixel slope, $d\Phi_{ej}^b$, is given by

$$d\Phi_{ej}^b = \Theta_{sj} E_s \cos i_{sj} \rho_{BSA\_high}(i_{sj}, \varphi_{sj}) dA_{tj} \tag{4}$$

where Θ represents the shadow factor, which is set to 1 for an illuminated slope and 0 otherwise [29–31] and can be calculated using the ray-tracing method. Thus, Θ_{sj} indicates whether or not the subpixel slope is sunlit. Variables i_{sj} and φ_{sj} represent the relative SZA and SAA, respectively, corresponding to the subpixel slope, and $\rho_{BSA\_high}(i_{sj}, \varphi_{sj})$ represents the BSA of the subpixel slope.

$\rho_{BSA\_high}(i_{sj}, \varphi_{sj})$ is affected by topographic obstructions because several parts of the reflected radiation from the outgoing hemisphere are obstructed. Variable $d\Phi_{ej}^b$ can be approximately calculated as follows:

$$d\Phi_{ej}^b = d\tilde{\Phi}_{ej}^b V_j \tag{5}$$

where $d\tilde{\Phi}_{ej}^b$ denotes the reflected radiation flux neglecting obstructions from the adjacent terrain, and V_j represents the sky view factor, which can be calculated using the DEM [32]. Sky view factor V_j represents the unobstructed portion of the sky at a given location and ranges between 0 and 1. A value of V close to 1 indicates that almost the entire hemisphere is unobstructed and visible, which is the case for exposed features, such as planes and peaks; values close to 0 are present in deep sinks and lower regions of deep valleys, where almost no sky is visible [33].

Therefore, $\rho_{BSA\_high}(i_{sj}, \varphi_{sj})$ is equal to:

$$\rho_{BSA\_high}(i_{sj}, \varphi_{sj}) = \frac{d\tilde{\Phi}_{ej}^b V_j}{d\Phi_{ij}^b} = \tilde{\rho}_{BSA\_high}(i_{sj}, \varphi_{sj}) V_j \tag{6}$$

where $\tilde{\rho}_{BSA\_high}(i_{sj}, \varphi_{sj})$ denotes the BSA neglecting obstructions from an adjacent terrain. Thus, the sum of the subpixel slope radiation fluxes over a rugged terrain, Φ_e^b is as follows:

$$\Phi_e^b = E_s \int_{A(s)} \cos i_{sj} \tilde{\rho}_{BSA\_high}(i_{sj}, \varphi_{sj}) V_j dA_{tj} \tag{7}$$

where $A_{(s)}$ denotes the subpixel slopes that are illuminated by the sun. Combining Equations (3) and (7), we obtain

$$\int_{A(s)} \cos i_{sj} \tilde{\rho}_{BSA\_high}(i_{sj}, \varphi_{sj}) V_j dA_{tj} = \cos i_s^e \rho_{BSA\_eq}(i_s^e, \varphi_s^e) A_e(\theta_s, \phi_s) \tag{8}$$

To focus on the topographic effects on BSA, land cover type within the rugged terrain is assumed to be homogeneous, and the differences among $\tilde{\rho}_{BSA\_high}(i_{sj}, \varphi_{sj})$ for different subpixel slopes are only caused by different solar illumination geometries and DEM characteristics. $\tilde{\rho}_{BSA\_high}(i_{sj}, \varphi_{sj})$ has an identical function form with that of $\rho_{BSA\_eq}(i_s^e, \varphi_s^e)$, except that the latter input angle parameters are different.

To unify the symbols and simplify the deduction, Equation (8) is substituted with

$$\int_{A(s)} Y(u_j, v_j) V_j dA_{tj} = Y(u_e, v_e) A_e(u, v) \tag{9}$$

where $Y(u_j, v_j) = \cos i_{sj} \tilde{\rho}_{BSA\_high}(i_{sj}, \varphi_{sj})$, $Y(u_e, v_e) = \cos i_s^e \rho_{BSA\_eq}(i_s^e, \varphi_s^e)$, $u_j = \cos i_{sj}$, $v_j = \cos \varphi_{sj}$, $u_e = \cos i_s^e$, $v_e = \cos \varphi_s^e$, $u = \cos \theta_s$, and $v = \cos \phi_s$. To construct the virtual slope, we obtain the specific formulas for u_e, v_e and A_e, which determine if the virtual slope exists. An alternative method based on the Taylor expansion, which is similar to the derivation strategy of the Hapke shadow function [34,35], was used to solve Equation (9). Specifically, Y is assumed to be mathematically well behaved, and Y is expanded on both sides of Equation (9) in a Taylor series about u and v:

$$A_e(u,v)[Y(u,v) + \tfrac{\partial Y}{\partial u}(u,v)(u_e - u) + \tfrac{\partial Y}{\partial v}(u,v)(v_e - v) + \ldots]$$
$$= Y(u,v) \int_{A(s)} V_j dA_{tj} + \tfrac{\partial Y}{\partial u}(u,v) \int_{A(s)} (u_j - u) V_j dA_{tj} + \tfrac{\partial Y}{\partial v}(u,v) \int_{A(s)} (v_j - v) V_j dA_{tj} + \ldots \tag{10}$$

Since u and v are independent variables, and Y is an arbitrary function of u and v, Equation (10) is satisfied only if the coefficients of Y and its partial derivatives are separately equal on both sides of the equation. Neglecting the higher order terms of the Taylor expansion for Y, we obtain Equation (11):

$$A_e(u,v) = \int_{A(s)} V_j dA_{tj}$$
$$A_e(u,v) = \int_{A(s)} (u_j - u) V_j dA_{tj}/(u_e - u) \tag{11}$$
$$A_e(u,v) = \int_{A(s)} (v_j - v) V_j dA_{tj}/(v_e - v)$$

Solving Equation (11), u_e, v_e, and A_e can be specifically and respectively be formulated as:

$$A_e(u,v) = \int_{A(s)} V_j dA_{tj}$$
$$u_e = \frac{\int_{A(s)} u_j V_j dA_{tj}}{\int_{A(s)} V_j dA_{tj}} \tag{12}$$
$$v_e = \frac{\int_{A(s)} v_j V_j dA_{tj}}{\int_{A(s)} V_j dA_{tj}}$$

By combining Equations (1), (2), and (12), we obtain:

$$\rho_{BSA\_low}(\theta_s, \phi_s) = \frac{\cos i_s^e \rho_{BSA\_eq}(i_s^e, \varphi_s^e) A_e(\theta_s, \phi_s)}{\cos \theta_s \int \cos a_j dA_{tj}} \tag{13}$$

By introducing Equation (12) into Equation (13), we obtain

$$\rho_{BSA\_low}(\theta_s, \phi_s) = \rho_{BSA\_eq}(i_s^e, \varphi_s^e) \frac{\int_{A(s)} \cos i_{sj} V_j dA_{tj}}{\cos \theta_s \int \cos a_j dA_{tj}} \tag{14}$$

Equation (14) shows that the BSA over a rugged terrain can be obtained by the product between the BSA of the virtual slope and a specified factor. Its discrete formula is written as

$$\rho_{BSA\_low}(\theta_s, \phi_s) = \rho_{BSA\_eq}(i_s^e, \phi_s^e) \frac{\sum\limits_{k=1}^{N} \Theta_{sk} \cos i_{sk} V_k / \cos a_k}{N \cos \theta_s} \tag{15}$$

where N represents the number of the subpixel slopes within the pixel, and k denotes the kth subpixel slope.

Equations (13) and (14) indicate that this method combines well with any single slope BSA estimation method to estimate the BSA over a rugged terrain. In this paper, the combined PROSPECT leaf optical property model and the SAIL canopy bidirectional reflectance model, also referred to as PROSAIL, is selected as a the single slope surface reflectance and the BSA estimation method. The PROSAIL model has been widely used to study plant canopy spectral reflectance [36] and is relatively mature and efficient. For an infinitely inclined homogeneous vegetation canopy, the topographic influences on the anisotropic reflectance simulated by the PROSAIL model can be categorized into two aspects: photon path length alteration inside the vegetation layer and the adjustment of the extinction coefficient. The first effect is handled by a simple geometric correction of the solar-terrain sensor and the consideration of vertical tree growth [18,24]. Second, a spherical leaf inclination distribution function (LIDF) is assumed, which allows the effective extinction coefficient for a unit path length to be fixed in all directions, where the topographic effects on the extinction coefficient can be neglected. Given the specific vegetation parameter and the terrain configuration, we can use Equations (13) and (14) and the PROSOIL model under topographic considerations to estimate the BSA over a rugged terrain.

2.3. Topographic Effect Analysis Methods

The topographic effects on BSA are influenced by the spatial distributions of the subpixel slopes, SZA and SAA. A local sensitivity analysis method is used in this paper to analyze the effects of these different factors. This method estimates the effect of a single factor on the outputs while maintaining the other factors at their nominal values [37]. The maximum absolute deviation (MAD) and the maximum relative deviation (MRD) are used to quantitatively analyze these sensitivities:

$$MAD = \underset{i=1}{\overset{N}{Max}}\{a_{model}^i\} - \underset{i=1}{\overset{N}{Min}}\{a_{model}^i\} \tag{16}$$

$$MRD = \left[\underset{i=1}{\overset{N}{Max}}\{a_{model}^i\} - \underset{i=1}{\overset{N}{Min}}\{a_{model}^i\}\right] / \underset{i=1}{\overset{N}{Max}}\{a_{model}^i\} \tag{17}$$

where N represents the amount of estimated BSAs, and i represents the BSA index. Variable a_{model}^i denotes the ith estimated BSA calculated by the proposed method in this paper.

3. Datasets

3.1. Simulated DEM Dataset

To simulate and evaluate the BSA over a rugged terrain, nine simulated DEMs with 30 m spatial resolution were generated to provide various magnitudes of roughness [19]. The statistical mean values of the elevation, slope and sky view factor are listed in Table 1. For this research, only the central 1×1 km area (i.e., 33×33 grid cells) of the aforementioned DEMs was used to avoid calculation errors at the edge of the DEM.

Table 1. Basic parameters for nine simulated DEMs.

Filter	Exaggeration	Mean Elevation (m)	Mean Slope (°)	Mean Sky View Factor
5 × 1	1	29.49	1.33	1.00
3 × 1	1	29.49	1.88	1.00
1 × 1	1	29.49	2.51	1.00
5 × 1	10	294.89	12.84	0.95
3 × 1	10	294.89	17.59	0.90
1 × 1	10	294.89	22.62	0.85
5 × 1	20	589.78	23.70	0.83
3 × 1	20	589.78	30.90	0.74
1 × 1	20	589.78	37.72	0.64

3.2. Global Digital Elevation Model (GDEM)

With an average elevation exceeding 4500 m, the Tibetan Plateau (Figure 2) is the largest and highest plateau in the world. This plateau is characterized by high spatial heterogeneity due to the presence of mountainous areas [38]. Therefore, it is an ideal region to compare and validate the developed method in this paper. A 100 km × 100 km test area, with elevations ranging from 2492 m to approximately 6769 m, is identified from the plateau. The DEM shown in the upper-right part of Figure 2, with a 30-m spatial resolution of the study area, was collected by the Advanced Spaceborne Thermal Emission and Reflection Radiometer (ASTER) of the global digital elevation model, version 2 (ASTER GDEM2). The terrain over the southern parts of the study area was relatively complex and rugged, whereas the northern terrain was relatively gentle. This study area, with a high spatial heterogeneity, was suitable for the study of topographic effects on BSA.

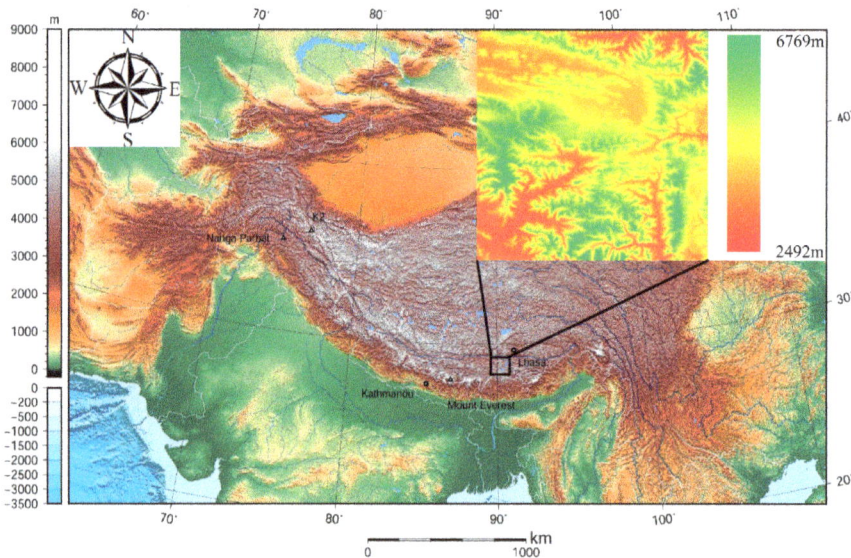

Figure 2. The Tibetan Plateau and the study area.

To effectively investigate the topographic effects on BSA, three typical categories for 1-km real DEMs, with typical surface fluctuations, were selected from the study area: gentle slope terrain (10°), moderate slope terrain (20°), and steep slope terrain (30°). Each DEM category was comprised of six representative DEMs with different subpixel aspect distributions. Their slopes and aspect distributions are shown in Figure 3, where the code names of the DEMs (e.g., dem-10-1) are marked in the legend.

The subpixel aspect distributions of the six DEMs from 1-km DEM category with a mean slope of 10° had the following characteristics, subsequently: dominant southwest-oriented distribution, relatively uniform distribution, dominant southwest-oriented distribution, relatively uniform distribution, dominant eastward-oriented distribution, and dominant northward-oriented distribution. The DEMs with a mean slope of 20° had the following characteristics, subsequently: dominant northward-oriented distribution, dominant eastward-oriented distribution, relatively uniform distribution, dominant southwest-oriented distribution, dominant northward-oriented distribution and dominant southward-oriented distribution. The DEMs with a mean slope of 30° had the following characteristics, subsequently: dominant northwest-oriented distribution, dominant southwest-oriented distribution, dominant southwest-oriented distribution, dominant southeast-oriented distribution, relatively uniform distribution and dominant eastward-oriented distribution.

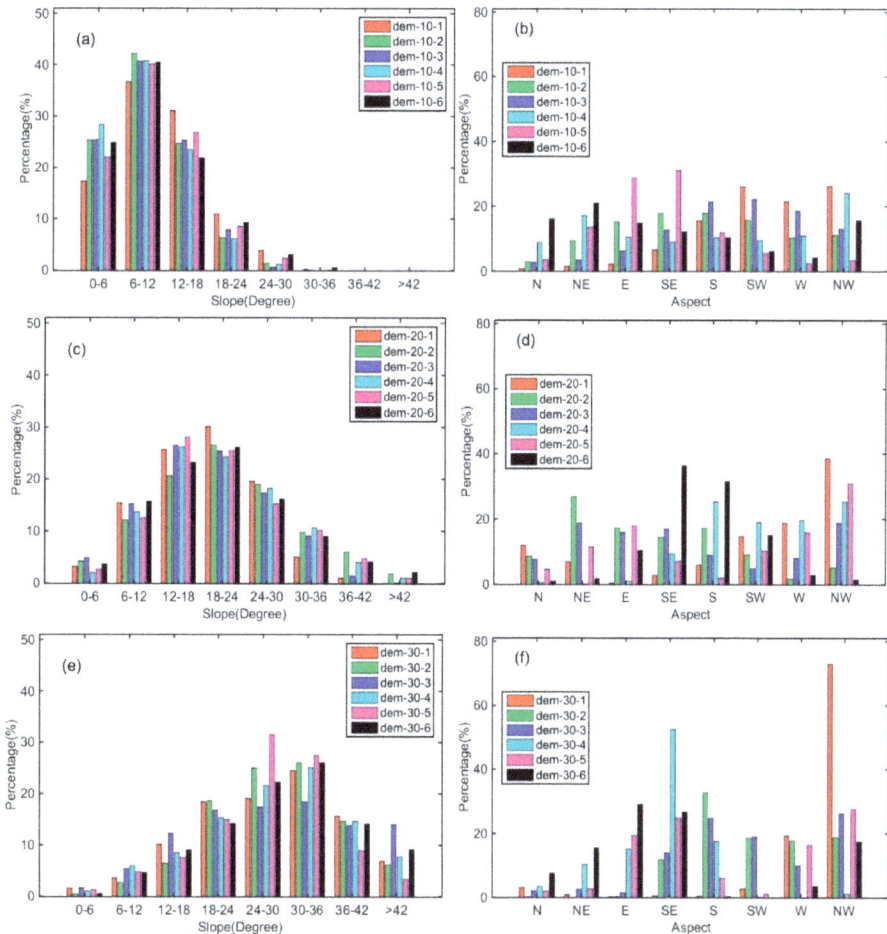

Figure 3. Distributions of slope (**a,c,e**) and aspect (**b,d,f**) within a 1-km pixel under real DEMs with different mean slopes: (**a,b**) 10°; (**c,d**) 20°; and (**e,f**) 30°. In the legends of (**b,d,f**), N, NE, E, SE, S, SW, W and NW stand for north, northwest, east, southeast, south, southwest, west and northwest, respectively. In the north, the SAA is 0°. The SAA gradually increases in a clockwise rotation of the determined direction, until the SAA is 360° (i.e., when the SAA rotated back to its original north position).

3.3. Reference BSA Dataset Simulation Based on the Radiative Approach

Considering that it is difficult to obtain the albedo reference over a rugged terrain, and current land albedo products have poor accuracies over a rugged terrain, the radiosity approach [39–41], which is a widely used computer simulation model, was used to generate the reference BSA to evaluate the performance of the proposed method. In this paper, the terrain was described by the DEM, and the land cover was assumed to be homogeneous in the simulation scenario.

Specifically, the procedure included three steps: the anisotropic reference reflectance simulation based on the radiosity approach, the spectral BSA calculation by integrating the reflectance over the hemispheric exitance given an illumination direction, and the broad-band shortwave BSA calculation by integrating the spectral BSA weighted by the incident radiation. In the first step, the reflectance characteristics of each subpixel slope were acquired based on the PROSAIL model under topographic consideration and upscaled to reflectance values over a rugged terrain based on the radiosity approach. In this paper, considering that the BSA of soil has a similar variation with topography as that of vegetation, only the vegetation BSA variation was analyzed; the input parameter specifications in the PROSAIL model are shown in Table 2 and Figure 4.

Table 2. Specification of input parameters in the PROSAIL model.

Model Parameters		Unit	Range
Leaf parameters	Leaf structure index	unitless	1.5
	Leaf chlorophyll content	$[\mu g/cm^2]$	40
	Leaf dry matter content	$[g/cm^2]$	0.009
	Leaf water content	$[cm]$	0.01
	Leaf brown pigment	$[g/cm^2]$	0.0
Soil parameters	Reflectance	——	Shown in Figure 4a
Canopy structure parameters	LAI	$[m^2/m^2]$	3
	Leaf inclination distribution function	——	Spherical
	Hot spot size parameter	$[m/m]$	0.01
Atmospheric condition	Incoming radiation	——	Shown in Figure 4b
Illumination view geometry	Solar zenith angle	$[°]$	0–90 at a 5° interval
	Solar azimuth angle	$[°]$	0–360 at a 5° interval
	View zenith angle	$[°]$	0–90 at a 5° interval
	View azimuth angle	$[°]$	0–360 at a 5° interval
Terrain	DEM	——	Nine simulated DEMs and real DEMs

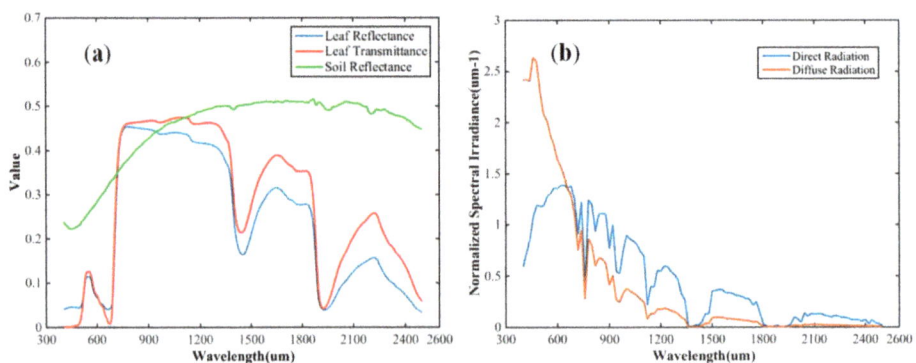

Figure 4. (**a**) Leaf reflectance, leaf transmittance and soil reflectance; (**b**) Normalized spectral irradiance curve.

4. Results and Discussion

4.1. Modeled BSA Accuracy Assessment

To analyze the accuracy of the modeled BSA with the proposed method, a simulation scenario was constructed based on both the nine simulated DEMs and three real DEM categories. For the simulation scenarios, the realistic tree shape parameters, component signatures, and other parameters are listed in Table 2. Then, the modeled BSAs are estimated by the proposed method in Section 2.2, and the reference BSAs are generated by the method in Section 3.3. The determination coefficient (R^2), root mean square errors (RMSE), mean absolute percentage error (MAPE) and mean bias (Bias) are adopted to evaluate the accuracy of the proposed method.

Figure 5a shows the comparison results of the modeled and reference BSAs over nine simulated DEMs. It is found that these two types of BSAs are close in magnitude, indicating that a majority of data points are distributed around the 1:1 line with a small RMSE (0.0060), MAPE (0.0038) and Bias (0.0038). This demonstrates that the modeled BSA is consistent with the reference BSA. Table 3 presents the error statistics of the BSA for different simulated terrains. The RMSE varies from 0.0002 to 0.0074, and the MAPE increases from 0.0001 to 0.0066 as the mean slope increases from 1.33° to 37.72°, which indicates that with an increase in mean slope, the accuracy of the proposed method gradually decreases but is still acceptable. This can be explained by the increase in mean slope, which causes the multi-scattering effects and the terrain obstruction effects for the reflected radiation to become increasingly obvious; however, the proposed method neglects multi-scattering effects and approximately considers the terrain obstruction effects. These results confirm the capability of the proposed method for estimating BSA with simulated DEMs.

The real DEMs provide a great number of terrains with different subpixel slope distributions, which are closer to the natural characteristics of the terrain than those of the simulated DEMs with normal distributions. Figure 5b indicates that the modeled BSA shows an adequate performance, with a high R^2 close to 1, a low RMSE of 0.0038, a MAPE of 0.0023 and a Bias of 0.0023. Table 4 presents the error statistics under different real terrain conditions. Overall, the results obtained from the comparison with the real DEMs are similar to those of the simulated DEMs. Therefore, the proposed method provides a high-quality BSA dataset and can be applied to investigate the topographic effects on BSA.

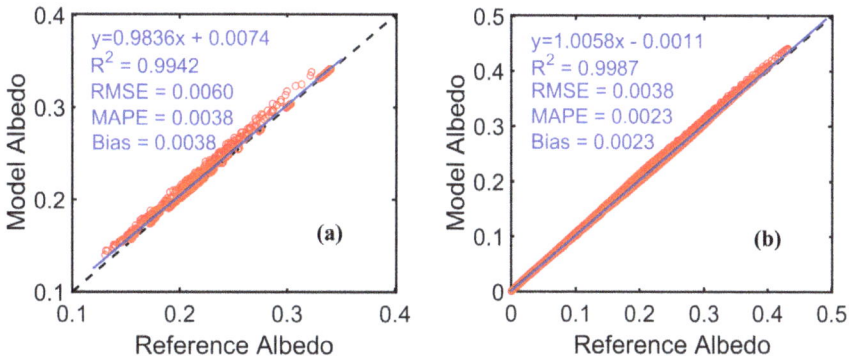

Figure 5. Scatterplots between the reference and the modeled BSAs over (**a**) simulated DEMs and (**b**) real DEMs.

Table 3. Accuracy statistics of the modeled BSA over different simulated DEMs.

Mean Slope (°)	R^2	RMSE	MAPE	Bias
1.33	1.0000	0.0002	0.0001	0.0001
1.88	1.0000	0.0003	0.0002	0.0002
2.51	1.0000	0.0006	0.0003	0.0003
12.84	0.9985	0.0059	0.0037	0.0037
17.59	0.9961	0.0071	0.0050	0.0050
22.62	0.9873	0.0078	0.0059	0.0059
23.70	0.9893	0.0078	0.0059	0.0059
30.90	0.9782	0.0078	0.0065	0.0065
37.72	0.9734	0.0074	0.0066	0.0066

Table 4. Accuracy statistics of the modeled BSA over different real DEMs.

Mean Slope (°)	R^2	RMSE	MAPE	Bias
10	0.9998	0.0020	0.0011	0.0011
20	0.9994	0.0030	0.0023	0.0023
30	0.9963	0.0056	0.0036	0.0036

4.2. Topographic Effects on BSA

4.2.1. Factors Influencing the BSA over a Rugged Terrain

Real DEMs were used to investigate the topographic effects over a rugged terrain because they provide a sufficient range of topographies that cover a wide variety of natural terrain characteristics. Figure 6 shows the hemispheric BSAs modeled with different solar illuminations over a flat terrain and the three typical real 1-km DEMs (i.e., dem-10-1, dem-20-1, and dem-30-1), which indicate that BSA distributions with solar illumination geometries are intensively affected by topography. Figure 6a shows that the BSA over a flat terrain monotonously increases with SZA regardless of the SAA value. When the terrain is relatively flat in dem-10-1, several minor changes occur in the shape of the BSA distribution, as shown in Figure 6b. The BSA increases with SZA when the SZA is substantially less than 90° and has a weak relationship with the SAA, but the BSA decreases when the SZA is close to 90° due to the terrain block of incident radiation. With an increase in mean slope for dem-20-1, the shape of the BSA distribution evidently changes and becomes asymmetric, as shown in Figure 6c. The north-facing slope terrains are dominant in dem-20-1, which indicates that the BSA is dependent on the SAA. The BSA increases monotonously with the SZA and is relatively large when the SAA is oriented north. In comparison, the BSA first increases then decreases with the SZA and becomes small when the SAA is oriented south due to the existence of shadows. As the terrain becomes more rugged in dem-30-1, the BSA generally decreases, and the shape of the BSA distribution clearly changes. This phenomenon can be explained by the fact that shadow effects are more obvious in dem-30-1 and substantially influence the BSA. Thus, we conclude that the subpixel slope distribution and the solar illumination geometry are two important factors influencing BSA, and BSA variations with varying illumination angles present different trends under different terrain conditions. Specifically, the mean slope, subpixel aspect distribution, SZA, and SAA are four main controlling factors for quantitatively analyzing the topographic effects on BSA.

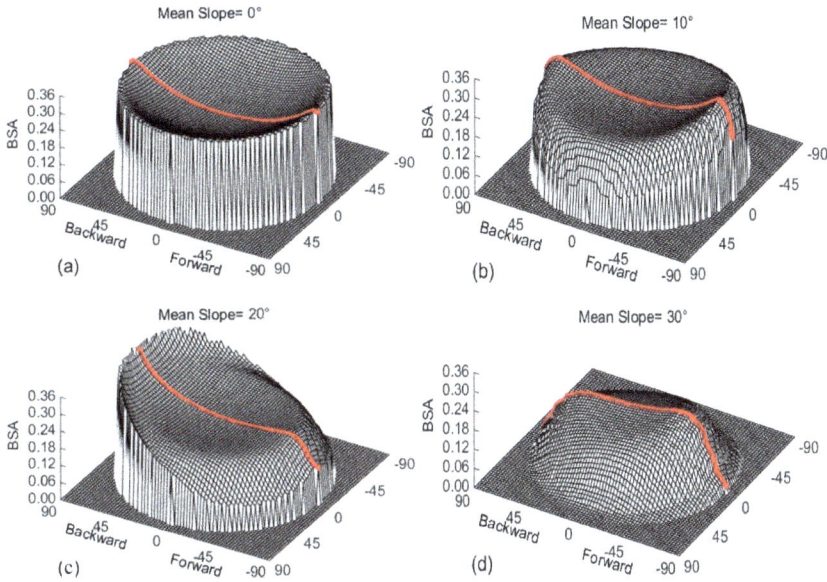

Figure 6. Hemispheric distribution of BSA under different SZAs and real 1-km DEMs: (**a**) flat terrain; (**b**) dem-10-1; (**c**) dem-20-1; and (**d**) dem-30-1. The radial coordinate is the SZA, and the angular coordinate is the SAA. The red line represents the north-south line; the backward side represents the northern aspect (i.e., where SAA is equal to 0°), and the forward side represents the southern aspect (i.e., where SAA is equal to 180°).

4.2.2. BSA Variation with Mean Slope

The 1-km DEMs over the study area were used to investigate the subpixel effects of the mean slope on the BSA (shown in Figure 7). When the SZA is 0°, a shadow does not exist, but the spatial variations in the subpixel slope and aspect within the rugged terrain affect the BSA. The MAD and MRD of the BSA reach 0.08 and 36%, respectively. However, when the SZA is 30°, shadows occur, and the topographic effects on the BSA are further enhanced. The MAD and MRD of the BSA increase to 0.12 and 50%, respectively. When the SZA is 60°, the MAD and MRD of the BSA reach 0.28 and approximately 85%, respectively, due to more obvious shadow effects and terrain obstruction effects.

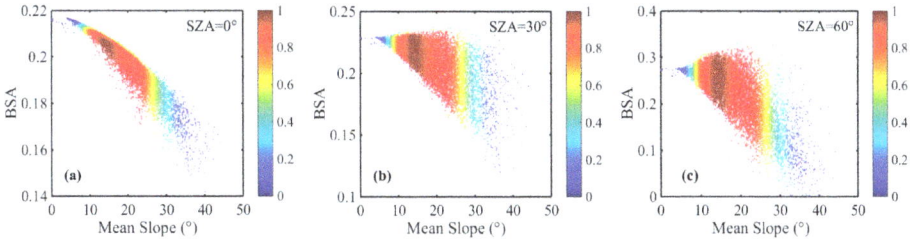

Figure 7. BSA variation with a mean slope under different SZAs: (**a**) 0°; (**b**) 30°; and (**c**) 60°. The colors refer to the density of points (from highest (red) to lowest (blue)).

Generally, BSA presents a decreasing trend with an increase in mean slope. When the SZA is 0°, the BSA clearly decreases with an increase in mean slope because several parts of the outgoing reflected radiation in the hemisphere are obstructed even though there is not a shadow present. However, when the SZA is 30°, and the SZA is 60°, the BSA variation with an increase in mean slope is maintained, but it is not as obvious. This is because BSA is affected by various factors, such as SZA, SAA, shadows, terrain obstruction, and the slope and aspect distribution of the subpixel slopes. When the terrain is gently rugged (20°), the shadow and occlusion effects of the terrain may be weak, but the alteration in regional illumination angle caused by the topography may result in an increase in BSA, as shown in Figure 7b,c. The regional SZA of the subpixel slope facing the sun is small, whereas that of the subpixel slope facing away from the sun can be relatively large, which indicates that the BSA is low when sunward subpixel slopes account for the majority of the data. When the terrain is steep (30°), the effects of shadows and terrain obstruction play a dominant role, which means that many subpixel slopes cannot be illuminated by the sun, and many parts of the outgoing reflected radiation in the hemisphere are obstructed. Therefore, the BSA decreases significantly regardless of the SZA, as shown in Figure 7.

4.2.3. BSA Variations with Sub-Pixel Aspect Distributions

Three typical 1-km DEM categories were used to investigate the effects of subpixel aspect distribution on BSA. Without loss of generality, the SAA is set to 150° in this analysis. For each category, the subpixel aspect distributions of the six DEMs differ significantly. Figure 8 shows the BSA variation with the subpixel aspect distribution. When the SZA is close to 0°, the influence of the subpixel aspect distribution on BSA under different terrains is minimal because shadowing does not occur in this scenario. The influence of the subpixel aspect distribution increases with SZA regardless of the mean slope of the terrain. When the mean slope is 30°, the MAD of the BSA increases from 0.01 to approximately 0.15 with an increase in SZA from 0° to 60°. This is because shadows gradually occur as the SZA increases, and the subpixel aspect distribution is related to the shadow ratio and distribution. When the terrain is relatively flat, the influence of the subpixel aspect distribution on the BSA is minimal, as shown in Figure 8a. With an increase in mean slope, the influence of the subpixel aspect distribution on the BSA gradually increases. The MADs of the BSAs over the terrain with mean slopes of 10°, 20°, and 30° are 0.06, 0.08, and 0.15, respectively. Overall, with an increase in SZA and mean slope, the influences of the subpixel aspect distribution gradually increase. In addition, the influence of the subpixel aspect distribution has a strong relationship with the SAA, which will be discussed in Section 4.2.5.

Based on specific solar illumination geometries and mean slopes, the topographic effects on the BSA are dependent on the spatial distribution characteristics of the subpixel aspect. When the proportion of the subpixel slope facing the sun is high, the BSA is relatively large, whereas the BSA becomes small when the proportion of the subpixel slope facing away from the sun is high. For instance, when the SZA is 60°, and the SAA is 150° (i.e., the sun is to the southeast), the BSA of the dominant eastward-oriented dem-10-5 is at a maximum in Figure 8a, the BSA of the dominant southward-oriented dem-20-6 is at maximum in Figure 8b, and the BSA of the dominant eastward-oriented dem-30-6 is at a maximum (and that of the dominant northwest-oriented dem-30-1) is at a minimum in Figure 8c.

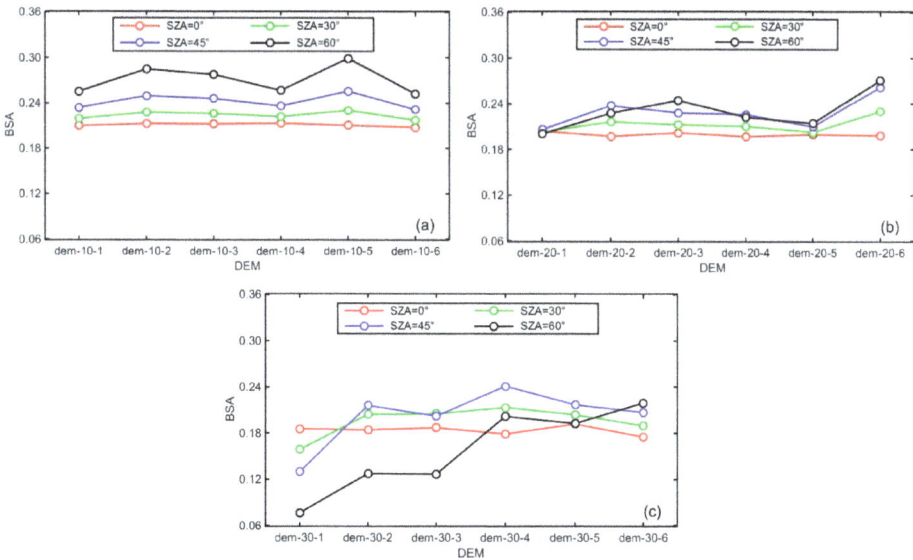

Figure 8. BSA variations with subpixel aspect distributions for terrains with different mean slopes:
(**a**) 10°; (**b**) 20°; and (**c**) 30°. The SAA is 150°.

4.2.4. BSA Variation with SZA

Both in situ measurements and remote sensing data from satellite and aircraft platforms have shown that BSA is strongly dependent on SZA [42–45]. Figure 9b,d show the BSA simulations with different SZAs, which exhibits obvious spatial variation characteristics corresponding to the mean slope distribution (shown in Figure 9a). As the SZA increases, the dynamic range of BSA becomes gradually large due to the modulation in the regional illumination angle and shadow effects. When the SZA is 0°, BSA varies from 0.14 to 0.22. In this case, shadows do not exist because the sun is at nadir, and the variation in BSA is mainly caused by the alteration in the regional illumination angle due to the inclined terrain. When the SZA is 30°, BSA varies from 0.12 to 0.24. The range increases slightly due to the existence of shadows and the increase in SZA. When the SZA is 60°, the BSA range, which becomes even larger, varies from 0.05 to 0.33. The low value of BSA (close to 0) can be explained by the fact that shadow effects are serious in some circumstances when a majority of the subpixel slopes face away from the sun. The high BSA value can be attributed to the high proportion of subpixel slopes facing toward the sun, where few shadows exist, even though the SZA is large.

Furthermore, three typical 1-km DEM categories are used to analyze the variations in topographic effects based on SZA. The variations in BSA based on SZA for different terrains are shown in Figure 10. When the terrain is flat (Figure 10a), BSA increases monotonously as the SZA increases. When the terrain is rugged (Figure 10b,d), BSA first increases then decreases with an increase in SZA. When the mean slope is 30°, the MAD and MRD of the BSA over rugged terrains caused by different SZAs exceed 0.22 and approximately 88%, respectively. This is because when the SZA is close to 90°, the shadow effect is obvious. The SZA at the inflection points gradually decreases from 70° to approximately 50° as the mean slope increases from 10° to 30°, respectively. These inflection points can be attributed to the increase in mean slope, which causes shadows to more easily appear when the SZA is relatively small. It is concluded that the increasing trend in BSA with SZA gradually slows as the mean slope increases.

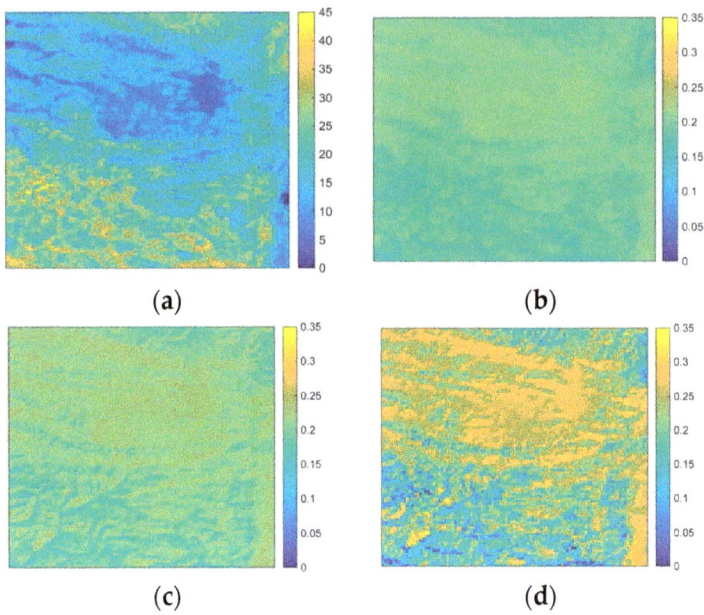

Figure 9. Maps of mean slope (**a**) within each 1-km pixel and the spatial distributions of BSA with different SZAs: (**b**) 0°; (**c**) 30°; and (**d**) 60°.

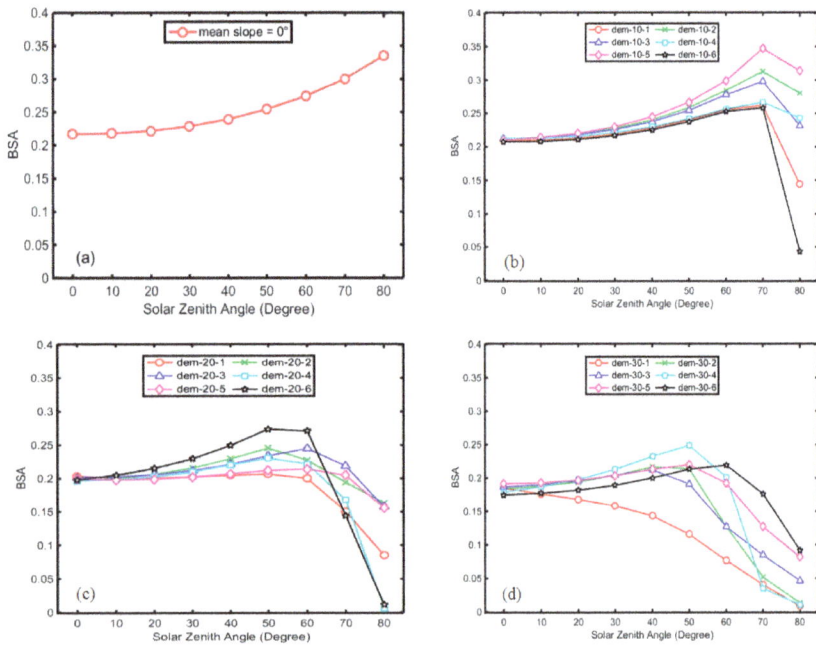

Figure 10. BSA variation with SZA for terrains with different mean slopes: (**a**) 0°; (**b**) 10°; (**c**) 20°; and (**d**) 30°. The SAA is 150°.

4.2.5. BSA Variation with SAA

Figure 11 shows the variation in BSA based on DEMs with different mean slopes with a specified SAA for different SZAs. These DEMs are dem-10-3, dem-20-3, and dem-30-3, whose subpixel aspect distributions exhibit a dominant southwest-orientation, a relatively uniform distribution, and a dominant southwest-orientation, respectively. The influence of the SAA on BSA increases with SZA. For a mean slope of 30°, the MAD of the BSA with a 30° SZA is approximately 0.03, while the MAD of the BSA with a 60° SZA exceeds 0.07. The SAA has little impact on BSA when the terrain is relatively flat, whereas the SAA considerably affects the BSA when the terrain is rugged. The influence of the SAA on BSA gradually increases with an increase in mean slope. Figure 11d shows that as the mean slope increases from 0° to 30°, the MAD of BSA varies from 0 to approximately 0.07, which can be explained because the shadow effects are obvious as the SZA and the mean slope increase simultaneously.

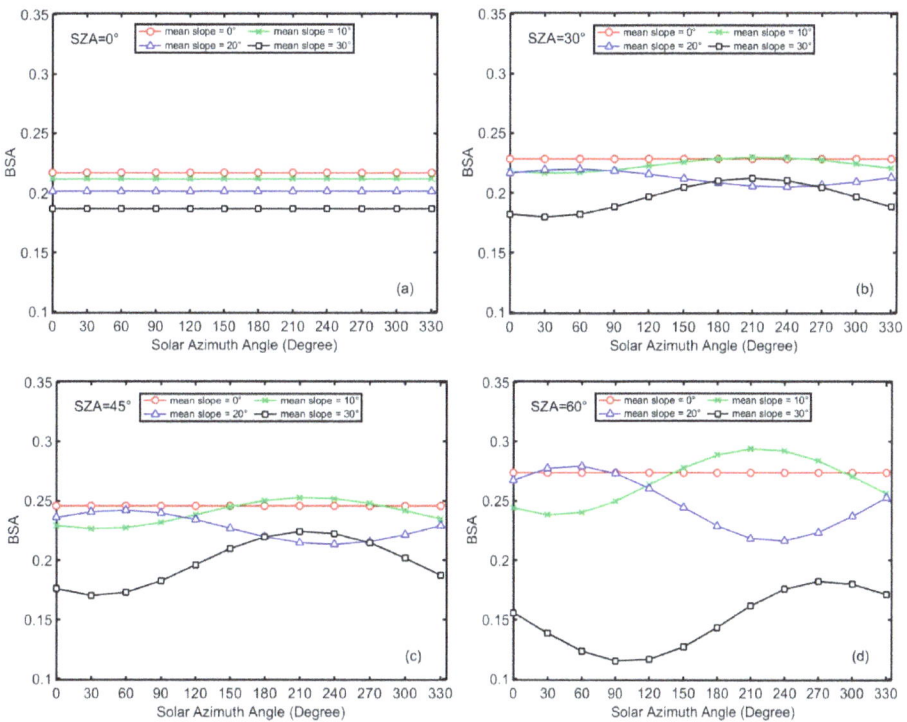

Figure 11. BSA variations with the SAA under different SZAs: (**a**) 0°; (**b**) 30°; (**c**) 45°; and (**d**) 60°.

Furthermore, the variations in BSA with SAA under different subpixel aspect distributions are also analyzed. Figure 12 shows the variation in BSA with SAA over the different terrains with different aspect distributions when the SZA is 30°. It is concluded that there is a parabolic distribution between BSA and the SAA. When the SAA is close to the predominant aspect of the subpixel slopes, the BSA is relatively large, as shown in the variation curves with SAA in Figures 11 and 12. This is because the shadow area is relatively small in this situation. When the SAA is opposite that of the predominant aspect at the end of the parabolic distribution, the BSA reaches a minimum due to the large shadow ratio. For terrains in dem-10-2, dem-10-4, dem-20-3 and dem-30-5, which have relatively uniform aspect distributions, the SAA has little effect on BSA. When the aspect is unevenly distributed, the parabolic distribution characteristics between the SAA and BSA are distinct. For dem-30-1, the northwest-facing slopes comprise the majority; therefore, the maximum and minimum BSAs appear in the northwestern and southeastern directions, respectively, which is where the sun is located. The BSA variations in dem-30-2 and dem-30-3 with SAA coincide due to the analogous aspect distributions between these two DEMs. These results demonstrate that BSA over a rugged terrain is sensitive to SAA, and the subpixel aspect distribution has considerable influence on the relationship between the SAA and the BSA.

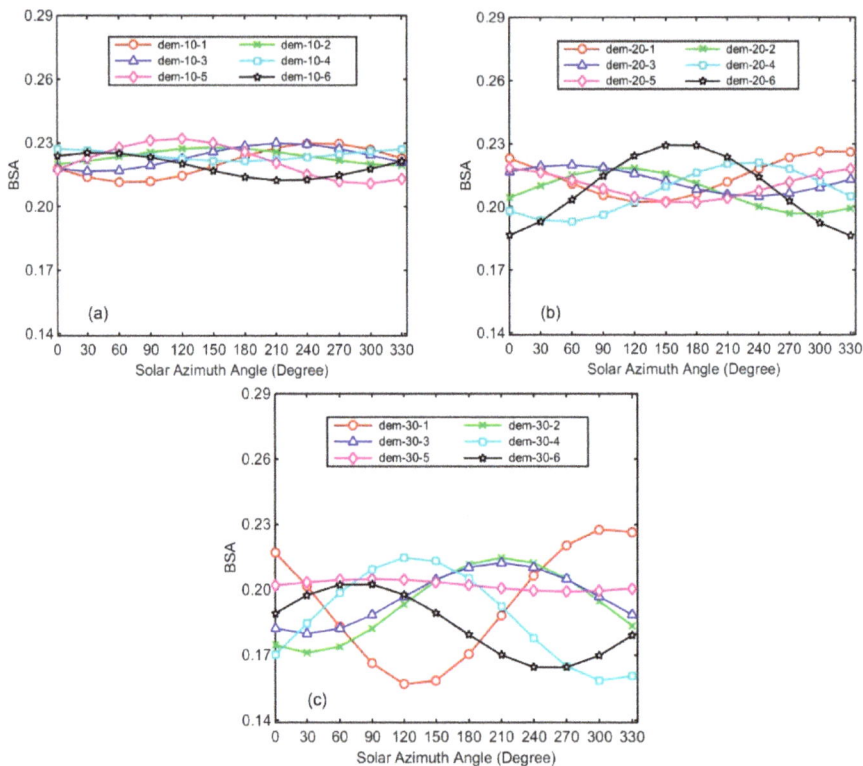

Figure 12. BSA variations with SAA for terrains with different mean slopes: (**a**) $10°$; (**b**) $20°$ and (**c**) $30°$. The SZA is $30°$.

4.3. Method Limitations

The accuracy assessment shows that with an increase in mean slope, BSA estimation errors gradually increase due to model simplification and approximations in problems such as multi-scattering and terrain obstruction. However, in general, the multi-scattering effect from adjacent terrains has a weaker contribution to BSAs, unless the slope lies in deep valleys or the adjacent terrains have high reflectance [28,46]. Therefore, considering that multi-scattering effect is negligible for snow-free land surface and the developed method has high precision in the validation experiments, the proposed method can have an adequate performance in vegetation and soil albedo estimations; these problems may have little effect on the analysis of topographic effects on BSA. Unfortunately, the developed method is not applicable to snow covered land surfaces with high reflectance. In addition, due to the model assumption of homogeneous land cover, the developed method faces a challenge in the mixed-pixel region. Subsequent research will focus on solving these problems.

This paper mainly uses modeled BSA data with the proposed method to analyze the topographic effects on BSA. The developed BSA estimation method was validated against the radiosity approach. Therefore, developing an efficient collection method for in situ albedo data over rugged terrains and validating the developed method against in situ data are urgent. In addition, in practical applications, DEM quality and the mismatching between the DEM and the remote sensing imagery also affect the accuracy of albedo estimation with the developed method. These basic data processing issues need to be further discussed and addressed, although they are beyond the scope of this paper.

5. Conclusions

Neglecting topographic effects may lead to significant bias when estimating land surface albedo. In this paper, we presented an efficient snow-free BSA estimation method over a rugged terrain. The proposed method was validated using simulated DEMs and real DEMs. The validation results showed that the modeled BSAs were consistent with the reference BSA, with an RMSE smaller than 0.01, which confirmed the ability of the proposed method to estimate BSA (with acceptable errors). By comparing the modeled BSA using the proposed method over the real DEM scenario, the topographic effects on BSA were investigated in detail.

The BSA over a rugged terrain is influenced by the subpixel slope distribution (mean slope and subpixel aspect distribution) and the solar illumination angle (SZA and SAA). These factors are related to the modulation in the regional illumination angle and shadow effects, which play key roles in the topographic effects on BSA. The more rugged the terrain and the larger the solar illumination angle, the more obvious topographic effects are on BSA. Specifically, for the subpixel slope distribution, the mean slope has a higher influence on BSA than that on the subpixel aspect distribution. For the mean slope, BSA generally presents a decreasing trend with an increase in mean slope. The larger the SZA, the more obvious the decreasing trend in BSA is with an increase in mean slope. When analyzing the subpixel aspect distribution, for an increase in mean slope, the influence of the subpixel aspect distribution on BSA gradually increases. Under specific solar illumination geometries and mean slopes, the influences of the subpixel aspect distribution on BSA are dependent on the proportion of subpixel slopes facing the sun. For the solar illumination angle, the SZA has a greater impact on BSA than that from the SAA. When the terrain is relatively flat, BSA increases monotonously with an increasing SZA. As the terrain becomes rugged, BSA first increases then decreases with an increase in SZA due to the incident radiation terrain block. For the SAA, BSA over a rugged terrain is sensitive to the SAA, and the influence of SAA on BSA increases with an increasing SZA and mean slope. A parabolic distribution is found between BSA and the SAA. When the SAA is close to the predominant aspect of the subpixel slopes, BSA is relatively large.

The motivation and findings in this study can benefit land surface albedo modeling and retrieval in the field of remote sensing. Subsequent research will focus on practical remote sensing applications and method improvements by considering multi-scattering effects.

Acknowledgments: This work was jointly supported by the Chinese Natural Science Foundation Project (No.: 41671363, 41331171) and the National Basic Research Program of China (No.: 2013CB733401).

Author Contributions: Dalei Hao and Jianguang Wen conceived and designed the method and the relative experiments; Dalei Hao, Jianguang Wen and Qing Xiao performed the experiments and were involved in the resulting analysis and discussion; Shengbiao Wu, Xingwen Lin, Baocheng Dou, Dongqin You and Yong Tang gave a considerable amount of helpful advice on the revised work; Dalei Hao wrote the paper.

Conflicts of Interest: The authors declare no conflict of interest.

References

1. Liang, S. *Quantitative Remote Sensing of Land Surfaces*; Wiley-Interscience: Hoboken, NJ, USA, 2003; pp. 413–415.
2. Qu, Y.; Liang, S.; Liu, Q.; He, T.; Liu, S.; Li, X. Mapping surface broadband albedo from satellite observations: A review of literatures on algorithms and products. *Remote Sens.* **2015**, *7*, 990–1020. [CrossRef]

3.   Martonchik, J.V.; Bruegge, C.J.; Strahler, A.H. A review of reflectance nomenclature used in remote sensing. *Remote Sens. Rev.* **2000**, *19*, 9–20. [CrossRef]

4.   Schaepman-Strub, G.; Schaepman, M.E.; Painter, T.H.; Dangel, S.; Martonchik, J.V. Reflectance quantities in optical remote sensing—Definitions and case studies. *Remote Sens. Environ.* **2006**, *103*, 27–42. [CrossRef]

5.   Schaepman-Strub, G.; Schaepman, M.E.; Martonchik, J.V.; Painter, T.H.; Dangel, S. Radiometry and reflectance: From terminology concepts to measured quantities. In *The SAGE Handbook of Remote Sensing*; Sage Publications: Thousand Oaks, CA, USA, 2009.

6.   Lucht, W.; Schaaf, C.B.; Strahler, A.H. An algorithm for the retrieval of albedo from space using semiempirical BRDF models. *IEEE Trans. Geosci. Remote Sens.* **2002**, *38*, 977–998. [CrossRef]

7.   Combal, B.; Isaka, H.; Trotter, C. Extending a turbid medium BRDF model to allow sloping terrain with a vertical plant stand. *IEEE Trans. Geosci. Remote Sens.* **2000**, *38*, 798–810. [CrossRef]

8.   Schaaf, C.; Li, X.; Strahler, A. Topographic effects on bidirectional and hemispherical reflectances calculated with a geometric-optical canopy model. *IEEE Trans. Geosci. Remote Sens.* **1994**, *32*, 1186–1193. [CrossRef]

9.   Strahler, A.H.; Muller, J.; Lucht, W.; Schaaf, C.; Tsang, T.; Gao, F.; Li, X.; Lewis, P.; Barnsley, M.J. MODIS BRDF/Albedo Product: Algorithm Theoretical Basis Document. Version 5.0. Available online: https://modis.gsfc.nasa.gov/data/atbd/atbd_mod09.pdf (accessed on 1 October 2017).

10. Liang, S. A direct algorithm for estimating land surface broadband albedos from MODIS imagery. *IEEE Trans. Geosci. Remote Sens.* **2003**, *41*, 136–145. [CrossRef]

11. Tasumi, M.; Allen, R.G.; Trezza, R. At-surface reflectance and albedo from satellite for operational calculation of land surface energy balance. *J. Hydrol. Eng.* **2008**, *13*, 51–63. [CrossRef]

12. Schaaf, C.B.; Gao, F.; Strahler, A.H.; Lucht, W.; Li, X.; Schaaf, C.B.; Gao, F.; Strahler, A.H.; Lucht, W.; Li, X. First Operational BRDF, Albedo Nadir Reflectance Products from MODIS. *Remote Sens. Environ.* **2002**, *83*, 135–148. [CrossRef]

13. Deschamps, P.Y.; Breon, F.M.; Leroy, M.; Podaire, A.; Bricaud, A.; Buriez, J.C.; Seze, G. The polder mission: Instrument characteristics and scientific objectives. *IEEE Trans. Geosci. Remote Sens.* **1994**, *32*, 598–615. [CrossRef]

14. Diner, D.J.; Beckert, J.C.; Reilly, T.H.; Bruegge, C.J.; Conel, J.E.; Kahn, R.A.; Martonchik, J.V.; Ackerman, T.P.; Davies, R.; Gerstl, S.A.W. Multi-angle imaging spectroradiometer (MISR) instrument description and experiment overview. *IEEE Trans. Geosci. Remote Sens.* **2005**, *36*, 1072–1087. [CrossRef]

15. Rutan, D.; Charlock, T.P.; Rose, F.; Kato, S.; Zentz, S.; Coleman, L. *Global Surface Albedo from Ceres/Terra Surface and Atmospheric Radiation Budget (SARB) Data Product*; ACM: New York City, NY, USA, 2006; pp. 2245–2255.

16. Ryu, Y.; Kang, S.; Moon, S.K.; Kim, J. Evaluation of land surface radiation balance derived from moderate resolution imaging spectroradiometer (MODIS) over complex terrain and heterogeneous landscape on clear sky days. *Agric. For. Meteorol.* **2008**, *148*, 1538–1552. [CrossRef]

17. Wen, J.; Zhao, X.; Liu, Q.; Tang, Y.; Dou, B. An improved land-surface albedo algorithm with DEM in rugged terrain. *IEEE Geosci. Remote Sens. Lett.* **2014**, *11*, 883–887.

18. Gao, B.; Jia, L.; Menenti, M. An improved method for retrieving land surface albedo over rugged terrain. *IEEE Geosci. Remote Sens. Lett.* **2014**, *11*, 554–558. [CrossRef]

19. Wen, J.G.; Qiang, L.; Liu, Q.H.; Xiao, Q.; Li, X.W. Scale effect and scale correction of land-surface albedo in rugged terrain. *Int. J. Remote Sens.* **2009**, *30*, 5397–5420. [CrossRef]

20. Fan, W.; Chen, J.M.; Ju, W.; Zhu, G. GOST: A geometric-optical model for sloping terrains. *IEEE Trans. Geosci. Remote Sens.* **2014**, *52*, 5469–5482.

21. Mousivand, A.; Verhoef, W.; Menenti, M.; Gorte, B. Modeling top of atmosphere radiance over heterogeneous non-lambertian rugged terrain. *Remote Sens.* **2015**, *7*, 8019–8044. [CrossRef]

22. Yin, G.; Li, A.; Zhao, W.; Jin, H.; Bian, J.; Wu, S. Modeling canopy reflectance over sloping terrain based on path length correction. *IEEE Trans. Geosci. Remote Sens.* **2017**, *55*, 4597–4609. [CrossRef]

23. Manninen, T.; Andersson, K.; Riihelä, A. Topography Correction of the CM-SAF Surface Albedo Product SAL. In Proceedings of the EUMETSAT Meteorological Satellite Conference, Oslo, Norway, 5–9 September 2011.

24. Kawata, Y.; Ueno, S.; Kusaka, T. Radiometric correction for atmospheric and topographic effects on landsat mss images. *Remote Sens.* **1988**, *9*, 729–748. [CrossRef]

25. Li, H.; Shen, Y.; Yang, P.; Zhao, W.; Allen, R.G.; Shao, H.; Lei, Y. Calculation of albedo on complex terrain using MODIS data: A case study in Taihang mountain of China. *Environ. Earth Sci.* **2015**, *74*, 1–10. [CrossRef]

26. Cherubini, F.; Vezhapparambu, S.; Bogren, W.; Astrup, R.; Strømman, A.H. Spatial, seasonal, and topographical patterns of surface albedo in Norwegian forests and cropland. *Int. J. Remote Sens.* **2017**, *38*, 4565–4586. [CrossRef]

27. Ashdown, I. *Radiosity: A Programmer's Perspective*; John Wiley & Sons, Inc.: Hoboken, NJ, USA, 1994; pp. 7–10.

28. Proy, C.; Tanre, D.; Deschamps, P.Y. Evaluation of topographic effects in remotely sensed data. *Remote Sens. Environ.* **1989**, *30*, 21–32. [CrossRef]

29. Chen, Y.; Hall, A.; Liou, K.N. Application of three-dimensional solar radiative transfer to mountains. *J. Geophys. Res. Atmos.* **2006**, *111*, 5143–5162. [CrossRef]

30. Richter, R.; Schläpfer, D. Geo-atmospheric processing of airborne imaging spectrometry data. Part 2: Atmospheric/topographic correction. *Int. J. Remote Sens.* **2002**, *23*, 2631–2649. [CrossRef]

31. Sandmeier, S.; Itten, K.I. A physically-based model to correct atmospheric and illumination effects in optical satellite data of rugged terrain. *IEEE Trans. Geosci. Remote Sens.* **1997**, *35*, 708–717. [CrossRef]

32. Dozier, J.; Frew, J. Rapid calculation of terrain parameters for radiation modeling from digital elevation data. *IEEE Trans. Geosci. Remote Sens.* **1990**, *28*, 963–969. [CrossRef]

33. Zaksek, K.; Ostir, K.; Kokalj, Z. Sky-view factor as a relief visualization technique. *Remote Sens.* **2011**, *3*, 398–415. [CrossRef]

34. Hapke, B. *Theory of Reflectance and Emittance Spectroscopy: Second Edition*; Cambridge University Press: Cambridge, UK, 2012; pp. 303–335.

35. Hapke, B. Bidirectional reflectance spectroscopy: 3. Correction for macroscopic roughness. *Icarus* **1984**, *59*, 41–59. [CrossRef]

36. Jacquemoud, S.; Verhoef, W.; Baret, F.; Bacour, C.; Zarcotejada, P.J.; Asner, G.P.; François, C.; Ustin, S.L.; Ustin, S.L.; Schaepman, M.E. Prospect+sail models: A review of use for vegetation characterization. *Remote Sens. Environ.* **2009**, *113*, S56–S66. [CrossRef]

37. Cariboni, J.; Gatelli, D.; Liska, R.; Saltelli, A. The role of sensitivity analysis in ecological modelling. *Ecol. Model.* **2007**, *203*, 167–182. [CrossRef]

38. Roupioz, L.; Nerry, F.; Jia, L.; Menenti, M. Improved surface reflectance from remote sensing data with sub-pixel topographic information. *Remote Sens.* **2014**, *6*, 10356–10374. [CrossRef]

39. Liang, S.; Lewis, P.; Dubayah, R.; Qin, W.; Shirey, D. Topographic effects on surface bidirectional reflectance scaling. *J. Remote Sens.* **1997**, *1*, 82–93.

40. Borel, C.C.; Gerstl, S.A.W.; Powers, B.J. The radiosity method in optical remote sensing of structured 3-d surfaces. *Remote Sens. Environ.* **1991**, *36*, 13–44.

41. Goel, N.S.; Rozehnal, I.; Thompson, R.L. A computer graphics-based model for scattering from objects of arbitrary shapes in the optical region. *Remote Sens. Environ.* **1991**, *36*, 73–104. [CrossRef]

42. Wang, K.; Wang, P.; Liu, J.; Sparrow, M.; Haginoya, S.; Zhou, X. Variation of surface albedo and soil thermal parameters with soil moisture content at a semi-desert site on the western Tibetan plateau. *Bound. Layer Meteorol.* **2005**, *116*, 117–129. [CrossRef]

43. Oguntunde, P.G.; Ajayi, A.E.; Giesen, N.V.D. Tillage and surface moisture effects on bare-soil albedo of a tropical loamy sand. *Soil Tillage Res.* **2006**, *85*, 107–114. [CrossRef]

44. Pinty, B.; Verstraete, M.M.; Dickinson, R.E. A physical model for predicting bidirectional reflectances over bare soil . *Remote Sens. Environ.* **1989**, *27*, 273–288. [CrossRef]

45. Pinker, R.T.; Laszlo, I. Modeling surface solar irradiance for satellite applications on a global scale. *J. Appl. Meteorol.* **1992**, *31*, 194–211. [CrossRef]

46. Sirguey, P. Simple correction of multiple reflection effects in rugged terrain. *Int. J. Remote Sens.* **2009**, *30*, 1075–1081.

remote sensing

MDPI

Article

Local Effects of Forests on Temperatures across Europe

Bijian Tang [1,2,3]**, Xiang Zhao** [1,2,]***** and Wenqian Zhao** [1,2]

[1] State Key Laboratory of Remote Sensing Science, Jointly Sponsored by Beijing Normal University and Institute of Remote Sensing and Digital Earth of Chinese Academy of Sciences, Bijian 100875, China; tangbj@mail.bnu.edu.cn (B.T.); wenqianzhao@mail.bnu.edu.cn (W.Z.)

[2] Beijing Engineering Research Center for Global Land Remote Sensing Products, Institute of Remote Sensing Science and Engineering, Faculty of Geographical Science, Beijing Normal University, Beijing 100875, China

[3] Division of Environment and Sustainability, The Hong Kong University of Science and Technology, Kowloon, Hong Kong, China

***** Correspondence: zhaoxiang@bnu.edu.cn; Tel.: +86-10-5880-0152

Received: 20 December 2017; Accepted: 27 March 2018; Published: 29 March 2018

Abstract: Forests affect local climate through biophysical processes in terrestrial ecosystems. Due to the spatial and temporal heterogeneity of ecosystems in Europe, climate responses to forests vary considerably with diverse geographic and seasonal patterns. Few studies have used an empirical analysis to examine the effect of forests on temperature and the role of the background climate in Europe. In this study, we aimed to quantitatively determine the effects of forest on temperature in different seasons with MODIS (MODerate-resolution Imaging Spectroradiometer) land surface temperature (LST) data and in situ air temperature measurements. First, we compared the differences in LSTs between forests and nearby open land. Then, we paired 48 flux sites with nearby weather stations to quantify the effects of forests on surface air temperature. Finally, we explored the role of background temperatures on the above forests effects. The results showed that (1) forest in Europe generally increased LST and air temperature in northeastern Europe and decreased LST and air temperature in other areas; (2) the daytime cooling effect was dominate and produced a net cooling effect from forests in the warm season. In the cold season, daytime and nighttime warming effects drove the net effect of forests; (3) the effects of forests on temperatures were mainly negatively correlated with the background temperatures in Europe. Under extreme climate conditions, the cooling effect of forests will be stronger during heatwaves or weaker during cold spring seasons; (4) the background temperature affects the spatiotemporal distribution of differences in albedo and evapotranspiration (forest minus open land), which determines the spatial, seasonal and interannual effects of forests on temperature. The extrapolation of the results could contribute not only to model validation and development but also to appropriate land use policies for future decades under the background of global warming.

Keywords: land surface temperature; satellite observations; flux measurements; latitudinal pattern; land cover change

1. Introduction

Forests cover more than ~42 million km^2 in the Northern Hemisphere (~30% of the land surface), and affects local climate mainly through biophysical processes [1–4]. The biophysical processes (e.g., albedo, evapotranspiration rate (ET) and surface roughness) all have effects on surface energy fluxes, which causes the effects of forests on local climates to be complicated [5]. Forests usually have a lower albedo than that for grasslands and croplands, especially in winter, when grass or crops are covered by snow. When open land, (i.e. grass or crops), are converted into forests, or even when a deciduous forest transforms into an evergreen forest, the albedo changes, resulting in a change in regional radiative forcing and surface temperature [6]. The surface roughness and ET of forests tend to be higher than those of open land, especially in summer, when forests have a high leaf area index (LAI). In the daytime, forests tend to cool the local temperature with a high latent heat flux. At nighttime, forests may be store heat and increase the local temperature by strengthening the nocturnal temperature inversion [7–9]. Therefore, the balance of these different processes determines the net effect of forest on local climate. A comparison of these biophysical processes between forests and open lands can help us determine the effects of forests on local temperatures, a topic which has been analyzed in previous studies [9–14].

Depending on the location, biophysical processes in forests may cause the cooling or warming of local temperatures [15]. In tropical regions, the cooling effect of ET, surface roughness and the larger land sink of CO_2 from forests dominates the warming effect induced by albedo; therefore, forest surface temperatures tend to be cooler than open land surface temperatures. In contrast, an opposite effect occurs in the boreal zone, where the warming effect is inferior to the cooling effect and is induced by a higher albedo and a land albedo–sea ice feedback mechanism [16]. Many studies have shown that the effects of forests on temperature show a gradient distribution that gradually changes from cooling to warming from the equator to the poles [17–19]. Deforestation significantly reduces air temperatures above 45°N in Asia and North America and leads to warming below 35°N [7,19]. Afforestation can obviously increase the land surface temperature (LST) north of 45°N and reduce the LST south of 35°N in the Northern Hemisphere (NH) [14,18,20].

In fact, biophysical processes vary not only with diverse geographic patterns but also with seasonal patterns, which results in different effects of forests on temperature during different seasons [21,22]. Albedo is affected by soil wetness, soil color and snow cover [23,24]. In winter, the difference in albedo between forests and open land is largest because open land tends to be covered with snow. There are also some differences among different forest types at different latitudes. At the high latitudes, the type of forest is mainly needleleaf, which has a ground and forest canopy that are both easily covered with snow. At lower latitudes in Europe, most forests are either mixed or broadleaf forests, where only the canopy is easily covered with snow. This characteristic causes the albedo difference of forest minus open land to vary in space and with the season [21]. The difference in ET is nearly zero because of the small leaf index and radiation limitation [25]. This result implies that the cooling effect, which result from ET and surface roughness, is weaker than the warming effect of albedo in winter. In contrast, in summer, the difference in albedo is smallest, and the difference in ET is largest in summer, which causes the cooling effect resulting that results from ET and surface roughness to be stronger than the warming effect for albedo. The net effects of periodically changed albedo, surface roughness and ET play roles in the seasonal changes of the net effects of forests on the climate at mid- and high- latitudes [20]. This phenomenon is supported by a recent study that found that the effects of deforestation have a south-north gradient from warming to cooling in winter due to snow-cover, which always warms the local temperature in summer in the eastern United States [10].

Moreover, biophysical processes may also change with a variable background climate in different years, enabling the effects of forests change with years [26]. From a cold year to a warm year, the difference in albedo between forests and open land may decrease because of the lower snow depth. The difference in ET may also change, which is related to the local soil moisture. The change in albedo and ET difference between forests and open land may affect the effects of forests, cooling or warming local temperature. Several model studies found that the cooling effect of deforestation decreased with a warmer background [26–28]. However, there are also some disagreements within these model results. Winckler et al. [27] and Armstrong et.al [28] found that this decrease would occur globally, while Pitman et al. [26] found that the cooling of deforestation increased in a lower latitude and decreased in a higher latitude. The effect of the background climate on forest effects must be comprehensively evaluated, particularly for the current decade, which is experiencing intense climate warming.

Europe has a temperate marine ecosystem, a Mediterranean ecosystem and a temperate continental climate. There are also many famous mountains in Europe, such as the Alps, the Apennines and the Pyrenees. Considerable heterogeneity of climate and topography has created complex spatial patterns regarding the effect of forest on temperature. In addition, a growing number of extreme climate events, such as heat waves and extreme precipitation, have been observed throughout Europe due to on-going climate change. These various background climates have also created complex spatial and temporal patterns regarding the effects of forests on temperature.

Over the past few decades, most of the known effects of forests on the climate in Europe have been derived from models, while global climate models were not suitable for the local impacts of forest on climate due to their coarse spatial resolutions and uncertainties in the physical processes, parameterization and input data [5,15]. Regional climate models in Europe are often based on comparisons among climate model outputs for different land cover conditions, with the major difference being that forests in one scenario are replaced by open land conditions in another scenario [29]. However, considerable heterogeneity in the climate and topography of Europe has created complex spatial patterns regarding the effects of forests on temperature. A growing number of extreme climate events, such as heat waves and extreme precipitation, have also created complex temporal patterns regarding the effects of forests on temperature [30]. It is not easy for regional climate models to simulate the complex spatial and temporal patterns of forest effects, and some model studies often show contradictory results [31,32].

There are two kinds of observations, that are widely used to explore the effects of forests on temperature: satellite and in-situ data [7,14,17–20]. Some studies used satellite observations found that afforestation in China cooled daytime temperatures and warmed the nocturnal temperatures [18,20]. Mi Zhang [19] and Lee et al [7] used forest flux site observations to study the effects of deforestation in North America and eastern Asia and found that deforestation warmed temperature at low latitudes and cooled temperatures at high latitudes. However, few studies have used observations to examine the effects of forests on climate in Europe, which has experienced a large increase in forests over the last two decades. Although Li et al. [14,17] used satellite observation to explore the effect of forest globally and found that forests decrease local temperatures in Europe, they focused on spatial patterns and did not consider the effects of background climate on forests during different yeas. Additionally, using two time observations to represent daily averages may also lead to some uncertainties [33]. Further studies should emphasize the effects of forests on the climate in Europe using both kinds of observations to explore the spatiotemporal patterns of forest effects and the effect of the background climate.

In this paper, we used the MODIS (MODerate resolution Imaging Spectroradiometer) land surface temperature (LST), ET, albedo and land cover classification, and FLUXNET site observations from European Fluxes Database Cluster to analyze the effect of forest on temperature in Europe (11°W - 40°E, 35°N - 70°N) and the effect of the background climate. The specific objectives of this study are as follows: (1) to identify the spatiotemporal pattern of forest effects on LST and air temperature and (2) to explore the impact of the background climate, such as extreme climate, on the forest effects to understand how it impacts temperature.

2. Data and Methods

2.1. Data

The MODIS/Aqua (MYD11A2, version 6) products, with temporal and spatial resolutions of 8 days and 1 km, respectively, from 2003 to 2016, were used in our study because the Aqua satellite passes over the region approximately 13:30 and 01:30, which is close to when the daily maximum and minimum temperatures occur. The MODIS LST data are retrieved from clear-sky conditions over each 8-day period with best quality [34]. Based on the quality control documentation, only temperatures with an emissivity error and an LST error less than or equal to 0.02 and 2 K, respectively, were selected for further study.

The MODIS 16-day/1 km albedo product (MCD43B3 version 5) from 2003 to 2016 was used to calculate the difference in surface albedo between forests and open land (Table 1). The product contains black-sky and white-sky albedo, which can be used to calculate the actual (blue-sky) albedo based on the ratio of direct to diffusive shortwave radiation [35]. In our study, we simply used the averages of black-sky and white sky albedo to represent the blue-sky albedo due to the small difference and high correlation between white-sky and black-sky albedo [14]. The albedo quality in our study was controlled by the MCD43B2 data set, and only pixels that were identified as 'best quality', 'good quality' and 'mixed quality' were chosen for further study. The main reason why we did not choose the version 6 data set was because the albedo product in version 6 had only daily data which was nearly 2 terabytes for our study area and was not convenient to process.

Table 1. List of data products used

Produce Name	Product Type	Resolution	Period Considered
MYD11A2 (V6)	LST	1 km	2003–2016
MCD43B3 (V5)	Albedo	1 km	2003–2016
MOD16A2 (V5)	ET	1 km	2003–2014
MCD12Q1 (V5)	Land cover type	500 m	2012
Forest flux sites	Air temperature		Valid year from 1996–2016

The MODIS ET product (MOD16A2 version 5) was used to quantify the changes in ET (Table 1). The ET product is the first regular 1 km land surface ET data set for the 109.03 million km^2 of global vegetation land areas at an 8-day interval [36]. The mean absolute bias of this ET product is approximately 0.33–0.39 mm/day compared with the ET in situ observations. The pixels were only identified as 'good quality' or 'other quality', but further examination identified the 'clouds NOT present 'classification, which was selected for further study. The MODIS ET version 5 product produced data until 2014. Here, we used MODIS ET data from 2003 to 2014. Our study area was located from mid- to high-latitudes. In the version 6 data set, there were many interpolated values at high latitudes in spring, autumn and winter, which were not suitable for our research.

Annual 500 m MODIS land cover data (MCD12Q1 version 5) in 2012 with the International Geosphere–Biosphere Programme (IGBP) classification, was used to classify forests and open land. Evergreen needleleaf forests, evergreen broadleaf forests, deciduous needleleaf forests, deciduous broadleaf forests and mixed forests in the IGBP land cover classification were merged into one forest type. Croplands and grasslands were combined into an open land classification.

Forest flux sites from the European Fluxes Database Cluster and meteorological stations from Global Summary of the Day (GSOD) database in Europe were used in this study. The GSOD database was based on data exchanged under the World Meteorological Organization (WMO) World Weather Watch Program according to WMO Resolution 40 (Cg-XII). In accordance with the requirements of the WMO, the surface meteorological stations need to be in open grasslands and far away from cities and water bodies. The effects of forests on surface air temperature could be analyzed by comparing the air temperature differences between forest flux sites and meteorological stations.

2.2. Data Processing

2.2.1. Data Aggregation Strategy

All MODIS data used in this study were re-projected into a 0.01° resolution. Each year, there were 46 LST, 46 ET, and 46 albedo images. The 46 albedo images were derived from the phased production strategy. First, we aggregated the 8-day or 16-day MODIS data into monthly means using 8-day or 16-day composites every month. Second, we aggregated the monthly averages into annual averages only if a pixel had 12 monthly averages. Third, we aggregated the annual averages into multiple-year averages if a pixel had at least one valid annual average. Fourth, the monthly averages for a single year were all aggregated into multiple-year monthly averages. Finally, we acquired multiple-year monthly and annual averages for LST, albedo and ET; in total, there were 14 years of annual averages for LST and albedo for the period 2003–2016, and 11 years of annual averages for ET for the period 2003–2014.

The temperature data from the forest flux sites were on a half-hour time scale. First, we aggregated the half-hour data set onto a daily time scale. Only if valid values for one day comprised more than 90% of all measurements (48) were the maximum (T_{max}), minimum (T_{min}) and mean temperatures (T_{mean}) of that day calculated. Second, we aggregated the daily T_{max}, T_{min} and T_{mean} into monthly and annual averages when the valid daily values comprised more than 90% of days in the whole year. Third, we aggregated the monthly and annual averages into multiple-year averages of T_{max}, T_{min} and T_{mean} for forests. We chose the 90% threshold as a compromise between accuracy and a sufficient number of valid years.

2.2.2. Window Searching Strategy

In our study, we applied a window searching strategy similar to that of Li et al. [14] to determine all of the available sample windows and compare forests and open land over Europe. Here, the sample windows as squares contained 40 × 40 pixels, which were approximately equal to 40 km × 40 km. Any two adjacent windows were overlapped by 15 pixels. If a window individually had more than 10% of the pixels for forests and open land, and the absolute average elevation difference between the forest and open land pixels was less than 100 m, it was a valid window that was used to calculate the mean differences in LST, albedo and ET between forests and open land. Additionally, to explore the effects of background temperature on forests, we calculated the mean LST for all pixels within a window and regarded it as the background temperature of that window. As a result, there were 3363 windows selected (Figure 1).

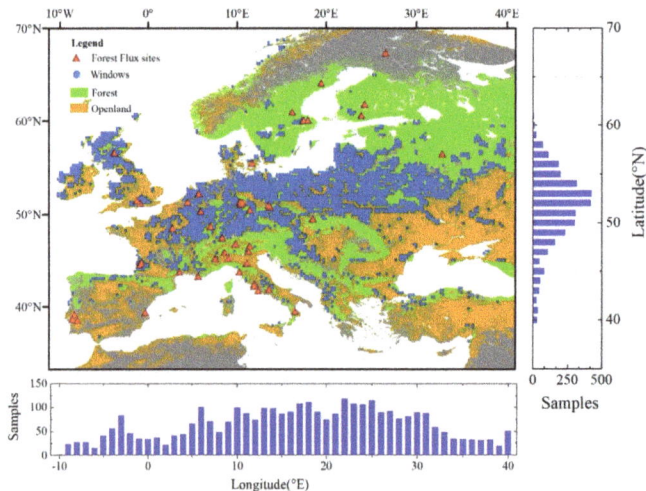

Figure 1. The spatial distributions of land cover types, paired sites and selected windows. The green and orange backgrounds refer to areas with forests and open lands, respectively. The paired sites are marked with red triangles. The blue points refer to the selected windows (0.4 × 0.4°), which have areas with more than 10% of forests and open land. The small panels in the below and right show the sample window numbers at each 1° (longitude and latitude) band.

2.2.3. Paired Sites Strategy

Here, we developed paired sites between the forest flux sites and the meteorological stations based on the following criteria. For a given forest flux site, we found all meteorological stations near the forest site within 1°. Then, we chose the meteorological station that had the smallest latitudinal distance between the paired site and the forest flux sites. Temperatures from the GSOD data were set at the daily time scale. We applied the same strategy as that for the forest flux sites to obtain the multiple-year monthly and annual averages of T_{max}, T_{min} and T_{mean} for open land. Only if the number of valid years in the forest and open land data sets were greater than one was the paired sites deemed valid to calculate the mean differences in T_{max}, T_{min} and T_{mean}. Finally, a total of 48 paired sites were obtained (Figure 1, Table S1). The average difference in elevation at the paired sites was 15.7 m, and the largest difference was 776 m, which was in Lavarone, Italy (the IT-Lav site). The average linear distance from the paired sites was approximately 34.3 km, and the longest distance was nearly 99.3 km, which was in Leinefelde, Germany (the DE-Lnf site).

2.2.4. Temperature Differences and Elevation Adjustment Strategy

The effects of forests on local land surface temperature or air temperature was expressed as the LST difference (ΔLST) and the T difference (ΔT) between forest and open land within a window and a paired site.

$$\Delta LST = LST_f - LST_o \tag{1}$$

$$\Delta T = T_f - T_o \tag{2}$$

where LST_f and LST_o represent the average LST of forests and open land pixels in a window, respectively, and T_f and T_o represent the air temperature at a paired site. Positive (negative) values of ΔLST or ΔT represent forests that are warmer (cooler) than open land. The ET difference and the albedo difference between forests and open land were defined similarly.

Even within a window or a paired site, the elevation difference between forest and open land might be large, which results in a systematic bias in ΔLST and ΔT due to the lapse rate. In our study, we applied the elevation adjustment method from Li et al. [14] to eliminate an elevation-induced bias from the original value. Here, we produced the correct term based the elevation difference (ΔELV) and the regression slope (k), which were calculated from the linear regression of ΔLST or ΔT versus ΔELV.

$$\Delta LST_a = \Delta LST - k*\Delta ELV \tag{3}$$

$$\Delta T_a = \Delta T - k*\Delta ELV \tag{4}$$

where ΔLST_a and ΔT_a represent the adjusted LST and air temperature differences, respectively. Here, k for daytime, nighttime, and daily ΔLST were 8.6 °C/km, 2.3 °C/km and 5.4 °C/km, respectively. The k values for the differences in T_{max}, T_{min} and T_{mean} were 7.2 °C/km, 4.6 °C/km, and 5.8 °C/km, respectively.

3. Results

3.1. Geographic Patterns in Temperature Difference

During the daytime, forests have a cooling effect on LST relative to the effect of open land in Europe, except for some areas in England and northern Norway (Figure 2a). The mean annual daytime cooling effect of forests is −1.06 ± 0.03 °C (at the 95% confidence interval and estimated by the *t*-test; the results hereafter was calculated in the same manner). The cooling effect shows a clear decreasing pattern moving towards higher latitudes and increases slowly with an increase in longitude. Forests cool daytime temperatures within 35°–63°N and warm daytime temperatures north of 66°N. At night, forests tend to have a higher LST than that of open land in Europe, except for a small area in southwest France (Figure 2b). The mean annual nighttime warming effect of forests is 0.58 ± 0.01 °C. The warming effect increases significantly from west to east and decreases slightly towards higher latitudes south of 63°N. Daytime cooling dominates nighttime warming, which results in a daily net cooling in most areas of Europe (Figure 2c). The mean annual daily cooling effect of forests is

−0.24 ± 0.01 °C. The cooling effect decreases with an increase in latitudes and longitudes from west to east and transitions into a warming effect over northeastern Europe. In northeastern Europe (north of 55°N and east of 25°E), the mean annual daily warming effect is 0.16 ± 0.03 °C. Similar results are obtained from sampling windows with different sizes or different threshold values of vegetation (figures not shown), suggesting that differences in vegetation type are a major cause of LST differences, which are independent of the vegetation threshold values and sample sizes.

Forests have a similar but slightly different effect on local air temperature. Forests tend to decrease the T_{max} and increase the T_{min} (Figure 3a,b). However, an adverse phenomenon is found near the Mediterranean area, where forests increase the T_{max} and decrease the T_{min}. The mean annual T_{max} cooling effect and the T_{min} warming effect of forests are −0.47 ± 0.28 °C and 0.52 ± 0.44 °C, respectively. The T_{max} cooling effect decreases slightly with an increase in longitude, and the T_{min} warming effect increases significantly towards higher latitudes, which results in a cooling effect from forests on T_{mean} in southwestern Europe and a warming effect from forests on T_{mean} in northwestern Europe (Figure 3c). In southwestern Europe (south of 55°N and west of 15°E), the mean annual cooling effect from forests on T_{mean} is −0.25 ± 0.23 °C. In northeastern Europe (north of 55°N and east of 15°E), the warming effect from forests on T_{mean} is 0.36 ± 0.28 °C. Additionally, several paired sites with high elevations have warming effects on T_{mean}, which is consistent with the results of previous studies [23]. However, some paired sites show an inconsistent phenomenon. For example, one paired site near London and several sites in Italy show a large warming effect on T_{mean}, while neighboring paired sites have a cooling effect on T_{mean}, which may result from the background climate and will be discussed in Section 3.3.

3.2. Seasonal Patterns in Temperature Differences

A clear, seasonal variation in the effect of forests on LST can be seen in Europe (Figure 4). Forests cool daytime LSTs during the warm season, and this cooling effect decreases with latitude and increases with longitude (Figure 4a,d). In the cold season, the daytime warming effect dominates all forests, and the effect increases towards high latitudes and with longitude from west to east. The forest nighttime warming effect occurs year-round and increases with longitude (Figure 4b,e). In the warm season, the daytime cooling effect dominates the nighttime warming effect, which results in a net daily cooling effect. The daily cooling effect on LST decreases from low to high latitudes and increases from west to east, which is mainly due to the latitudinal and longitudinal patterns of the daytime cooling effect (Figure 4c,f). In the cold season, forests have a warming effect on LST because of daytime and nighttime warmings. The warming effect increases towards eastern longitudes.

Figure 5 shows the seasonal variations in T_{max} and T_{min} between forests and open land and their T_{mean} differences in three different latitudinal and longitudinal zones. Forests tend to have a cooling effect on air temperature during the warm season and a warming effect during the cold season (Figure 5). The mean annual effects of forests on T_{mean} are −0.37 ± 0.30 °C (number of site pairs: n = 17) below 45°N, −0.07 ± 0.36 °C (n = 21) within 45–55°N and 0.25 ± 0.30 °C (n = 10) north of 55°N. From low latitudes to high latitudes, the cooling effect of forests on T_{max} and the warming effect of forests on T_{min} both increase. The magnitude of the increase in the cooling effect is weaker than that of the warming effect, which results in a cooling effect from forests on T_{mean} that decreases with latitude (Figure 5a,c,e). The mean annual effects of forests on T_{mean} are −0.25 ± 0.65 °C (n = 12) west of 5°E, −0.25 ± 0.20 °C (n = 26) within 5–15°E, and 0.43 ± 0.26 °C (n = 10) east of 15°E. From west to east, the cooling effect of forests on T_{max} decreases, and the warming effect of forests on T_{max} increases, which results in a cooling effect of forests on T_{mean} that decreases with an increase in longitude (Figure 5b,d,f).

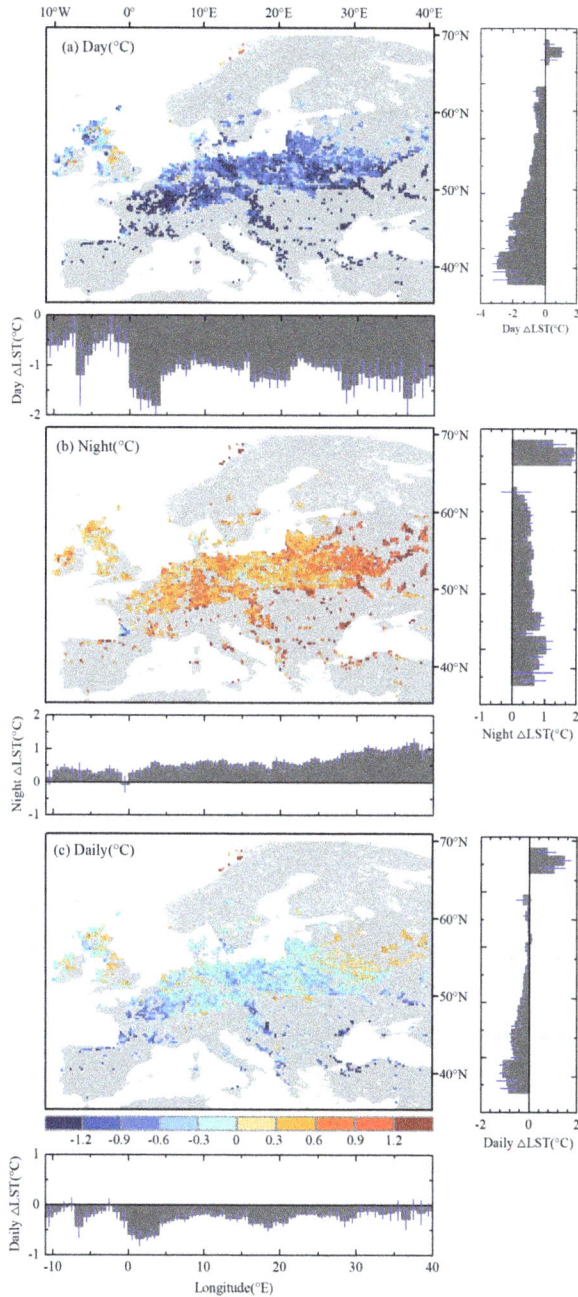

Figure 2. The spatial distributions of the annual mean (**a**) daytime, (**b**) nighttime and (**c**) daily average ΔLST (forest minus open land) in Europe during the period 2003–2016. The small panels in the below and right of each ΔLST show the longitudinal and latitudinal zonal average of each ΔLST for every 1° bin. The blue lines represent the 95% confidence interval (CI) estimated by the t-test.

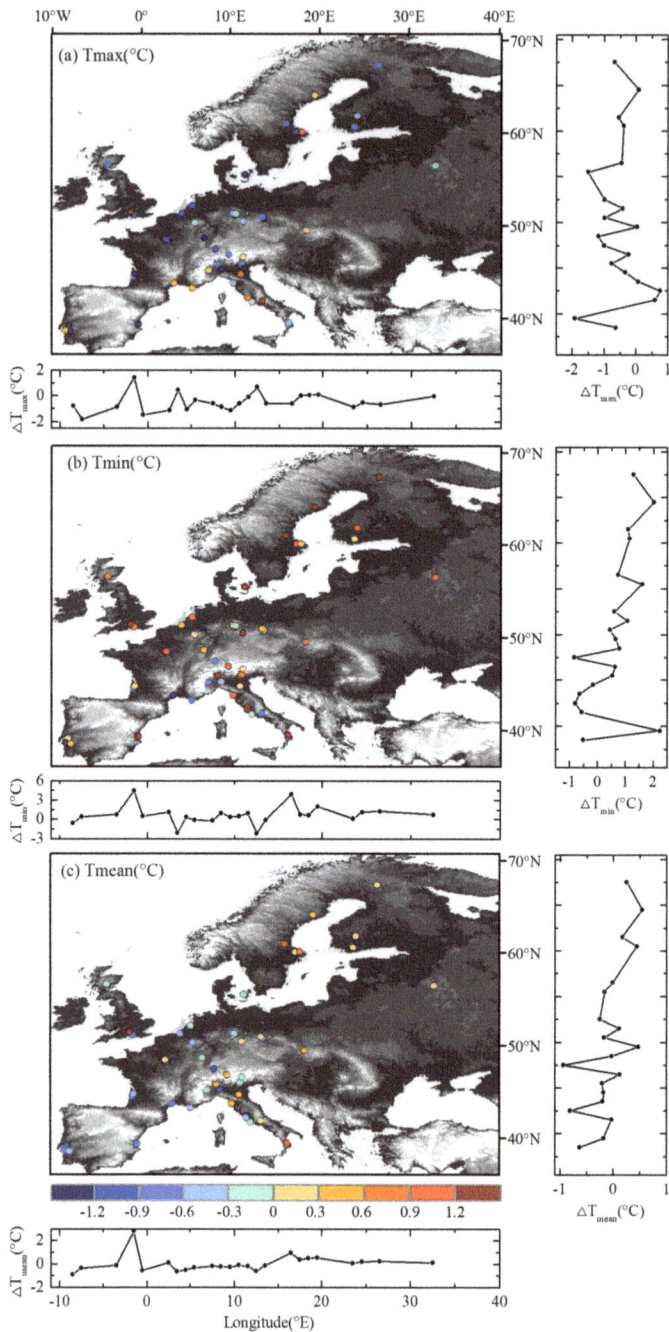

Figure 3. The spatial distributions of the annual mean (**a**) maximum, (**b**) minimum and (**c**) daily average ΔT (forest minus open land) in Europe. The small panels in the below and right show the longitudinal and latitudinal zonal average of each ΔT for every 1° bin. The background color refers to the elevation, which gradually increases from black to white.

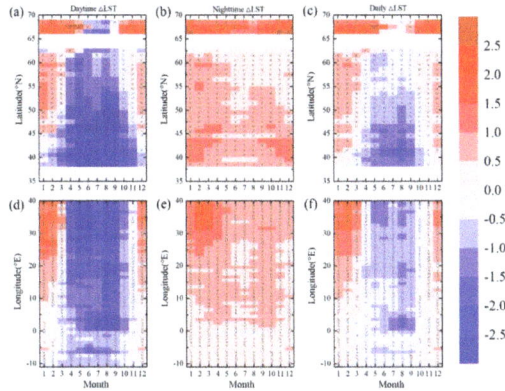

Figure 4. Spatiotemporal patterns of latitudinal variations in (**a**) daytime, (**b**) nighttime, (**c**) daily LST differences (forest minus open land) and longitudinal variations in (**d**) daytime, (**e**) nighttime and (**f**) daily LST differences (forest minus open land) during the period 2003–2016. Grids with cross symbols indicate that the LST differences are significant at the 95% CI by the *t*-test.

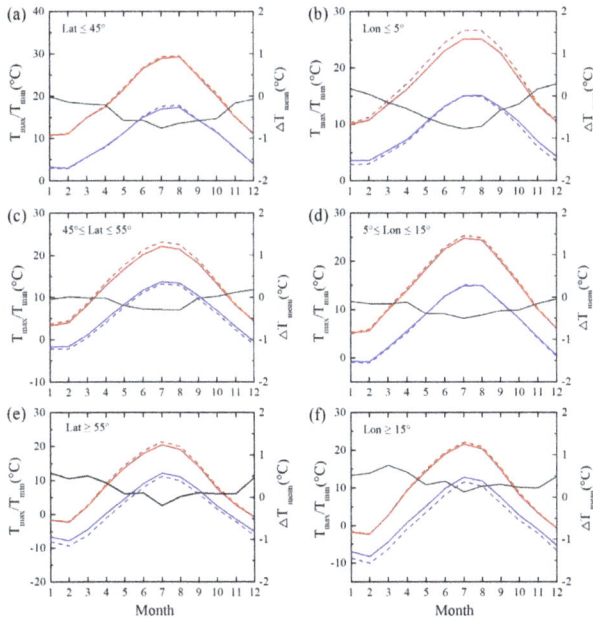

Figure 5. Comparison of seasonal variations in daily maximum, daily minimum and daily mean temperature differences in three latitudinal (**a**) south of 45°N, (**c**) between 45°N and 55°N, (**e**) north of 55°N, and longitudinal (**b**) west of 5°E, (**d**) between 5°E and 15°E, (**f**) east of 15°E ranges. The red solid and red dashed lines indicate Tmax for forests and open lands, respectively. The blue solid and blue dashed lines indicate Tmin for forests and open lands, respectively. The black solid line indicates the Tmean difference from forests minus open lands.

3.3. Effects of Background Temperatures on the Effects of Forests

The effects of forests on LST and air temperature show both spatial patterns and changes with latitude and longitude, which may be related to background temperatures. In this study, we calculated the average daily LST of all pixels within a window and referred to it as the background LST of a window. For paired sites, we simply regarded the air temperature of a forest site as the background air temperature. Figure 6 shows the relationship between the effects of forests and the background temperature. Forests tend to cool temperatures in warmer locations, such as tropical areas, and warm temperatures in cooler locations, such as arctic and high elevation areas. The same transitional temperature (near 6.5 °C) is found for both LST and air temperature. Forests have a cooling effect when the background temperature is higher than the transitional temperature. In contrast, forests have a warming effect if the background temperature is lower than the transitional temperature.

Figure 7 presents the relationship between forest effects and background temperatures for various years. Here, only when the valid years with a window were greater than five years was the relationship between background LST and daily LST differences (forest minus open land) calculated for various years. Since the number of valid years for the paired sites were usually limited, we focused on only the influence of background LSTs on the effects of forests on LST. The daily ΔLST was negatively related to the background LST in most areas of Europe, except for some areas in England and Germany (Figure 7a). This pattern indicated that the cooling effects of forests on LST increased as the background LST increased. Areas with significant changes were mainly located in eastern and southern Europe (Figure 7b).

The effect of background temperature on forest effects is also examined during different seasons (Figure 8). Although the effects of forests are mainly negatively related with background LSTs during the four seasons, there are several differences within the cold seasons (winter and spring) and the warm seasons (summer and autumn). The relationship between background LSTs and the effects of forests is more uniform in cold seasons than that in warm seasons. In spring, the cooling effect of forests tends to increase in eastern Europe and decrease in western Europe as the background LST increases (Figure 8a). In winter, the warming effect of forests decreases in eastern Europe and increases in western Europe as the background LST increases, especially in France (Figure 8d). In summer, the cooling effect of forests decreases in eastern Europe and increases in western Europe as the background LST increases. These relationships are much more complex in autumn, when the cooling effect either increases or decreases as the background LST changes (Figure 8b,c).

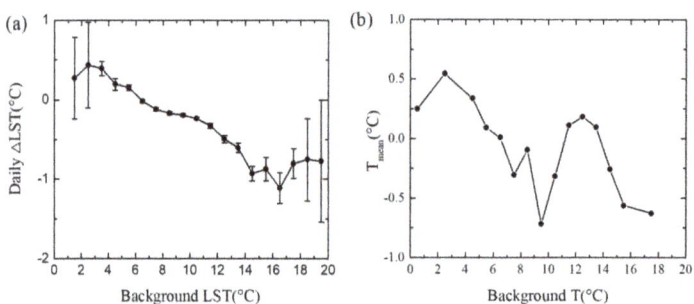

Figure 6. The relationships between (**a**) background LST and daily LST differences (forest minus open land) and (**b**) background air temperature and daily mean air temperature differences (forest minus open land). The daily LST differences are binned and averaged on 1° background LST intervals (i.e., the LST for all pixels within a window). The daily mean air temperature differences are binned and averaged on 1° grids for background air temperature (i.e., air temperatures of forest sites). The thin black bars represent the 95% confidence interval (CI) by the *t*-test.

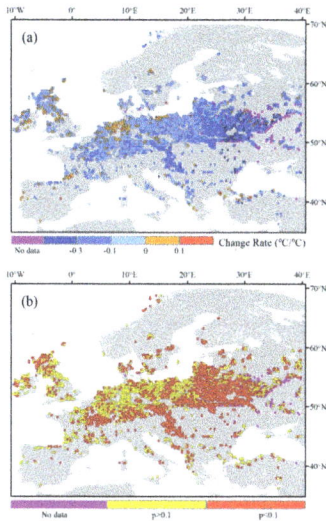

Figure 7. The relationship between background LST and daily LST differences (forest minus open land) during various years. (**a**) The rate of change for daily LST differences under different background LSTs. The number 0.1 indicates that the daily LST difference increases by 0.1 °C when the background LST increases by 1 °C. (**b**) Significance of the relationship between background LST and daily LST differences.

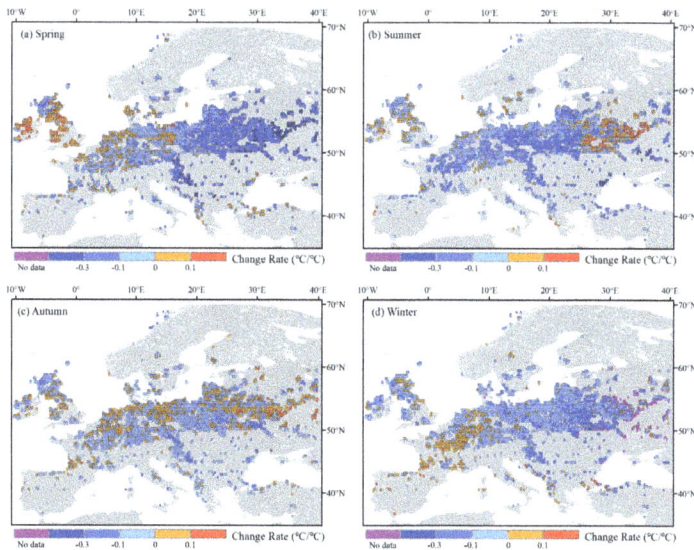

Figure 8. The relationship between background LST and daily LST differences (forest minus open land) in (**a**) spring, (**b**) summer, (**c**) autumn, and (**d**) winter during various years. Spring, summer, autumn and winter are defined by March and May, June and August, September and November, and December and February, respectively. The significance map of the four seasons is similar to Figure 7b and is not shown here.

3.4. Drivers of Temperature Difference

Albedo and ET are identified as key drivers in terms of forest effects. The spatial and seasonal distributions of ΔAlbedo and ΔET are shown in Figures 9 and 10, respectively, which are used to determine if the albedo in forests is lower than that on open lands in all seasons. The difference in albedo increases towards high latitudes and from west to east. Additionally, the difference in albedo is greater in the cold season than that in the warm season. Albedo is affected by soil color and soil wetness. ΔAlbedo is magnified by the presence of snow, as open land can be covered by snow. Snow is more likely to consistently and persistently occur in northern and eastern Europe [37]. Therefore, the ΔAlbedo in northern and eastern Europe is larger than that in western Europe, especially in the winter (Figure 10a,c). Forests have a higher ET than that of open land, especially in the warm season. The difference in ET (forest minus open land) decreases with latitude. For southern Europe, ET is strongly controlled by soil moisture availability, while it is constrained by radiation in northern Europe [25]. Forests can maintain a larger uptake of water because of deeper roots, while open land is more likely to be subjected to water limitations. However, incoming solar radiation over forests is like that of the nearby open land in northern Europe. Thus, the ΔET in southern Europe is greater than that in northern Europe (Figures 9b and 10b).

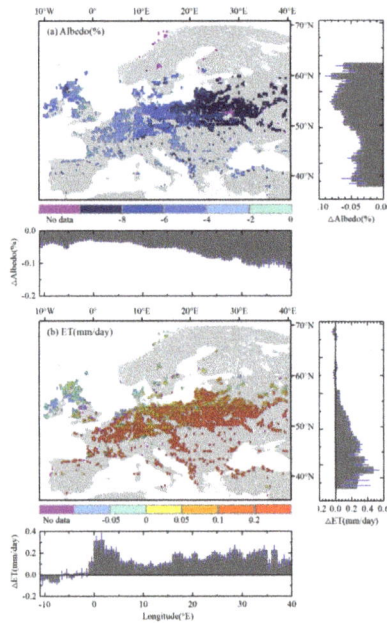

Figure 9. The spatial distributions of annual mean (**a**) albedo (%) and (**b**) ET differences (forest minus open land) in Europe. The periods used to analyze albedo and ET differences are 2003–2016 and 2003–2014, respectively. The below and right small panels for each difference show the longitudinal and latitudinal zonal averages for every 1° bin. The blue lines represent the 95% confidence interval (CI) estimated by the *t*-test.

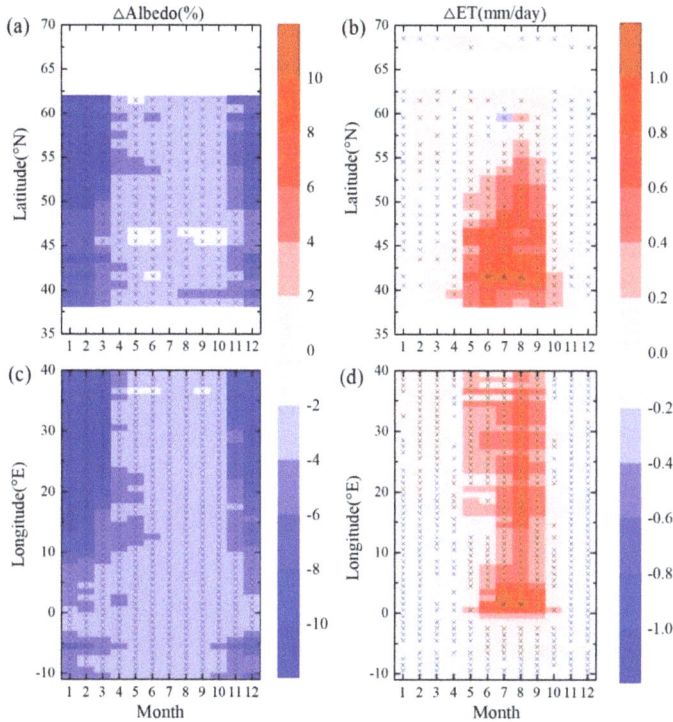

Figure 10. Spatiotemporal patterns of (**a**) latitudinal and (**c**) longitudinal variations in albedo (%) differences (forest minus open land) during the period 2003–2016 and (**b**) latitudinal and (**d**) longitudinal variations in ET differences (forest minus open land) during the period 2003–2014. Grids with cross symbols indicate differences that are significant at the 95% CI by the *t*-test.

The spatial and seasonal patterns of differences in albedo and ET (forest minus open land) determine the spatiotemporal differences in LST and T. From south to north, the cooling effect caused by ET differences decreases, while the warming effect caused by albedo differences increases. This results in a cooling effect from forests that decreases with latitude and gradually changes into a warming effect (Figures 2c and 3c). From west to east, the warming effects from albedo increase with longitude; therefore, the cooling effect of forests decreases with longitude in Europe (Figures 2c and 3c). In northeastern Europe, the annual incoming shortwave radiation energy per square meter that is received in the daytime ranges between 1246 and 1835 MJ. The annual mean of ΔAlbedo is $-10.39 \pm 0.25\%$. Therefore, forests absorb an extra 129–191 MJ of energy compared with adjacent open lands each year. Moreover, the annual mean ΔET is 0.05 ± 0.01 mm/day in northeastern Europe, which is roughly equivalent to the additional 45 ± 10 MJ of energy that is dissipated from forests compared to that dissipated by open lands each year. The additional energy absorbed by forests via albedo differences is greater than the energy dissipated by forests via ET differences in contrast to open lands, which results in a warming effect from forests in northeastern Europe (Figures 2c and 3c). From the warm season to the cold season, the cooling effect from ET differences decreases, and the warming effect from albedo increases, which results in a net cooling effect from forests in summer and net warming effect from forests in winter (Figures 4 and 5).

The background temperatures during various years may result in different forest effects on local temperature by changing the differences in albedo and ET. Figure 11 shows the relationship between the differences in albedo and ET and background LSTs in Europe. Although some windows have less than five years of valid data, we find that the differences in albedo increase with increasing background LSTs, indicating that the absolute value of the difference in albedo decreases (Figure 11a). The warming effect induced by the difference in albedo decreases when the background temperature increases, especially in eastern Europe, which is significantly affected by snow. The ET difference (forest minus open land) increases with the background LST in most areas of Europe. The cooling effect induced by the difference in ET increases when the background temperature increases. Therefore, the increase in the cooling effect caused by ET and the decrease in the warming effect caused by albedo result in an increase in the net cooling effect of forests on LST (Figure 7a). However, it is noticeable that the difference in ET decreases with an increase in LST in some areas. The change in the cooling effect caused by ET may be complicated under background temperature changes, which results in noise regarding the change in the net effect during warm seasons (Figure 8b,c). In Figure 3c, we observe a paired site that has a strong warming effect from forests near London. This was induced by a heat wave in 2006. In 2006, the annual temperature difference (forest minus open land) was nearly 6 °C, while in normal years, the temperature difference was 1.2 °C. The paired sites in Italy were also strongly affected by the local background temperature. This effect will require further examination in the future.

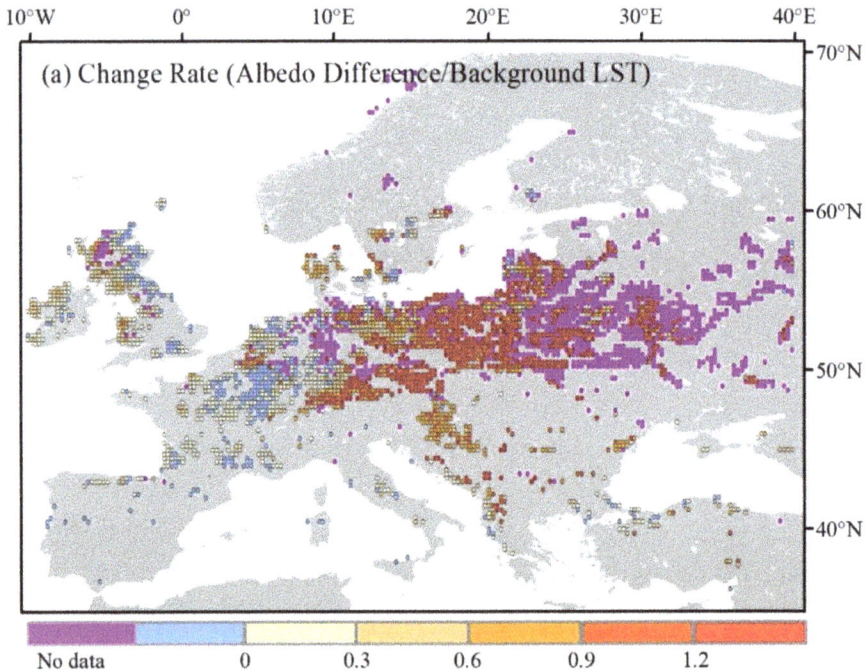

Figure 11. *Cont.*

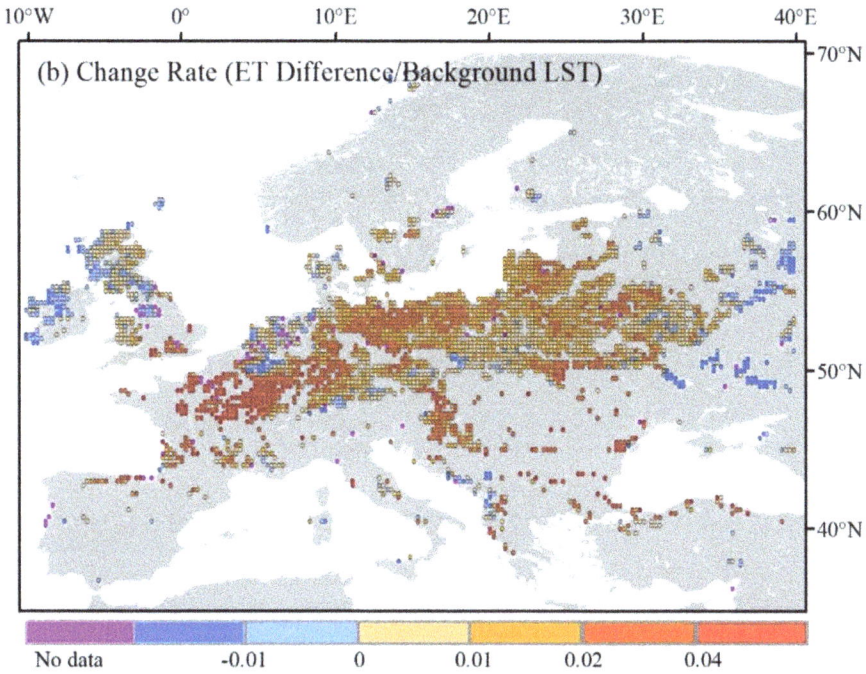

Figure 11. The relationship between the annual mean background LST and the annual mean (**a**) albedo (%) differences (forest minus open land) and (**b**) ET differences (forest minus open land) during various years. The number 0.1 in (**a**,**b**) indicates that the albedo and ET differences increase by 0.1% and 0.1 mm/day, respectively, when the background LST increases by 1 °C. Only when the valid year of a window was greater than five years was the relationship calculated.

4. Discussion

In this study, we use satellite observations and in situ measurements to examine the effects of forests on local surface temperatures. Overall, our results show that forests cool surface temperatures in Europe. Our findings agree with those of previous studies to a certain extent. Arora et al. proved that the bio-geophysical component of net temperature responses to 100% afforestation simulations was greater than zero in northeastern Europe and less than zero in other European regions [38]. Montenegro et al. studied the effect of afforestation on biochemical and biophysical processes using satellite data to investigate the effects of afforestation on temperature and proved that afforestation lowered the temperature in areas between 40°S and 60°N [39]. In this paper, an analysis based on measurements demonstrated that forests increase surface temperatures only in northeastern Europe and decrease temperatures in other European areas.

The diurnal asymmetry in temperature results from different energy balance processes, such as solar heating and radiative and dynamic cooling [40]. Daytime temperatures can be modified by incoming solar radiation, land surface properties (e.g., albedo and emissivity), latent and sensible heat fluxes, air mass advection and near-surface atmospheric boundary layer conditions [41]. Given the small size of a single grid cell, incoming solar radiation is likely to be similar in each grid cell. Thus, the amount of absorbed radiation is determined by surface albedo. The consumption of this energy by either latent and sensible heat fluxes or heat storage in soil and biomasses is controlled by vegetation activity and the soil moisture status [36]. In the daytime, due to deeper roots and larger leaf areas, forests have higher efficiencies in dissipating heat into the atmospheric boundary layer through turbulent diffusion than do open lands, which are aerodynamically smoother [42]. Forests absorb more solar energy and dissipate more energy than open land; these two processes determine the daytime ΔLST. Nighttime ET from vegetation is negligible; therefore, nighttime LST is mainly influenced by energy stored during the day and the near-surface atmospheric boundary layer. Lee et al (2011) hypothesized that forests tend to be warmer than open lands at night because forests are usually taller than open land, which enhances turbulence and draws heat from aloft towards the surface [7]. Several similar mechanisms have been proposed for wind farms and orchards, where machines are used to promote turbulence at night to warm the surface temperature [41,43]. Other factors (e.g., soil moisture, air humidity and boundary layer clouds) also help warm surface temperatures at night. The increase in soil moisture for forests tends to increase the surface heat capacity, which results in increases in daytime heat storage and nighttime heating. In addition, due to the higher ET from forests relative to open land, the increases in air humidity and boundary layer clouds increase the downward longwave radiation from the atmosphere and decrease the upward longwave radiation from the surface, which increases surface temperatures at night [9,15,17].

In our study, we found that background temperature was an important factor when determining the effects of forests not only in space but also at the interannual timescale. Forests tend to decrease the local temperature in warmer areas and increase the local temperature in colder areas, which generates a transitional latitude (56°N) and a transitional background temperature (near 6.5 °C) in Europe and causes the cooling effect to switch to a warming effect. Several previous studies that focused on North America and Asia also found a transitional temperature change latitudinally, which was different than that for Europe [7,14,18–20]. Latitudinal profiles of the zonal means were derived for air temperatures north of 10°N in North America, Asia, Europe and the Northern Hemisphere. These patterns were calculated via a monthly temperature data set with a spatial resolution of 0.5°, which was obtained from the Climate Research Unit (CRU), version TS3.22 [44–46]. We found that the annual mean air temperature (6.5 °C) occurs at 45°N in North America, 45°N in Asia, 56°N in Europe and 46°N in the Northern Hemisphere (Figure 12). Even though a different transitional latitude was observed in Europe than in those studies that observed different continents, the background temperatures were very close at these transitional latitudes.

There were record-breaking heatwaves in Europe in 2003 [30,47]. When calculating the average summer ΔLST in central Europe (45°–52°N, 5°–10°E) during the period 2003–2014, we found that the cooling effect of forests in the summer of 2003 was relatively larger than those in other years (Figure 13a), which was consistent with previous findings. For open lands, there was a significantly greater decrease in LAI, which is an important factor influencing ET that amplified the decrease in ET and resulted in a larger ΔET. This phenomenon was most remarkable in central Europe, where the largest number of severe heat waves have occurred. A long, continuous period with temperatures below 0 °C affected large parts of central Europe in 2005/2006, which resulted in a delayed snow melt in the spring of 2006 [37]. We also found that the average spring ΔLST in 2006 was higher than those in other years, which indicated that the cooling effect from forests decreased with colder background temperatures. The main reason for this phenomenon was that the difference in albedo (forest minus open land) became larger when there was a longer snow cover duration in spring, which led to a smaller cooling effect from forests (Figure 13b).

Background temperatures may be a major factor that can influence the effects of forests on temperature by altering biophysical processes, such as the differences in albedo and ET (forest minus open land). Under a warmer climate background, the warming effect caused by albedo is smaller due to reduced snow cover [15,27]. The difference in ET between forests and open lands may decrease in regions with a sufficient amount of soil moisture, but it increases when a region experiences soil moisture depletion [48]. Thus, the change in the cooling effect caused by ET has a greater uncertainty under a warmer background, and it may vary in space as the degree of the background temperature increases. In most areas of Europe, a decrease in the warming effect and an increase in the cooling effect from forest results in an increase in the total forest cooling effect under warmer background temperatures.

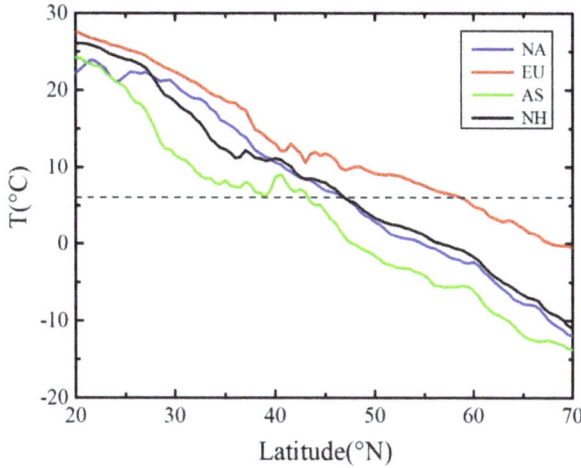

Figure 12. Annual mean air temperatures at different latitudes in the Northern Hemisphere (north of 10°N) during the period 2003–2016, where NA represents North America, EU represents Europe, AS represents Asia and NH represents the Northern Hemisphere. The dotted line represents 6.5 °C.

In our study, we combined evergreen forests and deciduous forests, which may result in several uncertainties. During the growing season, these two forests are not very different. During the non-growing season, transpiration from evergreen forests may be higher than that from deciduous forests [11]. This means that the cooling effect caused by ET in evergreen forests is greater than that in deciduous forests. However, compared with evergreen forests, deciduous forests are more easily covered with snow, which means that the difference in albedo between evergreen forests and open lands tends to be larger than that between deciduous forests and open lands [30]. The warming effect caused by albedo from evergreen forests is greater than that from deciduous forests. Although there may be different effects between evergreen forests and deciduous forests, the total difference in the effect on local temperature may be small due to the divergent effect (i.e., larger cooling from ET and larger warming from albedo in evergreen forests).

There are several other uncertainties and influential factors in our study. Although MODIS LST has relatively low uncertainties compared with most LST satellite data, there are still some unavoidable uncertainties (i.e., clouds) in the data set [49]. When we study the differences in LST between forests and open lands, uncertainties still exist. In our study, we ignore the measurement height differences between the forest flux sites and the meteorological stations, which may also cause some uncertainties. The height of the temperature measurement in the FLUXNET network varies from 2 to 15 m above the canopy, while the height of measurement at the meteorological stations is 2 m, which may change the annual mean temperature difference by a maximum of 0.008 °C [7]. In addition, we use temporal land cover data instead of long-term unchanged land cover data, which may cause some uncertainties in our results. In addition, numerous factors can affect biophysical processes, which determine the effects of forests on local temperature. Other factors, such as the leaf area index (LAI), root depth, soil moisture, and extreme climate and background temperatures can influence the effects of forests through biophysical processes.

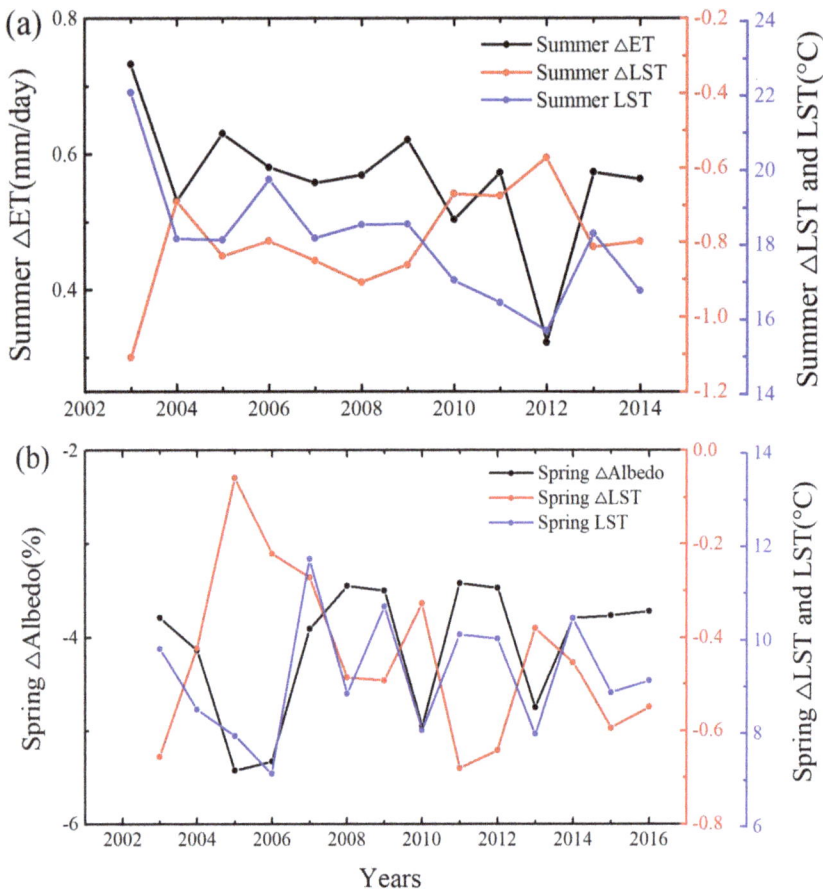

Figure 13. Interannual variability of (**a**) the seasonal mean LST in summer, the difference in ET (forest minus open land) in summer, and background LST in summer and (**b**) the seasonal mean LST in spring, the difference in albedo (forest minus open land) in spring, and the background LST in spring.

5. Conclusions

Throughout this work, we analyzed the effects of forests on LST and air temperature in Europe using remote sensing data and in situ measurements to reveal the geographic and seasonal patterns of this effect. Additionally, we explored the influence of background temperatures on the spatial and temporal patterns of forest effects on temperature in Europe.

Our results show that (1) forests generally cool the LST and air temperature in Europe, and the cooling effect varies in space and decreases with increases in latitude and longitude, which causes a switch to a warming effect in northeastern Europe; (2) daytime cooling dominates the effect of forests in the warm seasons. In the cold seasons, the daytime and nighttime warming effects drive the effects of forests; (3) background temperature plays a role in the effects of forests on local temperature, and there is a transitional temperature (6.5 °C) in Europe; and (4) the effects of forests are negatively correlated with the background temperature. For example, the cooling effect from forests are larger in heatwaves and smaller in cold springs. Furthermore, (5) background temperatures affect the spatiotemporal patterns of the differences in albedo and ET (forest minus open land), which determines the spatial, seasonal and interannual effects of forests on temperature. This study reveals that other factors (e.g., LAI, root depth, soil moisture and background temperature) may also have an influence on the effects of forests. Further studies should take more biophysical processes into account and combine satellite data, in situ measurements and model results to examine the effects of forests on surface temperature.

Overall, this study provides an empirical analysis to reveal the geographic and seasonal patterns caused by the effects of forests on surface temperatures and the role of background temperature in forest effects in Europe. Understanding the main drivers of climate responses to forests could provide essential information for adaptation strategies, especially under future climate conditions.

Supplementary Materials: The following are available online at http://www.mdpi.com/2072-4292/10/4/529/s1; Table S1: The information on the forest sites.

Acknowledgments: The authors thank Dr. Tao Zhou and Donghai Wu for their helpful comments, which improved this manuscript. This study was supported by the National Key Research and Development Program of China (No. 2016YFA0600103, No. 2016YFB0501404). We are grateful to the members of the FLUXNET community (http://www.fluxdata.org/DataInfo/default.aspx) and, in particular, the European Eddy Fluxes Database Cluster (http://gaia.agraria.unitus.it/home) for acquiring the forest flux data.

Author Contributions: Bijian Tang and Xiang Zhao conceived and designed the experiments. Bijian Tang and Wenqian Zhao processed and analyzed the data. All authors contributed to the design, writing, and discussion.

Conflicts of Interest: The authors declare no conflict of interest.

References

1. Bonan, G.B. Forests and climate change: Forcings, feedbacks, and the climate benefits of forests. *Science* **2008**, *320*, 1444–1449. [CrossRef] [PubMed]
2. Forzieri, G.; Alkama, R.; Miralles, D.G.; Cescatti, A. Satellites reveal contrasting responses of regional climate to the widespread greening of Earth. *Science* **2017**, *356*, 1180–1184. [CrossRef] [PubMed]
3. Zeng, Z.; Piao, S.; Li, L.Z.X.; Zhou, L.; Ciais, P.; Wang, T.; Li, Y.; Lian, X.; Wood, E.F.; Friedlingstein, P.; et al. Climate mitigation from vegetation biophysical feedbacks during the past three decades. *Nat. Clim. Chang.* **2017**, *7*, 432–436. [CrossRef]
4. Bright, R.M.; Davin, E.; O'Halloran, T.; Pongratz, J.; Zhao, K.; Cescatti, A. Local temperature response to land cover and management change driven by non-radiative processes. *Nat. Clim. Chang.* **2017**, *7*, 296–302. [CrossRef]
5. Davin, E.L.; de Noblet-Ducoudre, N. Climatic impact of global-scale Deforestation: Radiative versus nonradiative processes. *J. Clim.* **2010**, *23*, 97–112. [CrossRef]
6. Bright, R.M.; Antón-Fernández, C.; Astrup, R.; Cherubini, F.; Kvalevåg, M.; Strømman, A.H. Climate change implications of shifting forest management strategy in a boreal forest ecosystem of Norway. *Glob. Chang. Biol.* **2014**, *20*, 607–621. [CrossRef] [PubMed]
7. Lee, X.; Goulden, M.L.; Hollinger, D.Y.; Barr, A.; Black, T.A.; Bohrer, G.; Bracho, R.; Drake, B.; Goldstein, A.; Gu, L.; et al. Observed increase in local cooling effect of deforestation at higher latitudes. *Nature* **2011**, *479*, 384–387. [CrossRef] [PubMed]
8. Zhao, Q.; Wentz, E.A.; Murray, A.T. Tree shade coverage optimization in an urban residential environment. *Build. Environ.* **2017**, *115*, 269–280. [CrossRef]
9. Schultz, N.M.; Lawrence, P.J.; Lee, X. Global satellite data highlights the diurnal asymmetry of the surface temperature response to deforestation. *J. Geophys. Res. Biogeosci.* **2017**, *122*, 903–917. [CrossRef]
10. Burakowski, E.; Tawfik, A.; Ouimette, A.; Lepine, L.; Novick, K.; Ollinger, S.; Zarzycki, C.; Bonan, G. The role of surface roughness, albedo, and Bowen ratio on ecosystem energy balance in the Eastern United States. *Agric. For. Meteorol.* **2017**, *249*, 367–376. [CrossRef]
11. Juang, J.Y.; Katul, G.; Siqueira, M.; Stoy, P.; Novick, K. Separating the effects of albedo from eco-physiological changes on surface temperature along a successional chronosequence in the southeastern United States. *Geophys. Res. Lett.* **2007**, *34*, L21408. [CrossRef]

12. Luyssaert, S.; Jammet, M.; Stoy, P.C.; Estel, S.; Pongratz, J.; Ceschia, E.; Churkina, G.; Don, A.; Erb, K.; Ferlicoq, M.; et al. Land management and land-cover change have impacts of similar magnitude on surface temperature. *Nat. Clim. Chang.* **2014**, *4*, 389–393. [CrossRef]

13. Rigden, A.J.; Li, D. Attribution of surface temperature anomalies induced by land use and land cover changes. *Geophys. Res. Lett.* **2017**, *44*, 6814–6822. [CrossRef]

14. Li, Y.; Zhao, M.; Motesharrei, S.; Mu, Q.; Kalnay, E.; Li, S. Local cooling and warming effects of forests based on satellite observations. *Nat. Commun.* **2015**, *6*, 6603. [CrossRef] [PubMed]

15. Li, Y.; De Noblet-Ducoudré, N.; Davin, E.L.; Motesharrei, S.; Zeng, N.; Li, S.; Kalnay, E. The role of spatial scale and background climate in the latitudinal temperature response to deforestation. *Earth Syst. Dyn.* **2016**, *7*, 167–181. [CrossRef]

16. Longobardi, P.; Montenegro, A.; Beltrami, H.; Eby, M. Deforestation induced climate change: Effects of spatial scale. *PLoS ONE* **2016**, *11*, e0153357. [CrossRef] [PubMed]

17. Li, Y.; Zhao, M.; Mildrexler, D.J.; Motesharrei, S.; Mu, Q.; Kalnay, E.; Zhao, F.; Li, S.; Wang, K. Potential and Actual impacts of deforestation and afforestation on land surface temperature. *J. Geophys. Res. Atmos.* **2016**, *121*, 14372–14386. [CrossRef]

18. Peng, S.-S.; Piao, S.; Zeng, Z.; Ciais, P.; Zhou, L.; Li, L.Z.X.; Myneni, R.B.; Yin, Y.; Zeng, H. Afforestation in China cools local land surface temperature. *Proc. Natl. Acad. Sci. USA* **2014**, *111*, 2915–2919. [CrossRef] [PubMed]

19. Zhang, M.; Lee, X.; Yu, G.; Han, S.; Wang, H.; Yan, J.; Zhang, Y.; Li, Y.; Ohta, T.; Hirano, T.; et al. Response of surface air temperature to small-scale land clearing across latitudes. *Environ. Res. Lett.* **2014**, *9*, 34002. [CrossRef]

20. Ma, W.; Jia, G.; Zhang, A. Multiple satellite-based analysis reveals complex climate effects of temperate forests and related energy budget. *J. Geophys. Res. Atmos.* **2017**, *122*, 3806–3820. [CrossRef]

21. Wickham, J.; Wade, T.G.; Riitters, K.H. An isoline separating relatively warm from relatively cool wintertime forest surface temperatures for the southeastern United States. *Glob. Planet. Chang.* **2014**, *120*, 46–53. [CrossRef]

22. Wickham, J.D.; Wade, T.G.; Riitters, K.H. Comparison of cropland and forest surface temperatures across the conterminous United States. *Agric. For. Meteorol.* **2012**, *166–167*, 137–143. [CrossRef]

23. Tudoroiu, M.; Eccel, E.; Gioli, B.; Gianelle, D.; Schume, H.; Genesio, L.; Miglietta, F. Negative elevation-dependent warming trend in the Eastern Alps. *Environ. Res. Lett.* **2016**, *11*, 44021. [CrossRef]

24. Zhao, Q.; Myint, S.W.; Wentz, E.A.; Fan, C. Rooftop surface temperature analysis in an Urban residential environment. *Remote Sens* **2015**, *7*, 12135–12159. [CrossRef]

25. Seneviratne, S.I.; Corti, T.; Davin, E.L.; Hirschi, M.; Jaeger, E.B.; Lehner, I.; Orlowsky, B.; Teuling, A.J. Investigating soil moisture-climate interactions in a changing climate: A review. *Earth-Sci. Rev.* **2010**, *99*, 125–161. [CrossRef]

26. Pitman, A.J.; Avila, F.B.; Abramowitz, G.; Wang, Y.P.; Phipps, S.J.; de Noblet-Ducoudré, N. Importance of background climate in determining impact of land-cover change on regional climate. *Nat. Clim. Chang.* **2011**, *1*, 472–475. [CrossRef]

27. Winckler, J.; Reick, C.H.; Pongratz, J. Why does the locally induced temperature response to land cover change differ across scenarios? *Geophys. Res. Lett.* **2017**, *44*, 3833–3840. [CrossRef]

28. Armstrong, E.; Valdes, P.; House, J.; Singarayer, J. The role of CO_2 and dynamic vegetation on the impact of temperate land-use change in the HadCM3 coupled climate model. *Earth Interact.* **2016**, *20*. [CrossRef]

29. Gálos, B.; Hagemann, S.; Hänsler, A.; Kindermann, G.; Rechid, D.; Sieck, K.; Teichmann, C.; Jacob, D. Case study for the assessment of the biogeophysical effects of a potential afforestation in Europe. *Carbon Balance Manag.* **2013**, *8*, 3. [CrossRef] [PubMed]

30. Teuling, A.J.; Seneviratne, S.I.; Stöckli, R.; Reichstein, M.; Moors, E.; Ciais, P.; Luyssaert, S.; van den Hurk, B.; Ammann, C.; Bernhofer, C.; et al. Contrasting response of European forest and grassland energy exchange to heatwaves. *Nat. Geosci.* **2010**, *3*, 722–727. [CrossRef]

31. Deng, X.; Zhao, C.; Yan, H. Systematic modeling of impacts of land use and land cover changes on regional climate: A review. *Adv. Meteorol.* **2013**, *2013*. [CrossRef]

32. Naudts, K.; Chen, Y.; McGrath, M.J.; Ryder, J.; Valade, A.; Otto, J.; Luyssaert, S. Europes forest management did not mitigate climate warming. *Science* **2016**, *351*, 597–600. [CrossRef] [PubMed]

33. Williamson, S.; Hik, D.; Gamon, J.; Kavanaugh, J.L.; Flowers, G.E. Estimating temperature fields from MODIS land surface temperature and air temperature observations in a sub-arctic alpine environment. *Remote Sens.* **2014**, *6*. [CrossRef]

34. Wan, Z. New refinements and validation of the collection-6 MODIS land-surface temperature/emissivity product. *Remote Sens. Environ.* **2014**, *140*, 36–45. [CrossRef]

35. Schaaf, C.B.; Gao, F.; Strahler, A.H.; Lucht, W.; Li, X.; Tsang, T.; Strugnell, N.C.; Zhang, X.; Jin, Y.; Muller, J.P.; et al. First operational BRDF, albedo nadir reflectance products from MODIS. *Remote Sens. Environ.* **2002**, *83*, 135–148. [CrossRef]

36. Mu, Q.; Zhao, M.; Running, S.W. Improvements to a MODIS global terrestrial evapotranspiration algorithm. *Remote Sens. Environ.* **2011**, *115*, 1781–1800. [CrossRef]

37. Dietz, A.J.; Wohner, C.; Kuenzer, C. European snow cover characteristics between 2000 and 2011 derived from improved modis daily snow cover products. *Remote Sens.* **2012**, *4*, 2432–2454. [CrossRef]

38. Arora, V.K.; Montenegro, A. Small temperature benefits provided by realistic afforestation efforts. *Nat. Geosci.* **2011**, *4*, 514–518. [CrossRef]

39. Montenegro, A.; Eby, M.; Mu, Q.; Mulligan, M.; Weaver, A.J.; Wiebe, E.C.; Zhao, M. The net carbon drawdown of small scale afforestation from satellite observations. *Glob. Planet. Chang.* **2009**, *69*, 195–204. [CrossRef]

40. Hain, C.R.; Anderson, M.C. Estimating morning change in land surface temperature from MODIS day/night observations: Applications for surface energy balance modeling. *Geophys. Res. Lett.* **2017**, *44*, 9723–9733. [CrossRef] [PubMed]

41. Zhou, L.; Tian, Y.; Baidya Roy, S.; Thorncroft, C.; Bosart, L.F.; Hu, Y. Impacts of wind farms on land surface temperature. *Nat. Clim. Chang.* **2012**, *2*, 539–543. [CrossRef]

42. Rotenberg, E.; Yakir, D. Distinct patterns of changes in surface energy budget associated with forestation in the semiarid region. *Glob. Chang. Biol.* **2011**, *17*, 1536–1548. [CrossRef]

43. Tang, B.; Wu, D.; Zhao, X.; Zhou, T.; Zhao, W.; Wei, H. The Observed Impacts of Wind Farms on Local Vegetation Growth in Northern China. *Remote Sens.* **2017**, *9*, 332. [CrossRef]

44. Wu, D.; Zhao, X.; Liang, S.; Zhou, T.; Huang, K.; Tang, B.; Zhao, W. Time-lag effects of global vegetation responses to climate change. *Glob. Chang. Biol.* **2015**, *21*, 3520–3531. [CrossRef] [PubMed]

45. Zhao, W.; Zhao, X.; Zhou, T.; Wu, D.; Tang, B.; Wei, H. Climatic factors driving vegetation declines in the 2005 and 2010 Amazon droughts. *PLoS ONE* **2017**, *12*. [CrossRef] [PubMed]

46. Harris, I.; Jones, P.D.; Osborn, T.J.; Lister, D.H. Updated high-resolution grids of monthly climatic observations—The CRU TS3.10 Dataset. *Int. J. Climatol.* **2014**, *34*, 623–642. [CrossRef]

47. Bevan, S.L.; Los, S.O.; North, P.R.J. Response of vegetation to the 2003 European drought was mitigated by height. *Biogeosciences* **2014**, *11*, 2897–2908. [CrossRef]

48. Anderson, R.G.; Canadell, J.G.; Randerson, J.T.; Jackson, R.B.; Hungate, B.A.; Baldocchi, D.D.; Ban-Weiss, G.A.; Bonan, G.B.; Caldeira, K.; Cao, L.; et al. Biophysical considerations in forestry for climate protection. *Front. Ecol. Environ.* **2011**, *9*, 174–182. [CrossRef]

49. Hulley, G.C.; Hughes, C.G.; Hook, S.J. Quantifying uncertainties in land surface temperature and emissivity retrievals from ASTER and MODIS thermal infrared data. *J. Geophys. Res. Atmos.* **2012**, *117*. [CrossRef]

remote sensing

MDPI

Article

New Scheme for Validating Remote-Sensing Land Surface Temperature Products with Station Observations

Wenping Yu [1,*], Mingguo Ma [1,*] , Zhaoliang Li [2], Junlei Tan [3] and Adan Wu [3]

1   Chongqing Engineering Research Center for Remote Sensing Big Data Application, School of Geographical Sciences, Southwest University, No. 2 Tiansheng Road, Beibei District, Chongqing 400715, China
2   ICube, Uds, CNRS (UMR7357), 300 Bld Sebastien-Brant, CS10413, 67412 Illkirch, France; lizl@unistra.fr
3   Heihe Remote Sensing Experimental Research Station, Northwest Institute of Eco-Environment and Resources, Chinese Academy of Sciences, 320 Donggang West Road, Lanzhou 730000, China; tanjunlei@lzb.ac.cn (J.T.); wuadan@lzb.ac.cn (A.W.)
*   Correspondence: ywpgis2005@swu.edu.cn (W.Y.); mmg@swu.edu.cn (M.M.)

Received: 27 September 2017; Accepted: 20 November 2017; Published: 24 November 2017

Abstract: Continuous land-surface temperature (LST) observations from ground-based stations are an important reference dataset for validating remote-sensing LST products. However, a lack of evaluations of the representativeness of station observations limits the reliability of validation results. In this study, a new practical validation scheme is presented for validating remote-sensing LST products that includes a key step: assessing the spatial representativeness of ground-based LST measurements. Three indicators, namely, the dominant land-cover type (DLCT), relative bias (RB), and average structure scale (ASS), are established to quantify the representative levels of station observations based on the land-cover type (LCT) and LST reference maps with high spatial resolution. We validated MODIS LSTs using station observations from the Heihe River Basin (HRB) in China. The spatial representative evaluation steps show that the representativeness of observations greatly differs among stations and varies with different vegetation growth and other factors. Large differences in the validation results occur when using different representative level observations, which indicates a large potential for large error during the traditional T-based validation scheme. Comparisons show that the new validation scheme greatly improves the reliability of LST product validation through high-level representative observations.

Keywords: spatial representativeness; heterogeneity; validation; land-surface temperature products (LSTs); observations; HiWATER; remote sensing

---

1. Introduction

Land-surface temperature (LST) is an important parameter related to the surface energy and water balance at local and global scales and has principal significance for applications such as monitoring the climate, hydrological cycle, and vegetation [1]. Satellite remote sensing provides a repetitive synoptic view in short intervals of the global land surface and is a vital tool for monitoring the LST of the Earth. With the development of remote-sensing technology, many LST products have been provided by different groups based on retrieval from different satellite data [2–5]. The first long-term global sensing LST dataset, the NOAA/NASA Pathfinder AVHRR Land dataset (PAL) [2], was released in 1994. The second generation AVHRR Land Pathfinder II (PALII) was a refinement product from the PAL released in 2000 [3]. Sun and Pinker estimated LST products from a Geostationary Operational Environmental Satellite (GEOS) in 2003 [4]. The LSTs from Spinning Enhanced Visible and Infrared Imager (SEVIRI) onboard the Meteosat Second Generation (MSG) was retrieved in 2008 [6]. As a part of NASA Earth Observing System (EOS) project, MODIS LSTs have played an important role in recent studies, especially in regional studies, because of the suitable temporal and spatial resolution, acceptable accuracy, and accessibility of these LSTs. Therefore, MODIS daily LST/LSE products with 1-km spatial resolution are validated in this study.

Remotely sensed LSTs must be appropriately and precisely evaluated to ensure effective application [7]. Mainly two types of methods exist for validating LST products retrieved from thermal-infrared satellite data: temperature-based methods (T-based) and radiance-based methods (R-based) [8–14]. The main advantage of R-based methods is that they work during both the daytime and nighttime because in situ LST observations are not required, and finding validation sites with small spatial variations in land-surface emissivity is relatively easy [14]. However, the atmospheric and water vapor profiles at validation sites from radiosonde balloons that are synchronously launched with the satellite are a necessary dataset, which limits the implementation of this method for long-term and large-region validation. Therefore, T-based methods remain common, and ground-based measurements are still the primary source of datasets to directly validate remotely sensed LSTs. However, we cannot perform a direct comparison with a pixel grid, especially for a coarse-resolution product over heterogeneous areas, because of the spatial heterogeneity and different scales between ground-based observations and remotely sensed LST pixels.

A generally accepted method is a systematic site-to-network method, which deeply develops an in situ sampling strategy and upscaling approaches to acquire the truth at the pixel scale over a heterogeneous surface based on multiscale, multi-platform and multi-source observations [15]. This approach employs both field measurements from nodes of a wireless sensor network (WSN) and high-resolution remote-sensing data from synchronous high-resolution satellites or airborne sensors to establish a site-specific relationship and generate high-resolution LST reference maps over the validation area [15]. These LST reference maps are then treated as benchmarks to obtain multiscale validation datasets by upscaling methods [16–18]. However, only a few high-resolution LST reference maps can be synchronously obtained for a given region because of cost limitations and the revisiting cycles of satellites with high resolution, which is the greatest challenge towards the global validation of LST products, especially in terms of temporal consistency in product validation.

In contrast to more complicated R-based, site-to-network methods with limited LST reference maps, simple T-based methods are directly based on existing global, long-term ground LST measurements and are an important supplement for validation. Simple T-based methods have been widely used to validate remotely sensed LST products at homogeneous stations [8,13,19]. When directly validating LST products with spatial resolutions above hundreds or even thousands of meters by ground-based measurements, the error from the scale mismatch changes with the land-cover type (LCT) and the proportions of mixtures in pixel grids reduce the reliability of the validation and hinder the application of ground-based LST measurements during the validation of remotely sensed LST products. Coll et al. have pointed out that during the day, LST can vary by 10 K or more over a few meters in a heterogeneous surface [9]. Ground-based LST measurements from two types LST observation instruments with different field of view (FOV) were selected to discuss the scale mismatch implications for validation of remote sensing LST products in the study by Yu et al., and the validation results show that there is an extra 26.9% in the error >3 K range caused by the 41.5 FOV difference [20]. Therefore, we must assess the spatial representativeness of station observations at a given spatial resolution to reliably validate remotely sensed LSTs. Recently, several methods have been used to assess the spatial representativeness of different land-surface parameters, such as the leaf area index [21], surface solar radiation [22], bidirectional reflectance distribution function (BRDF)/albedo [23], air temperature [24] and air quality [25], which are observed by ground stations. These methods are based on two factors: the point-to-area consistency and the spatial heterogeneity [21]. The point-to-area consistency indicator can be calculated through two methods. The first involves directly comparing the footprint of the ground-station observations to the corresponding product pixels [26] or the average value of the corresponding area [27]. In the second approach, the observational representativeness is determined by the average difference between a given station and its neighboring stations based on multi-temporal observations from multiple stations [9,28]. The semi-variance is usually selected to describe the spatial representativeness by analyzing the spatial heterogeneity around the stations [29]. A first-order statistical algorithm is an important spatial heterogeneity indicator, for example, using window-size analysis to assess the spatial variation of the landscape around a given station [30]. Spatial representativeness assessments have been widely implemented to validate satellite-albedo, evapotranspiration, and LAI products [21,23,26,31,32]. However, few representativeness assessments exist for station LST observations, which increases the uncertainty of the validation of LST products, particularly for simple T-based implementations, and hinders the application of station observations.

This paper presents a new methodology for validating LST products that focuses on quantifying the spatial representativeness of station observations to improve the accuracy and reliability of LST product validation. The term "spatial representativeness" refers to measurements of the degree to which ground-based observations can resolve the surrounding LST by extending to the satellite footprint. This validation technique assesses the spatial characteristics of the LST, and the seasonal representativeness changes within a statistical framework. A scheme that is based on spatial representativeness indicators is presented in this paper, and then the grading criteria are outlined in detail. All the stations from the Heihe Watershed Allied Telemetry Experimental Research (HiWATER) [33] are selected for applying the validation strategy. The study area and data-processing procedure are introduced in Section 3. In Section 4, the representativeness of the given station observations is assessed to validate MODIS V5 daily LST products (MOD/MYD11A1). The representativeness assessment and LST product validation are also analyzed and discussed in this Section. Finally, the conclusions are summarized.

2. Methodology

2.1. New Validation Scheme

The LST is a land-surface parameter with great spatial and temporal heterogeneity, which creates many challenges for "point-to-pixel" comparisons. Local changes in the surface temperature within and between different ecosystems introduce scale mismatch errors. Moreover, these patterns change seasonally and are particularly difficult to identify during periods of rapidly changing surface conditions. Therefore, "point" measurements alone are not sufficient to validate satellite-derived LST retrievals, especially remote-sensing LST products with moderate and low resolution (illustrated in Figure 1). This temporal mismatch can be solved by increasing the observation-acquisition frequency of stations. The scheme that is developed here to validate remote-sensing LST products (see Figure 1) attempts to solve spatial mismatch effects during validation. The key in this scheme is to assess the spatial representativeness, which is based on remote-sensing data with high spatial resolution that are closely related to the LST. Indicators are proposed to quantify the spatial representativeness, and then the grading criteria are designed before selecting appropriate ground-based measurements for validation.

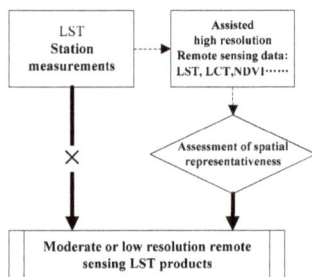

Figure 1. New scheme for land surface temperature (LST) validation based on the assessment of local spatial representativeness (site level). In the scheme, LCT is the abbreviation of land-cover type, and NDVI is the abbreviation of normalized difference vegetation index.

2.2. Indicators for Assessing Spatial Representativeness

Three indicators are proposed to describe the spatial characteristics for a specific parameter, mainly including the consistency from points to pixels and the spatial heterogeneity within pixels. These indicators are calculated on high-resolution images, which are much easier to access, thus simplifying the process. The LST is a direct parameter for assessing the representativeness of a given station's LST observations. However, obtaining temporal high-spatial-resolution LST matches for LST products with lower resolution is difficult. Therefore, high-resolution land-cover type (LCT) and the normalized difference vegetation index (NDVI), which can indicate the surface conditions and their changes, are chosen as additional supporting parameters to evaluate ground-based LST observations in addition to LST data with a high spatial resolution.

If the LCTs observed by stations do not match the dominant LCTs in the pixels, these station LST observations cannot represent the LST of the LCTs in the pixels. Thus, the dominant LCT (DLCT), which is given by the percentage of the observed LCT throughout the pixel's area, is defined as

$$\text{DLCT} = \frac{M(s)}{N(s)} \times 100 \tag{1}$$

where s is the product pixel, $M(s)$ is the area with the LCT that is observed by the given station, and $N(s)$ is the total area of the LCTs in the LST product's pixel grid. When using a high-resolution LCT map, the DLCT can also be described as the percentage of fine pixel numbers covered by station-observed LTC to total LCT pixel numbers in the LST product's pixel range. A high DLCT indicates that the LCT observed by a station is consistent with that in the LST product's pixel and low heterogeneity within the product pixel because the mixing rate of LCTs in the pixel is not large.

We developed a relative bias (RB) indicator to assess how close a ground-based LST measurement is to the value of the corresponding pixel area. According to the high-spatial-resolution LST reference images, the relative bias is used to describe the difference between the LST value $T(s)$ at a station and the average LST value $\overline{T}(s)$ in the product pixel's area. If we consider the comparability between different ranges of LST values, the RB is defined as

$$\text{RB} = \frac{|T(s) - \overline{T}(s)|}{\overline{T}(s)} \times 100 \tag{2}$$

where s is the product pixel and RB depends on both its resolution and the resolution of the reference LST image. This indicator can quantify the certainty of the ground-based measurements to the product pixel area LST values. A smaller RB indicates more spatially representative observations for the corresponding pixel at the specific spatial resolution of s.

The two above indicators mainly measure the point-to-area value consistency, so the heterogeneity of the spatial distribution of the LST within a pixel, which is correlated with the vegetation growth, must be seasonally quantified. In terms of the spatial autocorrelation of LST parameters, semivariogram is one of the most commonly used and efficient geostatistical analysis tools for quantitatively evaluating spatial variations. Semi-variance and related geostatistical kriging were developed from mining research during the late 1950s and have been widely used after a publication by Journel and Huijbregts in 1978 [34,35]. These geostatistical techniques have been used in many scientific projects, such as in describing the distribution and density of plants and animals [36,37] and in determining the spatial scales of variation and sampling strategies in remote sensing [38–40]. In regionalized variable theory, the semi-variance measures the dissimilarity of a spatial variable observed at different locations. The semi-variance is calculated by the average squared difference between observations $Z(x_i)$ and $Z(x_j)$, which are separated by distance h, as described below:

$$r(h) = \frac{1}{2N(h)} \sum_{||x_i - x_j|| = h} (Z(x_i) - Z(x_j))^2 \tag{3}$$

where $2N(h)$ is the number of observation pairs, which are separated by a distance h, or lag, as intervals to calculate the semi-variance. In this study, the variogram estimator $r(h)$ is computed on discretized point values from high-spatial resolution LST pixels, and then a variogram model is established as a parametric functional approximation based on these semi-variance values. Several theoretical variogram model types exist, including linear models, spherical models, exponential models, and Gaussian models. Among these models, spherical models are the most widely used variogram models for their strong fitting and generalization capabilities and are recommended for assessing the spatial representativeness of observations [23,32]. The isotropic spherical variogram that is used to estimate the variogram is as follows:

$$ r_{sph}(h) = \begin{cases} c_0 + c \times \left(1.5 \times \frac{h}{a} - 0.5 \times \left(\frac{h}{a}\right)^3 \right) & for\ 0 \le h \le a \\ c_0 + c\ for\ h > a \end{cases} \tag{4} $$

In Equation (4), it is obvious that the $r_{sph}(h)$ generally increases from a nonzero value to a relatively stable constant value with h. When $h = 0$, the $r_{sph}(h)$ value is a nonzero value c_0, namely, $r_{sph}(0) = c_0$, and c_0 is the nugget of the variogram. The stable constant value is the sill $(c_0 + c)$ of the variogram. When $r_{sph}(h)$ reaches the sill, the value of the variable h is a, namely, the range of the variogram. The key parameters range (a), nugget (c_0), and sill $(c_0 + c)$ can be obtained from fitting Equation (4). a is the maximal distance between the two correlated points and indicates the average structural scale (ASS) of the given area. The nugget c_0 indicates the level of $Z(x_i)$'s randomness, which may be caused by internal variations in $Z(x_i)$ over a smaller distance h than the sampling distance or may be derived from the sampling error. The sill $(c_0 + c)$ represents the largest extent of the regionalized variation. Li and Reynolds [41] introduced the proportion of structural variation, which is based on subtracting the variogram nugget (c_0) from the sill (c) and then dividing by the sill, to discuss the definition and quantification of ecological heterogeneity. However, the heterogeneity of LST parameters considerably varies over time when compared to that of ecological parameters, and reference high-resolution LST maps may not be completely synchronous with the validated LST products. Therefore, the ASS is introduced based on the range of the variogram model, which reflects the average structural scale of the given area and the size of the homogeneous area. A large ASS value indicates that the station observations represent a large homogeneous area.

3. Data Instruction and Preparation

3.1. Ground-Based LST Measurements

In this study, we selected the Heihe River Basin (HRB), which is the second largest inland valley in China's arid regions, to evaluate MODIS LSTs with 1-km resolution. The HRB is located in the northern arid region within 97.1°E–102.0°E and 37.7°N–42.7°N. Glaciers, frozen soils, alpine meadows, forests, irrigated crops, riparian ecosystems, deserts, and gobi are distributed from upstream to downstream regions (see Figure 2). The HRB was selected as an experimental watershed to reveal the processes and mechanisms of the ecohydrological system in an inland river basin. Allied telemetry experiments such as the Heihe Basin Field Experiment (HEIFE) [42] and Watershed Allied Telemetry Experimental Research (WATER) [43] have been conducted in the HRB, and the HiWATER [33] project has been ongoing since 2012. The stations that collect watershed hydrological observations cover a wider range than those in previous studies and provide a large amount of ecohydrological data for evaluation. Eighteen atmospheric stations from HiWATER are scattered around the HRB region. Longwave-radiation data for eighteen stations from 2013 to 2014 were selected to obtain ground-based LSTs to evaluate the MODIS LSTs. The locations of the stations are shown in Figure 2. The information for the stations is listed in Table 1, and environmental photos of these sites are shown in Figure 3. The ARC, ARS, ARY, DSL, JYL, HZS, HCG, and EBZ stations are located in the upstream area; the DMZ, GBZ, HZZ, SDZ, and SSW stations are located in the midstream area; and the downstream area contains the SDQ, HJL, HYZ, NTZ, and LTZ stations. The LCTs of these stations are all typical types in the three significantly different areas.

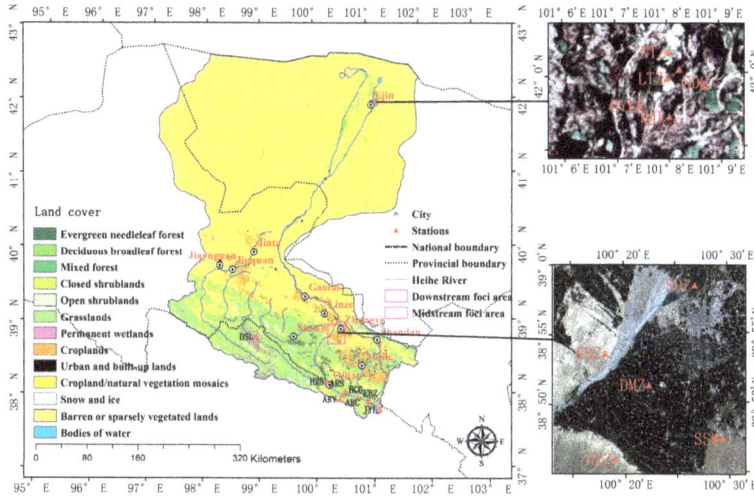

Figure 2. Study area and locations of the validation stations.

Table 1. Information for the stations in this study.

Station Name	Longitude/°E	Latitude/°N	Altitude/m	Height/m	Footprint [1]/m	Landscapes
A'rou superstation (ARC)	100.46	38.05	3033	5	37.32	Alpine meadow
A'rou sunny slope station (ARS)	100.52	38.09	3559	6	44.78	Alpine grassland
A'rou shade station (ARY)	100.42	37.99	3538	6	44.78	Alpine grassland
Dashalong station (DSL)	98.95	38.83	3775	6	44.78	Swamp meadow
Jinyangling station (JYL)	101.11	37.85	3700	6	44.78	Alpine meadow
Huangzangsi station (HZS)	100.19	38.23	2660	6	44.78	Cropland (wheat)
Huangcaogou station (HCG)	100.73	38.00	3186	6	44.78	Alpine grassland
E'bo station (EBZ)	100.94	37.96	3407	6	44.78	Alpine grassland
Daman superstation (DMZ)	100.37	38.86	1519	12	89.57	Cropland (maize)
Gobi station (GBZ)	100.30	38.89	1571	6	44.78	Gobi Desert
Huazhaizi desert station (HZZ)	100.32	38.77	1726	2.5	18.66	Desert steppe
Wetland station (SDZ)	100.45	38.97	1460	6	44.78	Wetland
Shenshawo desert station (SSW)	100.49	38.79	1582	6	44.78	Desert
Populus forest station (HYZ)	101.12	41.99	927	24	179.14	Populus forest
Cropland station (NTZ)	101.13	42.01	919	6	44.78	Cropland
Barren-land station (LTZ)	101.13	41.99	931	6	44.78	Bare soil
Sidaoqiao station (SDQ)	101.12	41.99	935	10	74.64	Euphrates poplar olive and Tamarix mixed forest
Mixed forest station (HJL)	101.13	41.99	929	24	179.14	Euphrates poplar olive and Tamarix mixed forest

[1] "Footprint" refers to the diameter of the footprint; "Height" indicates the installation height.

Eighteen meteorological towers are located in the HRB region, consisting of three superstations and fifteen ordinary stations. Pyrgeometers are deployed at these 10-m to 35-m high meteorological station towers to measure longwave radiation (see Figure 3 and Table 1). At least two pyrgeometers are positioned on a single tower: one facing upward and the other facing downward. The field-of-view

(FOV) of the upward-facing pyrgeometer is nearly 180°, while that of the downward-facing pyrgeometer is 150°. Therefore, the effective diameter of the FOV of the pyrgeometers on a 10-m to 35-m tower is approximately 2.5–24 m with a 6-m average mounting height, and the diameters of the ground-observation footprints are shown in Table 1. The pyrgeometers are sensitive to the spectral range from 4.5 to 42 μm in the longwave band. All the instruments at each station were calibrated before and after field deployment. Field-routing exams were implemented once per month. Assurance and quality control provided the best possible data for the level-2 daily data. All these data and related information can be found at the HiWATER website [44]. All the ground-based measured data from the eighteen HiWATER sites were 10-min averaged values. The longwave radiation data were selected according to the field viewing time for the MODIS LST. The LST is related to land-surface longwave radiation according to the Stefan-Boltzmann law [1,45]:

$$L_\uparrow = \varepsilon_b \delta T^4 + (1 - \varepsilon_b) \times L_\downarrow \tag{5}$$

where $L_\uparrow$ is the surface upwelling longwave radiation, ε_b is the surface broadband emissivity, δ is Stefan-Boltzmann's constant (5.67×10^{-8} W·m^{-2}·K^{-4}), and $L_\downarrow$ is the atmospheric downwelling longwave radiation at the surface. Therefore, the ground-measured LST can be estimated from station longwave-radiation observations by the following equation:

$$T_s = \left[\frac{L_\uparrow - (1 - \varepsilon_b) \times L_\downarrow}{\varepsilon_b \delta} \right]^{\frac{1}{4}} \tag{6}$$

In Equation (6), $L_\uparrow$ and $L_\downarrow$ are obtained from the ground-based measurements. Seven narrowband emissivities exist in MOD/MYD11B1 LST/LSE products. ε_b can be estimated from these MODIS narrowband emissivities [45]:

$$\varepsilon_b = 0.2122 \times \varepsilon_{29} + 0.3859 \times \varepsilon_{31} + 0.4029 \times \varepsilon_{32} \tag{7}$$

where ε_b is the broadband emissivity, and ε_{29}, ε_{31}, and ε_{32} are the narrow emissivity products of MODIS bands 29, 31, and 32 that are retrieved from the MODIS day/night LST algorithm (MOD/MYD11B1 LST/LSE).

Figure 3. *Cont.*

252

Figure 3. Photos of the 18 Heihe Watershed Allied Telemetry Experimental Research (HiWATER) stations. The two pyrgeometers that were used to record the longwave radiation were deployed at an average height of 6 m, with one facing upwards and the other facing downwards.

3.2. Remote-Sensing Data

3.2.1. MODIS Data

As a component of NASA's Earth Observing System (EOS) project, two MODIS instruments were placed onboard the Terra and Aqua satellite platforms to provide information for global atmosphere-, land- and oceanic-process studies [46]. MOD/MYD22_L2, MOD/MYD11A1, MOD/MYD11B1 and MOD/MYD07_L2 are all daily LST products based on thermal-infrared data from MODIS. MOD/MYD11_L2 was retrieved by a generalized split-window algorithm with 1-km spatial resolution [47]. MOD/MYD11A1 is tile-based and gridded in the sinusoidal projection from MOD/MYD11_L2. MOD/MYD11B1 was obtained using the day/night LST algorithm at 5-km spatial resolution [48]. MOD/MYD07_L2 was retrieved by the atmospheric team using statistical regression methods [49]. In this study, we focus on the collection of 5 MOD/MYD11A1 products, which are more widespread. Uncertainties from the satellite measurements and improvements in the original

MODIS LSTs for cloudy days are beyond the scope of this paper. To eliminate effects from the inversion algorithm and clouds, only pixels with high-quality MODIS LSTs were selected for the evaluation based on a quality control flag value of 0. The narrow emissivity products from MOD/MYD11B1 LST/LSE were selected to estimate the land-surface broadband emissivity ε_b and obtain ground-based LSTs [45]. The narrow emissivities from MOD/MYD11B1 LST/LSE at 5-km resolution were resampled to 1-km spatial resolution to match the evaluated MODIS LSTs.

3.2.2. High-Spatial-Resolution Images

Land-cover images with a spatial resolution of 30 m (data doi:10.3972/hiwater.155.2014.db, downloaded from the HiWATER land cover datasets [50], which were produced by Zhong et al. [51,52], were selected to obtain the DLCT in a 1-km LST product pixel. This dataset was mainly based on charge-coupled device (CCD) data from the Huan Jing 1 (HJ-1) satellite, which was launched on 6 September 2008, by the China Center for Resources Satellite Data and Application (CRESDA). HJ-1/CCD has three visible bands and one near-infrared band [53]. This dataset provides monthly land-cover maps of the HRB from 2011 to 2015 with 30-m spatial resolution. LCTs change with the seasons, and these changes are similar across consecutive years, so the DLCT products in 2013 were collected to calculate the DLCT indicator.

Landsat 8, which is called the Landsat Data Continuity Mission, is extending the distinguished 40-year records of Landsat-series satellites and has enhanced capabilities, such as adding new spectral bands in the visible and thermal-infrared wavelengths and improving the signal-to-noise ratio and radiometric resolution of the sensor [54]. The Landsat 8 satellite includes two instruments: an Operational Land Imager (OLI) and Thermal Infrared Sensor (TIRS). High-resolution LST maps were retrieved from the Landsat 8 TIRS data and OLI data. Before retrieving the LST maps, the Landsat OLI and TIRS images were preprocessed, including radiometric calibration and atmospheric correction based on the correction model in the ENVI software. Monthly NDVI maps were obtained to assess the relationships between the monthly changes in the indicators and vegetation growth. The NDVI maps were based on the visible red band (R, band 4) and the near-infrared band (NIR, band 5) according to the following equation:

$$\text{NDVI} = \frac{NIR - R}{NIR + R} \tag{8}$$

In this study, the LST data were estimated from the TIRS aboard Landsat 8 based on a practical split-window (SW) algorithm developed by Du et al. [55]. The SW algorithm can be expressed as follows:

$$T = b_0 + \left(b_1 + b_2\frac{1-\varepsilon}{\varepsilon} + b_3\frac{\Delta\varepsilon}{\varepsilon^2}\right)\frac{T_i + T_j}{2} + \left(b_4 + b_5\frac{1-\varepsilon}{\varepsilon} + b_6\frac{\Delta\varepsilon}{\varepsilon^2}\right)\frac{T_i - T_j}{2} + b_7\left(T_i - T_j\right)^2 \tag{9}$$

where T is LST, T_i and T_j are the TOA brightness temperatures in the thermal-infrared channels i and j, respectively; ε is the average emissivity of the two channels (i.e., $\varepsilon = 0.5(+\varepsilon_j)$); $\Delta\varepsilon$ is the channel emissivity difference (i.e., $\varepsilon = 0.5(\varepsilon_i - \varepsilon_j)$); and $b_k(k = 0, 1, \ldots 7)$ are the algorithm coefficients from the simulated dataset. In this algorithm, the coefficients were determined based on atmospheric water-vapor subranges, which were obtained through a modified split-window covariance-variance ratio method. The channel emissivities were acquired from newly released global land-cover products at 30 m and a fraction of the calculated vegetation cover from visible and near-infrared images that were obtained by Landsat 8.

The effect of heterogeneity changes depending on the season, so we selected Landsat 8 data from September 2013 to August 2014 with a 16-day temporal resolution to calculate the RB, and ASS. A total of 92 Landsat 8 images for all the stations in the HRB were downloaded from the following USGS website [56]. The statistical results were based on per-month averages to eliminate invalid data from cloud cover.

4. Results and Discussion

4.1. Spatial Representativeness Classification

Since the DLCT and RB can measure point-to-pixel value consistency, and ASS can assess the spatial patterns for a given station, respectively. Therefore, all three indicators were combined to describe the representativeness including point-to-pixel consistency and spatial heterogeneity. The spatial representativeness was also classified based on these three indicators. The DLCT indicator determines the representativeness of the station's LCT in the product pixel. When the LCT in the view footprint of a station is not dominant within the product pixel, the station observations cannot be representative of the pixel, and the LST value of all other vegetation-cover types may be ignored, even if the point-to-area LST consistency is high at the station sometime. When a pixel has a large DLCT value, the RB and ASS subsequently would determine the spatial representativeness together. Presumably, the station-observed LSTs represent ideal data for LST product validation if the RB value is small and the ASS value is large. If the RB and are ASS value are both small, the station may have some spatial representation. In some extreme cases, the station observing area is an average heterogeneity sub-areas in the products pixel, which means the station observations are representative for pixels, although the surface is heterogeneous in these products pixels. Finally, if the RB is large but the ASS is small, the observations probably differ from the values of the pixel.

Reasonable thresholds for the DLCT, RB and ASS are required to determine the representativeness level of a given station's observations. The emissivity products of the version 5 collection of MODIS LST/LSE products are retrieved based on the LCT of a pixel, and the LCT is classified as the land cover of this pixel based on the classification rule for MODIS land-cover products (MCD12Q1) if the area percent of one LCT in a pixel is higher than 60% [57]. Thus, 60% was defined as the threshold of the DLCT in this study. The RB was used to evaluate the difference between the land-based measurements within the view footprint at each station (in Table 1, the view footprints are shown in the sixth column) and the mean pixel value at the station locations. The ideal RB value is close to zero. However, the threshold of the RB is not unique and depends on the spatial resolution of the LST products, the view footprint of the station measurements, and the retrieval accuracy of the high-spatial-resolution LST maps that are used to evaluate the representativeness. Thus, a reasonable threshold for the RB was 0.5% in this study for MODIS LST products with 1-km spatial resolution, a station view footprint above 30 m and a 1-K retrieval error from the high-resolution LST map itself [55]. The ASS is calculated from variogram models based on the semi-variance in a 3-km × 3-km area that is centered at a given station and can indicate the greatest distance over which the value at a point on the surface is related to the value at another point. The ASS defines the maximum neighborhood over which control points should be selected to estimate a grid node to take advantage of the statistical correlation among observations and can describe the spatial distribution of LSTs and quantify the average LST spatial structures in the given area. In this study, the measurements from stations were used to evaluate the MODIS LSTs with a 1-km spatial resolution. Therefore, a reasonable ASS indicator should be larger than the spatial resolution of the LST products, that is, 1 km in this study.

The spatial representativeness of the station's LST observations was classified into five different levels based on the difference-constraining degrees of the three indicators and their thresholds. The levels and their descriptions are presented in Table 2.

Table 2. Grading of the spatial representativeness.

Level	DLCT > 60%	RB < 0.5%	ASS > 1 km	Description
1	√	√	√	Best representativeness level with strict point-to-pixel LST and LCT value consistency, and a homogeneous LST distribution
2	√	√	×	High representativeness level with high point-to-pixel LST and LCT value consistency but an incomplete homogeneous LST distribution
3	√	×	√	Moderate representativeness level with high LCT consistency and a relatively homogeneous LST distribution but low point-to-area LST consistency
4	√	×	×	Low representativeness level with low point-to-area LST consistency and a heterogeneous LST distribution
5	×	-	-	Minimal representativeness level with various LCTs in pixels and mismatch land-cover observations

4.2. Representativeness Grading of the HiWATER Station Observations

The representativeness of the LST observations for the eighteen HiWATER stations was graded based on the three indicators and level in Table 2. Figure 4 shows the DLCT indicator results for the stations, and the red dashed line shows the threshold value of the DLCT in the figure. In the upstream area, all the stations were covered by grass or meadows, except for HZS, which was covered by wheat. The DLCT values within a 1-km pixel indicated that five stations were covered by a single LCT and one station (DSL) was covered by a dominant grass type, with a DLCT value above 80%. In the upstream area of the HRB, only one station (HZS) had a DLCT value below 60%. Among the five stations in the midstream area of the HRB, SSW, HZZ, and GBZ were covered by a single LCT and SDZ was dominated by one type within a 1-km pixel. The LCTs within a 1-km pixel were diverse and broken at DMZ, and the DLCT value of the station observations was less than 60%. The LCTs of the five downstream stations were more complex within a 1-km pixel, and their DLCT values were all less than 60%. According to the DLCT threshold, the observations of seven stations, including HZS, DMZ, SDQ, HJL, HYL, NTZ, and LTZ, were graded at Level 5 (Table 2). Eleven remaining stations exhibited vegetation-type cover, matching the types in the corresponding MODIS LST pixels, but their representative levels varied depending on the RB and ASS values.

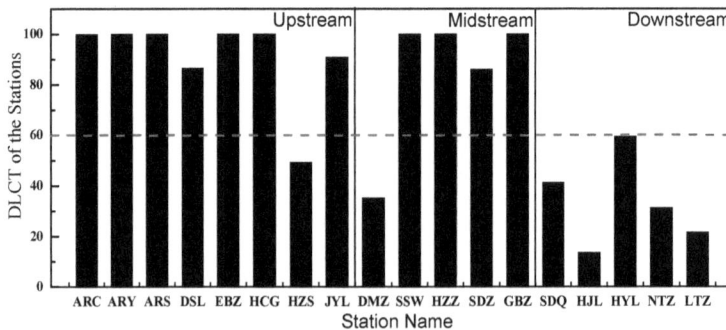

Figure 4. The dominant land cover type (DLCT) for the HiWATER stations.

The annual vegetation growth at stations can severely affect changes in their spatial representativeness. When stations have homogeneous vegetation types and density, the representativeness of stations may not vary with vegetation growth. Instead, the representativeness changes with the level of the point-to-area LST match, vegetation-density homogeneity and spatial contexture of the pixel during different seasons. The spatial representativeness of station observations can vary with vegetation growth in a year. When the vegetation type is uniform and the density is homogeneous, the representativeness at different growth stages may not significantly change. Otherwise, the representativeness may change because of changes in the point-to-area consistency and spatial heterogeneity. Seasonal changes in the spatial representativeness are similar in consecutive years when not interfered with by outside influences. Therefore, the RB and ASS values of the station observations were calculated monthly and are shown in Figures 5 and 6.

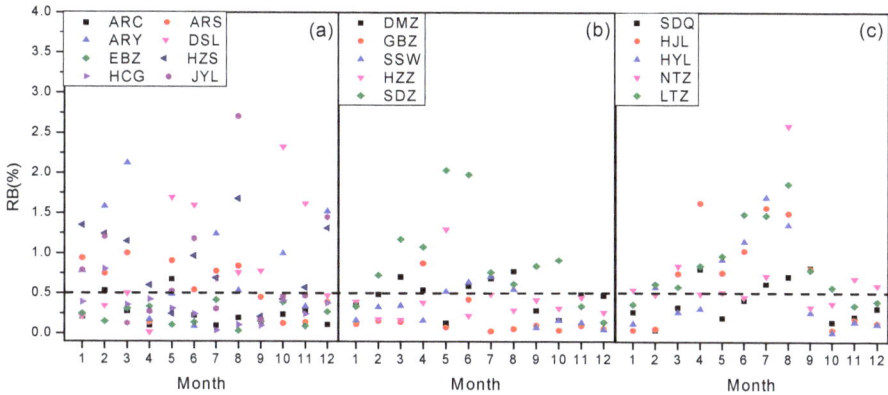

Figure 5. Monthly relative bias (RB) of the observations from the 18 stations within 1-km MODIS LST pixels. (**a**), (**b**) and (**c**) respectively show the monthly RB of the stations in the upstream, midstream and downstream of the Heihe River Basin (HRB).

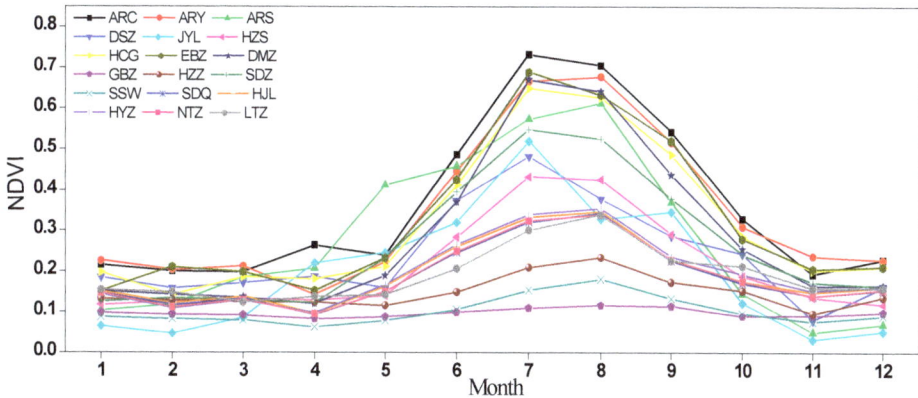

Figure 6. Monthly normalized difference vegetation index (NDVI) at the eighteen sites from 2013 to 2014.

The RB values of the stations in the upstream area, midstream area, and downstream area and their monthly changes are shown in Figure 5a–c, respectively. The eight upstream stations did not show a consistent trend with vegetation growth. As shown in Table 1, the LCTs of these stations were mostly grass or meadows. The DLCT values of these stations were all greater than 80%, with the exception of HZS, and the area of the heat-shadow effect in the thermal-infrared image was wider than the actual area. The RB depended on the homogeneity of the vegetation density for the stations that were covered by only one vegetation type and its changes with the seasons. Many factors can influence the density's homogeneity and its changes with vegetation growth, such as non-uniform rain or snow, random grazing in the grass, and uneven growth. Thus, complex factors are likely the main reasons for the inconsistency and volatility of the stations' changing RB trends. The chance of an RB value below 0.5% was greatest for stations with a DLCT above 60%.

The monthly changes in the RB at the stations in the midstream and downstream areas of the HRB could be divided into two types (Figure 5b,c). The first type included GBZ, SSW and HZZ, with DLCT values of 100%. Their RB values showed similar trends to those of the upstream stations, with inconsistency and volatility. The RB values of these stations were mostly smaller than 0.5%. The RB values of the other midstream and downstream stations, which had smaller DLCT values, increased with the NDVI (shown in Figure 6), although the RB's maximum peak point appeared within, before or after the month of the maximum NDVI peak. The RB values of these stations were all larger than 0.5% during the growing season. The land within a 1-km pixel area at these stations was covered by more than two LCTs, and the LST difference increased with vegetation growth and increasing solar radiation, especially at the maximum peak during summer. For instance, the midstream SDZ station was mainly covered by water and seeds. During winter, the station was predominantly covered by water, resulting in a small RB value below 0.5%. As the seeds began to grow, the seed area became larger, and the LST difference between seed and water increased as the solar radiance became stronger, creating a larger RB value. After reaching its maximum, the RB value decreased as vegetation growth and the solar radiance simultaneously decreased. The above analysis shows that the RB is relatively consistent with the DLCT and that an RB threshold is reasonable to effectively distinguish each level for a 1-km pixel.

The monthly ASS values of all the stations in the HRB are shown in Figure 7, with the ASS plots of stations in the upstream area in Figure 7a, the ASS plots of stations in the midstream area in Figure 7b, and the ASS plots of stations in the downstream area in Figure 7c. The black dashed lines in Figure 7 are the threshold of the ASS. The ASS plots of each station in Figure 7 did not show consistent trends as the seasons changed. However, when the DLCT values of the stations were less than 60%, the corresponding ASS values were less than 1000 m. Even for stations with DLCT values of 100%, the ASS value could be smaller than 1000 m, such as JYL in January and February (see Figures 4a and 6a). The ASS results for the stations were determined by the heterogeneity in the 1-km LST pixel. The heterogeneity of stations with a DLCT below 60% was mainly caused by the LST differences among different LCTs in the pixel. By contrast, the heterogeneity of stations with greater DLCT values, including values of 100%, was determined by the density heterogeneity of the dominant vegetation in the pixel, which is influenced by many factors, including uneven vegetation growth, human activity, such as random grazing, and localized weather, such as non-uniform rainfall. Moreover, the results were affected by the combination of these factors, leading to fluctuating ASS plot trends for stations with DLCT values of 100%.

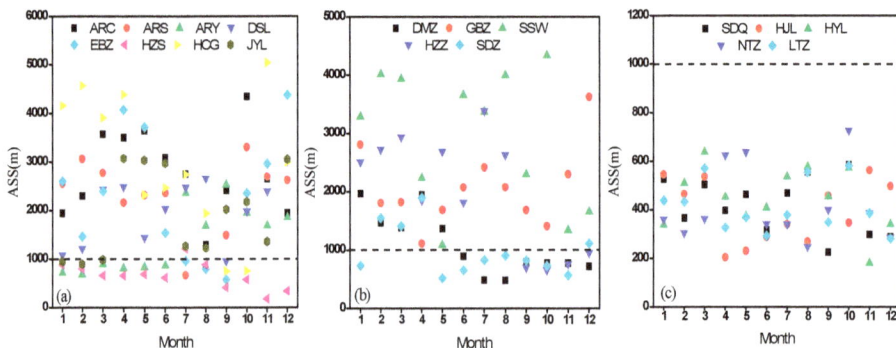

Figure 7. Monthly average structure scale (ASS) of the stations. (**a**), (**b**) and (**c**) respectively show the monthly ASS of the stations in the upstream, midstream and downstream of the HRB.

The monthly representativeness levels of the multi-temporal LST observations of all the stations in the HRB are shown in Figure 8 based on the DLCT values and the monthly RB and ASS values, which were calculated using the high-resolution LST images from the Landsat 8 images. Eight stations were present in the upstream area, so the upstream stations were divided into two figures, namely, Figure 8a,b, to clearly show the monthly representativeness. The monthly representativeness levels of the midstream and downstream stations are shown in Figure 8c,d. All the downstream stations had DLCT values below 60%, so all the stations are listed as Level 5 in Figure 8d. The spatial representativeness varied by month for the different stations. Among the 216 months of observations (18 stations in 12 months from September 2013 to September 2014), only 73 months of observations were graded as Level 1, so the number of station observations that perfectly represented the pixel were limited.

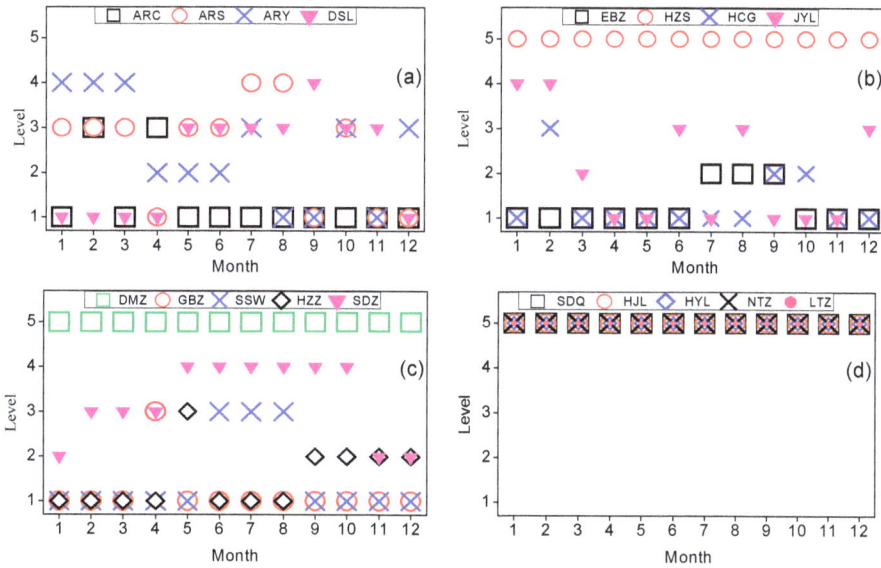

Figure 8. Grading results of the stations. (**a,b**) show the monthly representativeness level of the upstream stations; (**c,d**) show the monthly representativeness level of midstream and downstream stations

4.3. Traditional Validation Results without Spatial Representative Assessment

Traditional initial validation results, which neglect the evaluation of the spatial representativeness of the station observations, are presented first to demonstrate the effectiveness of our proposed scheme. The scatter plots in Figures 9 and 10 show the daytime and nighttime comparison results between the MODIS LST and ground-based LST measurements. As shown in Figure 8, the scatters in half of the sub-plots were close to the 1:1 line (the black line in the scatter plot). By contrast, the scatters for ARY, JYL, SDZ, SSW, SDQ, HJL, HYZ, NTZ, and LTZ were more decentralized. In Figure 10, the scatters of these sites were closer to the 1:1 line than those in Figure 9. Table 3 presents a statistical comparison of the daytime and nighttime results between the ground-based measurements and MODIS LST products that were retrieved from the MODIS data onboard Terra and Aqua, respectively. The bias (ground-based LST measurements—MODIS LST) at eighteen sites during the daytime ranged between -4.72 K and 6.46 K for the MOD11A1 LST/LSE products and between -3.44 K and 6.24 K for the MYD11A1 LST/LSE products; the root mean square error (RMSE) during the daytime ranged from 1.74 K to 5.50 K for the Terra MODIS LSTs (MOD11A1 LST/LSE products) and from 1.75 K to 5.94 K for the Aqua MODIS LSTs (MYD11A1 LST/LSE products). By contrast, the bias and RMSE between the Terra MODIS LST and ground-based measurements at nighttime ranged from −0.51 K to 3.86 K and from 1.19 K to 2.64 K, respectively; for the Aqua MODIS LST, the bias ranged from −0.39 K to 3.48 K and the RMSE ranged from 1.94 K to 3.82 K.

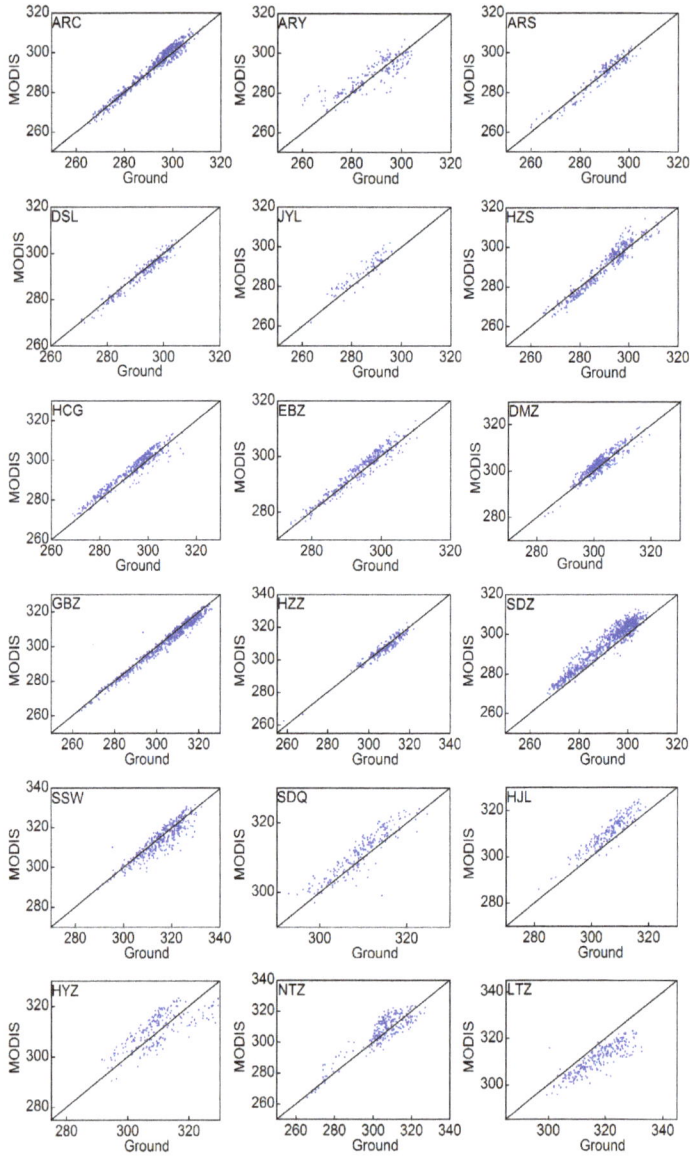

Figure 9. Scatterplots of the daytime comparison results of the LSTs from MODIS and the LST measurements at the eighteen HiWATER sites.

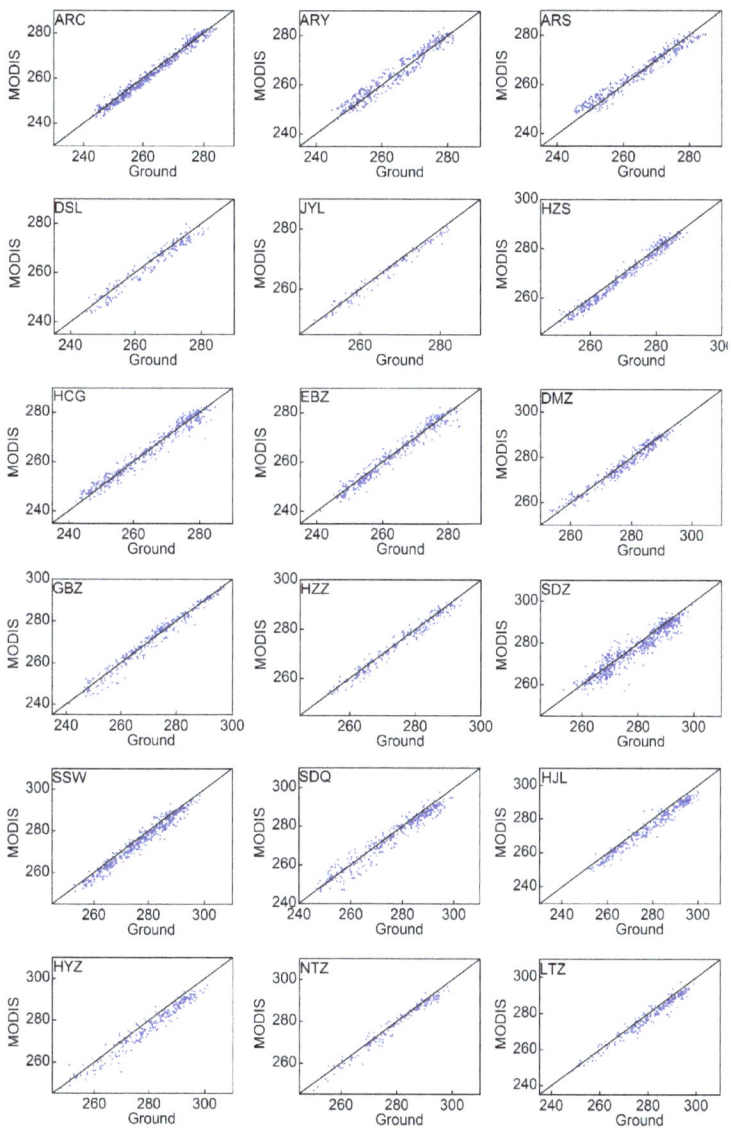

Figure 10. Scatterplots of the nighttime comparison results of the LSTs from MODIS and the LST measurements at the eighteen HiWATER sites.

The initial results revealed that the MODIS LSTs fit better with MODIS at nighttime than during the daytime, and the comparison results greatly differed between the daytime and nighttime for some stations, especially HZS, HYZ, NTZ, and LTZ, which are all Level-5 stations. The LST had stronger heterogeneity during the daytime than at nighttime. During the day, LSTs can vary by 10 K or more over a few meters depending on the nature of the surface and the local meteorological conditions, with the variability being lower at night [9]. When comparing the unrepresentative observations of these stations to the MODIS LSTs, the stronger heterogeneity resulted in greater differences in the bias and RMSEs of these stations between the daytime and nighttime because of the constant accuracy of the MODIS LST products within the same pixel. If these results were considered the final results for the accuracy of the MODIS LST, an obvious error would be introduced into the validation because of the unrepresentativeness of the station observations when compared to the corresponding scale of the remote-sensing observations. The stronger heterogeneity during the daytime affected the Aqua MODIS LST and Terra MODIS LST validation process in a similar manner, so no significant differences in accuracy existed between the LST products from Terra MODIS and the LST products from Aqua MODIS when compared to the ground-measured LSTs at these sites, as shown in Table 3.

Table 3. Summary of the statistical comparison results between the MODIS LSTs and the LST measurements from the HiWATER sites.

| Station | MOD [a] | | | | MYD [b] | | | |
| | Day | | Night | | Day | | Night | |
	Bias	RMSE	Bias	RMSE	Bias	RMSE	Bias	RMSE
ARC	−0.55	2.16	0.98	2.15	−0.52	1.92	0.67	1.94
ARY	−0.79	5.50	−0.51	2.25	−0.96	4.48	−0.28	2.56
ARS	0.31	2.56	0.01	2.64	0.56	2.54	−0.39	2.57
DSL	1.07	1.74	1.25	1.22	0.89	1.87	1.27	2.16
JYL	−2.79	3.14	0.96	1.41	−2.94	3.09	0.47	2.26
HZS	0.10	3.49	1.65	1.19	−0.13	3.73	0.96	1.94
HCG	−1.31	2.36	0.62	1.96	−1.68	2.92	−0.25	2.33
EBZ	−0.78	2.07	0.77	1.61	−0.97	2.39	0.23	2.56
DMZ	−0.88	2.81	1.23	1.45	−0.96	2.77	0.21	2.21
GBZ	0.17	2.25	−0.01	1.52	1.08	1.75	0.66	2.99
HZZ	0.74	1.86	0.93	1.28	0.25	2.46	0.69	2.08
SDZ	−4.12	2.93	1.90	2.74	−3.44	2.71	1.26	3.78
SSW	1.34	4.18	2.49	1.60	1.57	2.95	1.55	2.51
SDQ	−1.84	3.18	1.95	1.21	−1.65	2.35	1.50	3.80
HJL	−4.23	2.70	3.56	1.90	−4.55	2.81	3.48	3.82
HYZ	−0.72	5.18	3.86	2.03	−0.71	5.94	2.75	3.14
NTZ	−4.72	4.08	1.43	1.47	−0.17	5.44	1.15	2.91
LTZ	6.46	4.21	2.45	1.15	6.24	3.68	1.77	3.02

[a] The MOD columns show the results of a comparison between the LSTs from the MODIS onboard Terra and the ground-based measurements at the eighteen sites; [b] The MYD columns show the results of a comparison between the LSTs from the MODIS onboard Aqua and the ground-based measurements at the eighteen sites.

4.4. MODIS LST Product Validation Considering the Influence of Spatial Representativeness Evaluation

All the station observation validation results and representative station observation validation results were calculated and compared to discuss the effect of representativeness on the validation of satellite-retrieved LSTs. In total, 7527 station observations from eighteen stations were evaluated based on the Landsat 8 LST reference maps according to the satellite-overpass times and quality control flags of the MODIS LSTs. The RMSE and bias of the MOD/MYD11A1 LST products based on all the observations and the different levels of observations are listed in Table 4. For all the observations, the bias of the MODIS LST ranged from −0.27 K to 2.39 K, and the RMSE ranged from 3.32 K to 4.93 K. The RMSEs were the smallest for Level 1, which contained 2472 station observations. During both the daytime and nighttime, the bias between the station observations and MODIS LSTs was better than 1 K and the RMSE was less than 2 K. No significant differences existed between the daytime and nighttime comparison results for the Level 1 observations. These results using Level 1 observations are inconsistent with those from the traditional validation scheme in Section 4.3 and are more reasonable based on the retrieval accuracy of the MODIS LSTs during the daytime and nighttime for the same pixel. The comparison results between the MODIS LSTs and LST observations for Levels 2 and 3 had a similar bias and RMSE, which were not larger than those for all the observations. The results for the Level-4 observations had a larger bias and RMSE than those for Levels 3 and 2. The largest RMSE values were obtained when using Level 5 observations. The RMSE value increased from Level 1 to

Level 5. Large differences existed in the bias and RMSE between the daytime and nighttime at all levels, except for Level 1, because of the greater spatial heterogeneity during the daytime. Table 4 indicates that the highly representative validated observations usually had good results. Thus, highly representative observations can more accurately describe satellite-retrieved LST values and limit spatial mismatch errors from point-station observations. Therefore, validating all the stations without grading their representativeness usually underestimated the accuracy of the satellite-retrieved LSTs; these underestimated accuracy values were larger for the daytime validation than the nighttime validation.

Table 4. Validation results for MOD/MYD11A1 based on the spatial representativeness levels of the station observations.

	MOD [a]				MYD [b]			
Level	Day		Night		Day		Night	
	Bias	RMSE	Bias	RMSE	Bias	RMSE	Bias	RMSE
Level 1	−0.19	1.75	0.8	1.58	0.30	1.79	0.43	1.84
Level 2	−1.13	4.58	1.24	2.53	−1.48	3.24	0.32	2.60
Level 3	1.45	4.93	1.44	2.23	0.47	3.67	0.77	2.67
Level 4	−2.94	5.93	2.51	3.89	−2.25	4.56	1.33	4.15
Level 5	2.03	6.27	3.60	4.52	−0.36	5.88	2.28	4.29
All	0.59	4.93	2.39	3.55	−0.27	4.58	1.28	3.32

[a] The MOD columns show the results of a comparison between the LSTs from the MODIS onboard Terra and the ground-based measurements at the eighteen sites; [b] The MYD columns show the results of a comparison between the LSTs from the MODIS onboard Aqua and the ground-based measurements at the eighteen sites.

4.5. Other Potential Factors during the LST Validation Process

Other factors may potentially influence the accuracy of the validation results. In this study, we only focused on the sensor view zenith angle (VZA) and broadband sensitivity issues, which are the main factors that affect T-based validation for MODIS LSTs in addition to the representativeness of the station observations. The effects of these two factors are discussed in detail below.

4.5.1. Dependence of the LST Error on the Sensor VZA

The relationship between the error (the absolute difference between MODIS and ground-based LSTs) and the sensor VZA was first investigated at all HiWATER sites to analyze the potential factors that create large errors at certain sites. The VZAs differed between ground-based instruments and the MODIS sensor: the ground-measured LSTs were obtained at a VZA of 0°, whereas the MODIS observations were acquired over a large range of VZAs (0–65°). Sufficient ground-based measurements were available to statistically analyze multiple VZAs using data from the eighteen sites. Scatterplots of the errors and sensor VZAs are presented in Figure 11. The average errors did not significantly depend on the VZAs during the daytime (see Figure 11a). By contrast, the probability of a larger error increased with the VZA for the nighttime scatterplot (Figure 11b). The average absolute bias for observations that were acquired with a smaller VZA (≤30°) was 0.2 K lower than that for those at greater VZAs (>30°). Greater errors for LSTs observed under larger sensor VZAs were also observed in a prior validation study [8,13], and the remote sensor may see different percentages of shadows at different VZAs during the daytime, when larger heterogeneities occur [58]. The larger LST heterogeneity at certain sites was the most likely cause of the daytime scatterplot (Figure 11a), rather than the obvious dependence of the error of the LST on the sensor VZA.

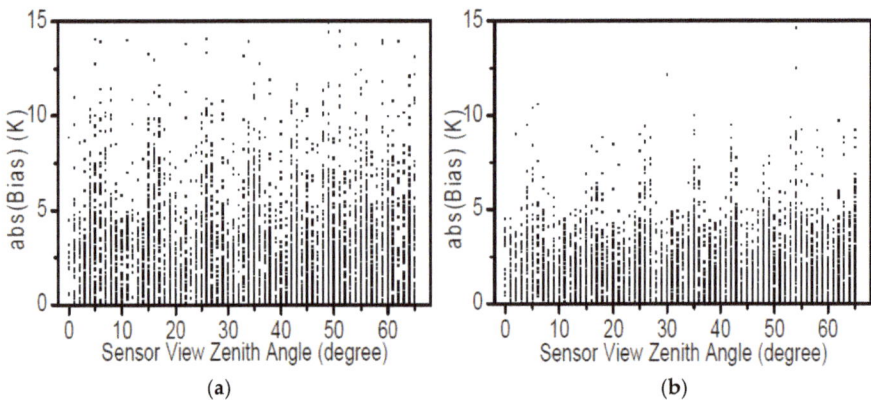

Figure 11. Relationships between the bias and zenith angle of MODIS when viewing the pixels: (**a**) relationship during the daytime; and (**b**) relationship at nighttime.

4.5.2. Broadband Sensitivity Issue

The accuracy of LSTs from ground-based longwave-radiation measurements is another important factor that influences the evaluation of MODIS LSTs. The accuracy of longwave-radiation measurements and the accuracy of the broadband emissivity, which is used to calculate LSTs from ground-based longwave-radiation measurements, are the two major factors for ground-based LST estimation. The accuracy of longwave-radiation measurements depends on the sensor calibration. Calibration is typically performed before and after field deployment and during monthly field-routing exams [59], so the estimation accuracy of the broadband emissivity is a key parameter that influences the estimation accuracy of ground-based LST observations. Our previous study indicated that the maximum mean absolute error (MAE) of the LST was less than 0.3 K when the estimation emissivity MAE was ≤0.01, which was estimated by adding a series bias of ±0.001, ±0.002, ±0.003, ±0.004, ±0.005, and ±0.01 based on the ε_b values at the stations in the northern arid region of China, including ARC, HZZ, and DMZ (also called YK) [60]. In this study, the satellite-estimated broadband-emissivity method was used based on the MODIS narrowband emissivities. A suitable set of coefficients for all the data was used to calculate the broadband emissivity, as outlined in Equation (7), according to the diversity of the land cover at the sites. In a study by Wang et al., the broadband emissivity that was estimated by this method was compared to three years of ground-based emissivity measurements at Gaize (32.30°N and 84.06°E, with an elevation of 4420 m) in the western Tibetan Plateau. The results showed that the broadband-emissivity calculation from the MODIS narrowband emissivities reasonably matched the ground measurements, with a standard deviation of 0.0085 and a bias of 0.0015 [45]. Therefore, the average estimation error of the broadband emissivity based on this method was less than 0.037 K, which is the maximum average absolute error value of all the stations in northern arid China for an emissivity bias of 0.02.

5. Conclusions

In traditional validation schemes, as discussed in Section 4.3, station observations are usually considered as the true LST values of the corresponding pixels. Therefore, a source of uncertainty for the validation results is introduced when the representativeness of station observations is not evaluated. The spatial representativeness assessment of station observations is a key step to reliably validate satellite-retrieved LST products, which has barely been discussed by previous researchers. In this study, a new validation scheme was proposed that adds a key step to evaluate the spatial representativeness of station LST observations. Three indicators, namely, the DLCT, RB, and ASS, were constructed in this new scheme to assess and grade the representativeness of station observations. The representativeness of the station observations was divided into five levels for a 1-km spatial resolution pixel according to the values of these three indicators. Then, the proposed method was used to evaluate the spatial representativeness of HiWATER LST observations for 1-km-pixel daily MODIS LST products. The analysis showed that the new validation scheme can effectively limit the error that is introduced by spatial mismatches between the station observations and remote-sensing products. This spatial representative assessment synthesized the three indicators, providing quantitative and accurate descriptions and reliably evaluating the observations' representativeness, although reasonable thresholds for the three indicators should consider the LST products' spatial resolution, the theoretical accuracy of the retrieval algorithm, and instrument error. Therefore, several conclusions can be drawn.

First, the traditional validation results when using all the station observations showed obvious errors and RMSE values between the daytime and nighttime validations, especially when using observations from the downstream stations. The retrieval method for the MODIS LST products could obtain LST products with similar accuracy during the daytime and nighttime. Therefore, the error was likely introduced by unrepresentative station observations. According to the HRB traditional validation sample, this effect caused a maximum bias of 8.03 K and a maximum RMSE of 3.25 K according to the difference between the daytime and nighttime validation results from the same stations.

Second, the monthly representativeness analysis illustrated that the spatial contexture heterogeneity and point-to-pixel consistency of the LST changed with variations in plant growth and other factors, such as human activity, the solar-radiation intensity, and the local climate, which created irregular changes in the monthly spatial representativeness of the stations. The monthly changes in all the factors that influenced the representativeness were not always similar between consecutive years, so the spatial representativeness in the same month of different years may have greatly changed for a given station. Therefore, the spatial representativeness for consecutive years could be reassessed through a rigorous LST product-validation process.

Third, the RMSE of the MOD/MYD11A1 products increased from 1.58 K to 6.27 K from Level 1 to Level 5, and the RMSE values from Level 1 to Level 3 were smaller than those across all the observations. For all levels except Level 1, the RMSE values were larger during the daytime than those at nighttime because of the stronger LST heterogeneity during the daytime. The obvious differences in the bias and RMSE values among the levels indicated that the representativeness method could sufficiently differentiate the spatial representativeness level within 1-km pixels. Moreover, the Level-1 LST observations were acceptable validation data for the MODIS 1-km LST products. While the RMSE difference between all observations (using in the traditional validations) and Level 1 observations validations are 3.0 K in daytime and 1.7 K in nighttime, which indicates the error introduced by the traditional validation without the representativeness assessment. Therefore, the error of the MOD/MYD11A1 products was better than 1 K and the RMSE was less than 2 K. Moreover, no obvious differences existed in the accuracy of the MODIS LST products among the four daily times that Terra and Aqua passed.

Finally, the bias increased from -0.19 K at Level 1 to 3.6 K at Level 5, and the RMSE increased from 1.58 to 6.27. Thus, a bias difference of 3.79 K and a RMSE difference of 4.69 K existed between Level 1, which was the most representative level, and Level 5, which was the least representative level. The dependence of the absolute biases on the sensor VZA during the daytime was strongly influenced by the land-surface heterogeneity at heterogeneous sites. During the nighttime, the LSTs of these sites were more homogeneous and any surface-heterogeneity effects were smaller, so the average absolute bias for observations that were acquired under lower VZA ($\leq 30°$) was 0.2 K lower than those that were acquired at greater VZAs ($>30°$). The estimated broadband emissivities from the narrowband MODIS LST/LSE products retrieved by the day/night algorithm varied with the land-surface conditions, such as vegetation growth and land cover, with a bias of 0.0015. Therefore, the estimation error of the broadband emissivities was less than 0.004 K. Compared to the effects of the representativeness, the errors from these last two factors were very small. Thus, the evaluation of spatial representativeness is a key step to reliably validate LST products with special spatial resolution, which is the greatest advantage of our proposed validation scheme.

A more reasonable and accurate scheme to validate remotely sensed LST products was proposed in this study. However, areas with more heterogeneous LSTs are not suitable for validation with single-point ground-based measurements and require conditional upscaling by multi-point ground-based measurements. Therefore, this scheme can be further improved for different conditions, and can be actually used to study potential new sites for LST validation. Furthermore, the thresholds of the three indicators were suitable for grading the levels of station observations in a 1-km pixel. This scheme can also be used to evaluate other LST products by renewing the thresholds and evaluating the spatial representativeness in other pixel grids.

Acknowledgments: This work was supported by the Natural Science Foundation of China (Grant No. 41601448), the National Key Technology R&D Program of China (Grant No. 2016YFC0500106), the Fundamental Research Funds for the Central Universities of China (Grant No. XDJK2016C094) and the Doctoral Fund of Southwest University (Grant No. SWU115063). This study was also supported by the China Scholarship Council. The authors would like to thank all the institutes and universities in the North Arid and Semi-Arid Area Cooperative Experimental Observation Integrated Research Program for providing the meteorological data.

Author Contributions: Mingguo Ma outlined the research topic and assisted with manuscript writing. Zhaoliang Li gave much helpful advice on the revising work. Junlei Tan and Adan Wu were involved in the data collection and preprocessing step. Wenping Yu processed the data, analyzed the results and wrote the manuscript.

Conflicts of Interest: The authors declare no conflict of interest.

References

1. Liang, S.L. *Quantitative Remote Sensing of Land Surface*; John Wiley & Sons: Hoboken, NJ, USA, 2004.
2. Smith, P.M.; Kalluri, S.N.V.; Prince, S.; Defries, R. The NOAA/NASA pathfinder AVHRR 8-km land data set. *Photogramm. Eng. Remote Sens.* **1997**, *63*, 12–27.
3. El Saleous, N.; Vermote, E.; Justice, C.; Townshend, J.; Tucker, C.; Goward, S. Improvements in the global biospheric record from the Advanced Very High Resolution Radiometer (AVHRR). *Int. J. Remote Sens.* **2000**, *21*, 1251–1277. [CrossRef]
4. Sun, D.; Pinker, R.T. Estimation of land surface temperature from a Geostationary Operational Environmental Satellite (GOES-8). *J. Geophys. Res. Atmos.* **2003**, *108*. [CrossRef]
5. Sobrino, J.A.; Jiménez-Muñoz, J.C.; Sòria, G.; Romaguera, M.; Guanter, L.; Moreno, J.; Plaza, A.; Martínez, P. Land surface emissivity retrieval from different VNIR and TIR sensors. *IEEE Trans. Geosci. Remote Sens.* **2008**, *46*, 316–327. [CrossRef]
6. Trigo, I.F.; Peres, L.F.; DaCamara, C.C.; Freitas, S.C. Thermal land surface emissivity retrieved from SEVIRI/Meteosat. *IEEE Trans. Geosci. Remote Sens.* **2008**, *46*, 307–315. [CrossRef]
7. Li, Z.-L.; Tang, B.-H.; Wu, H.; Ren, H.; Yan, G.; Wan, Z.; Trigo, I.F.; Sobrino, J.A. Satellite-derived land surface temperature: Current status and perspectives. *Remote Sens. Environ.* **2013**, *131*, 14–37. [CrossRef]
8. Coll, C.; Caselles, V.; Galve, J.M.; Valor, E.; Niclos, R.; Sanchez, J.M.; Rivas, R. Ground measurements for the validation of land surface temperatures derived from AATSR and MODIS data. *Remote Sens. Environ.* **2005**, *97*, 288–300. [CrossRef]
9. Coll, C.; Wan, Z.; Galve, J.M. Temperature-based and radiance-based validations of the v5 MODIS land surface temperature product. *J. Geophys. Res. Atmos.* **2009**, *114*. [CrossRef]
10. Wan, Z. New refinements and validation of the collection-6 MODIS land-surface temperature/emissivity product. *Remote Sens. Environ.* **2014**, *140*, 36–45. [CrossRef]
11. Wan, Z.; Zhang, Y.; Zhang, Q.; Li, Z.L. Quality assessment and validation of the MODIS global land surface temperature. *Int. J. Remote Sens.* **2004**, *25*, 261–274. [CrossRef]
12. Wang, K.; Liang, S. Evaluation of ASTER and MODIS land surface temperature and emissivity products using long-term surface longwave radiation observations at SURFRAD sites. *Remote Sens. Environ.* **2009**, *113*, 1556–1565. [CrossRef]
13. Wang, W.; Liang, S.; Meyers, T. Validating MODIS land surface temperature products using long-term nighttime ground measurements. *Remote Sens. Environ.* **2008**, *112*, 623–635. [CrossRef]

14. Wan, Z. New refinements and validation of the MODIS land-surface temperature/emissivity products. *Remote Sens. Environ.* **2008**, *112*, 59–74. [CrossRef]
15. Wang, S.; Li, X.; Ge, Y.; Jin, R.; Ma, M.; Liu, Q.; Wen, J.; Liu, S. Validation of regional-scale remote sensing products in China: From site to network. *Remote Sens.* **2016**, *8*, 980. [CrossRef]
16. Gupta, R.K.; Prasad, T.S.; Krishna Rao, P.V.; Bala Manikavelu, P.M. Problems in upscaling of high resolution remote sensing data to coarse spatial resolution over land surface. *Adv. Space Res.* **2000**, *26*, 1111–1121. [CrossRef]
17. Liu, Y.; Hiyama, T.; Yamaguchi, Y. Scaling of land surface temperature using satellite data: A case examination on ASTER and MODIS products over a heterogeneous terrain area. *Remote Sens. Environ.* **2006**, *105*, 115–128. [CrossRef]
18. Liu, Y.B.; Yamaguchi, Y.; Ke, C.Q. Reducing the discrepancy between aster and MODIS land surface temperature products. *Sensors* **2007**, *7*, 3043–3057. [CrossRef] [PubMed]
19. Hook, S.J.; Vaughan, R.G.; Tonooka, H.; Schladow, S.G. Absolute radiometric in-flight validation of mid infrared and thermal infrared data from ASTER and MODIS on the terra spacecraft using the Lake Tahoe, CA/NV, USA, automated validation site. *IEEE Trans. Geosci. Remote Sens.* **2007**, *45*, 1798–1807. [CrossRef]
20. Yu, W.; Ma, M. Scale mismatch between in situ and remote sensing observations of land surface temperature: Implications for the validation of remote sensing LST products. *IEEE Geosci. Remote Sens. Lett.* **2015**, *12*, 497–501.
21. Xu, B.; Li, J.; Liu, Q.; Huete, A.R.; Yu, Q.; Zeng, Y.; Yin, G.; Zhao, J.; Yang, L. Evaluating spatial representativeness of station observations for remotely sensed leaf area index products. *IEEE J. Sel. Top. Appl. Earth Obs. Remote Sens.* **2016**, *9*, 3267–3282. [CrossRef]
22. Hakuba, M.Z.; Folini, D.; Sanchez-Lorenzo, A.; Wild, M. Spatial representativeness of ground-based solar radiation measurements. *J. Geophys. Res. Atmos.* **2013**, *118*, 8585–8597. [CrossRef]
23. Román, M.O.; Schaaf, C.B.; Woodcock, C.E.; Strahler, A.H.; Yang, X.; Braswell, R.H.; Curtis, P.S.; Davis, K.J.; Dragoni, D.; Goulden, M.L. The MODIS (Collection V005) BRDF/albedo product: Assessment of spatial representativeness over forested landscapes. *Remote Sens. Environ.* **2009**, *113*, 2476–2498. [CrossRef]
24. Janis, M.J.; Robeson, S.M. Determining the spatial representativeness of air-temperature records using variogram-nugget time series. *Phys. Geogr.* **2004**, *25*, 513–530. [CrossRef]
25. Henne, S.; Brunner, D.; Folini, D.; Solberg, S.; Klausen, J.; Buchmann, B. Assessment of parameters describing representativeness of air quality in-situ measurement sites. *Atmos. Chem. Phys.* **2010**, *10*, 3561–3581. [CrossRef]
26. Jia, Z.; Liu, S.; Xu, Z.; Chen, Y.; Zhu, M. Validation of remotely sensed evapotranspiration over the Hai River Basin, China. *J. Geophys. Res. Atmos.* **2012**, *117*. [CrossRef]
27. Janssen, S.; Dumont, G.; Fierens, F.; Deutsch, F.; Maiheu, B.; Celis, D.; Trimpeneers, E.; Mensink, C. Land use to characterize spatial representativeness of air quality monitoring stations and its relevance for model validation. *Atmos. Environ.* **2012**, *59*, 492–500. [CrossRef]
28. Chan, C.-C.; Hwang, J.-S. Site representativeness of urban air monitoring stations. *J. Air Waste Manag. Assoc.* **1996**, *46*, 755–760. [CrossRef] [PubMed]
29. Wang, Z.; Schaaf, C.B.; Chopping, M.J.; Strahler, A.H.; Wang, J.; Roman, M.O.; Rocha, A.V.; Woodcock, C.E.; Shuai, Y. Evaluation of Moderate-resolution Imaging Spectroradiometer (MODIS) snow albedo product (MCD43A) over tundra. *Remote Sens. Environ.* **2012**, *117*, 264–280. [CrossRef]
30. Chen, B.; Coops, N.C.; Fu, D.; Margolis, H.A.; Amiro, B.D.; Black, T.A.; Arain, M.A.; Barr, A.G.; Bourque, C.P.-A.; Flanagan, L.B. Characterizing spatial representativeness of flux tower eddy-covariance measurements across the Canadian Carbon Program Network using remote sensing and footprint analysis. *Remote Sens. Environ.* **2012**, *124*, 742–755. [CrossRef]
31. Cescatti, A.; Marcolla, B.; Vannan, S.K.S.; Pan, J.Y.; Román, M.O.; Yang, X.; Ciais, P.; Cook, R.B.; Law, B.E.; Matteucci, G. Intercomparison of MODIS albedo retrievals and in situ measurements across the global FLUXNET network. *Remote Sens. Environ.* **2012**, *121*, 323–334. [CrossRef]
32. Wang, Z.; Schaaf, C.B.; Strahler, A.H.; Chopping, M.J.; Román, M.O.; Shuai, Y.; Woodcock, C.E.; Hollinger, D.Y.; Fitzjarrald, D.R. Evaluation of MODIS albedo product (MCD43A) over grassland, agriculture and forest surface types during dormant and snow-covered periods. *Remote Sens. Environ.* **2014**, *140*, 60–77. [CrossRef]

33. Li, X.; Cheng, G.; Liu, S.; Xiao, Q.; Ma, M.; Jin, R.; Che, T.; Liu, Q.; Wang, W.; Qi, Y.; et al. Heihe watershed allied telemetry experimental research (HiWATER): Scientific objectives and experimental design. *Bull. Am. Meteorol. Soc.* **2013**, *94*, 1145–1160. [CrossRef]

34. Journel, A.G.; Huijbregts, C.J. *Mining Geostatistics*; Academic Press: London, UK, 1981.

35. Cressie, N. The origins of kriging. *Math. Geol.* **1990**, *22*, 239–252. [CrossRef]

36. Rossi, R.E.; Mulla, D.J.; Journel, A.G.; Franz, E.H. Geostatistical tools for modeling and interpreting ecological spatial dependence. *Ecol. Monogr.* **1992**, *62*, 277–314. [CrossRef]

37. Robertson, G.P. Geostatistics in ecology: Interpolating with known variance. *Ecology* **1987**, *68*, 744–748. [CrossRef]

38. Curran, P.J.; Atkinson, P.M. Geostatistics and remote sensing. *Prog. Phys. Geogr.* **1998**, *22*, 61–78. [CrossRef]

39. Atkinson, P.M.; Tate, N.J. Spatial scale problems and geostatistical solutions: A review. *Prof. Geogr.* **2000**, *52*, 607–623. [CrossRef]

40. Burcsu, T.K.; Robeson, S.M.; Meretsky, V.J. Identifying the distance of vegetative edge effects using Landsat TM data and geostatistical methods. *Geocarto Int.* **2001**, *16*, 61–70. [CrossRef]

41. Li, H.; Reynolds, J. On definition and quantification of heterogeneity. *Oikos* **1995**, *73*, 280–284. [CrossRef]

42. Hu, Y.D.; Gao, J.M.; Wang, J.M.; Ji, G.L.; Shen, Z.B.; Cheng, L.S.; Chen, J.Y.; Li, S.Q. Some achievements in scientific research during HEIFE. *Plateau Meteorol.* **1994**, *13*, 225–236.

43. Li, X.; Li, X.; Li, Z.; Ma, M.; Wang, J.; Xiao, Q.; Liu, Q.; Che, T.; Chen, E.; Yan, G.; et al. Watershed allied telemetry experimental research. *J. Geophys. Res.* **2009**, *114*. [CrossRef]

44. Heihe Watershed Allied Telemetry Experimental Research (HiWATER) Home. Available online: http://card.westgis.ac.cn/hiwater (accessed on 17 March 2017).

45. Wang, K.; Wan, Z.; Wang, P.; Sparrow, M.; Liu, J.; Zhou, X.; Haginoya, S. Estimation of surface long wave radiation and broadband emissivity using Moderate Resolution Imaging Spectroradiometer (MODIS) land surface temperature/emissivity products. *J. Geophys. Res. Atmos.* **2005**, *110*, D11109. [CrossRef]

46. Salomonson, V.V.; Barnes, W.; Maymon, P.W.; Montgomery, H.E.; Ostrow, H. Modis: Advanced facility instrument for studies of the earth as a system. *IEEE Trans. Geosci. Remote Sens.* **1989**, *27*, 145–153. [CrossRef]

47. Wan, Z.M.; Dozier, J. A generalized split-window algorithm for retrieving land-surface temperature from space. *IEEE Trans. Geosci. Remote Sens.* **1996**, *34*, 892–905.

48. Wan, Z.M.; Li, Z.L. A physics-based algorithm for retrieving land-surface emissivity and temperature from EOS/MODIS data. *IEEE Trans. Geosci. Remote Sens.* **1997**, *35*, 980–996.

49. Seemann, S.W.; Li, J.; Gumley, L.E.; Strabala, K.I.; Menzel, W.P. Operational retrieval of atmospheric temperature, moisture, and ozone from MODIS infrared radiances. *J. Appl. Meteorol.* **2003**, *42*, 1072–1091. [CrossRef]

50. HiWATER: Land Cover Map of Heihe River Basin. Available online: http://card.westgis.ac.cn/hiwater/rsproduct (accessed on 21 March 2017).

51. Zhong, B.; Ma, P.; Nie, A.; Yang, A.; Yao, Y.; Lü, W.; Zhang, H.; Liu, Q. Land cover mapping using time series HJ-1/CCD data. *Sci. China Earth Sci.* **2014**, *57*, 1790–1799. [CrossRef]

52. Zhong, B.; Yang, A.; Nie, A.; Yao, Y.; Zhang, H.; Wu, S.; Liu, Q. Finer resolution land-cover mapping using multiple classifiers and multisource remotely sensed data in the Heihe River Basin. *IEEE J. Sel. Top. Appl. Earth Obs. Remote Sens.* **2015**, *8*, 4973–4992. [CrossRef]

53. Zhong, B.; Zhang, Y.; Du, T.; Yang, A.; Lv, W.; Liu, Q. Cross-calibration of HJ-1/CCD over a desert site using Landsat ETM + Imagery and ASTER GDEM product. *IEEE Trans. Geosci. Remote Sens.* **2014**, *52*, 7247–7263. [CrossRef]

54. Roy, D.P.; Wulder, M.; Loveland, T.; Woodcock, C.; Allen, R.; Anderson, M.; Helder, D.; Irons, J.; Johnson, D.; Kennedy, R. Landsat-8: Science and product vision for terrestrial global change research. *Remote Sens. Environ.* **2014**, *145*, 154–172. [CrossRef]

55. Du, C.; Ren, H.; Qin, Q.; Meng, J.; Zhao, S. A practical split-window algorithm for estimating land surface temperature from Landsat 8 data. *Remote Sens.* **2015**, *7*, 647–665. [CrossRef]

56. USGS Home. Available online: https://glovis.usgs.gov/ (accessed on 4 February 2017).

57. MODIS Land Cover Product Algorithm Theoretical Basis Document (ATBD) Version 5.0. Available online: http://modis.gsfc.nasa.gov/data/atbd/land_atbd.php (accessed on 2 May 2016).

58. Ermida, S.L.; Trigo, I.F.; DaCamara, C.C.; Göttsche, F.M.; Olesen, F.S.; Hulley, G. Validation of remotely sensed surface temperature over an oak woodland landscape—The problem of viewing and illumination geometries. *Remote Sens. Environ.* **2014**, *148*, 16–27. [CrossRef]

59. Xu, Z.W.; Liu, S.M.; Li, X.; Shi, S.J.; Wang, J.M.; Zhu, Z.L.; Xu, T.R.; Wang, W.Z.; Ma, M.G. Intercomparison of surface energy flux measurement systems used during the HiWATER-MUSOEXE. *J. Geophys. Res.* **2013**, *118*, 13140–13157. [CrossRef]

60. Yu, W.; Ma, M.; Wang, X.; Geng, L.; Tan, J.; Shi, J. Evaluation of MODIS LST products using longwave radiation ground measurements in the northern arid region of China. *Remote Sens.* **2014**, *6*, 11494–11517. [CrossRef]

remote sensing

MDPI

Article

Estimating Land Surface Temperature from Feng Yun-3C/MERSI Data Using a New Land Surface Emissivity Scheme

Xiangchen Meng [1,2], Jie Cheng [1,2,*] and Shunlin Liang [1,2,3]

[1] State Key Laboratory of Remote Sensing Science, Jointly Sponsored by Beijing Normal University and Institute of Remote Sensing and Digital Earth of Chinese Academy of Sciences, Beijing 100875, China; xiangchenmeng@yeah.net (X.M.); sliang@umd.edu (S.L.)

[2] Institute of Remote Sensing Science and Engineering, Faculty of Geographical Science, Beijing Normal University, Beijing 100875, China

[3] Department of Geographical Sciences, University of Maryland, College Park, MD 20742, USA

* Correspondence: Jie_Cheng@bnu.edu.cn

Received: 2 November 2017; Accepted: 29 November 2017; Published: 1 December 2017

Abstract: Land surface temperature (LST) is a key parameter for a wide number of applications, including hydrology, meteorology and surface energy balance. In this study, we first proposed a new land surface emissivity (LSE) scheme, including a lookup table-based method to determine the vegetated surface emissivity and an empirical method to derive the bare soil emissivity from the Global LAnd Surface Satellite (GLASS) broadband emissivity (BBE) product. Then, the Modern Era Retrospective-Analysis for Research and Applications (MERRA) reanalysis data and the Feng Yun-3C/Medium Resolution Spectral Imager (FY-3C/MERSI) precipitable water vapor product were used to correct the atmospheric effects. After resolving the land surface emissivity and atmospheric effects, the LST was derived in a straightforward manner from the FY-3C/MERSI data by the radiative transfer equation algorithm and the generalized single-channel algorithm. The mean difference between the derived LSE and field-measured LSE over seven stations is approximately 0.002. Validation of the LST retrieved with the LSE determined by the new scheme can achieve an acceptable accuracy. The absolute biases are less than 1 K and the STDs (RMSEs) are less than 1.95 K (2.2 K) for both the 1000 m and 250 m spatial resolutions. The LST accuracy is superior to that retrieved with the LSE determined by the commonly used Normalized Difference Vegetation Index (NDVI) threshold method. Thus, the new emissivity scheme can be used to improve the accuracy of the LSE and further the LST for sensors with broad spectral ranges such as FY-3C/MERSI.

Keywords: FY-3C/MERSI; GLASS; Land surface temperature; Land surface emissivity

1. Introduction

Land surface temperature (LST) is one of the key parameters in the land surface physical processes at regional and global scales, integrating the interactions between the surface and atmosphere and all energy exchanges between the atmosphere and the land [1,2]. LST plays a significant role in many research fields, such as weather forecasting, global ocean circulation and climate change research [3]. Remote sensing is a unique way of obtaining the LST at regional and global scales. Three kinds of algorithms have been proposed in the past decades to derive the LST from satellite data [4,5], i.e., the single-channel algorithm [6–8], the split-window (SW) algorithm [1,9–11] and the multi-channel algorithm [12,13]. With these versatile algorithms, many LST products have been produced from different satellite data, such as the Advanced Spaceborne Thermal Emission and Reflection Radiometer (ASTER) [14,15], Moderate Resolution Imaging Spectroradiometer (MODIS) [1,16], Visible Infrared

Imaging Radiometer Suite (VIIRS) [10,17], Geostationary Operational Environmental Satellites (GOES) [18,19] and Spinning Enhanced Visible and Infrared Imager (SEVIRI) [20]. Those LST products have been widely used for monitoring urban heat islands [21,22] and volcanoes [23–25], detecting forest fires [26,27], and so on.

FengYun-3C (FY-3C) satellites are China's second-generation polar-orbiting meteorological satellites. The Medium Resolution Spectral Imager (MERSI) is the instrument onboard the FY-3C, with 4 of 19 visible/shortwave channels and 1 thermal infrared (TIR) channel are set for 250 m spatial resolution, with other channels for 1 km spatial resolution. FY-3C/MERSI also provides 1000 m radiance data interpolated from the original 250 m data. Thus, we can obtain 250 m and 1000 m spatial resolution thermal infrared data from FY-3C/MERSI. FY-3C/MERSI provides a new data source for the retrieval of LST and meteorology monitoring. However, to our knowledge, an operational FY-3C/MERSI LST product is unavailable.

For the sensor with only one TIR channel, a set of LST retrieval algorithms, such as the radiative transfer equation algorithm [8,28], the mono-window algorithm [6] and the generalized single-channel algorithm [7], have been developed for estimating the LST. According to the validation results obtained from four Surface Radiation Budget Network (SURFRAD) sites by Yu et al. [29], the radiative transfer equation algorithm has the highest accuracy, and the root mean square error is less than 1 k. Additionally, the study by Windahl and Beurs [30] verified the precision of the radiative transfer equation method, mono-window algorithm and generalized single-channel algorithm. The accuracy of the three algorithms decreased with the increase of water vapor content, and the radiative transfer equation method has a higher precision under a high water vapor content. The research of Jiménez-Muñoz et al. [31] indicated that the accuracy of the generalized single-channel algorithm is below 1 K when the water vapor content was lower than 2 g/cm^2. The effective mean atmospheric temperatures in the mono-window algorithm are often estimated from the empirical formula with the near-surface air temperature, and this may be not suitable for some special study areas [32]. Thus, both the radiative transfer equation algorithm and the generalized single-channel algorithm are the potential LST retrieval algorithms for FY-3C/MERSI.

The accuracy of the single channel algorithm depends on the accuracy of the atmospheric correction and land surface emissivity (LSE). At present, the Normalized Difference Vegetation Index (NDVI) threshold method [33], the Vegetation Cover Method (VCM) [3] and the classification-based method [34] are widely used in the single channel algorithms. However, these techniques present several limitations. LSE based on the classification-based method cannot reflect the land cover changes [35]. For example, the MODIS LST products (collection 5) underestimate the LST in an arid area of northwest China, due to an overestimation of the LSE by the classification-based method [36]. Therefore, an accurate LSE scheme is the prerequisite of accurate LST retrieval. In addition to the determination accuracy of the land surface emissivity, the precision of the atmospheric correction also affects the results of the LST. Most of the single channel algorithms were developed without considering the effect of the view zenith angle; this will introduce a large error into the results of the sensor with a large view zenith angle [32]. Given that the view zenith angle of FY-3C/MERSI can reach up to 55 degrees, we will conduct an angular dependent atmospheric correction.

This study aims to accurately estimate the LST from the FY-3C/MERSI data using a more realistic LSE scheme. The structure of this paper is arranged as follows: Section 2 introduces the data used in this study and the estimation of the ground LST. Section 3 describes the methodology used in this study, including a new LSE scheme, atmospheric correction and the determination of the LST. The results and analysis are presented in Section 4. A discussion is provided in Section 5, and the main conclusions are summarized in Section 6.

2. Data

2.1. Satellite Products

The FY-3C/MERSI images, Global LAnd Surface Satellite (GLASS) broadband emissivity (BBE) product, GLASS leaf area index (LAI) product, MODIS Surface Reflectance data (MOD09GQ) and MODIS Land Cover data (MCD12Q1) were used to estimate the land surface temperature. Nine FY-3C/MERSI images ranging from 17 July 2014 to 6 October 2014 were obtained in this study. For the convenience of registration, the calibrated MERSI 1000 m resolution earth viewing data was used after calibration. The land surface reflectance, land cover or LAI products from FY-3C/MERSI are unavailable; therefore, the MOD09GQ, MCD12Q1 and GLASS LAI were used for the LST retrieval from FY-3C/MERSI. The FY-3C/MERSI images were re-projected to the projection of MOD09GQ. All the satellite products used were resampled to 1 km resolution to match the spatial resolution of the FY-3C/MERSI data.

The MOD09GQ Version 6 product provides an estimate of the surface spectral reflectance of the Terra MODIS 250 m data corrected for the atmospheric conditions such as gases, aerosols, and Rayleigh scattering. The MOD09GQ were used to calculate the NDVI and then identify the vegetated surfaces. The MODIS Land Cover product (MCD12Q1) is an annual land cover dataset with a 500 m spatial resolution, which contains five classification schemes. The International Geosphere Biosphere Program (IGBP) global vegetation classification scheme was selected in this paper to determine the leaf emissivity of vegetated surfaces.

The GLASS BBE product [37] was derived from the Advanced Very High Resolution Radiometer (AVHRR) and MODIS data using the newly developed algorithms [38,39]. The GLASS LAI product was generated using a general regression neural network (GRNN) from the MODIS surface reflectance data [40]. Both GLASS BBE and LAI products have spatial and temporal resolutions of 1 km and eight days, respectively. Detailed information for the GLASS product can be found in Liang et al. [37]. The GLASS BBE and the GLASS LAI products were used for estimating the LSE for FY-3C/MERSI.

2.2. Ground Measurements

The Heihe Watershed Allied Telemetry Experimental Research (HiWATER) [41,42] was performed in the Heihe River Basin, which is a typical inland river basin in northwest China. In this study, three datasets were selected from this experiment: the dataset of the thermal infrared spectrum observed by BOMEM MR304 in the middle reaches of the Heihe River Basin [43], the dataset of the hydrometeorological observation network (automatic weather station, 2014) [44] and the dataset of infrared temperature in the Zhanye Airport desert [45].

The surface-leaving radiance of the different components of land surfaces (soil, sand, corn leaf, soybean leaf, apple leaf, etc.) and atmospheric downward radiance were measured by ABB BOMEM MR304 spectroradiometers and a diffuse gold plate [36]. The emissivity spectra in the range of 8~14 μm with a spectral resolution of 1 cm^{-1} were retrieved using the Iterative Spectrally Smooth Temperature and Emissivity Separation (ISSTES) algorithm [46]. The emissivity spectra in the ASTER and the MODIS spectral library were also used to determine the leaf emissivity in Section 3.3 and to derive the LSE for the FY-3C/MERSI from the GLASS BBE for bare soils.

There are five automatic weather stations in this study: Bajitan Gobi Desert station (GB), Shenshawo sandy desert station (SSW), Huazhaizi desert steppe station (HZZ), Zhangye wetland station (SD) and Daman Superstation (CJZ) and the automatic weather stations were installed with Kipp and Zonen CNR1 net radiometers at a 6-m-height or with the SI-111 radiometer at a 2.65-m-height. The datasets of the five automatic weather stations and infrared temperatures in the Zhanye Airport desert (JCHM) were used to validate the LST estimated from the FY-3C/MERSI. The spatial distribution of the six field sites is shown in Figure 1.

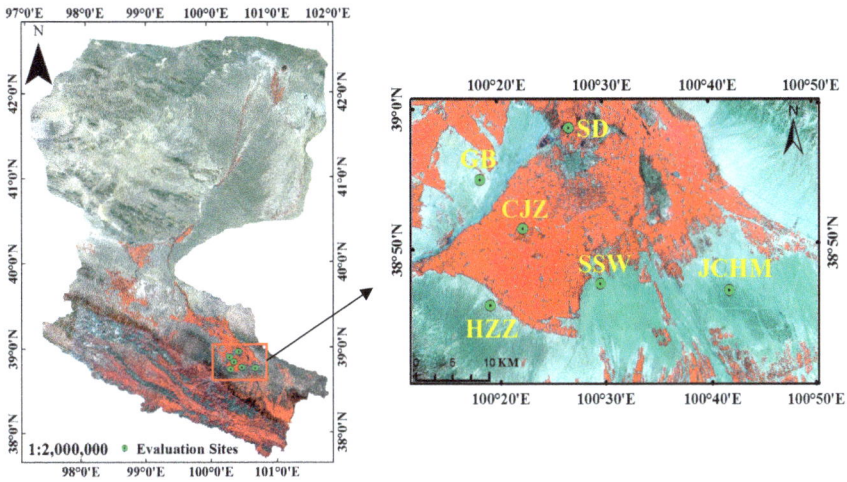

Figure 1. Spatial distribution of the six in situ sites. (The base map is from HJ-1 CCD false color composite image and the RGB components are channels 4, 3 and 2, respectively).

For the CJZ, GB and SSW sites, the ground LSTs were estimated from the upward and downward longwave radiation, which were observed at nadir by Kipp and Zonen CNR1 net radiometers, using the following equation:

$$T_s = [\frac{F^\uparrow - (1 - \varepsilon_b) \cdot F^\downarrow}{\varepsilon_b \cdot \sigma}]^{1/4} \tag{1}$$

where T_s is the LST, $F^\uparrow$ is the surface upward longwave radiation, ε_b is the BBE, σ is the Stefan-Boltzmann constant (5.67×10^{-8} Wm^{-2} K^{-4}), and $F^\downarrow$ is the atmospheric downward longwave radiation at the surface. The BBE is estimated from the ASTER narrowband emissivity using the following linear equation derived by Cheng et al. [47]:

$$\varepsilon_{bb} = 0.197 + 0.025\varepsilon_{10} + 0.057\varepsilon_{11} + 0.237\varepsilon_{12} + 0.333\varepsilon_{13} + 0.146\varepsilon_{14} \tag{2}$$

where ε_{bb} is the surface broadband emissivity at a spectral range of 8 ~13.5 μm, and ε_{10} ~ε_{14} are the ASTER narrowband emissivity of five channels. The average ε_{bb} values of 0.944 ± 0.009 and 0.914 ± 0.009 were obtained from all 20 scenes of the AST_05 product (May of 2012 to August of 2015) for the GB and SSW sites, respectively. The average ε_{bb} values of the CJZ sites were calculated from the GLASS BBE of each eight-day period, as the ASTER LSE was inaccurate over vegetated surface [14,48] and the GLASS BBE achieves an acceptable accuracy over vegetated surfaces [49].

For the SD, HZZ and JCHM sites, the ground LSTs were calculated from radiometric temperatures measured at nadir by the SI-111 radiometer, using the following equation:

$$B(T_s) = [B(T_r) - (1 - \varepsilon)L_{sky}]/\varepsilon \tag{3}$$

where B is the Planck function weighted for the spectral response function of the SI-111 radiometer, ε is the surface emissivity of the SI-111 channel with a spectral range of 8–14 μm and L_{sky} is the downward longwave radiation measured by the SI-111 radiometers installed at the JCHM site, which is aimed at the sky at approximately 55° from the zenith. The average ε was estimated from the five ASTER narrowband emissivity using the following linear equation derived from the spectral library: the average ε values of 0.964 ± 0.006 and 0.955 ± 0.007 were adopted for the HZZ and JCHM sites, respectively. The average ε_{bb} values of the SD sites were also calculated from the GLASS BBE of each eight-day period for the same reason as the CJZ sites.

$$\varepsilon_{SI-111} = 0.1309 + 0.0918\varepsilon_{10} + 0.0701\varepsilon_{11} + 0.1069\varepsilon_{12} + 0.5456\varepsilon_{13} + 0.0515\varepsilon_{14} \tag{4}$$

3. Methods

3.1. Algorithms Used for Estimating the LSTs

3.1.1. The Radiative Transfer Equation (RTE) Algorithm

According to the radiative transfer equation, the blackbody radiance under clear sky conditions can be expressed by the following formula:

$$B_i(T_s) = \frac{L_i^{sen} - L_i^{\uparrow}(\theta)}{\tau_i(\theta)\varepsilon_i} - \frac{1 - \varepsilon_i}{\varepsilon_i}L_i^{\downarrow} \tag{5}$$

where L_i^{sen} is the at-sensor radiance of channel i, T_s is the land surface temperature, θ is the view zenith angle (VZA), $B_i(T_s)$ is the blackbody radiance of channel i, ε_i is the land surface emissivity of channel i, $\tau_i(\theta)$ and $L_i^{\uparrow}(\theta)$ are the atmospheric transmittance and atmospheric upward radiance of channel i at VZA θ, and $L_i^{\downarrow}$ is the downward atmospheric irradiance of channel i. Provided with LSE and three atmospheric parameters, the LST calculation is straightforward.

3.1.2. The Generalized Single-Channel (GSC) Algorithm

Jiménez-Muñoz and Sobrino [7] developed a generalized single-channel method to retrieve the land surface temperature from a single thermal sensor. The land surface temperature is expressed by the following formula:

$$T_s = \gamma\left[\frac{1}{\varepsilon_i}(\psi_1 L_i + \psi_2) + \psi_3\right] + \delta \tag{6}$$

where ε_i is the land surface emissivity of channel i, γ and δ given by the following:

$$\gamma = \left\{\frac{c_2 L_i}{T_i^2}\left[\frac{\lambda_i^4}{c_1}L_i + \lambda_i^{-1}\right]\right\}^{-1}; \delta = -\gamma L_i + T_i \tag{7}$$

where T_i refers to the at-sensor brightness temperature; L_i is the radiance received by channel i of the sensor; c_1 is $1.19104 \times 10^8 \, W \times \mu m^4 \times m^{-2} \times sr^{-1}$ and c_2 is $14387.7 \, \mu m \times K$; λ_i is the effective band wavelength for band i; and ψ_1, ψ_2 and ψ_3 are the atmospheric functions, given by the following:

$$\psi_1 = \frac{1}{\tau_i(\theta)}; \psi_2 = -L_i^{\downarrow} - \frac{L_i^{\uparrow}(\theta)}{\tau_i(\theta)}; \psi_3 = L_i^{\downarrow} \tag{8}$$

where $\tau_i(\theta)$ and $L_i^{\uparrow}(\theta)$ are the atmospheric transmittance and atmospheric upward radiance of channel i at VZA θ, and $L_i^{\downarrow}$ is the downward atmospheric irradiance of channel i. If the atmospheric parameters $\tau_i(\theta)$, $L_i^{\downarrow}$ and $L_i^{\uparrow}(\theta)$ are known, the atmospheric functions can be calculated from (8).

3.2. Angular Dependent Atmospheric Correction

It is well known that $\tau_i(\theta)$ and $L_i^{\uparrow}(\theta)$ vary with VZA [32], so the effect of VZA should be considered in the atmospheric correction, because the VZA of the FY-3C/MERSI can reach up to a maximum value of 55 degrees. In the RTE algorithm, many studies have indicated that various atmospheric reanalysis products, such as NCEP/FNL, MERRA and ERA-Interim can obtain good atmospheric correction results [8,50–53]. Benefitting from its high vertical resolution (42 pressure levels from 1000 hpa to 0.1 hpa) and high spatial resolution (2/3° longitude × 1/2° latitude) [52], the MERRA reanalysis data in conjunction with the fast radiative transfer model RTTOV 11.3 [54] are utilized for atmospheric correction in the RTE algorithm. The precipitable water vapor product of FY-3C/MERSI, which is longitude/latitude projected with a 0.05° resolution, were used to calculate the atmospheric functions in the GSC algorithm.

For a particular FY-3C/MERSI scene, the atmospheric transmittance and upward radiance at nadir view ($\tau_i(0)$ and $L_i^\uparrow(0)$) were calculated using RTTOV and MERRA in the RTE algorithm and calculated from the water vapor content in the GSC algorithm. To minimize the computational time, the downward radiance was modeled as a non-linear function of the upward radiance at nadir view [55]. For a given VZA, $\tau_i(\theta)$ and $L_i^\uparrow(\theta)$ can be fitted as the non-linear function of $\tau_i(0)$ and $L_i^\uparrow(0)$. MODTRAN 5.2 and SeeBor V5.0 training database of global profiles [56,57] (SeeBor V5.0 profiles for simplicity as follows) were used for establishing the non-linear relationship.

According to the study of Galve et al. [58], 2762 SeeBor V5.0 profiles acquired on land under clear sky conditions were chosen for the simulation. Given that the VZA of the FY-3C/MERSI can reach up to 55 degrees, the VZA are designed with a range from 0 to 65 degrees in a 5-degree step. The result of atmospheric transmittance and upward radiance under various VZAs are depicted in Figure 2. Clearly, the atmospheric transmittance or upward radiance differences increase with VZA, due to the rise of the atmospheric path with the angle. For a given VZA, $\tau_i(\theta)$ and $L_i^\uparrow(\theta)$ can be expressed by $\tau_i(0)$ and $L_i^\uparrow(0)$ through the following quadratic equation and $L_i^\downarrow$ can also be expressed by $L_i^\uparrow(0)$ through the following expression. The coefficients will be given in Section 4.1.

$$Y = aX^2 + bX + c \tag{9}$$

where X is $\tau_i(0)$ or $L_i^\uparrow(0)$, Y is $\tau_i(\theta)$ or $L_i^\uparrow(\theta)$.

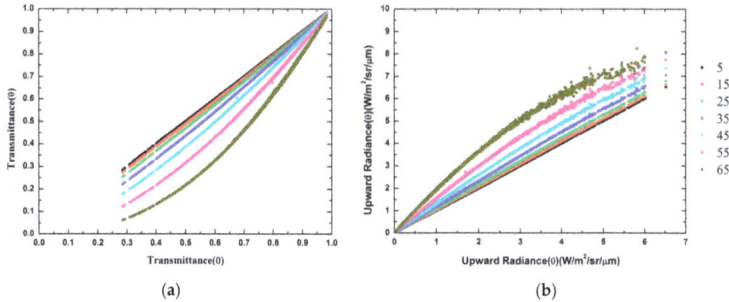

Figure 2. Plot of the atmospheric transmittance (**a**) or upward radiance (**b**) at a given VZA against the atmospheric transmittance or upward radiance at nadir view.

3.3. A New Scheme for Determining Land Surface Emissivity

The land surface was divided into vegetated surfaces and bare soils to calculate their LSEs, respectively. Figure 3 shows the spectral response for band 5 of the FY-3C/MERSI. We can see that the MERSI has a broadband spectral range of 9.5–13 μm, which is inside the spectral range of GLASS BBE (8–13.5 μm). The study of Ren and Cheng [59] indicated that a linear relationship between MERSI emissivity and GLASS BBE existed for bare soils using the soil emissivity spectra in the ASTER spectral library. As the accuracy of GLASS BBE is better than 0.02 [60], the performance of this linear relationship is certainly better than the constant assumption or a linear function fitting of red reflectance adopted by the NDVI threshold and the VCM methods in predetermining LSE for LST estimation.

Thus, the GLASS BBE was used to determine the LSE for LST estimation from the FY-3C/MERSI. The emissivity of the bare soils was estimated with the following equations:

$$\varepsilon_{soil} = 0.8731 * \varepsilon_{BB} + 0.1269 \tag{10}$$

where ε_{soil} is the soil emissivity and ε_{BB} is the GLASS BBE. The regression coefficients are determined by a total of 45 emissivity spectra from the ASTER spectral library, MODIS spectral library and the measured soil emissivity from Wang et al. [43].

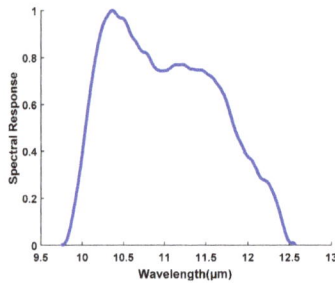

Figure 3. Spectral response for band 5 of FY-3C/MERSI.

Regarding the vegetated surfaces, we followed the method proposed by Cheng et al. [49], which can reflect the abundance of the vegetation and also has an accuracy of better than 0.005 over the fully covered vegetated surface. We used the 4Scattering by Arbitrary Inclined Leaves (4SAIL) model to construct a lookup table (LUT) of the LSE for vegetated surfaces. The variation ranges for three principal model inputs were set as follows: the leaf emissivity ranges from 0.935 to 0.995 and has an interval of 0.01; the soil emissivity varies from 0.71 to 0.99 and has an interval of 0.01; and the LAI ranges from 0 to 6.0 and has an interval of 0.5. After that, we can derive the emissivity of vegetated surfaces by interpolating the LUT using three inputs: leaf emissivity, soil emissivity, and LAI. Vegetated surfaces were identified using the NDVI calculated from MOD09GQ. LAI was extracted from the GLASS LAI product. Leaf emissivity was calculated from the measurements by Pandya et al. [61], Wang et al. [43], and the ASTER and MODIS spectral library for five composited vegetation land cover types based on MCD12Q1; these are shown in Table 1. The soil background emissivity underneath the vegetation canopy was derived from the mean GLASS BBE from month 10 of this year to month 4 of the following year. The land surface in this period is covered by soil rather than vegetation, and the emissivity in this period is the emissivity of the soil background for vegetated surfaces.

Table 1. Leaf emissivity values for five composited vegetation land cover types.

IGBP Class	Composite Type	Leaf Emissivity	Sources
1~7	Forest and Shrubland	0.967	Mean of conifer and deciduous emissivity from ASTER spectral library and 24 leaf emissivities from MODIS spectral library and 10 measured leaf emissivities from Wang et al. [43]
8, 9	Savanna	0.966	50% forest + 50% grassland
10	Grassland	0.965	Mean of green grass emissivity from ASTER spectral library and elephant grass emissivity from Pandya et al. [61]
12, 14	Cropland	0.966	Mean of 9 leaf emissivities from Pandya et al. [61], 4 wheat emissivities of Li et al. [36] and 39 measured leaf emissivities from Wang et al. [43]
16, 254	Other types	0.966	Mean value of above four types

4. Results and Analysis

4.1. Coefficients for Atmospheric Correction

Values for $\tau_i(0)$ and $L_i^\uparrow(0)$ in Equation (9) can be derived after running the RTTOV, or they can be calculated from the precipitable water vapor through the following expression:

$$\tau_i(0) = 0.9703 - 0.0563w - 0.02059w^2 + 0.00208w^3, \ R^2 = 0.977$$
$$L_i^\uparrow(0) = 0.07306 + 0.41283w + 0.20374w^2 - 0.01948w^3, \ R^2 = 0.965 \tag{11}$$

Different coefficients are obtained for each VZA when regressions of $\tau_i(\theta)$ and $L_i^\uparrow(\theta)$ are completed against $\tau_i(0)$ and $L_i^\uparrow(0)$. To obtain a uniform angular dependent atmospheric correction expression, a linear function of $sec(\theta) - 1$ is used [62], and the quadratic equation can be written as follows:

$$Y = (a1*S^2 + a2*S + a3)X^2 + (b1*S^2 + b2*S + b3)X + (c1*S^2 + c2*S + c3) \tag{12}$$

where $S = sec(\theta) - 1$, $a1$, $a2$, $a3$, $b1$, $b2$, $b3$, $c1$, $c2$ and $c3$ are the coefficients of the formula. The angular dependent atmospheric correction coefficients for FY-3C/MERSI are given in Table 2 and the coefficient of determination of 0.99 was obtained by fitting all data to Equation (12).

Table 2. Angular Dependent Atmospheric correction coefficients of Equation (12) for FY3C/MERSI.

Parameters	a1	a2	a3	b1	b2	b3	c1	c2	c3
$L_i^\uparrow(\theta)$	−0.0111	−0.0846	0.0007	−0.0955	0.9205	0.9997	0.0189	−0.0198	0.0003
$\tau_i(\theta)$	0.1077	0.721	−0.0055	−0.2987	−0.4775	1.0104	0.1885	−0.2376	−0.005

4.2. Evaluation with the In Situ LSE

In this paper, the measured emissivities on 4 September 2014 or 5 September 2014 were selected to validate the estimated LSE on 4 September 2014. The emissivities were measured near the in situ sites by the 102F Portable Fourier Transform Infrared (FTIR) Spectrometer [63]. There are six measurements for the vegetation surfaces, namely, corn leaf, Chinese cabbage leaf, *Alhagi sparsifolia*, sparse vegetation, large cluster sparse vegetation, and one measurement for bare soil. The leaf emissivity of corn and Chinese cabbage were measured on 5 September 2014, the emissivities of other types were measured on 4 September 2014.

Table 3 summarizes the values of measured emissivity, estimated LSE and the bias between them for the different sites. Figure 4 shows the photos of measurement sites. Compared to the measured emissivity values, the LSE at the CJZ01, SSW02, SSW03 and GB02 sites were overestimated by 0.012, 0.025, 0.003 and 0.003, respectively; and LSE at CJZ02, GB01 and SSW01 sites were underestimated by 0.002, 0.016 and 0.015, respectively. What the spectrometer actually measured at the CJZ01 site is the emissivity of corn leaf, and so it is lower than the estimated LSE because the multi-scattering is not considered. The bias is very small at a relatively homogeneous site such as GB02. Regarding the sparsely vegetated surfaces, the spatial scaling effect is very strong because it is difficult to determine exactly what the spectrometer has seen in its narrow field of view. We should be very cautious when using the measured data in sparse vegetation. For example, the field of view of the spectrometer may be totally soil or vegetation canopy if the vegetation coverage is quite low. Figure 4c can also illustrate this phenomenon. The measured object is *Alhagi sparsifolia*, whose area ratio is quite low in the whole pixel, and so the estimated LSE of the whole pixel is close to the measured soil emissivity (Figure 4f). The larger bias that appears over GB01 and SSW02 is not difficult to understand. Assuming that the measured emissivity is accurate enough, the large bias between the measured emissivity and estimated LSE may come from the uncertainty of a new emissivity determination method or the spatial variability of field stations, both of which will be discussed in the following sections.

Table 3. List of measured emissivities, estimated LSEs and the bias between them for different types.

Station Name	Type Name	Measured Emissivity	Estimated LSE Using the Method in Section 3.3	Bias
CJZ01	Corn leaf	0.973	0.985	0.012
CJZ02	Chinese cabbage leaf	0.987	0.985	−0.002
GB01	*Alhagi sparsifolia*	0.976	0.960	−0.016
SSW01	Sparse vegetation	0.986	0.971	−0.015
SSW02	Large cluster sparse vegetation	0.960	0.985	0.025
SSW03	Large cluster sparse vegetation	0.959	0.962	0.003
GB02	Bare soil	0.957	0.960	0.003

Figure 4. Photos of measurement sites. (**a**) CJZ01; (**b**) CJZ02; (**c**) GB01; (**d**) SSW01; (**e**) SSW02,03; (**f**) GB02.

4.3. Sensitivity Analysis of Emissivity

To analyze the effects of leaf emissivity, soil emissivity and LAI on the estimated LSE over vegetated surfaces, we adhered to the following rules to conduct the simulation: (1) the leaf emissivity ranges from 0.92 to 0.98 with an interval of 0.02, the LAI ranges from 0.1 to 6.0 with an interval of 0.5 from 0.5 to 6.0, and the soil emissivity was set to 0.96; (2) the soil emissivity ranges from 0.90 to 0.98 with an interval of 0.02, the LAI ranges from 0.1 to 6.0 with an interval of 0.5 from 0.5 to 6.0, and the leaf emissivity was set to 0.96. Figure 5 shows the emissivity calculated from the 4SAIL model. When the LAI values change from 0.1 to 2.0, the variation of soil emissivity has more influence on the estimated LSE than the variation of the leaf emissivity. In this range, when the leaf emissivity changes from 0.92 to 0.98, the difference between the maximum and minimum estimated LSE changed from 0.033 to 0.046, but when the soil emissivity changed from 0.90 to 0.98, the difference between the maximum and minimum estimated LSE changed from 0.075 to 0.007. This indicated that the soil emissivity might cause great errors to the estimated LSE of the vegetated surface when the LAI is less than 2.0. The emissivity shows little change when the LAI is greater than 3.0, when the leaf emissivity changes from 0.92 to 0.98, the emissivity remains from 0.978 to 0.994, no matter how the soil emissivity and LAI change. The emissivity shows little change when the LAI is greater than 3.0, when the soil emissivity changes from 0.90 to 0.98 and the leaf emissivity is equal to 0.96, the emissivity remains 0.989, no matter how much the LAI changes.

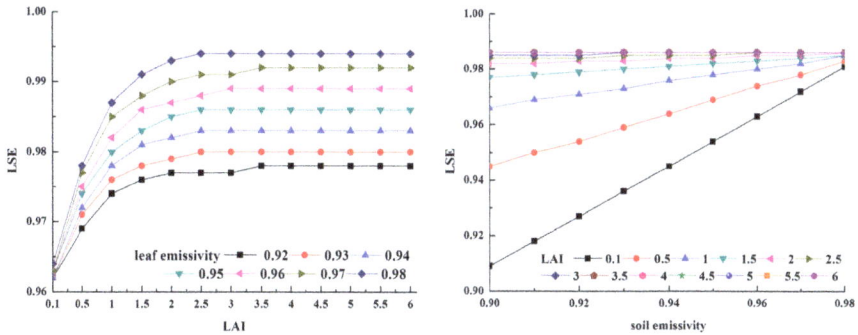

Figure 5. LSE simulated by 4SAIL model varying LAI at a fixed soil emissivity of 0.96 (**left**) or a fixed leaf emissivity of 0.96 (**right**).

We also analyzed the stability of the soil background emissivity underneath the vegetation canopy and the applicability of the regression formula over bare soils. First, we plot the BBE of each eight-day period from 2001 to 2014 for different stations, as shown in Figure 6. As described in Section 3.3, the emissivity from month 10 of this year to month 4 of next year is the emissivity of the soil background for vegetated surfaces. We can infer from this that the soil background emissivity underneath the vegetation canopy is stable, based on Figure 6. According to the validation using four field trials by Cheng et al. [39], the accuracy of the GLASS BBE of bare soil is 0.016, and so the mean soil emissivity can be used to represent the soil background emissivity underneath the vegetation canopy. Second, using only one formula to regress the relationship between the GLASS BBE and estimated LSE may introduce some errors. The BBE variation is various for the different soil types, different seasons and different areas [49]. This trend is also obvious in Figure 6, as the BBE are slightly different at the various stations. Because of this, we chose soil and sand samples from the spectral library to calculate the regression coefficient between the GLASS BBE and estimated LSE, respectively. The regression results from soil and sand samples are obviously different, so it is noteworthy when using the regression formula. If we used the regression formula for soil samples to calculate the LSE of the sand surface, it may be inaccurate.

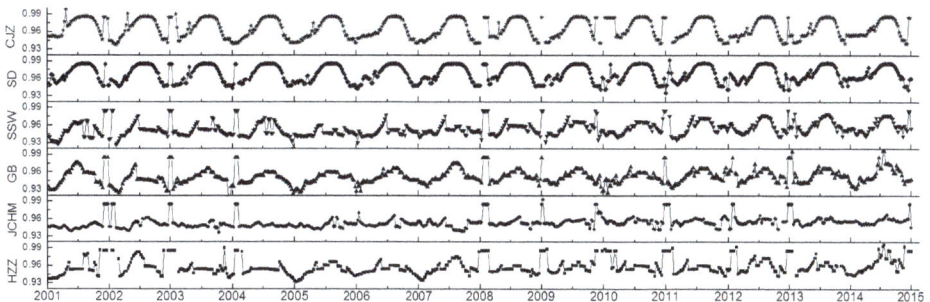

Figure 6. The BBE of each eight-day period from 2001 to 2014 for six stations.

4.4. Validation of Retrieved LST from FY-3C/MERSI with In Situ LST

Both the radiative transfer equation algorithm and the generalized single-channel algorithm were used for estimating the LST from the FY-3C/MERSI images acquired in 2014. Figure 7 shows the statistical results between the estimated LST from the RTE algorithm or the GSC algorithm and the in situ LST. The results derived from the RTE algorithm, using the LSE calculated by the new emissivity scheme, was denoted as RTE1. The NDVI threshold method developed by Sobrino et al. [33] was also used to retrieve the LST by the RTE algorithm. The results were called RTE2 accordingly. The results derived from the GSC algorithm, which the LSE calculated by new scheme, was denoted as GSC1, whereas the result derived from the GSC algorithm using the LSE calculated by the NDVI threshold method was denoted as GSC2. Note that the LST of the GSC algorithm on 17 July 2014 was empty due to a lack of atmospheric water vapor data.

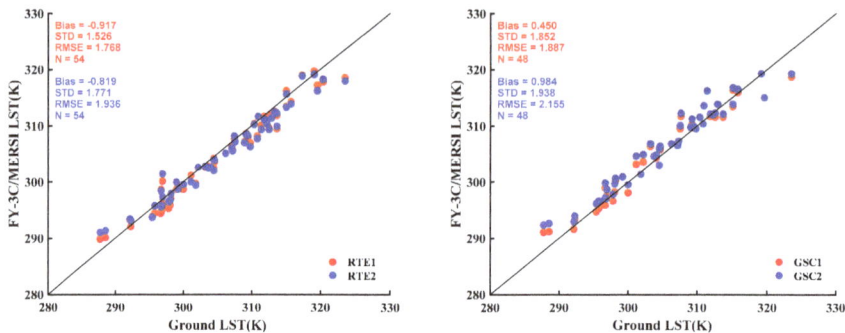

Figure 7. Scatterplots between estimated LST from RTE algorithm (**left**) or GSC algorithm (**right**) and in situ LST.

From Figure 7, we can see that there is a high correlation between the estimated LST and the in situ LSTs for both the RTE algorithm and the GSC algorithm. The bias and STD (RMSE) of the RTE algorithm with the LSE derived from the new scheme are −0.92 K and 1.53 K (1.77 K). The bias and STD (RMSE) of the RTE algorithm using the LSE calculated by the NDVI threshold method are −0.82 K and 1.77 K (1.94 K). The bias and STD (RMSE) of the GSC algorithm with the LSE derived from the new LSE scheme are 0.45 K and 1.85 K (1.89 K). The bias and STD (RMSE) of the GSC algorithm using the LSE calculated by the NDVI threshold method are 0.98 K and 1.94 K (2.16 K). The validation results showed that the estimated LSTs are obviously underestimated at most stations for the two RTE algorithms, while most of the LSTs derived from the GSC algorithm was overestimated. Overall, we can conclude that the new LSE scheme can achieve a higher precision of LST determination.

Comparing the LSE derived by the NDVI threshold method with the LSE derived by the new scheme, the two methods have nearly similar LSTs, because the soil emissivity used here is the same as the new scheme, rather than the constant assumption adopted by the original NDVI threshold. According to the MODIS precipitable water product (MOD05), most of the total water vapor content of the validation stations in 2014 was less than 2.0 g/cm^2, and so the RTE and GSC algorithms all have obtained a high level of accuracy. The results are consistent with the study by Jiménez-Muñoz et al. [31], who claimed that the GSC algorithm has a high accuracy in a low water vapor content area. Although a high precision of the LST can be achieved with the LSE determined by the new scheme, we should be very cautious when using the atmospheric profiles for atmospheric correction. When the surroundings of study area were covered by clouds, the LST of the RTE algorithm is underestimated, e.g., JCHM station on 24 August 2014.

5. Discussion

5.1. Comparison with LST Derived from ASTER Emissivity Product

In this section, we used the LSE calculated from the ASTER Surface Emissivity product (AST_05) to evaluate the accuracy of the LSE estimated by the new scheme, as well as the estimated LSTs. The AST_05 is an on-demand product generated using the Temperature/Emissivity Separation (TES) algorithm [12] and combined with the Water Vapor Scaling (WVS) atmospheric correction method [55] for the five thermal infrared (TIR) 90 m resolution bands. The study by Hully et al. [64] indicated that the accuracies in retrieving the spectral emissivity for ASTER were below 0.016. The AST_05 product on 23 July 2014 was used to evaluate the emissivity determined by the new scheme.

First, the FY-3C/MERSI LSE can be calculated from the linear transformation formula, as shown in Equation (13). We adopted the least-squares fitting method to establish the transformation formula, using the 251-emissivity spectra in the ASTER spectral library and the spectral response function of the ASTER band 13, 14 and FY-3C/MERSI.

$$LSE_{aster} = 0.7045\varepsilon_{13} + 0.2381\varepsilon_{14} + 0.055 \qquad (13)$$

where LSE_{aster}, ε_{13} and ε_{14} are the land surface emissivity of the FY-3C/MERSI and ASTER channels 13 and 14, respectively. Then, we compared the estimated LSE on 17 and 26 July 2014 to the LSE calculated from AST_05 on 23 July 2014, assuming that the land surface emissivity of the study area is stable during a short time. Figure 8 shows the images of the LSE difference on 17 and 26 July 2014 and the corresponding images of the LST difference. The LSE difference was calculated using the LSE estimated from the scheme presented in this paper (LSE_{mersi}) minus the LSE estimated from the AST_05 product (LSE_{aster}). The LSTs were derived from the RTE algorithm. The histogram of the LSE difference on 17 and 26 July 2014 and the corresponding histogram of the LST difference ($LST_{mersi} - LST_{aster}$) are also provided.

The average bias of the LSE difference over vegetated surfaces on 17 July 2014 is 0.009 and the RMSE is 0.013. The average bias and RMSE of the LSE difference over the vegetated surface on 26 July 2014 are the same as the values on 17 July 2014. The new scheme provides accurate values over bare soil surface on 17 July 2014 and 26 July 2014, with an average bias of less than 0.002 and a RMSE of less than 0.011, respectively. The LSE estimated from the new scheme over the vegetated surface on 17 July 2014 and on 26 July 2014 have been overestimated by 0.01, when compared with the LSE estimated from the AST_05 product. This result is consistent with the experiments of Gillespie et al. [65] and the research of Jiménez-Muñoz et al. [66]. They found that the TES algorithm has larger uncertainties over low spectral contrast surfaces, such as vegetation, and it provides accurate values for soil and rocks [49].

The average bias of the LST difference over vegetated surfaces on 17 July 2014 is −0.581 K and the RMSE is 0.808 K. The average bias and RMSE of the LST difference over vegetated surfaces on 26 July 2014 are −0.577 K and 0.784 K, respectively. The new scheme provides accurate values over the bare soil surfaces on 17 July 2014 and 26 July 2014, with an average bias of less than 0.06 K and RMSE less than 0.7 K, respectively. The statistical results show that the LST bias over the vegetated and bare soil surfaces have small statistical errors, most of the absolute bias values are within 2 K. Table 4 shows the results of the LSE estimated from AST_05 and that determined by the new scheme on 17 and 26 July 2014, as well as the corresponding LST at the validation stations. The LSEs of the GB and HZZ sites provided accurate values on 17 and 26 July 2014, with biases of less than 0.003 and the biases of the corresponding LSTs are within 0.3 K. Compared with the LSEs estimated from AST_05, the LSEs of the SSW site are overestimated by 0.01 and 0.013, respectively, and the LSTs are underestimated by 0.71 K and 0.87 K, respectively. The LSTs of the SD and CJZ sites are all within 2 K with in situ measured LSTs. From the above analyses, we can conclude that the LSE determined by the new scheme is acceptable over a bare soil surface compared with the LSE directly calculated from the

AST_05 product. Therefore, a high precision of the LST can be achieved from the RTE algorithm with the LSE determined by the new scheme.

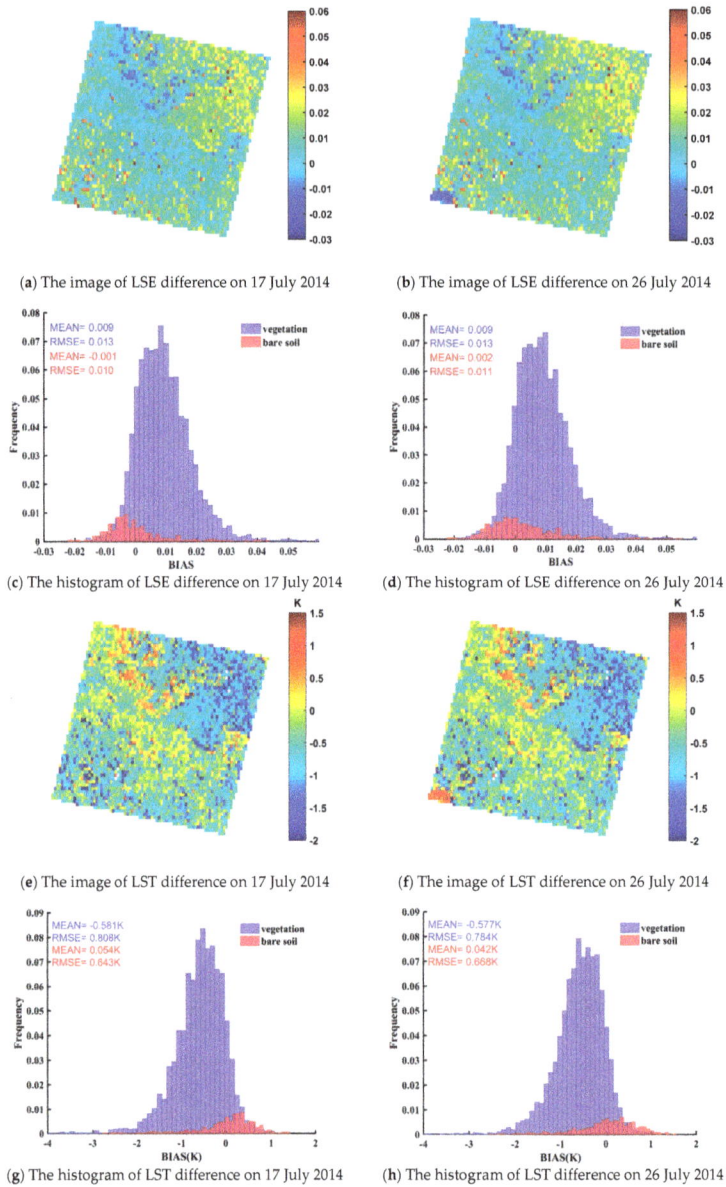

(a) The image of LSE difference on 17 July 2014

(b) The image of LSE difference on 26 July 2014

(c) The histogram of LSE difference on 17 July 2014

(d) The histogram of LSE difference on 26 July 2014

(e) The image of LST difference on 17 July 2014

(f) The image of LST difference on 26 July 2014

(g) The histogram of LST difference on 17 July 2014

(h) The histogram of LST difference on 26 July 2014

Figure 8. The images and histograms of LSE difference between the AST_05 product and the new method on 17 and 26 July 2014; the corresponding images and histograms of LST difference. (a) The image of LSE difference on 17 July 2014; (b) The image of LSE difference on 26 July 2014; (c) The histogram of LSE difference on 17 July 2014; (d) The histogram of LSE difference on 26 July 2014; (e) The image of LST difference on 17 July 2014; (f) The image of LST difference on 26 July 2014; (g) The histogram of LST difference on 17 July 2014; (h) The histogram of LST difference on 26 July 2014.

Table 4. LSE estimated from AST_05 or the new method on 17 and 26 July 2014, as well as the corresponding LST.

	17 July 2014				26 July 2014			
Sites	LSE_{aster}	LSE_{mersi}	LST_{aster} (K)	LST_{mersi} (K)	LSE_{aster}	LSE_{mersi}	LST_{aster} (K)	LST_{mersi} (K)
SD	0.955	0.984	300.566	298.887	0.955	0.984	302.115	300.447
CJZ	0.963	0.985	302.865	301.524	0.963	0.985	300.360	299.062
GB	0.954	0.955	312.475	312.401	0.954	0.957	315.341	315.154
SSW	0.954	0.964	317.890	317.179	0.954	0.967	318.487	317.616
HZZ	0.971	0.968	314.886	315.099	0.971	0.968	314.020	314.224

5.2. Effects of Spatial Scale on the LST and LSE Evaluation

The LST validation sites must be homogeneous from the point scale to several kilometers [67,68]. The spatial variability of the validation stations has large effects on the validation results. To analyze the spatial variability of the six validation stations, we extracted the LSE or LST of the 3 × 3 (270 m), 5 × 5 (450 m), 7 × 7 (630 m), 9 × 9 (810 m), and 11 × 11 (990 m) pixels centered on each validation station from the twenty scenes of the ASTER LSE or LST products, respectively. The standard deviation of the different window sizes calculated from the twenty scenes of the ASTER LST or ASTER LSE products reflects the spatial variability of the validation stations. The average standard deviations of the different window sizes at the six stations are shown in Table 5. The average standard deviation of the ASTER LSE at the six stations changes little or remains the same with the various window sizes, with a range from 0.003 to 0.006. The average standard deviation of the ASTER LSE over the bare soil surface, e.g., HZZ and JCHM station changes are smaller than the vegetation surface, e.g., CJZ and SD stations. The average standard deviation of the ASTER LST at the six stations changes variously with the different window sizes. The HZZ, JCHM and GB stations show the lower standard deviation at all window sizes, with a range from 0.34 K to 0.89 K. The standard deviation of the SD and CJZ sites were higher than 1.0 K, except the 3 × 3 window size, which had a range from 1.04 K to 1.98 K. It appears to be that the two vegetation sites have higher heterogeneity than other sites.

Table 5. The average STD of the 3 × 3, 5 × 5, 7 × 7, 9 × 9, and 11 × 11 pixels extracted from 20 scenes of ASTER LST or ASTER LSE products at six stations.

	LST STD (K)					LSE STD				
Sites	3 × 3	5 × 5	7 × 7	9 × 9	11 × 11	3 × 3	5 × 5	7 × 7	9 × 9	11 × 11
SD	0.929	1.086	1.242	1.470	1.664	0.006	0.006	0.006	0.006	0.006
CJZ	0.804	1.041	1.387	1.801	1.977	0.005	0.005	0.005	0.005	0.005
GB	0.435	0.625	0.776	0.842	0.887	0.003	0.004	0.004	0.005	0.005
SSW	0.671	0.829	0.942	1.121	1.485	0.004	0.004	0.005	0.005	0.005
HZZ	0.396	0.468	0.530	0.601	0.684	0.003	0.004	0.004	0.004	0.005
JCHM	0.342	0.426	0.491	0.506	0.523	0.003	0.004	0.004	0.004	0.004

To illustrate the heterogeneous surface's effects on the estimated LST, we estimated the land surface temperature using the 250 m FY-3C/MERSI data. The GLASS BBE and GLASS LAI products were resized to a 250 m resolution based on the nearest neighbor interpolation model. Figure 9 shows the boxplots between the estimated LSTs from the RTE algorithm (left) or the GSC algorithm (right) at a 250 m resolution and ground LSTs. The trend is similar to the results of the 1000 m spatial resolution. The estimated LSTs are obviously underestimated in the RTE algorithm and most of the LSTs derived from the GSC algorithm were larger than the in situ LSTs.

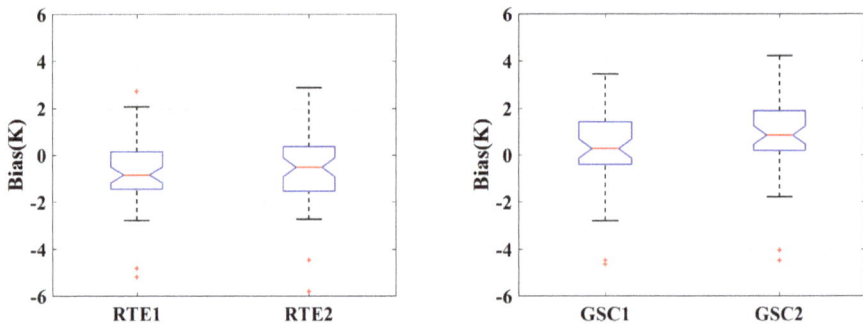

Figure 9. Boxplots between estimated LSTs from RTE algorithm (**left**) or GSC algorithm (**right**) at 250 m resolution and in situ LSTs.

Compared with the results of the two algorithms at 1000 m resolution, the deviation of the two algorithms at 250 m resolution has improved to a certain extent. The bias of the RTE algorithm has changed from -0.92 K to -0.73 K for RTE1 and from -0.82 K to -0.53 K for RTE2. The bias of the GSC algorithm has changed from 0.45 K to 0.23 K for GSC1 and from 0.98 K to 0.87 K for GSC2. In addition, the STD (RMSE) of the RTE and GSC algorithms are all less than 1.76 K (1.95 K). From the above analyses, we can conclude that the two LST algorithms with the LSE derived from the new scheme are all suitable for estimating land surface temperatures. The estimated LSTs with the new LSE scheme have a higher precision than the estimated LST with the NDVI threshold method. Although the scaling effect on the ASTER LST is significant, the heterogeneous surface has little effect on the estimated LST for the coarser spatial resolution (250 m and 1000 m).

6. Conclusions

In this study, we proposed a new scheme to predetermine the LSE for estimating LST from the FY-3C/MERSI with only one thermal infrared channel. The new scheme first divides the land surface into bare soils and vegetated surfaces, then takes advantage of the 4SAIL model's ability to derive the land surface emissivity for vegetated surfaces and establishes the linear relationship between the GLASS BBE and the land surface emissivity of the FY-3C/MERSI data for bare soil surfaces. After determining the LSE, the LST was retrieved by using the RTE and GSC algorithms.

The LSE derived by the new scheme was validated by the field measurements collected at seven stations during the HiWATER experiment. The mean difference between the derived LSE and field-measured LSE is approximately 0.002. When the LSE determined by the new scheme was used for the LST retrieval from the FY-3C/MERSI data using the RTE and GSC algorithms, an acceptable accuracy was achieved, i.e., at 1000 m resolution, the absolute bias of the two algorithms were less than 1 K, and the STD (RMSE) values were all less than 1.95 K (2.2 K). At 250 m resolution, the absolute bias, STD and RMSE of the two algorithms were all less than 0.87 K, 1.76 K and 1.95 K, respectively. Compared to the LST derived by the same algorithm but with the commonly used NDVI threshold method, the new land surface emissivity scheme can achieve better results. Additionally, the new scheme was evaluated by the ASTER emissivity product. The new scheme can provide an accurate LSE estimate, with an average bias of less than 0.009 and RMSE of less than 0.013. Both the ASTER LSE and LSE determined by the new scheme were used to retrieve the LST from the FY-3C/MERSI data, and good agreement was obtained. The average bias and RMSE of the corresponding LST differences are -0.6 K and 0.8 K, respectively. Regarding the validation and evaluation results, we can conclude that the new emissivity scheme can be used to improve the accuracy of the LSE and further the LST for sensors with a broad spectral range such as the FY-3C/MERSI.

Acknowledgments: The ground data used in this study is provided by Cold and Arid Regions Science Data Center at Lanzhou (http://westdc.westgis.ac.cn). The GLASS BBE and LAI products are downloaded from GLCF (ftp://ftp.glcf.umd.edu/glcf/GLASS/). The MODIS products are downloaded from NASA (https://search.earthdata.nasa.gov/). This work was partly supported by the National Natural Science Foundation of China via grant 41771365, the National Key Research and Development Program of China via grant 2016YFA0600101, and the National Natural Science Foundation of China via grant 41371323.

Author Contributions: Jie Cheng conceived and designed the algorithm; Xiangchen Meng performed the algorithm, analyzed the data and wrote the paper. All authors participated in the editing of the paper.

Conflicts of Interest: The authors declare no conflict of interest.

References

1. Wan, Z.; Dozier, J. A generalized split-window algorithm for retrieving land-surface temperature from space. *IEEE Trans. Geosci. Remote Sens.* **1996**, *34*, 892–905.

2. Mannstein, H. Surface Energy Budget, Surface Temperature and Thermal Inertia. In *Remote Sensing Applications in Meteorology and Climatology*; Vaughan, R.A., Ed.; Springer Netherlands: Dordrecht, The Netherlands, 1987; pp. 391–410.

3. Valor, E.; Caselles, V. Mapping land surface emissivity from NDVI: Application to european, African, and South American areas. *Remote Sens. Environ.* **1996**, *57*, 167–184. [CrossRef]

4. Cheng, J.; Liang, S.; Wang, J.; Li, X. A stepwise refining algorithm of temperature and emissivity separation for hyperspectral thermal infrared data. *IEEE Trans. Geosci. Remote Sens.* **2010**, *48*, 1588–1597. [CrossRef]

5. Li, Z.-L.; Wu, H.; Wang, N.; Qiu, S.; Sobrino, J.A.; Wan, Z.-M.; Tang, B.-H.; Yan, G.-J. Land surface emissivity retrieval from satellite data. *Int. J. Remote Sens.* **2013**, *34*, 3084–3127.

6. Qin, Z.; Karnieli, A.; Berliner, P. A mono-window algorithm for retrieving land surface temperature from landsat tm data and its application to the Israel-Egypt border region. *Int. J. Remote Sens.* **2001**, *22*, 3719–3746.

7. Jiménez-Muñoz, J.C. A generalized single-channel method for retrieving land surface temperature from remote sensing data. *J. Geophys. Res. Atmos.* **2003**, *108*. [CrossRef]

8. Ellicott, E.; Vermote, E.; Petitcolin, F.; Hook, S.J. Validation of a new parametric model for atmospheric correction of thermal infrared data. *IEEE Trans. Geosci. Remote Sens.* **2008**, *47*, 295–311. [CrossRef]

9. Becker, F.; Li, Z.L. Towards a local split window method over land surfaces. *Int. J. Remote Sens.* **1990**, *11*, 369–393. [CrossRef]

10. Yu, Y.; Privette, J.L.; Pinheiro, A.C. Evaluation of split-window land surface temperature algorithms for generating climate data records. *IEEE Trans. Geosci. Remote Sens.* **2008**, *46*, 179–192. [CrossRef]

11. Yunyue, Y.; Tarpley, D.; Privette, J.L.; Goldberg, M.D.; Rama Varma Raja, M.K.; Vinnikov, K.Y.; Hui, X. Developing algorithm for operational GOES-R land surface temperature product. *IEEE Trans. Geosci. Remote Sens.* **2009**, *47*, 936–951. [CrossRef]

12. Gillespie, A.; Rokugawa, S.; Matsunaga, T.; Cothern, J.S.; Hook, S.; Kahle, A.B. A temperature and emissivity separation algorithm for advanced spaceborne thermal emission and reflection radiometer (ASTER) images. *IEEE Trans. Geosci. Remote Sens.* **1998**, *36*, 1113–1126. [CrossRef]

13. Masiello, G.; Serio, C.; De Feis, I.; Amoroso, M.; Venafra, S.; Trigo, I.F.; Watts, P. Kalman filter physical retrieval of surface emissivity and temperature from geostationary infrared radiances. *Atmos. Meas. Tech.* **2013**, *6*, 3613–3634. [CrossRef]

14. Coll, C.; Caselles, V.; Valor, E.; Niclòs, R.; Sánchez, J.M.; Galve, J.M.; Mira, M. Temperature and emissivity separation from ASTER data for low spectral contrast surfaces. *Remote Sens. Environ.* **2007**, *110*, 162–175. [CrossRef]

15. Sobrino, J.; Jimenezmunoz, J.; Balick, L.; Gillespie, A.; Sabol, D.; Gustafson, W. Accuracy of ASTER level-2 thermal-infrared standard products of an agricultural area in Spain. *Remote Sens. Environ.* **2007**, *106*, 146–153. [CrossRef]

16. Wan, Z. New refinements and validation of the collection-6 modis land-surface temperature/emissivity product. *Remote Sens. Environ.* **2014**, *140*, 36–45. [CrossRef]

17. Liu, Y.; Yu, Y.; Yu, P.; Göttsche, F.; Trigo, I. Quality assessment of S-NPP VIIRS land surface temperature product. *Remote Sens.* **2015**, *7*, 12215–12241. [CrossRef]

18. Sun, D.; Pinker, R.T. Estimation of land surface temperature from a geostationary operational environmental satellite (GOES-8). *J. Geophys. Res. Atmos.* **2003**, *108*. [CrossRef]

19. Sun, D.; Yu, Y. Land Surface Temperature (LST) Retrieval from Goes Satellite Observations. In *Satellite-based Applications on Climate Change*; Qu, J., Powell, A., Sivakumar, M., Eds.; Springer Netherlands: Dordrecht, The Netherlands, 2013; pp. 289–334.

20. Masiello, G.; Serio, C.; Venafra, S.; Liuzzi, G.; Göttsche, F.; Trigo, I.F.; Watts, P. Kalman filter physical retrieval of surface emissivity and temperature from seviri infrared channels: A validation and intercomparison study. *Atmos. Meas. Tech.* **2015**, *8*, 2981–2997. [CrossRef]

21. Weng, Q.; Lu, D.; Schubring, J. Estimation of land surface temperature-vegetation abundance relationship for urban heat island studies. *Remote Sens. Environ.* **2004**, *89*, 467–483. [CrossRef]

22. Imhoff, M.L.; Zhang, P.; Wolfe, R.E.; Bounoua, L. Remote sensing of the urban heat island effect across biomes in the continental USA. *Remote Sens. Environ.* **2010**, *114*, 504–513. [CrossRef]

23. Fee, D.; Matoza, R.S. An overview of volcano infrasound: From hawaiian to plinian, local to global. *J. Volcanol. Geotherm. Res.* **2013**, *249*, 123–139. [CrossRef]

24. Diker, C.; Ulusoy, I. Monitoring Thermal Activity of Eastern Anatolian Volcanoes Using Modis Images. In Proceedings of the European Geosciences Union General Assembly, Vienna, Austria, 27 April–2 May 2014; pp. 375–381.

25. Ulusoy, İ. Temporal monitoring of radiative heat flux from the craters of tendürek volcano (East Anatolia, Turkey) using ASTER satellite imagery. In Proceedings of the European Geosciences Union General Assembly, Vienna, Austria, 27 April–2 May 2014.

26. Guangmeng, Q.; Mei, Z. Using MODIS land surface temperature to evaluate forest fire risk of Northeast China. *IEEE Geosci. Remote Sens. Lett.* **2004**, *1*, 98–100. [CrossRef]

27. ManzoDelgado, L.; SánchezColón, S.; Álvarez, R. Assessment of seasonal forest fire risk using NOAA-AVHRR: A case study in central mexico. *Int. J. Remote Sens.* **2009**, *30*, 4991–5013. [CrossRef]

28. Meng, X.; Li, H.; Du, Y.; Liu, Q.; Zhu, J.; Sun, L. Retrieving land surface temperature from landsat 8 TIRS data using RTTOV and ASTER GED. In Proceedings of the Geoscience and Remote Sensing Symposium, Beijing, China, 10–15 July 2016; pp. 4302–4305.

29. Yu, X.; Guo, X.; Wu, Z. Land surface temperature retrieval from landsat 8 TIRS—Comparison between radiative transfer equation-based method, split window algorithm and single channel method. *Remote Sens.* **2014**, *6*, 9829–9852. [CrossRef]

30. Windahl, E.; Beurs, K.D. An intercomparison of landsat land surface temperature retrieval methods under variable atmospheric conditions using in situ skin temperature. *Int. J. Appl. Earth Obs. Geoinf.* **2016**, *51*, 11–27. [CrossRef]

31. Jimenez-Munoz, J.C.; Cristobal, J.; Sobrino, J.A.; Soria, G.; Ninyerola, M.; Pons, X.; Pons, X. Revision of the single-channel algorithm for land surface temperature retrieval from landsat thermal-infrared data. *IEEE Trans. Geosci. Remote Sens.* **2009**, *47*, 339–349. [CrossRef]

32. Zhou, J.; Zhan, W.; Hu, D.; Zhao, X. Improvement of mono-window algorithm for retrieving land surface temperature from HJ-1B satellite data. *Chin. Geogr. Sci.* **2010**, *20*, 123–131. [CrossRef]

33. Sobrino, J.A.; Raissouni, N.; Li, Z.L. A comparative study of land surface emissivity retrieval from NOAA data. *Remote Sens. Environ.* **2001**, *75*, 256–266. [CrossRef]

34. Snyder, W.C.; Wan, Z.; Zhang, Y.; Feng, Y.Z. Classification-based emissivity for land surface temperature measurement from space. *Int. J. Remote Sens.* **1998**, *19*, 2753–2774. [CrossRef]

35. Hulley, G.; Veraverbeke, S.; Hook, S. Thermal-based techniques for land cover change detection using a new dynamic MODIS multispectral emissivity product (MOD21). *Remote Sens. Environ.* **2014**, *140*, 755–765. [CrossRef]

36. Li, H.; Sun, D.; Yu, Y.; Wang, H.; Liu, Y.; Liu, Q.; Du, Y.; Wang, H.; Cao, B. Evaluation of the VIIRS and MODIS lst products in an arid area of Northwest China. *Remote Sens. Environ.* **2014**, *142*, 111–121. [CrossRef]

37. Liang, S.; Zhao, X.; Liu, S.; Yuan, W.; Cheng, X.; Xiao, Z.; Zhang, X.; Liu, Q.; Cheng, J.; Tang, H.; et al. A long-term global land surface satellite (GLASS) data-set for environmental studies. *Int. J. Digit. Earth* **2013**, *6*, 5–33. [CrossRef]

38. Cheng, J.; Liang, S. Estimating global land surface broadband thermal-infrared emissivity from the advanced very high resolution radiometer optical data. *Int. J. Digit. Earth* **2013**, *6*, 34–49. [CrossRef]

39. Cheng, J.; Liang, S. Estimating the broadband longwave emissivity of global bare soil from the MODIS shortwave albedo product. *J. Geophys. Res. Atmos.* **2014**, *119*, 614–634. [CrossRef]

40. Xiao, Z.; Liang, S.; Wang, J.; Chen, P.; Yin, X.; Zhang, L.; Song, J. Use of general regression neural networks for generating the glass leaf area index product from time-series MODIS surface reflectance. *IEEE Trans. Geosci. Remote Sens.* **2013**, *52*, 209–223. [CrossRef]

41. Xu, Z.; Liu, S.; Li, X.; Shi, S.; Wang, J.; Zhu, Z.; Xu, T.; Wang, W.; Ma, M. Intercomparison of surface energy flux measurement systems used during the HIWATER-MUSOEXE. *J. Geophys. Res. Atmos.* **2013**, *118*, 13140–13157. [CrossRef]

42. Li, X.; Cheng, G.; Liu, S.; Xiao, Q.; Ma, M.; Jin, R.; Che, T.; Liu, Q.; Wang, W.; Qi, Y.; et al. Heihe watershed allied telemetry experimental research (HIWATER): Scientific objectives and experimental design. *Bull. Am. Meteorol. Soc.* **2013**, *94*, 1145–1160. [CrossRef]

43. Wang Heshun, L.H.; Cao, B.; Du, Y.; Xiao, Q.; Liu, Q. *HIWATER: Dataset of Thermal Infrared Spectrum Observed by BOMEM MR304 in the Middle Reaches of the Heihe River Basin*; Institute of Remote Sensing Applications, Chinese Academy of Sciences: Beijing, China, 2012.

44. Liu, S.M.; Xu, Z.W.; Wang, W.Z.; Jia, Z.Z.; Zhu, M.J.; Bai, J.; Wang, J.M. A comparison of eddy-covariance and large aperture scintillometer measurements with respect to the energy balance closure problem. *Hydrol. Earth Syst. Sci.* **2011**, *15*, 1291–1306. [CrossRef]

45. Tan Junlei, M.M. *HIWATER: Dataset of Infrared Temperature in Zhanye Airport Desert*; Cold and Arid Regions Environmental and Engineering Research Institute, Chinese Academy of Sciences: Beijing, China, 2012.

46. Ingram, P.M.; Muse, A.H. Sensitivity of iterative spectrally smooth temperature/emissivity separation to algorithmic assumptions and measurement noise. *IEEE Trans. Geosci. Remote Sens.* **2001**, *39*, 2158–2167. [CrossRef]

47. Cheng, J.; Liang, S.; Yao, Y.; Ren, B.; Shi, L.; Liu, H. A comparative study of three land surface broadband emissivity datasets from satellite data. *Remote Sens.* **2013**, *6*, 111–134. [CrossRef]

48. Jimenez-Munoz, J.C.; Sobrino, J.A. Feasibility of retrieving land-surface temperature from ASTER TIR bands using two-channel algorithms: A case study of agricultural areas. *IEEE Geosci. Remote Sens. Lett.* **2007**, *4*, 60–64. [CrossRef]

49. Cheng, J.; Liang, S.; Verhoef, W.; Shi, L.; Liu, Q. Estimating the hemispherical broadband longwave emissivity of global vegetated surfaces using a radiative transfer model. *IEEE Trans. Geosci. Remote Sens.* **2016**, *54*, 905–917. [CrossRef]

50. Barsi, J.A.; Butler, J.J.; Schott, J.R.; Palluconi, F.D.; Hook, S.J. Validation of a web-based atmospheric correction tool for single thermal band instruments. In Proceedings of the Optics and Photonics, San Diego, CA, USA, 31 July–4 August 2005; Volume 5882, pp. 136–142.

51. Li, H.; Liu, Q.; Du, Y.; Jiang, J.; Wang, H. Evaluation of the NCEP and MODIS atmospheric products for single channel land surface temperature retrieval with ground measurements: A case study of HJ-1B IRS data. *IEEE J. Sel. Top. Appl. Earth Obs. Remote Sens.* **2013**, *6*, 1399–1408. [CrossRef]

52. Cook, M.; Schott, J.; Mandel, J.; Raqueno, N. Development of an operational calibration methodology for the landsat thermal data archive and initial testing of the atmospheric compensation component of a land surface temperature (LST) product from the archive. *Remote Sens.* **2014**, *6*, 11244–11266. [CrossRef]

53. Rivalland, V.; Tardy, B.; Huc, M.; Hagolle, O.; Marcq, S.; Boulet, G. A useful tool for atmospheric correction and surface temperature estimation of landsat infrared thermal data. In Proceedings of the European Geosciences Union General Assembly, Vienna, Austria, 17–22 April 2016; Volume 8, p. 696.

54. Matricardi, M.; Chevallier, F.; Kelly, G.; Thépaut, J.N. An improved general fast radiative transfer model for the assimilation of radiance observations. *Q. J. R. Meteorol. Soc.* **2010**, *130*, 153–173. [CrossRef]

55. Tonooka, H. Accurate atmospheric correction of ASTER thermal infrared imagery using the WVS method. *IEEE Trans. Geosci. Remote Sens.* **2005**, *43*, 2778–2792. [CrossRef]

56. Jiang, G.M.; Zhou, W.; Liu, R. Development of split-window algorithm for land surface temperature estimation from the VIRR/FY-3A measurements. *IEEE Geosci. Remote Sens. Lett.* **2013**, *10*, 952–956. [CrossRef]

57. Borbas, E.E.; Seemann, S.W.; Huang, H.L.; Li, J.; Menzel, P.W. Global profile training database for satellite regression retrievals with estimates of skin temperature and emissivity. In Proceedings of the International TOVS Study Conference, Beijing, China, 25–31 May 2005.

58. Galve, J.M.; Coll, C.; Caselles, V.; Valor, E. An atmospheric radiosounding database for generating land surface temperature algorithms. *IEEE Trans. Geosci. Remote Sens.* **2008**, *46*, 1547–1557. [CrossRef]

59. Ren, B.; Cheng, J. *Land Surface Temperature Retrieval Algorithm of Single Channel for FY-3A/MERSI*; Beijing Normal University: Beijing, China, 2015.

60. Dong, L.X.; Hu, J.Y.; Tang, S.H.; Min, M. Field validation of glass land surface broadband emissivity database using pseudo-invariant sand dunes sites in Northern China. *Int. J. Digit. Earth* **2013**, *6*, 96–112. [CrossRef]

61. Pandya, M.R.; Shah, D.B.; Trivedi, H.J.; Lunagaria, M.M.; Pandey, V.; Panigrahy, S.; Parihar, J.S. Field measurements of plant emissivity spectra: An experimental study on remote sensing of vegetation in the thermal infrared region. *J. Indian Soc. Remote Sens.* **2013**, *41*, 787–796. [CrossRef]

62. Niclòs, R.; Caselles, V.; Coll, C.; Valor, E. Determination of sea surface temperature at large observation angles using an angular and emissivity-dependent split-window equation. *Remote Sens. Environ.* **2007**, *111*, 107–121. [CrossRef]

63. Yu Wenping, M.M.; Ren, Z.; Tan, J.; Li, Y.; Wang, H. *HIWATER: Dataset of Emissivity in the Heihe River Basin*; Cold and Arid Regions Science Data Center at Lanzhou: Lanzhou, China, 2013.

64. Hulley, G.C.; Hughes, C.G.; Hook, S.J. Quantifying uncertainties in land surface temperature and emissivity retrievals from ASTER and MODIS thermal infrared data. *J. Geophys. Res. Atmos.* **2012**, *117*. [CrossRef]

65. Guillevic, P.C.; Privette, J.L.; Coudert, B.; Palecki, M.A.; Demarty, J.; Ottlé, C.; Augustine, J.A. Land surface temperature product validation using NOAA'S surface climate observation networks—Scaling methodology for the visible infrared imager radiometer suite (VIIRS). *Remote Sens. Environ.* **2012**, *124*, 282–298. [CrossRef]

66. Jiménez-Muñoz, J.C.; Sobrino, J.A.; Gillespie, A.; Sabol, D.; Gustafson, W.T. Improved land surface emissivities over agricultural areas using ASTER NDVI. *Remote Sens. Environ.* **2006**, *103*, 474–487. [CrossRef]

67. Coll, C.; Valor, E.; Galve, J.M.; Mira, M.; Bisquert, M.; García-Santos, V.; Caselles, E.; Caselles, V. Long-term accuracy assessment of land surface temperatures derived from the advanced along-track scanning radiometer. *Remote Sens. Environ.* **2012**, *116*, 211–225. [CrossRef]

68. Cheng, J.; Liang, S.; Dong, L.; Ren, B.; Shi, L. Validation of the moderate-resolution imaging spectrometer (MODIS) land surface emissivity products over the taklimakan desert. *J. Appl. Remote Sens.* **2014**, *8*. [CrossRef]

remote sensing

MDPI

Article

Estimation of High Spatial-Resolution Clear-Sky Land Surface-Upwelling Longwave Radiation from VIIRS/S-NPP Data

Shugui Zhou [1,2] and Jie Cheng [1,2,3,*]

[1] State Key Laboratory of Remote Sensing Science, Jointly Sponsored by Beijing Normal University and Institute of Remote Sensing and Digital Earth of Chinese Academy of Sciences, Beijing 100875, China; zhoushugui1990@gmail.com

[2] Institute of Remote Sensing Science and Engineering, Faculty of Geographical Science, Beijing Normal University, Beijing 100875, China

[3] U. S. Department of Agriculture, Agricultural Research Service, Hydrology and Remote Sensing Laboratory, Beltsville, MD 20705, USA

* Correspondence: Jie_Cheng@bnu.edu.cn

Received: 12 December 2017; Accepted: 2 February 2018; Published: 7 February 2018

Abstract: Surface-upwelling longwave radiation (LWUP) is an important component of the surface radiation budget. Under the general framework of the hybrid method, the linear models and the multivariate adaptive regression spline (MARS) models are developed to estimate the 750 m instantaneous clear-sky LWUP from the top-of-atmosphere (TOA) radiance of the Visible Infrared Imaging Radiometer Suite (VIIRS) channels M14, M15, and M16. Comprehensive radiative transfer simulations are conducted to generate a huge amount of representative samples, from which the linear model and the MARS model are derived. The two models developed are validated by the field measurements collected from seven sites in the Surface Radiation Budget Network (SURFRAD). The bias and root-mean-square error (RMSE) of the linear models are -4.59 W/m^2 and 16.15 W/m^2, whereas those of the MARS models are -5.23 W/m^2 and 16.38 W/m^2, respectively. The linear models are preferable for the production of the operational LWUP product due to its higher computational efficiency and acceptable accuracy. The LWUP estimated by the linear models developed from VIIRS is compared to that retrieved from the Moderate Resolution Imaging Spectroradiometer (MODIS). They agree well with each other with bias and RMSE of -0.15 W/m^2 and 25.24 W/m^2 respectively. This is the first time that the hybrid method has been applied to globally estimate clear-sky LWUP from VIIRS data. The good performance of the developed hybrid method and consistency between VIIRS LWUP and MODIS LWUP indicate that the hybrid method is promising for producing the long-term high spatial resolution environmental data record (EDR) of LWUP.

Keywords: longwave upwelling radiation (LWUP); Visible Infrared Imaging Radiometer Suite (VIIRS); surface radiation budget; hybrid method; remote sensing

---

1. Introduction

The surface radiation budget (SRB) is an important indicator in the study of climate formation and change and environmental prediction, which plays a key role in the global matter, energy cycles and interactions between the surface and the atmosphere system [1–3]. SRB is dominated by longwave radiation in the night and during most of the year in the polar regions [4,5]. Surface-upwelling longwave radiation (LWUP, 4.0–100 µm), the sum of thermal radiation emitted by the surface and reflected atmospheric downward longwave radiation, is the main cause of surface cooling in the clear night sky and is also an indirect indicator of surface temperature [5]. An accurate estimate of LWUP is one of the prerequisites for obtaining accurate weather forecasts, climate simulations, and land-surface process simulations.

Generally, we can obtain the LWUP using three approaches: field measurement, satellite remote sensing and model prediction. LWUP can be accurately measured with field instruments. However, field networks are sparsely distributed globally. Furthermore, field measurement can only represent a limited area. The spatial resolution of model prediction is relatively coarse [6]. Remote sensing can provide various kinds of products with global coverage and horizontal spatial continuity, but the temporal resolution is always limited. High spatial-resolution LWUP is an important diagnostic parameter for mesoscale land surface and atmosphere models [7] and can also serve as a bridge for validating coarse resolution data [8]. The Visible Infrared Imaging Radiometer Suite (VIIRS) is one of the key instruments onboard the Suomi National Polar-orbiting Partnership (S-NPP) satellite system and was successfully launched in 2011. The spatial resolution of VIIRS thermal-infrared channels is 750 m. Ignoring the relatively low temporal resolution of VIIRS, it is the preferred data source for estimating high spatial-resolution LWUP.

Over the past few decades, substantial efforts have been devoted to estimating LWUP at the surface using remotely sensed observations from space-borne platforms. These methods can be primarily divided into two categories: the temperature-emissivity method [7–10] and the hybrid method [11–13]. Current operational satellite land-surface temperature and emissivity products facilitate the estimation of LWUP [14–17], but the large uncertainties in the land-surface temperature and emissivity products limit its accuracy [18,19]. For example, the study of Wang et al. indicated that the accuracy of the temperature-emissivity method is much lower than that of the hybrid method [7]. As shown in Figure 1, the weighting function of thermal infrared channels located in narrow bands that are semi-transparent to atmospheric gases and thus sensitive primarily to emission from the surface ("atmospheric windows") peaks at the surface and contains the surface-emission information, so LWUP can be derived from top-of-atmosphere (TOA) radiance or brightness temperature directly [13,20]. The hybrid method links the thermal-infrared TOA radiances or brightness temperatures with LWUP through comprehensive radiative transfer modeling and statistical regression. It is physically based and at the same time has a high computational efficiency. Furthermore, it can bypass the problem of temperature and emissivity separation and achieve an acceptable accuracy in practice. The hybrid method has been successfully used to produce high spatial-resolution regional [7,11,21] and global [12] LWUP recently.

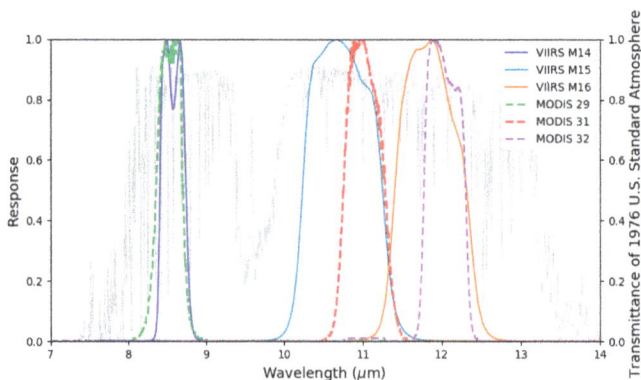

Figure 1. Relative spectral responses of Visible Infrared Imaging Radiometer Suite (VIIRS) channels M14, M15 and M16, and Moderate Resolution Imaging Spectroradiometer (MODIS) channels 29, 31 and 32. The gray line represents the transmittance of the 1976 U.S. Standard Atmosphere.

VIIRS was developed based on the heritage of Moderate Resolution Imaging Spectroradiometer (MODIS) instruments and has become a key bridge to ensure long-term continuity of the environmental data records (EDRs). VIIRS provides a large number of EDRs, including aerosol optical thickness [22], vegetation index [23], land-surface albedo [24], land-surface temperature [25], sea-surface temperature [26], etc. To our knowledge, an operational LWUP product from VIIRS is not available at regional and global scales. VIIRS do not provide operational products of surface broadband emissivity (BBE) and surface downward longwave radiation. Thus, it is difficult to calculate the LWUP with the temperature-emissivity method. In addition, no operational algorithm that can be used to retrieve LWUP from VIIRS has been reported in the literature.

We have developed a hybrid method for retrieving clear-sky LWUP from MODIS data and produced two years' global LWUP product recently [12]. It is possible to produce long-term high spatial-resolution EDR of LWUP by combining MODIS and VIIRS. The first step is to develop a hybrid method to estimate the global 750-m instantaneous clear-sky LWUP from VIIRS data. The article is arranged as follows. The data including satellite data, field measurements, and atmospheric profiles are described in Section 2. The method and validation results are provided in Sections 3 and 4. A brief discussion and conclusion are given in Sections 5 and 6.

2. Data

2.1. Visible Infrared Imaging Radiometer Suite (VIIRS) Data

VIIRS has been developed based on the heritage of legacy instruments, including AVHRR and MODIS, and extends and improves on them [27]. The VIIRS has 22 spectral channels with wavelengths ranging from 0.41 μm to 12.5 μm, which can be used for environmental monitoring and numerical weather forecasting. The on-orbit verification and intensive calibration and validation using a ground target show that VIIRS is working very well [28,29]. More than 20 environmental data records have been produced operationally from VIIRS data, including clouds, land-surface temperature and sea-surface temperature, vegetation index, aerosol optical thickness, active fire, snow/ice, surface albedo, etc.

The VIIRS data utilized in this study include the VIIRS sensor data records (SDR) and VIIRS Cloud Mask intermediate product (VCM). The VIIRS SDR contains the day–night band, imagery band, moderate resolution band and geolocation data. Three thermal-infrared channels, M14, M15 and M16, which are located in the "atmospheric windows" and are sensitive to the LWUP, are finally selected. These three channels have similar channel characteristics to MODIS channels 29, 31 and 32. The relative spectral responses of VIIRS channels M14, M15 and M16 as well as MODIS channels 29, 31 and 32 are displayed in Figure 1. The VIIRS VCM product is used to identify whether a pixel is clear-sky or cloudy [30]. The pixels with confident clear flag are identified as clear-sky pixels. In addition, the longitude, latitude and the sensor view zenith angle data are also provided in the SDR.

2.2. Atmospheric Profiles

In this study, two years of the Atmospheric Infrared Sounder (AIRS) level 2 standard atmospheric profiles are collected to construct the atmospheric profile database. Launched aboard the Earth Observing System (EOS) second satellite Aqua, the AIRS instrument has 2378 infrared spectral channels covering 3.74–15.39 μm with a high spectral resolution ($\lambda/\Delta\lambda$) of 1200 [31]. The AIRS infrared band is very stable, and the offset is less than 10 mK/yr [32]. The accuracy of temperature measurement of AIRS is better than 250 mK, and the absolute calibration accuracy of most bands can reach 100 mK. It is the most accurate and stable hyperspectral infrared detector up until now (http://daac.gsfc.nasa.gov/AIRS/). AIRS can be used to obtain the global atmospheric three-dimensional physical state (atmospheric temperature, water vapor, clouds, etc.) and the distribution of trace gases (ozone, carbon dioxide and methane, etc.) daily [33]. The temperature profile of AIRS has 28 layers, and the corresponding atmospheric pressure varies from 1100 hPa to 0.1 hPa. Meanwhile, the AIRS water-vapor profile has 14 layers, and the corresponding atmospheric pressure ranges from 1100 hPa to 50 hPa.

2.3. Surface Radiation Budget Network (SURFRAD)

The Surface Radiation Budget Network (SURFRAD) network was established in 1993 to supply accurate, longstanding and persistent measurements of the surface radiation budget for climate change studies [34]. SURFRAD provides quality-controlled field measurements including downward and upwelling solar irradiance and longwave infrared irradiation, along with other meteorological observations, such as wind speed, atmospheric pressure and relative humidity, etc. SURFRAD measurements are widely used for the validation of satellite-derived products and the details about the SURFRAD site and related instruments can be found in [34]. Daily one or three-minute SURFRAD data are organized into daily ASCII text and can be freely downloaded from ftp://aftp.cmdl.noaa.gov/data/radiation/surfrad. In addition, the basic data are routinely sent to several archives including the World Radiation Data Center (WRDC), National Climatic Data Center (NCDC), and Baseline Surface Radiation Network (BSRN). The location, elevation and land cover of the seven SURFRAD sites are listed in Table 1.

Site upwelling and downward thermal infrared irradiance are measured using two precision infrared radiometers (PIR). The PIRs are sensitive to the spectral range from 3 μm to 50 μm. Three standard PIRs, which are annually calibrated by world-reputable organizations, are used to calibrate these two PIRs adopted in SURFRAD network. The spectral range of the measurements can be extended to 4–100 μm by calibration [35]. The overall accuracy of PIR ground measurement is approximately ± 9 W/m^2 [34], and is reported to be about ± 5 W/m^2 recently [36]. The VIIRS moderate resolution band (M-band) has a resolution of 750 m at nadir. The spatial matching issue and scale effect need be considered when validating satellite-derived LWUP using SURFRAD ground measurements. The footprint of SURFRAD PIRs is much smaller compared to that of the VIIRS. Fortunately, SURFRAD sites were selected at the locations where the surrounding land cover of the site was homogeneous. For example, Wang et al. [7] compared the brightness temperature of the ASTER pixel that contains the SURFRAD site to other neighboring pixels within 1 km × 1 km and 2 km × 2 km windows. They found that the discrepancies between the central pixel and surrounding pixels were less than 1 K in general, and the standard deviations of center pixel and surrounding pixels were less than 2 K under most conditions. Therefore, the SURFRAD measurements were used to validate the derived VIIRS LWUP directly.

Table 1. Information of sites in the Surface Radiation Budget Network (SURFRAD) network.

Name	Location	Elevation (m)	Land Cover	Time Period of Used Data
Bondville_IL	40.0519°N, 88.3731°W	230	Cropland	2014–2017
Boulder_CO	40.1249°N, 105.2368°W	1689	Grassland	2014–2017
Desert_Rock_NV	36.6237°N, 116.0195°W	1007	Desert	2014–2017
Fort_Peck_MT	48.3078°N, 105.1017°W	634	Grassland	2014–2017
Goodwin_Creek_MS	34.2547°N, 89.873°W	98	Grassland	2014–2017
Penn_State_PA	40.7201°N, 77.9309°W	376	Cropland	2014–2017
Sioux_Falls_SD	43.7340°N, 96.6233°W	473	Grassland	2014–2017

3. Method

As shown in Figure 1, the channel characteristics of VIIRS channels M14, M15 and M16 are similar to those of MODIS channels 29, 31, and 32, which have been successfully used to derive LWUP using hybrid method at regional and global scales [7,12]. In this section, we developed the hybrid method for VIIRS using TOA radiance of channels M14, M15 and M16 under the general framework of the hybrid method. First, a huge amount of representative samples are generated by extensive radiative transfer modeling; then the linear model and MARS model are established to predict LWUP using TOA radiance of channels M14, M15 and M16.

3.1. Radiative Transfer Modeling

The surface upwelling longwave radiation consists of two components: longwave radiation emitted by the surface and surface-reflected atmospheric downward longwave radiation, which can be written as:

$$F_{up}^l = \varepsilon \int_{\lambda_1}^{\lambda_2} \pi B(T_s) d\lambda + (1 - \varepsilon) F_{down}^l \tag{1}$$

where the ε is the surface broadband emissivity. T_s is the surface temperature. F_{down}^l is the downward longwave radiation. λ_1 and λ_2 are the spectral integration range (4–100 μm).

A representative simulation database is required to train the hybrid method, such as the linear model and the MARS model. The moderate resolution atmospheric transmission (MODTRAN) software [37], which is widely used by researchers all over the world, can be used to simulate the radiation transmission and interaction between the atmosphere and the land surface under various atmospheric and surface conditions. Allowing for the difference between land-surface temperature and atmospheric temperature, the global land surface is divided into three regions: the low-latitude region (30°S–30°N), middle-latitude region (30°S–60°S, 30°N–60°N), high-latitude region (60°S–90°S, 60°N–90°N) [12]. At the same time, the atmospheric profiles are also divided into these three categories according to the latitude. For each sub-region, we extracted the atmospheric profiles from AIRS Level 2 standard atmosphere product and constructed the atmospheric profile database. To avoid excessive computation and to alleviate the similarity of the profiles, a screening process was applied, and the criteria proposed in [12] adopted to measure the similarity of atmospheric profiles. In total, 2842, 35,487 and 41,724 atmospheric profiles of high-latitude, mid-latitude and low-latitude region were obtained. MODTRAN 5.2 was used to simulate the spectral downward longwave radiance, thermal path radiance and spectral transmittance for each atmospheric profile and sensor view zenith angle. During the simulation, the sensor view zenith angles were set from 0° to 60° with an interval of 15°. We calculated the difference between surface temperature and the bottom layer temperature of atmospheric profiles for each region using two-year AIRS standard L2 product. The surface temperature was determined based on this difference. For example, the surface temperature of mid-latitude is equal to the bottom layer temperature plus a range of [−10, 15] K with a step of 5 K. Eighty-four representative spectral emissivity spectra including vegetation, soil, snow/ice, water selected from ASTER spectral library [38] and MODIS UCSB spectral library [39] and their combinations were used to characterize the land-surface conditions. Since there was no spectral emissivity value for wavelengths larger than 14 μm, the emissivity value was supposed to be the same as that at wavelength 14 μm in the following simulations. When the surface and atmospheric parameters were determined, the VIIRS TOA channel radiances are calculated using the following simplified equation:

$$L_i = \frac{\int_{\lambda_1}^{\lambda_2} ((\varepsilon_\lambda B(T_s) + (1 - \varepsilon_\lambda) L_{\downarrow\lambda})\tau_\lambda + L_{\uparrow\lambda}) f_i(\lambda) d\lambda}{\int_{\lambda_1}^{\lambda_2} f_i(\lambda) d\lambda} \tag{2}$$

where L_i is the TOA radiance for channel M14, M15 and M16 of VIIRS; $L_{\uparrow\lambda}$ and $L_{\downarrow\lambda}$ are path thermal radiance and spectral downward longwave radiance, respectively; $f_i(\lambda)$ is the spectral response function for channels M14, M15 and M16; λ_1 and λ_2 are the spectral range of VIIRS TIR channel. The flowchart of the hybrid method is displayed in Figure 2.

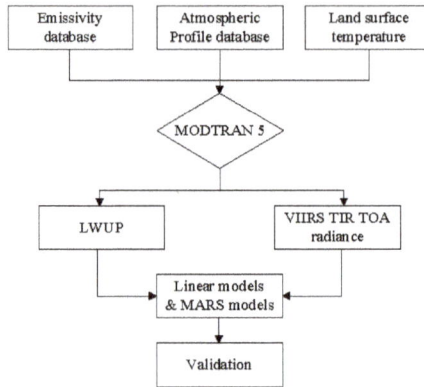

Figure 2. Flowchart for developing the hybrid method.

3.2. Linear Model

A linear model with the following expression was developed using the generated samples in Section 3.1:

$$LWUP = a_0 + a_1 M14 + a_2 M15 + a_3 M16 \tag{3}$$

where a_0, a_1, a_2 and a_3 are coefficients; and M14, M15 and M16 are TOA radiance for moderate resolution bands 14, 15 and 16 of the VIIRS. The linear model is fitted for each sub-region and view zenith angle.

3.3. The Multivariate Adaptive Regression Spline (MARS) Model

To probe the non-linear relationship between TOA spectral radiance and LWUP, we also use the multivariate adaptive regression spline (MARS) to model the non-linear relationship between TOA spectral radiance and LWUP with the same samples as those used by the linear model. MARS was proposed by Jerome H. Friedman in 1991 [40]. MARS is a highly generalized and highly specialized regression method for high-dimensional data. The regression method takes the tensor product of spline functions as the basis functions, while the determination of the basis functions and their number are automatically completed by the data without manual selection. In the multi-dimensional case, how to divide the space has become a critical problem, but the MARS model can solve this problem well. The MARS model is defined as follows:

$$
\begin{aligned}
\hat{f}(x) &= a_0 + \sum_{m=1}^{M} a_m B_m(x) \\
&= a_0 + \sum_{m=1}^{M} a_m \prod_{k=1}^{k_m} [S_{km}(x_{v(k,m)} - t_{km})]_+
\end{aligned}
\tag{4}
$$

where $B_m(x)$ is the mth basis function; a_0 is coefficient; a_m is the coefficient of the mth basis function; M is the number of the basis function; K_m is the number of knots in the mth basis function; S_{km} is 1 or -1, and indicates the spline function on the right or left side;$v(k,m)$ labels the independent variables; and t_{km} is a knot location. The basis function of MARS is a single spline function, or is the result of the interaction of multiple spline functions. The spline functions on the right and left sides are defined as follows:

$$
[+(x_{v(k,m)} - t)]_+^q = \begin{cases} (x - t_{km})^q, & x \geq t_{km} \\ 0, & \text{others} \end{cases}
$$
$$
[-(x_{v(k,m)} - t)]_+^q = \begin{cases} (t_{km} - x)^q, & x \leq t_{km} \\ 0, & \text{others} \end{cases}
\tag{5}
$$

where t is the position of the node; $x - t_{km}$ and $t_{km} - x$ are used to describe spline functions on the right and left regions when t is given; q is the power (>0) to which the splines are raised; and "+" takes 0 for negative values.

MARS is built in two phases: the forward-basis function selection and backward-pruning process. In the first stage, MARS begins with just the intercept term and iterations are conducted to add pairs of basis functions, which can minimize the training error to the utmost extent. In order to avoid overfitting, a backward-pruning process is needed. In the backward-pruning phase, the basis function, which contributes the least to the reduction of training error, is deleted at a time. The model with the lowest GCV (generalized cross-validation) is finally selected. The GCV, as an estimator for effective mean residual error, can achieve a balance between goodness of fit and model complexity. The GCV is calculated as follows:

$$GCV = \frac{\frac{1}{N} \sum_{i=1}^{N} (y_i - \hat{f}(x_i))^2}{(1 - \frac{enp}{N})^2} \tag{6}$$

where N is the number of samples in the training data; $\hat{f}(x_i)$ is the estimation of y_i; and enp is the effective number of parameters:

$$enp = k + c * (k - 1)/2 \tag{7}$$

where c is the penalty parameter; and k is the number of non-constant basis functions.

The training of MARS is based on the ARESLab package of the MATLAB platform, in which the MARS is implemented according to Friedman's original papers [40] and all parameters in the ARESLab package are automatically determined. The source code of the ARESLab toolbox can be downloaded from the following website: www.cs.rtu.lv/jekabsons/regression.html.

4. Results

4.1. Training Results of the Linear Model and MARS Model

The linear model can account for more than 97.7%, 98.5% and 99.1% of the variation of the LWUP in the simulation database for the low-latitude, middle-latitude and high-latitude regions, respectively. The bias is zero, and the RMSE ranges from 5.27 to 13.02 W/m². The RMSE in the low-latitude region is larger than that in the middle-latitude and high-latitude region. In addition, the RMSE for the larger view zenith angle is greater than that with the smaller view zenith angle. Details about fitting the results of the linear models can be found in Table 2.

The training results of the MARS model are also displayed in Table 2. The MARS model can account for more than 98.2%, 98.7% and 99.2% of the variation in the simulation database for the low-latitude, middle-latitude and high-latitude regions, respectively. The biases of all MARS models are zero, and the RMSEs range from 8.53 to 10.13 W/m², 6.64 to 8.24 W/m², and 4.8 to 6.47 W/m² for the low-latitude, middle-latitude and high-latitude regions, respectively. The RMSEs of the MARS models are slightly less than those of the linear models.

Table 2. Fitting results for the linear models and the Multivariate Adaptive Regression Spline (MARS) models.

				Low-Latitude Region						
			Linear Model					MARS		
Angle	a0	a1	a2	a3	R²	Bias	RMSE	R²	Bias	RMSE
0°	124.404	2.687	119.530	−93.350	0.989	0.00	8.82	0.990	0.00	8.53
15°	126.927	2.833	121.603	−95.997	0.988	0.00	8.97	0.993	0.00	8.54
30°	135.126	3.434	128.092	−104.459	0.988	0.00	9.13	0.989	0.00	8.71
45°	151.431	5.290	139.829	−120.664	0.985	0.00	10.46	0.986	0.00	9.03
60°	182.429	12.293	157.379	−149.538	0.977	0.00	13.02	0.982	0.00	10.13

				Middle-Latitude Region						
			Linear Model					MARS		
Angle	a0	a1	a2	a3	R²	Bias	RMSE	R²	Bias	RMSE
0°	99.959	1.747	104.644	−73.428	0.993	0.00	6.94	0.993	0.00	6.64
15°	101.853	1.769	106.772	−75.933	0.992	0.00	7.04	0.993	0.00	6.84
30°	108.090	1.922	113.550	−84.018	0.992	0.00	7.39	0.992	0.00	7.18
45°	120.822	2.647	126.401	−99.870	0.990	0.00	8.11	0.991	0.00	7.88
60°	146.517	6.157	148.690	−129.866	0.985	0.00	9.79	0.987	0.00	8.24

				High-Latitude Region						
			Linear Model					MARS		
Angle	a0	a1	a2	a3	R²	Bias	RMSE	R²	Bias	RMSE
0°	77.525	0.915	87.049	−50.963	0.995	0.00	5.27	0.996	0.00	4.8
15°	79.219	1.588	88.103	−52.734	0.995	0.00	5.16	0.996	0.00	4.85
30°	82.928	1.339	94.582	−59.759	0.995	0.00	5.41	0.996	0.00	5.11
45°	90.741	1.020	107.407	−73.892	0.994	0.00	5.91	0.994	0.00	5.80
60°	107.699	1.298	132.253	−102.344	0.991	0.00	6.95	0.992	0.00	6.47

4.2. Validation with Field Measurements

4.2.1. The Linear Model

The field measurements collected from the SURFRAD network were used to validate the linear models developed. When the view zenith angle was not equal to 0°, 15°, 30°, 45°, 60°, LWUP was linearly interpolated from LWUPs predicted by the linear model with an adjacent view zenith angle. View zenith angle exceeding 60° was not considered. The cloud-mask information was extracted from the VIIRS Cloud Mask intermediate product. Clear sky was identified when 3×3 neighboring pixels of the site are clear to ensure that the field measurements at the site were not affected by clouds and cloud shadows. In total, 2901 validation samples were finally obtained.

Figure 3 shows the comparison between the estimated LWUP and the ground measurements. We can find that the average bias and RMSE are −4.59 and 16.15 W/m², respectively, at seven SURFRAD sites. For further analysis, the validation samples were divided into two groups: daytime and night time according to the local solar time. The bias ranged from −16.94 to 7.84 W/m², and RMSE ranges from 13.84 to 30.43 W/m² in the daytime. The average bias and RMSE were −1.40 W/m² and 21.57 W/m². If Desert_Rock_NVt is not considered, the average bias and RMSE were −1.12 W/m² and 21.71 W/m². During the night time, the bias ranged from −20.48 W/m² to 2.97 W/m², and RMSE ranged from 8.76 W/m² to 21.22 W/m². The average bias and RMSE were −8.3 W/m² and 13.46 W/m². If Desert_Rock_NVt is not considered, the average bias and RMSE were −5.33 W/m² and 10.73 W/m². The LWUP estimated at night was less divergent than that in the daytime.

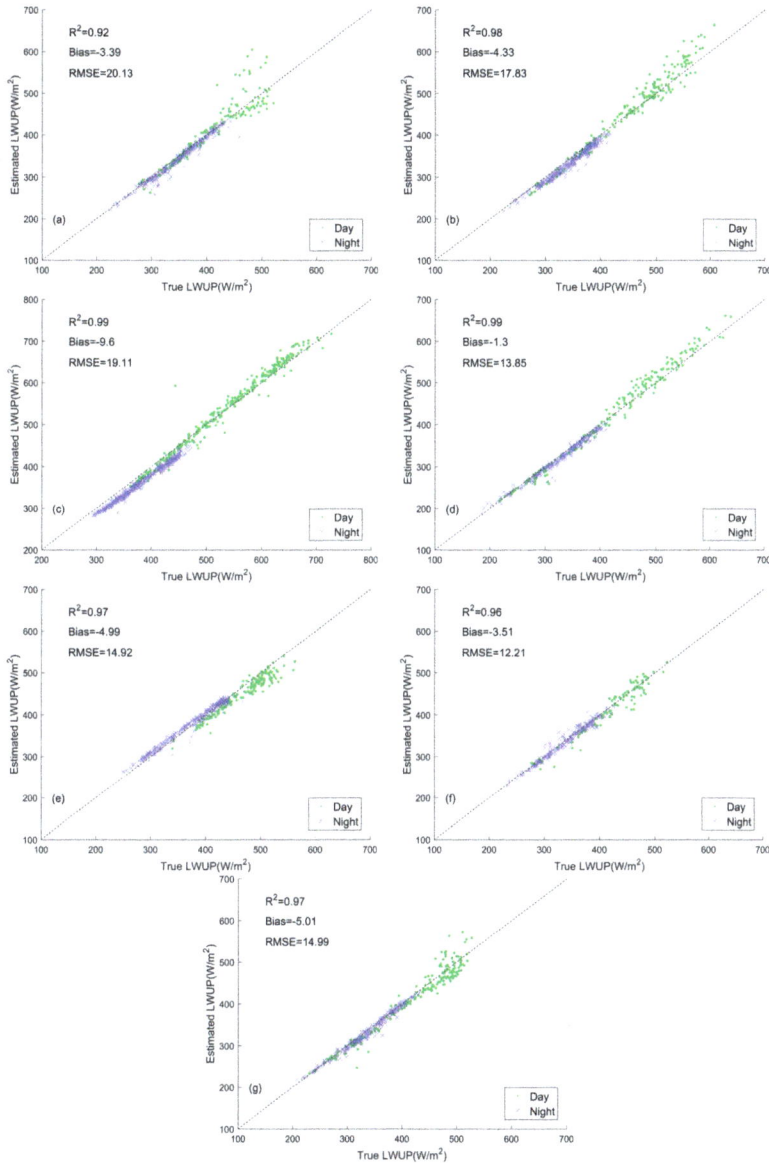

Figure 3. Validation results of the linear model at SURFRAD sites. (**a**) Bondville_IL; (**b**) Boulder_CO; (**c**) Desert_Rock_NV; (**d**) Fort_Peck_MT; (**e**) Goodwin_Creek_MS; (**f**) Penn_State_PA; (**g**) Sioux_Falls_SD.

4.2.2. MARS Models

The LWUP estimated by MARS models was validated with the same ground measurements. As shown in Figure 4, we can find that the average bias and RMSE were −5.23 and 16.38 W/m². The bias ranged from −16.15 to 8.15 W/m², and RMSE ranged from 12.12 to 29.96 W/m² in the daytime. The average bias and RMSE were −1.27 W/m² and 19.79 W/m². If Desert_Rock_NVt is not considered, the average bias and RMSE were −1.13 W/m² and 21.64 W/m². During the nighttime, the bias ranged from -20.46 W/m^2 to 2.25 W/m^2, and RMSE ranged from 8.66 W/m^2 to 15.26 W/m^2. The average bias and RMSE were -8.93 W/m^2 and 13.8 W/m^2. If Desert_Rock_NVt is not considered, the average bias and RMSE were -5.93 W/m^2 and 11.03 W/m^2. The distribution of the points in Figure 4 is similar to that in Figure 3.

Regarding the satellite-derived surface radiative fluxes, an accuracy of approximately ± 20 W/m^2 for instantaneous footprint values is required by the hydrological, meteorological, and agricultural research communities [41]. According to the validation results in this study, both the linear and MARS models can meet this requirement. Overall, the linear models are slightly better than the MARS models, although MARS can model the non-linearity between LWUP and TOA spectral radiance well during the training stage. Two reasons may account for this phenomenon. First, the zenith angles of the validation samples are all less than 40°, and the differences between the linear models and the MARS models are slight when the zenith angle is less than 45°, as shown in Table 2. Second, various kinds of satellite measurement errors are likely to be magnified by the non-linear model such as MARS. Due to the higher computational efficiency of the linear models, it is more adaptable to produce operational LWUP product.

A few studies have been devoted to estimating high spatial-resolution LWUP from MODIS and VIIRS. For example, Wang et al. [7] developed the linear models for estimating North American LWUP using MODIS data. The average bias and RMSE over SURFRAD site were -10.97 W/m^2 and 18.35 W/m^2. Jiao et al. [11] developed the neural network models for estimating LWUP from MODIS and VIIRS data over the Tibet Plateau. The average bias and RMSE of the validation results for MODIS were 11.24 W/m^2 and 26.78 W/m^2. The accuracy of LWUP from VIIRS was not validated. Cheng and Liang developed the linear models for estimating LWUP from MODIS data at global scale. The bias and RMSE of the validation over SURFRAD site were -4.49 W/m^2 and 13.47 W/m^2. Compared to these studies, the accuracies of the newly developed hybrid method are comparable or superior to those of the related published references.

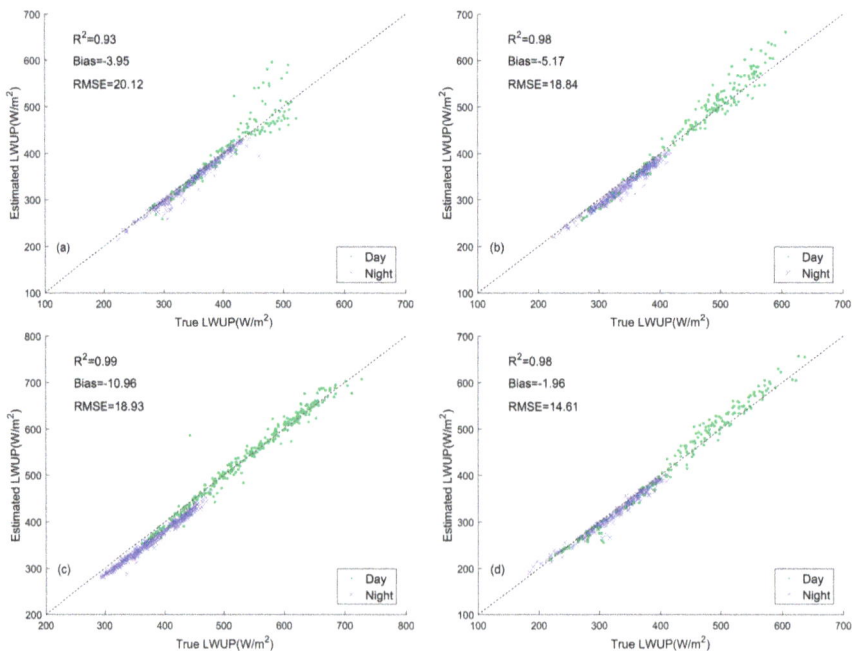

Figure 4. *Cont.*

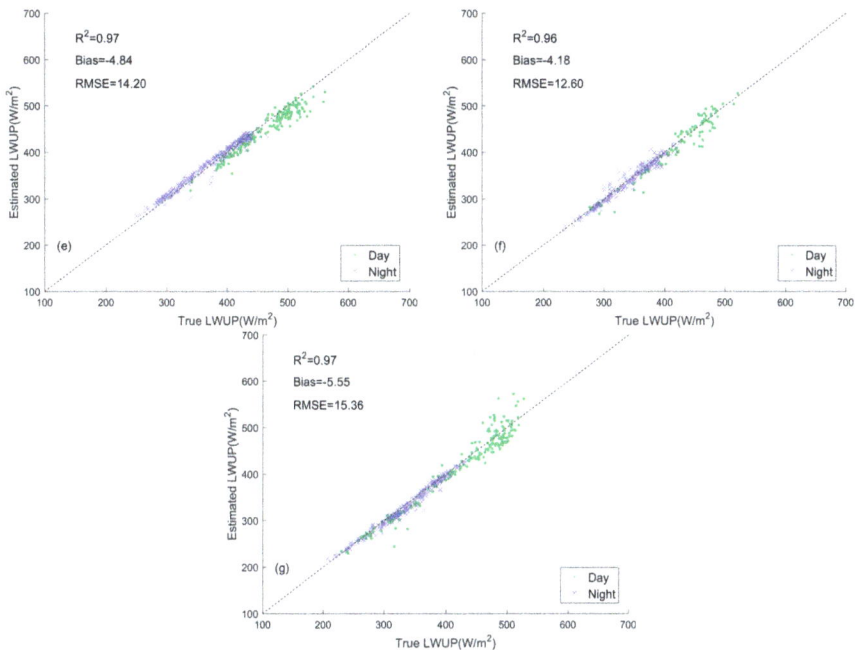

Figure 4. Validation results of the MARS model at SURFRAD sites. (**a**) Bondville_IL. (**b**) Boulder_CO. (**c**) Desert_Rock_NV. (**d**) Fort_Peck_MT. (**e**) Goodwin_Creek_MS. (**f**) Penn_State_PA. (**g**) Sioux_Falls_SD.

4.3. Comparison between Moderate Resolution Imaging Spectroradiometer (MODIS) and VIIRS Surface-Upwelling Longwave Radiation (LWUP)

Remote-sensing instruments such as MODIS and VIIRS are critical for providing continual and reliable measurements of the atmosphere, ocean and land-surface variables at the global scale. Furthermore, VIIRS is developed based on the heritage of MODIS instruments and has become a key bridge to ensure long-term continuity of the climate data records. LWUP estimated from VIIRS is compared with that retrieved from MODIS using the linear models of Cheng et al. [12] to check the consistency between the two instruments. The overpass time of the MODIS granule image is at 20:40 (UTC), 17 August 2014 and the VIIRS granule image is acquired from 20:46 to 20:51 (UTC) on the same day. The time difference between the two images is less than 10 min. Therefore, it is supposed that there are no significant differences in atmospheric conditions and land-surface properties within the 10-min period. In addition, the corresponding MOD35 and VIIRS Cloud Mask intermediate products are used to eliminate the influence of clouds. The linear hybrid method proposed by Cheng et al. [12] is used to calculate the LWUP from MODIS data, which has been validated by three measurement networks with a bias and RMSE of -0.31 W/m^2 and 19.92 W/m^2 in total.

The VIIRS image was aggregated to 1 km with bi-linear resampling method to match the spatial resolution of MODIS. The LWUP derived from both MODIS and VIIRS data were compared without considering the pixels that covered by invalid data or were contaminated by cloud. The spatial distributions of LWUP for MODIS and VIIRS are displayed in Figure 5, from which we can find that the LWUP distribution pattern of VIIRS is similar to that of the MODIS. In addition, the density plot of the LWUP is shown in Figure 6. Most of the LWUP values are concentrated between 400 and 800 W/m^2 with R^2 of 0.88. The bias and RMSE are -0.15 W/m^2 and 25.24 W/m^2, respectively. Thus, the LWUP of VIIRS is consistent with that of MODIS. This result is consistent with the study of Jiao et al. [11].

They compared the LWUP retrieved from MODIS and VIIRS at Tibet Plateau. The R^2, bias and RMES were 0.52 W/m^2, 2.87 W/m^2 and 26.02 W/m^2, respectively.

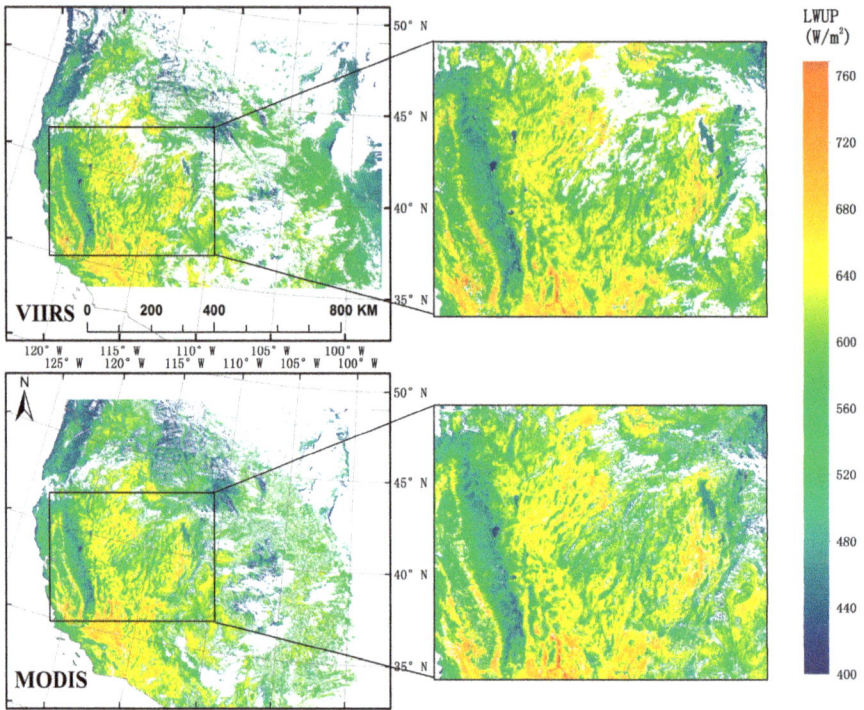

Figure 5. Distribution of LWUPs derived from VIIRS and MODIS images.

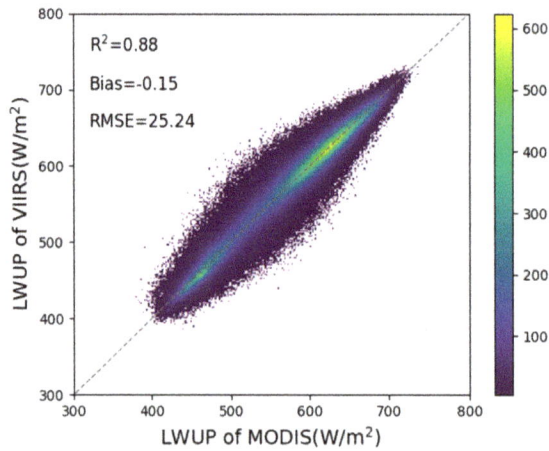

Figure 6. The scatterplot of retrieved LWUP from VIIRS and that retrieved from MODIS.

5. Discussion

5.1. Cloud Effect

In Section 4, the established hybrid methods were validated by the field measurements. Generally, most of the estimated LWUPs were closer to the one-to-one line. The points were less divergent in the night time than in the daytime. This may be because that the land surface is more homogeneous in the night time than in the daytime [42]. Compared to other sites, obvious underestimation is found in the desert site. Wang and Liang [7] thought air traffic out of Los Angeles produces many cirrus clouds over this site, and cloud contamination may be a significant source of error at this site. Thus, one of the reasons may be that the VIIRS Cloud Mask intermediate product cannot identify all kinds of cloud types, especially the cirrus cloud.

In order to test this assumption, we downloaded three years (2014–2017) field measurements from Atmospheric Radiation Measurement (ARM) program Southern Great Plains (SGP) site C1 (https://www.arm.gov). The downloaded data include the LWUP, cloud-base height and other auxiliary data. The cloud-base height information is derived from the Micropulse Lidar (MPL). MPL is a ground-based remote-sensing instrument that can be used to measure the altitude of clouds by transmitting pulses of light and using the receiver to detect the light scattered back by clouds and precipitation. From the time delay between each outgoing pulse and the backscattered signal, the distance to the scatterer is inferred [43]. The VIIRS pixels are, first, identified by the VIIRS Cloud Mask intermediate product, and then the extracted clear-sky pixels are further screened by the field measurements. The pixels with more than one detected cloud base are considered as cloudy, while those with no significant backscatter detected are regarded as true clear-sky pixels. The developed liner models are used to retrieve LWUP. The validation results of LWUP at site C1 are displayed in Figure 7. The bias and RMSE of cloudy pixels are -6.65 W/m^2 and 16.04 W/m^2, while bias and RMSE of clear-sky pixels are -2.94 W/m^2 and 15.62 W/m^2. The VIIRS Cloud Mask intermediate product actually cannot identify all kinds of cloud types, and undetected cloud will cause the underestimation of LWUP.

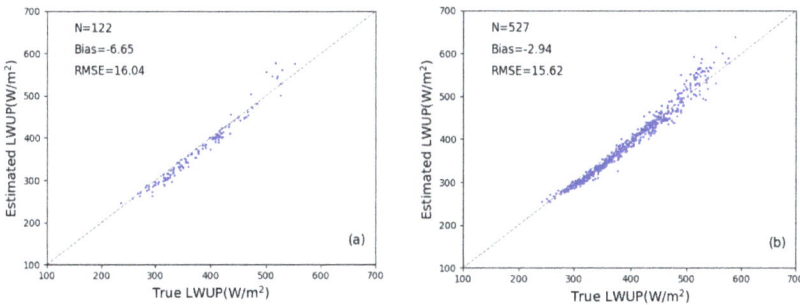

Figure 7. Validation results of the linear model at the Atmospheric Radiation Measurement (ARM) program Southern Great Plains (SGP) C1 site. (**a**) Pixels that were identified as clear sky by the VIIRS cloud mask whereas cloudy by ground-based Lidar; (**b**) pixels that were identified as clear sky by both the VIIRS Cloud Mask and the Lidar.

5.2. Broadband Emissivity (BBE) Effect

The obvious underestimation of LWUP at Desert_Rock_NV may also be related to BBE of the surface. Taking the developed linear model for the mid-latitude region at nadir view as an example, we investigated the effects of surface BBE on the accuracy of LWUP estimation. The bias and RMSE of the fitting liner model at nadir are zero and 6.94 W/m^2. We calculated the average bias and RMSE for each emissivity spectrum as well as the corresponding BBE. The relationship between BBE versus bias is shown in Figure 8; 78% samples have a negative bias when their BBE is less than 0.966 and 88% BBE has a positive bias when BBE is equal to or larger than 0.966. It is very likely to produce negative bias at Desert_Rock_NV from the point of the developed linear models, because its BBE is less than 0.966. We have also investigated the relationship between the fitting residual and land-surface temperature (LST) with the same data, and no significant overestimated or underestimated trend is found. Thus, cloud and surface BBE are two primary factors that affect the accuracy of the LWUP estimate.

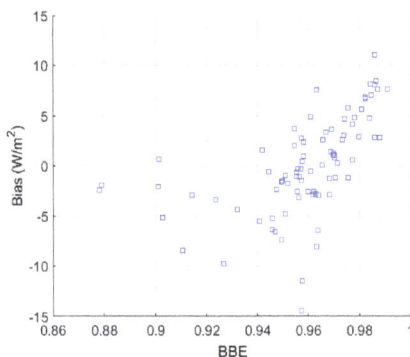

Figure 8. The relationships between Broadband Emissivity (BBE) versus bias of retrieved LWUP.

6. Conclusions

LWUP is an important component of the land-surface radiation budget. We developed two hybrid methods, namely, the linear model and MARS model, to estimate the 750-m instantaneous clear-sky LWUP from the VIIRS TOA channel radiances of M14, M15, and M16 at the global scale.

To consider the difference between the land-surface temperature and air temperature, the global land surface was divided into 3 regions based on latitude. Extensive radiative transfer modeling was conducted to produce a huge amount of representative samples for each sub region. The linear models and the MARS models were established at $0°$, $15°$, $30°$, $45°$ and $60°$ viewing zenith angles for each sub region. According to the statistical results, the linear models can account for more than 97.7% of the variation in the simulation database, the bias is zero, and the RMSEs range from 5.27 to 13.02 W/m^2; the MARS models can account for more than 98.2% of the variation in the simulation database, the bias is zero, and the RMSEs range from 4.8 to 10.13 W/m^2. Then, the two models were validated using three years (2014–2017) of ground measurements collected from seven SURFRAD sites. The average bias and RMSE of the linear models were -4.59 W/m^2 and 16.15 W/m^2, whereas the average bias and RMSE of the MARS models were -5.23 W/m^2 and 16.38 W/m^2. The difference between the linear models and the MARS models was not significant. The linear models have higher computational efficiency and are easy to implement, so it is a good choice for producing the operational LWUP product. The LWUP retrieved from VIIRS by the developed linear models was compared to that retrieved from MODIS by the previous linear models developed. Their spatial distribution pattern agreed well with each other; the bias and RMSE were -0.15 W/m^2 and 25.24 W/m^2.

This is the first time that the hybrid method has been applied to global estimates of clear-sky LWUP from VIIRS data. Limited validation results indicate that the accuracy of the hybrid method can meet the accuracy requirement of the hydrological, meteorological, and agricultural research communities. The good performance of the developed hybrid method and consistency between VIIRS LWUP and MODIS LWUP indicate that the hybrid method shows promise for producing a long-term high spatial-resolution environmental data record (EDR) of LWUP. The field measurements used for validation are collected from the SURFRAD network, which is located in the middle-latitude region. More validation data from low-latitude and high-latitude regions with other land-cover types need to be collected for further validation in the future. Also, more extensive comparison should be conducted before the generation of LWUP EDR using MODIS and VIIRS data.

Acknowledgments: The ground data used in this study are collected from Surface Radiation Budget Network (ftp: //aftp.cmdl.noaa.gov/data/radiation/surfrad) and ARM Climate Research Facility (https://www.arm.gov/data). The AIRS atmosphere profile product is downloaded from NASA (http://daac.gsfc.nasa.gov/AIRS/). The VIIRS SDR and Cloud mask product are downloaded from NOAA (https://www.class.ngdc.noaa.gov/saa/products/ welcome). The source code of the ARESLab toolbox is downloaded from the website: www.cs.rtu.lv/jekabsons/ regression.html. This work was partly supported by the National Natural Science Foundation of China via grant 41771365, the National Key Research and Development Program of China via grant 2016YFA0600101, and the Special Fund for Young Talents of the State Key Laboratory of Remote Sensing Sciences via grant 17ZY-02.

Author Contributions: Jie Cheng conceived and designed the algorithm; Jie Cheng and Shugui Zhou performed the algorithm and analyzed the data; Shugui Zhou wrote the paper. Both authors participated in the editing of the paper.

Conflicts of Interest: The authors declare no conflicts of interest.

References

1. Diak, G.R.; Mecikalski, J.R.; Anderson, M.C.; Norman, J.M.; Kustas, W.P.; Torn, R.D.; DeWolf, R.L. Estimating land surface energy budgets from space: Review and current efforts at the university of wisconsin—madison and usda–ars. *Bull. Am. Meteorol. Soc.* **2004**, *85*, 65–78. [CrossRef]
2. Ellingson, R.G. Surface longwave fluxes from satellite observations: A critical review. *Remote Sens. Environ.* **1995**, *51*, 89–97. [CrossRef]
3. Schmetz, J. Towards a surface radiation climatology: Retrieval of downward irradiances from satellites. *Atmos. Res.* **1989**, *23*, 287–321. [CrossRef]
4. Curry, J.A.; Schramm, J.L.; Rossow, W.B.; Randall, D. Overview of arctic cloud and radiation characteristics. *J. Clim.* **1996**, *9*, 1731–1764. [CrossRef]
5. Mlynczak, P.E.; Smith, G.L.; Wilber, A.C.; Stackhouse, P.W. Annual cycle of surface longwave radiation. *J. Appl. Meteorol. Climatol.* **2011**, *50*, 1212–1224. [CrossRef]
6. Ma, Q.; Wang, K.; Wild, M. Evaluations of atmospheric downward longwave radiation from 44 coupled general circulation models of cmip5. *J. Geophys. Res. Atmos.* **2014**, *119*, 4486–4497. [CrossRef]
7. Wang, W.; Liang, S.; Augustine, J.A. Estimating high spatial resolution clear-sky land surface upwelling longwave radiation from modis data. *IEEE Trans. Geosci. Remote Sens.* **2009**, *47*, 1559–1570. [CrossRef]
8. Bisht, G.; Venturini, V.; Islam, S.; Jiang, L. Estimation of the net radiation using modis (moderate resolution imaging spectroradiometer) data for clear sky days. *Remote Sens. Environ.* **2005**, *97*, 52–67. [CrossRef]
9. Long, D.; Gao, Y.; Singh, V.P. Estimation of daily average net radiation from modis data and dem over the baiyangdian watershed in north china for clear sky days. *J. Hydrol.* **2010**, *388*, 217–233. [CrossRef]
10. Wang, C.; Tang, B.-H.; Huo, X.; Li, Z.-L. New method to estimate surface upwelling long-wave radiation from modis cloud-free data. *Opt. Express* **2017**, *25*, A574–A588. [CrossRef] [PubMed]
11. Jiao, Z.; Yan, G.; Zhao, J.; Wang, T.; Chen, L. Estimation of surface upward longwave radiation from modis and viirs clear-sky data in the tibetan plateau. *Remote Sens. Environ.* **2015**, *162*, 221–237. [CrossRef]
12. Cheng, J.; Liang, S. Global estimates for high-spatial-resolution clear-sky land surface upwelling longwave radiation from modis data. *IEEE Trans. Geosci. Remote Sens.* **2016**, *54*, 4115–4129. [CrossRef]
13. Lee, H.-T.; Ellingson, R.G. Development of a nonlinear statistical method for estimating the downward longwave radiation at the surface from satellite observations. *J. Atmos. Ocean. Technol.* **2002**, *19*, 1500–1515. [CrossRef]
14. Liang, S.; Zhao, X.; Liu, S.; Yuan, W.; Cheng, X.; Xiao, Z.; Zhang, X.; Liu, Q.; Cheng, J.; Tang, H. A long-term global land surface satellite (glass) data-set for environmental studies. *Int. J. Digit. Earth* **2013**, *6*, 5–33. [CrossRef]
15. Cheng, J.; Liang, S.; Yao, Y.; Ren, B.; Shi, L.; Liu, H. A comparative study of three land surface broadband emissivity datasets from satellite data. *Remote Sens.* **2013**, *6*, 111–134. [CrossRef]
16. Vogt, J.V. Land surface temperature retrieval from noaa avhrr data. In *Advances in the Use of NOAA AVHRR Data for Land Applications*; Springer: Berlin/Heidelberg, Germany, 1996; pp. 125–151.
17. Wan, Z.; Dozier, J. A generalized split-window algorithm for retrieving land-surface temperature from space. *IEEE Trans. Geosci. Remote Sens.* **1996**, *34*, 892–905.

18. Wan, Z.; Zhang, Y.; Zhang, Q.; Li, Z.-L. Quality assessment and validation of the modis global land surface temperature. *Int. J. Remote Sens.* **2004**, *25*, 261–274. [CrossRef]

19. Cheng, J.; Liang, S.; Verhoef, W.; Shi, L.; Liu, Q. Estimating the hemispherical broadband longwave emissivity of global vegetated surfaces using a radiative transfer model. *IEEE Trans. Geosci. Remote Sens.* **2016**, *54*, 905–917. [CrossRef]

20. Tang, B.; Li, Z.-L. Estimating of instananeous net surface longwave radiation from modis cloud-free data. *Remote Sens. Environ.* **2008**, *112*, 3482–3492. [CrossRef]

21. Wang, T.; Yan, G.; Chen, L. Consistent retrieval methods to estimate land surface shortwave and longwave radiative flux components under clear-sky conditions. *Remote Sens. Environ.* **2012**, *124*, 61–71. [CrossRef]

22. Jackson, J.M.; Liu, H.; Laszlo, I.; Kondragunta, S.; Remer, L.A.; Huang, J.; Huang, H.C. Suomi-npp viirs aerosol algorithms and data products. *J. Geophys. Res. Atmos.* **2013**, *118*, 12673–12689. [CrossRef]

23. Vargas, M.; Miura, T.; Shabanov, N.; Kato, A. An initial assessment of suomi npp viirs vegetation index edr. *J. Geophys. Res. Atmos.* **2013**, *118*, 12301–312316. [CrossRef]

24. Wang, D.; Liang, S.; He, T.; Yu, Y. Direct estimation of land surface albedo from viirs data: Algorithm improvement and preliminary validation. *J. Geophys. Res. Atmos.* **2013**, *118*, 12577–12586. [CrossRef]

25. Yu, Y.; Privette, J.L.; Pinheiro, A.C. Analysis of the npoess viirs land surface temperature algorithm using modis data. *IEEE Trans. Geosci. Remote Sens.* **2005**, *43*, 2340–2350.

26. Petrenko, B. Evaluation and selection of sst regression algorithms for jpss viirs. *J. Geophys. Res. Atmos.* **2014**, *119*, 4580–4599. [CrossRef]

27. Schueler, C.F.; Clement, J.E.; Ardanuy, P.E.; Welsch, C.; DeLuccia, F.; Swenson, H. Npoess viirs sensor design overview. In Earth Observing Systems VI, Proceedings of the International Society for Optics and Photonics, San Diego, CA, USA, 29 July–4 August 2001; pp. 11–24.

28. Xiong, X.; Butler, J.; Chiang, K.; Efremova, B.; Fulbright, J.; Lei, N.; McIntire, J.; Oudrari, H.; Sun, J.; Wang, Z. Viirs on-orbit calibration methodology and performance. *J. Geophys. Res. Atmos.* **2014**, *119*, 5065–5078. [CrossRef]

29. Madhavan, S.; Brinkmann, J.; Wenny, B.N.; Wu, A.; Xiong, X. Evaluation of viirs and modis thermal emissive band calibration stability using ground target. *Remote Sens.* **2016**, *8*, 158. [CrossRef]

30. Godin, R. Joint polar satellite system (jpss) viirs cloud mask (vcm) algorithm theoretical basis document (atbd). *JPSS* **2014**, *474*, 474-00033.

31. Aumann, H.; Chanhine, M.T.; Gautier, C. Airs/amsu/hsb on the aqua mission: Design, science objectives, data products, and processing systems. *IEEE Trans. Geosci. Remote Sens.* **2003**, *41*, 253–264. [CrossRef]

32. Aumann, H.; Elliott, D.; Strow, L. Validation of the radiometric stability of the atmospheric infrared sounder. In Earth Observing Systems XVII, Proceedings of the International Society for Optics and Photonics, San Diego, CA, USA, 14–16 August 2012; p. 85100T.

33. Susskind, J.; Barnet, C.D.; Blaisdell, J.M. Retrieval of atmoshperic and surface parameters from airs/amsu/hsb data in the presence of clouds. *IEEE Trans. Geosci. Remote Sens.* **2003**, *41*, 390–409. [CrossRef]

34. Augustine, J.A.; DeLuisi, J.J.; Long, C.N. Surfrad—A national surface radiation budget network for atmospheric research. *Bull. Am. Meteorol. Soc.* **2000**, *81*, 2341–2357. [CrossRef]

35. Wang, K.; Dickinson, R.E. Global atmospheric downward longwave radiation at the surface from ground-based observations, satellite retrievals, and reanalyses. *Rev. Geophys.* **2013**, *51*, 150–185. [CrossRef]

36. Augustine, J.A.; Dutton, E.G. Variability of the surface radiation budget over the united states from 1996 through 2011 from high-quality measurements. *J. Geophys. Res. Atmos.* **2013**, *118*, 43–53. [CrossRef]

37. Berk, A.; Anderson, G.; Acharya, P.; Shettle, E. *Modtran5. 2.0. 0 User's Manual*; Spectral Sciences Inc.: Burlington, MA, USA; Air Force Research Laboratory: Hanscom, MA, USA, 2008.

38. Baldridge, A.; Hook, S.; Grove, C.; Rivera, G. The aster spectral library version 2.0. *Remote Sens. Environ.* **2009**, *113*, 711–715. [CrossRef]

39. Snyder, W.C.; Wan, Z.; Zhang, Y.; Feng, Y.-Z. Classification-based emissivity for land surface temperature measurement from space. *Int. J. Remote Sens.* **1998**, *19*, 2753–2774. [CrossRef]

40. Friedman, J.H. Multivariate adaptive regression splines. *Ann. Stat.* **1991**, 1–67. [CrossRef]

41. Gupta, S.K.; Kratz, D.P.; Wilber, A.C.; Nguyen, L.C. Validation of parameterized algorithms used to derive trmm–ceres surface radiative fluxes. *J. Atmos. Ocean. Technol.* **2004**, *21*, 742–752. [CrossRef]

42. Wang, W.; Liang, S.; Meyers, T. Validating modis land surface temperature products using long-term nighttime ground measurements. *Remote Sens. Environ.* **2008**, *112*, 623–635. [CrossRef]

43. Zhao, C.; Wang, Y.; Wang, Q.; Li, Z.; Wang, Z.; Liu, D. A new cloud and aerosol layer detection method based on micropulse lidar measurements. *J. Geophys. Res. Atmos.* **2014**, *119*, 6788–6802. [CrossRef]

remote sensing

MDPI

Article

Comparative Analysis of Chinese HJ-1 CCD, GF-1 WFV and ZY-3 MUX Sensor Data for Leaf Area Index Estimations for Maize

Jing Zhao [1], Jing Li [1,*], Qinhuo Liu [1,2,3,*], Hongyan Wang [4], Chen Chen [4], Baodong Xu [1,2] and Shanlong Wu [1]

[1] State Key Laboratory of Remote Sensing Science, Institute of Remote Sensing and Digital Earth, Chinese Academy of Sciences, Beijing 100101, China; zhaojing1@radi.ac.cn (J.Z.); xubd@radi.ac.cn (B.X.); wsl0579@163.com (S.W.)

[2] College of Resources and Environment, University of Chinese Academy of Sciences, Beijing 100049, China

[3] Joint Center for Global Change Studies, Beijing 100875, China

[4] Satellite Surveying and Mapping Application Center, National Administration of Surveying Mapping and Geo-Information of China, Beijing 100048, China; wanghy@sasmac.cn (H.W.); chenc@sasmac.cn (C.C.)

* Correspondence: lijing01@radi.ac.cn (J.L.); liuqh@radi.ac.cn (Q.L.); Tel./Fax: +86-010-6485-1880 (J.L.)

Received: 6 November 2017; Accepted: 3 January 2018; Published: 5 January 2018

Abstract: In recent years, China has developed and launched several satellites with high spatial resolutions, such as the resources satellite No. 3 (ZY-3) with a multi-spectral camera (MUX) and 5.8 m spatial resolution, the satellite GaoFen No. 1 (GF-1) with a wide field of view (WFV) camera and 16 m spatial resolution, and the environment satellite (HJ-1A/B) with a charge-coupled device (CCD) sensor and 30 m spatial resolution. First, to analyze the potential application of ZY-3 MUX, GF-1 WFV, and HJ-1 CCD to extract the leaf area index (LAI) at the regional scale, this study estimated LAI from the relationships between physical model-based spectral vegetation indices (SVIs) and LAI values that were generated from look-up tables (LUTs), simulated from the combination of the PROSPECT-5B leaf model and the scattering by arbitrarily inclined leaves with the hot-spot effect (SAILH) canopy reflectance model. Second, to assess the surface reflectance quality of these sensors after data preprocessing, the well-processed surface reflectance products of the Landsat-8 operational land imager (OLI) sensor with a convincing data quality were used to compare the performances of ZY-3 MUX, GF-1 WFV, and HJ-1 CCD sensors both in theory and reality. Apart from several reflectance fluctuations, the reflectance trends were coincident, and the reflectance values of the red and near-infrared (NIR) bands were comparable among these sensors. Finally, to analyze the accuracy of the LAI estimated from ZY-3 MUX, GF-1 WFV, and HJ-1 CCD, the LAI estimations from these sensors were validated based on LAI field measurements in Huailai, Hebei Province, China. The results showed that the performance of the LAI that was inversed from ZY-3 MUX was better than that from GF-1 WFV, and HJ-1 CCD, both of which tended to be systematically underestimated. In addition, the value ranges and accuracies of the LAI inversions both decreased with decreasing spatial resolution.

Keywords: LAI; ZY-3 MUX; GF-1 WFV; HJ-1 CCD; maize; PROSPECT-5B+SAILH (PROSAIL) model

1. Introduction

Leaf area index (LAI) is defined as one-half the total foliage area per unit ground surface area [1], and it is an important parameter for monitoring vegetation growth conditions [2,3]. LAI is a common variable that is used for regional and global climate, ecological, and hydrological models [4,5]. LAI has been widely used in global primary productivity measurements [6], agricultural yield estimations [7,8], and ecological and environmental assessments [9]. High-spatial-resolution LAI products play important roles in monitoring regional vegetation changes and evaluating the accuracy of low- resolution LAI products [10–12].

From the Landsat-5 satellite launched in 1984 to the present, there are many moderate- to high-resolution satellite sensors that are available in the world, such as the Landsat thematic mapper (TM)/enhanced thematic mapper plus (ETM+)/operational land imager (OLI), Terra advanced spaceborne thermal emission and reflection radiometer (ASTER), SPOT high resolution geometrical (HRG), IKONOS multi-spectral (MS), Sentinel-2 multispectral imager (MSI), and the Chinese environment satellite (HJ-1) charge coupled device (CCD), etc. Generally, LAI extracted from this moderate- to high-resolution imagery largely depends on the empirical relationships. Empirical expressions were established between LAI field measurements and the spectral vegetation indices (SVIs) from isolated dates [13–19]. The most commonly used SVIs include the normalized difference vegetation index (NDVI), the simple ratio index (SR), and the enhanced vegetation index (EVI) [14,15,20–23]. In addition, the reduced simple ratio (RSR), the soil-adjusted vegetation index (SAVI), and the perpendicular vegetation index (PVI) are widely used for LAI extraction [21,24–26]. When compared with LAI field measurements, the accuracies of satellite LAI estimates based on linear and non-linear regressions of the SVI-LAI relationships have coefficient of determination (R^2) values from approximately 0.37 to 0.98 and root mean square error (RMSE) values from approximately 0.17 m^2/m^2 to 1.14 m^2/m^2 for both crops (e.g., winter wheat, maize, and soybean) and forests (e.g., coniferous and deciduous) [16,21,23,27,28]. In addition, the SVI-LAI relationships are stronger for crop canopies than for coniferous forests, and are weakest for deciduous forests [15,22,29]. Empirical methods are computationally efficient when using remote sensing datasets at regional or large scales. However, empirical relationships that typically depend on unique vegetation types and regions are often constructed and used locally.

The physical model method, which is suitable for a variety of vegetation types, is also used to extract LAI from moderate- to high-resolution imagery [30–37]. Canopy reflectance models simulate the physical relationship between the canopy reflectance and the LAI in the forward direction. The scattering by arbitrarily inclined leaves with the hot-spot effect (SAILH) model [30,32,38], the Markov chain reflectance model (MCRM) [31], and the Li-Strahler geometric-optical model [34] have been used to extract LAI from moderate- to high-resolution imagery. LAI has also been estimated by indirect methods based on the inversion of canopy reflectance models, such as look-up tables (LUTs) and hybrid methods. The hybrid methods include decision tree learning, artificial neural networks, kernel methods, and Bayesian networks [39]. Additionally, the currently used indirect methods of the radiative transfer model (RTM) for Landsat ETM+ and Sentinel-2 MSI data are the LUT and neural networks [31,32,35,36]. The accuracy of satellite LAI inversions is better than SVI-LAI empirical relationships, with R^2 values from 0.54 to 0.82 and RMSE values from 0.17 m^2/m^2 to 0.71 m^2/m^2 for crops (e.g., maize and soybean), shrubs, and planted forests [30,31,33,34].

LAI field measurements were acquired via direct and indirect methods [40]. Direct LAI measurements, including leaf collection from deciduous forests and the destructive sampling of crops or low shrubs, are time consuming and difficult to collect at larger areas [40]. Indirect LAI measurement, including optical sensor-based method (such as those using the Licor LAI-2200 Plant Canopy Analyzer (LI-COR, Lincoln, NE, USA) [41], Tracing Radiation and Architecture of Canopies (TRAC) (3rd Wave, Nepean, ON, Canada) [42], AccuPar (Decagon Devices, Inc., Pullman, WA, USA) [43]), digital hemispherical photography (DHP), and new smartphone camera sensor technology (such as LAISmart [44,45] and PocketLAI [46]). These indirect methods are generally convenient, especially those that allow for LAI estimation using a smartphone, and generally efficient over larger spatial scales.

In recent years, China has developed and launched several satellites, such as the HJ-1A/B with a CCD sensor, the GaoFen No. 1 satellite (GF-1) with a wide field of view (WFV) camera and the resources satellite No. 3 (ZY-3) with a multi-spectral camera (MUX). Currently, the satellite data of

ZY-3 MUX, GF-1 WFV, and HJ-1 CCD have been applied for vegetation monitoring. HJ-1 CCD data have been widely used for LAI extraction at the regional scale based on SVI-LAI relationships or physical models [21,30,33,34,47]. In addition, GF-1 WFV data have been used to extract the fractional vegetation cover (FVC) [48,49] and estimate LAI from the NDVI-LAI empirical relationship [21,50,51] at the regional scale. However, there have not been any studies that have reported the extraction of LAI from ZY-3 MUX data. One of the objectives of this study is to analyze the potential use of three Chinese satellites, especially ZY-3 MUX, to extract LAI at the regional scale.

The study area was selected in the Huailai experiment station, Hebei Province, China. The satellite data from ZY-3 MUX, GF-1 WFV, and HJ-1 CCD for the study area were collected close to the date of LAI field measurements. The LAI extraction method was a physical model-based SVI- LAI relationship that was generated from the LUT based on the PROSPECT-5B + SAILH (PROSAIL) model with specific input parameters for each sensor (Section 2). In order to assess the surface reflectance quality of ZY-3 MUX, GF-1 WFV, and HJ-1 CCD after data preprocessing, the well- processed surface reflectance products of Landsat-8 OLI with a convincing data quality are used to compare the performances of these sensors, both in theory and reality. Furthermore, the accuracy of the LAI estimation results from ZY-3 MUX, GF-1 WFV, and HJ-1 CCD was validated based on LAI field measurements from the Huailai experiment station (Section 3). The up-scaled LAI inversions for ZY-3 MUX and GF-1 WFV were also compared with the HJ-1 CCD data at the same spatial resolutions for pure and mixed pixels (Section 3). The discussion and conclusions are presented in Sections 4 and 5, respectively.

2. Materials and Methods

2.1. Study Area and Field Measurements

The study area was at the Huailai experimental field (40°20′55.093″N, 115°46′59.569″E, altitude 488 m) in Hebei Province, China. This field is affiliated with the Chinese Academy of Sciences (CAS) (Figure 1). The study area is in a temperate and semi-arid region with four distinct seasons, abundant sunshine, simultaneous heat and moisture, and large temperature differences. The annual average temperature and average precipitation are approximately 9.5 °C and 392 mm, respectively. The primary vegetation type around the Huailai experimental field is farmland, and other land cover types are water, wetland beach, and residential. Maize is the dominant crop type in this study area. The soil type of the study area is brown soil.

Several vegetation structure parameters (e.g., leaf reflectance and transmittance, average leaf angle, canopy spectral measurements, and LAI) and biophysical and biochemical parameters (e.g., leaf chlorophyll-*a* and -*b* content (C_{ab}), leaf water content (C_w), and leaf dry matter content (C_m)) were acquired at the Huailai experimental field. The leaf reflectance and transmittance from 400 nm to 2500 nm were measured using a UV molecular spectroscopy (Lambda 900, PerkinElmer Inc., Waltham, MA, USA). The leaf inclination angle was the angle between a leaf and its normal direction, and was measured using a protractor. The average leaf angle is the mean leaf inclination angle for an entire plant. The canopy and soil spectra were measured using a spectroradiometer (Analytical Spectral Devices, ASD, Longmont, CO, USA) covering wavelengths from 400 nm to 1100 nm with a 5° field of view at noon on sunny days. C_{ab} was based on an average of six points on each leaf three times using the SPAD 502DL plus Chlorophyll meter (Spectrum technologies, Inc., Bridgend, UK). The fresh leaves were weighed and placed into an envelope bag. The envelope bag was then put in the oven at 105 °C for 30 min and then at 85 °C for 24 h. C_m is the weight of leaves after oven drying, and C_w is the proportion of leaf water (fresh weight minus dry weight) to the dry weight.

In this study, LAI field measurements from maize sample plots were acquired on 31 July 2014 (yellow dots in Figure 1b). The sample plots were selected based on NDVI values from Landsat-8 OLI according to Zeng et al. [52]. Then, the sample plots were selected in the field with a single plant type and uniform growth base on the coordinates from global positioning system (GPS). Finally, 17 sample plots were acquired in the study area. LAI field measurements were acquired using an LAI-2200 Plant

Canopy Analyzer (LI-COR, Inc., Lincoln, NE, USA). The LAI of each sample plot was measured from one above-canopy reading and nine below-canopy readings with a 45° view cup. The measurements were obtained from 06:30 to 10:00 and from 16:30 to 19:30 to avoid measurement errors caused by the direct sunlight, and the LAI value was measured twice at each site. To reduce the observer effects and other sources of error during LAI field measurements, except for the records of the fifth view angle (centered at 68°) acquired from the LAI-2200 instrument, all of the LAI field measurements were calculated based on a standard error LAI (SEL) of less than 0.5.

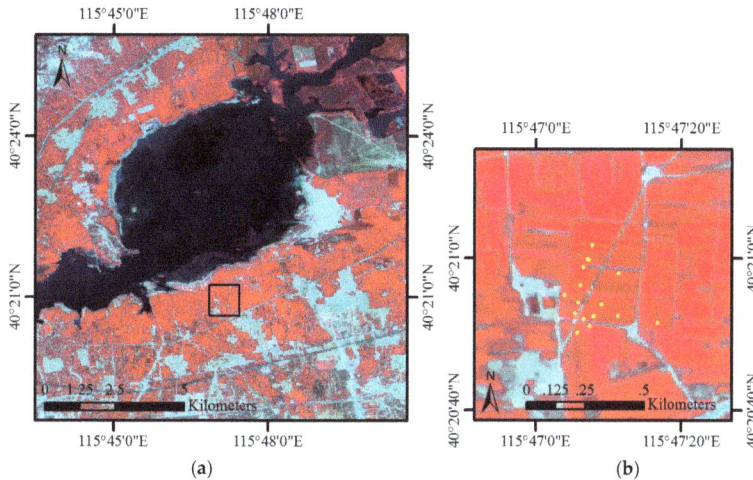

Figure 1. The geographic location of the study area of ZY-3 MUX based on the false color composite (NIR-red-green) (**a**), and the subset study area with leaf area index (LAI) field measurements for maize (yellow dots) (**b**).

2.2. Remote Sensing Data and Preprocessing

The HJ-1 satellite was launched on 6 September 2008, and the ZY-3 satellite was launched on 9 January 2012. Both the HJ-1 satellite and the ZY-3 satellite were launched from the Taiyuan Satellite Launch Center, Shanxi Province, China. The GF-1 satellite was launched on 26 April 2013 from the Jiuquan Satellite Launch Center, Gansu Province, China. The HJ-1, GF-1 and ZY-3 satellites are in sun synchronous orbits at altitudes of 649 km, 645 km, and 505 km, respectively. The technical specifications for the ZY-3 MUX, GF-1 WFV, and HJ-1 CCD sensors are shown in Table 1. Four spectral channels that are distributed in the visible and near-infrared (NIR) spectral domain ranging from 450 nm to 900 nm are identical in these three sensors. The radiometric resolutions of the GF-1 WFV and ZY-3 MUX sensors are higher than that of HJ-1 CCD by 2 bits, which improves the detectability of changes in the feature characteristics. The HJ-1 CCD data have a spatial resolution of 30 m, and the revisit time is approximately four days over China due to the combination of two satellites (HJ-1A and HJ-1B) with two cameras (CCD1 and CCD2) on each satellite. The GF-1 WFV data have a spatial resolution of 16 m and a revisit time of four days among the four combined cameras. The ZY-3 MUX data have the highest spatial resolution of 5.8 m, but the 51 km swath width is much narrower than that of the other two cameras.

Table 1. Technical specification of the ZY-3 MUX, GF-1 WFV, and HJ-1 CCD cameras.

Sensor		ZY-3 MUX		GF-1 WFV		HJ-1 CCD	
	Bands	Wavelength (µm)	Bands	Wavelength (µm)	Bands	Wavelength (µm)	
	1	0.45–0.52	1	0.45–0.52	1	0.43–0.52	
Spectral characteristics	2	0.52–0.59	2	0.52–0.59	2	0.52–0.60	
	3	0.63–0.69	3	0.63–0.69	3	0.63–0.69	
	4	0.77–0.89	4	0.77–0.89	4	0.76–0.90	
Spatial resolution (m)		5.8		16		30	
Radiometric resolution (Bit)		10		10		8	
Swath width (km)		51		200 (single); 800 (4 cameras)		360 (single); 700 (two)	
Revisit time (days)		5		4		4	

2.2.1. Remote Sensing Data Acquisition

The ZY-3 MUX data were acquired from the Satellite Surveying and Mapping Application Center (SASMAC) of the National Administration of Surveying, Mapping and Geo-information of China (NASG) [53]. The GF-1 WFV data were acquired from the Gaofen satellite data and information service system (GFDIS) [54], and the HJ-1 CCD data were acquired from the China Centre for Resources Satellite Data and Application (CRESDA) [55]. The surface reflectance data of Landsat-8 OLI for the study area was acquired from the United States Geological Survey (USGS) [56]. The Landsat-8 OLI data was used to evaluate the data stability of different sensors among ZY-3 MUX, GF-1 WFV and HJ-1 CCD. The acquisition information for these satellite data sources is provided in Table 2. All of the images over the study area were cloudless.

Table 2. Satellite data acquisition information for ZY-3 MUX, GF-1 WFV, HJ-1 CCD and Landsat-8 OLI.

Sensor	Date	Local Time	Solar Zenith Angle	Solar Azimuth Angle	View Zenith Angle (Mean)	View Azimuth Angle (Mean)
ZY-3 MUX	20140727	11:15:23	25.28°	216.38°	0°	216.38°
GF-1 WFV	20140727	11:50:19	22.08°	201.62°	29.97°	74.57°
HJ-1 CCD	20140728	10:00:09	34.97°	295.49°	25°	53.59°
Landsat-8 OLI	20140725	10:59:34	27.67°	131.88°	0°	95.31°

2.2.2. Remote Sensing Data Preprocessing

Preprocessing of the remote sensing data included radiometric calibration, geometric correction, and atmospheric correction. First, radiometric calibration converted the digital number value of the raw image to radiance based on Equation (1) [55].

$$L_e(\lambda_e) = Gain \cdot DN + Offset \tag{1}$$

where $L_e(\lambda_e)$ is the radiance, and Gain and Offset are the calibration coefficients. The unit is $W \times m^{-2} \times sr^{-1} \times \mu m^{-1}$ The Gain and Offset values for ZY-3 MUX, GF-1 WFV, and HJ-1 CCD were obtained from the CRESDA and are shown in Table 3 [55].

However, due to the unstable radiation performance of HJ-1 CCD data, a cross-radiometric calibration was conducted using a method that considers the characteristics of the surface bidirectional reflectance distribution function (BRDF) [47,57]. The calibration accuracy of the four spectral bands of the HJ-1 CCD sensor was 5%, which meets the requirements for absolute radiometric calibration accuracy [58].

The radiance was then converted to top of atmosphere (TOA) reflectance based on Equation (2) [16].

$$\rho = \frac{\pi \cdot L \cdot d^2}{ESUN_\lambda \cdot \cos \theta_s} \tag{2}$$

where ρ is TOA reflectance, π is 3.1415, L is the sensor spectral radiance, d is the earth-sun distance in astronomical units, $ESUN_\lambda$ is the extraterrestrial solar irradiance, and θ_s is the solar zenith angle. The $ESUN_\lambda$ values for ZY-3 MUX, GF-1 WFV, and HJ-1 CCD are shown in Table 3 [55].

Because of the high quality of the raw ZY-3 MUX and GF-1 WFV data on clear days, the atmospheric correction of these images was conducted using FLAASH in Environment and Visualizing Images (ENVI) software. However, due to the unstable data quality of HJ-1 CCD, the cross-radiometric calibration and atmospheric correction were accomplished simultaneous using the method proposed by Zhong et al. [57,59]. The atmospheric correction for the HJ-1 CCD image was conducted using a space-based aerosol optical depth (AOD) retrieval method [59]. The differences between the AOD retrieved from the HJ-1 CCD data and that from the Aerosol Robotic Network (AERONET) measurements ranged from –0.14 to 0.31. Approximately 50% of the derived AOD values correlated with AERONET AOD values with low discrepancy (less than 0.15), and the RMSE values for Xianghe and Beijing were 0.18 and 0.21, respectively.

Table 3. The calibration coefficients and extraterrestrial solar irradiance values for ZY-3 MUX, GF-1 WFV, and HJ-1 CCD.

Sensor	HJ-1 CCD			GF-1 WFV			ZY-3 MUX		
	Gain	Offset	*ESUN-*	Gain	Offset	*ESUN-*	Gain	Offset	*ESUN-*
Band 1	1.1451	4.6344	1929.81	0.1713	0.0000	1968.12	0.2509	0.0000	1958.30
Band 2	1.1660	4.0982	1831.14	0.1600	0.0000	1841.69	0.2338	0.0000	1855.71
Band 3	0.7647	3.7360	1549.82	0.1497	0.0000	1540.30	0.1885	0.0000	1548.72
Band 4	0.7558	0.7385	1078.32	0.1435	0.0000	1069.53	0.2035	0.0000	1085.60

The ZY-3 MUX image was corrected by its own coordinate file (*.rpc) that was acquired from NASG [53]. The geometric corrections of the HJ-1 CCD and GF-1 WFV images were conducted in ENVI software using the geometric correction image of ZY-3 MUX as the reference, and a second-order polynomial transformation with bilinear interpolation was used in the resampling. There were approximately 20 control points that were manually selected from the images, and the geometric registration error was less than one pixel of the images. The remote sensing data for the selected study area were projected to the World Geodetic System of 84 (WGS-84).

2.3. Generating the Forward Simulations

The widely used turbid medium model, i.e., SAILH, which is corrected by the hot-spot parameter, was selected due to its ease of use and consistent performance in validation practices [60,61]. The input parameters in the SAILH model included LAI, average leaf inclination angle (ALA), hot-spot, soil reflectance, leaf reflectance, leaf transmittance, diffuse fraction, solar zenith angle (SZA), view zenith angle (VZA), and relative azimuth angle (RAA). The leaf reflectance and transmittance values were simulated by the PROSPECT-5B model using several biochemical and biophysical parameters, including leaf mesophyll structure (N), C_{ab}, C_m, C_w, carotenoid content (C_{ar}), and brown pigment content (C_{brown}) [62]. The PROSPECT-5B+SAILH (PROSAIL) model has been used for more than twenty years for the retrieval of vegetation biophysical properties [63,64]. Previous studies have demonstrated that LAI, ALA, and C_{ab} have significant influences on canopy reflectance in the visible and NIR bands. However, other parameters, e.g., N, C_m, and C_w, are less sensitive to the canopy reflectance corresponding to the satellite bands [63,65]. Therefore, the parameters, e.g., N, C_m, C_w C_{ar}, and C_{brown}, in the PROSAIL model were fixed during the simulation to reduce the complexity and improve the efficiency of the LAI inversion. In this study, the soil reflectance for the PROSAIL model was acquired from field measurements. The parameters for maize, e.g., N, C_m, C_w C_{ar}, and C_{brown}, were fixed according to LOPEX'93. ALA varied from 40° to 70° at intervals of 10°; C_{ab} varied from 40 to 60 at intervals of 10; SZA varied from 0° to 85° at intervals of 1°; VZA varied from 0° to 35° at intervals of 1°; and, LAI varied from 0 to 8 at intervals of 0.1 (Table 4).

Table 4. The input variables for the PROSAIL model used to generate the forward simulation.

Parameters	Abbreviations	Units	Value Range	Interval
Leaf mesophyll structure	N	-	1.518	-
Leaf chlorophyll-*a* and -*b* content	C_{ab}	µg/cm	40–60	10
Leaf dry matter content	C_m	g/cm	0.003662	-
Leaf water content	C_w	cm	0.0131	-
Carotenoid content	C_{ar}	µg/cm	10	-
Brown pigment content	C_{brown}	-	0.05	-
Average leaf inclination angle	ALA	°	40–70	10
Hot-spot	Hot-spot	-	0.1	-
Leaf area index	LAI	m^2/m^2	0–8	0.1
Solar zenith angle	SZA	°	0–85	1
View zenith angle	VZA	°	0–35	1

The PROSAIL model was then run to simulate the actual satellite observations of canopy reflectance based on the spectral response curves of ZY-3 MUX, GF-1 WFV, and HJ-1 CCD, which were acquired from CRESDA [55], and of Landsat-8 OLI, which was acquired from USGS [56] (Figure 2). The reflectance simulations of each band based on each satellite spectral response can be calculated from Equation (3) [16].

$$\rho_s(\lambda) = \frac{\int_{\lambda_{min}}^{\lambda_{max}} \rho_s(\lambda_i)\varphi(\lambda_i)d\lambda}{\int_{\lambda_{min}}^{\lambda_{max}} \varphi(\lambda_i)d\lambda} \tag{3}$$

where $\rho_s(\lambda)$ is the simulated band reflectance of the satellite sensor, λ_{min} and λ_{max} are the lower and upper band wavelength limits, $\rho_s(\lambda_i)$ is the simulated hyperspectral reflectance for the *i*th wavelength, and $\varphi(\lambda_i)$ is the spectral response coefficient of the different sensors for the *i*th wavelength.

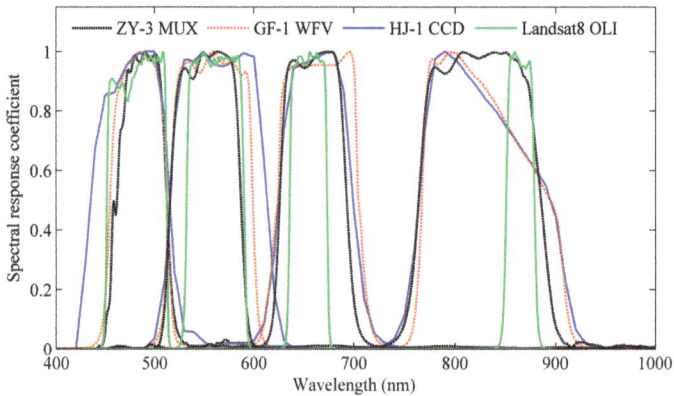

Figure 2. Spectral response curves for ZY-3 MUX, GF-1 WFV, and HJ-1 CCD from CRESDA, and for Landsat-8 OLI from USGS.

2.4. LAI Inversion Procedures

Before implementing the LAI inversion, the NDVI was used to separate the vegetation and non-vegetation pixels, and the non-vegetation pixels were removed. Generally, the NDVI values were less than 0.05 for bare soil [66,67]. Therefore, the pixels with NDVI values of less than 0.05 were identified as non-vegetation and were set to a filled value (marked as 0).

The LAI inversion method was chosen from the SVI-LAI empirical relationship based on the PROSAIL physical model to reduce the influences from the differences between the various sensors and spectral response curves and the residual errors from data preprocessing. The selected SVIs of this study include the NDVI and an NDVI combined with the NIR reflectance of vegetation (NIRv). The NDVI is a key parameter that is used to improve the accuracy of yield prediction for sugar beets, spring wheat, corn, and sunflower based on the NDVI relationships with optical signals under different nitrogen (N) and sulfur (S) contents [68–76]. Notably, the NDVI has been widely used for LAI extraction from high spatial resolution imagery [21,22,24,25,29,77]. NIRv is a new index that directly reflects the proportion of photons intercepted by chlorophyll, and it has a stronger linear relationship with LAI than does the NDVI [78].

Based on the simulation input parameters in Table 4, there are 2,972,160 records in the LUTs for each sensor. During the process of LAI inversion, the specific relationship between SVI and LAI was established according to the values of VZA and SZA for each sensor. Under the illumination and observation conditions of each sensor in the study area, the specific NDVI-LAI and NIRv-LAI exponential relationships for ZY-3 MUX, GF-1 WFV, and HJ-1 CCD were established from the PROSAIL LUT simulations for each sensor (details shown in Table 5). The NDVI and NIRv both had strong relationships with the LAI, and the R^2 values for ZY-3 MUX, GF-1 WFV, and HJ-1 CCD were all higher than 0.91. Moreover, all of the R^2 values for the NIRv-LAI exponential relationship were higher than those of the NDVI-LAI exponential relationship. LAI estimated from the specific NDVI-LAI and NIRv-LAI exponential relationships based on the image NDVI and NIRv calculations for ZY-3 MUX, GF-1 WFV, and HJ-1 CCD. The flowchart of the LAI inversion method based on ZY-3 MUX, GF-1 WFV, and HJ-1 CCD imagery is shown in Figure 3.

Table 5. The SVI-LAI exponential relationships for ZY-3 MUX, GF-1 WFV, and HJ-1 CCD.

SVIs	NDVI-LAI Relationship		NIRv-LAI Relationship	
	Expression	R^2	Expression	R^2
ZY-3 MUX	LAI = 0.0484 exp(5.2397 * NDVI)	0.91	LAI = 0.1725 exp(6.4087 * NIRv)	0.98
GF-1 WFV	LAI = 0.0385 exp(5.4728 * NDVI)	0.92	LAI = 0.1578 exp(5.6711 * NIRv)	0.98
HJ-1 CCD	LAI = 0.0380 exp(5.4241 * NDVI)	0.91	LAI = 0.1543 exp(5.6960 * NIRv)	0.98

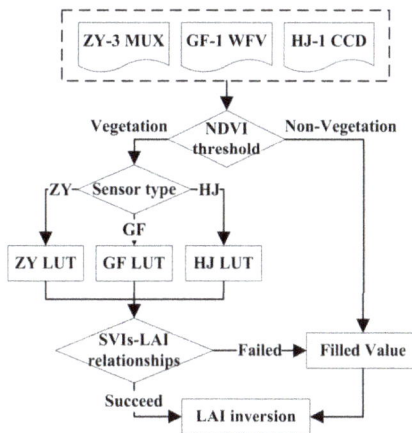

Figure 3. Flowchart of the LAI inversion method based on ZY-3 MUX, GF-1 WFV, and HJ-1 CCD.

2.5. Assessment of LAI Inversions for a Heterogeneous Surface

The performances of the LAI inversions from ZY-3 MUX, GF-1 WFV, and HJ-1 CCD data were evaluated using the LAI field measurements. Three indices, R^2, RMSE, and bias (BIAS), were used to evaluate the absolute discrepancies between the LAI inversions and LAI field measurements. R^2 describes the entire correlation between the LAI inversions and LAI field measurements. RMSE represents the standard deviation between the LAI inversions and LAI field measurements. Bias is a systematic variation that results from a random sampling or estimation process that does not give accurate results on average.

The heterogeneity of the surface vegetation has a strong effect on the LAI inversions at different spatial resolutions. Therefore, the spatial representativeness of the LAI field measurements was assessed based on the LAI inversions from the NDVI-LAI relationships for ZY-3 MUX data within a 30×30 m subpixel region. The spatial representativeness was evaluated using the relative absolute error (RAE) and the coefficient of sill (CS), according to Xu et al. [79]. In this study, the thresholds of RAE and CS were 10% for the representativeness evaluation based on LAI. Values of RAE and CS that were higher than 10% represented significant heterogeneity of the surface and lower spatial representativeness of the LAI field measurements. Then, the LAI inversions for ZY-3 MUX, GF-1 WFV, and HJ-1 CCD were evaluated based on the high and low spatial representativeness of the LAI field measurements.

Finally, a 3×3 km² range of LAI inversions was extracted from the NDVI-LAI relationships to further compare the LAI inversions from ZY-3 MUX, GF-1 WFV, and HJ-1 CCD at a consistent spatial resolution. After inversion of LAI from the remote sensing data, the LAI result from ZY-3 MUX was resampled to 16 and 30 m spatial resolutions, and the LAI result from GF-1 WFV was resampled to a 30 m spatial resolution using an upscaling method that considered the surface heterogeneity that described the variance of NDVI. Then, we compared the LAI inversions at the corresponding spatial resolutions. The LAI upscaling function is expressed in Equation (4) [80].

$$\text{LAI}_{upscaling} = \text{LAI}_{mean} - \frac{1}{2}\sigma_{\text{NDVI}} \times g''(m\text{NDVI}) \tag{4}$$

where $\text{LAI}_{upscaling}$ is the LAI value after upscaling from the high spatial resolution to the low spatial resolution, LAI_{mean} is the mean LAI value at the high spatial resolution within a pixel of low spatial resolution, and σ_{NDVI} and $g''(m\text{NDVI})$ are the variance and the second order differential of the mean NDVI value at the high spatial resolution within a pixel with low spatial resolution that was calculated from the NDVI-LAI exponential function, respectively.

3. Results and Analysis

The LAI in the study area was inversed based on the proposed LAI method for ZY-3 MUX, GF-1 WFV, and HJ-1 CCD, and the accuracy of the LAI inversions was validated based on the LAI field measurements of maize crops. The degree of the influence of the spatial resolution on the LAI inversion was analyzed using an upscaling method to determine the LAI differences among different sensors with varied spatial resolutions. The reflectance in the red and NIR bands was compared between ZY-3 MUX, GF-1 WFV, HJ-1 CCD, and Landsat-8 OLI to illustrate the feasibility of using LAI inversion with three Chinese satellite sensors.

3.1. LAI Validation for ZY-3 MUX, GF-1 WFV and HJ-1 CCD

The LAI inversions for ZY-3 MUX, GF-1 WFV, and HJ-1 CCD based on the proposed LAI method are shown in Figure 4. The LAI inversions from (a) to (c) were inversed from the NDVI-LAI exponential relationships in Table 5, and the LAI inversions from (d) to (f) were inversed from the NIRv-LAI exponential relationships in Table 5. The high LAI values were in the forest and cropland regions around the reservoir, whereas the low LAI values were in the wetland beach and grassland regions in the right corner and bottom of the images. Overall, the spatial variations in these three LAI inversions were similar. However, the values of the LAI inversions for the GF-1 WFV and HJ-1 CCD images were much lower, as is especially apparent in Figure 4c,e,f.

The performances of the proposed LAI estimation method for the ZY-3 MUX, GF-1 WFV, and HJ-1 CCD data were validated based on the LAI field measurements from 17 maize samples (Figure 5). Although the R^2 values of the field measurements, compared with the NDVI-LAI inversions from GF-1 WFV (R^2 = 0.66) and HJ-1 CCD (R^2 = 0.74), were higher than that of ZY-3 MUX (R^2 = 0.53), the NDVI-LAI inversions from GF-1 WFV and HJ-1 CCD tended to be systematically underestimated, especially for the higher LAI values. The RMSE and BIAS of the NDVI-LAI inversions from ZY-3 MUX (RMSE = 0.94 and BIAS = −0.73) were lower than those from GF-1 WFV (RMSE = 1.30 and BIAS = −1.19) and HJ-1 CCD (RMSE = 1.35 and BIAS = −1.22). Overall, the performance of the LAI inversed from the NDVI-LAI relationship for ZY-3 MUX was better than that from GF-1 WFV and HJ-1 CCD (Figure 5a–c). In addition, the performance of the method for LAI inversed from the NIRv-LAI relationships for GF-1 WFV and HJ-1 CCD was not sufficient, because the NIRv-LAI relationships largely depended on the reflectance of the NIR band, which was significantly underestimated in the NDVI-LAI inversions (Figure 5e,f). The performance of the method for LAI inversed from the NIRv-LAI relationships was better for ZY-3 MUX (R^2 = 0.62, RMSE = 0.54 and BIAS = −0.02) than for GF-1 WFV and HJ-1 CCD (Figure 5d). Because the LAI field measurements were all observed on the same day (31 July 2014), apart from one LAI value of 1.8, the other LAI values varied from 2.5 to 4.5. The concentrated LAI measurements led to small R^2 values between the LAI inversions and field measurements, as shown in Figure 5. Moreover, the RMSE and BIAS reflected the systematical underestimation of LAI estimations from GF-1 WFV and HJ-1 CCD.

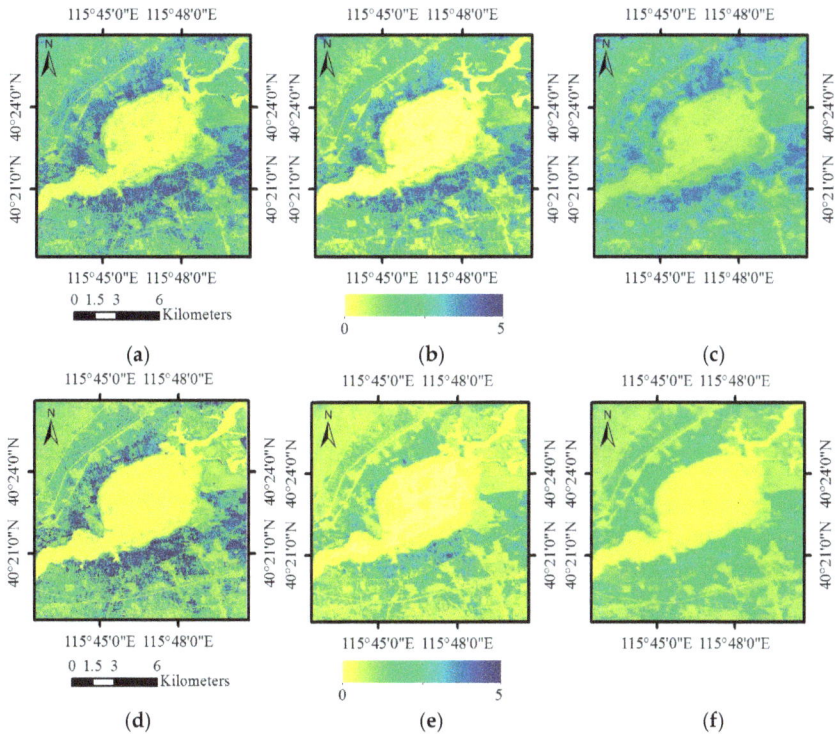

Figure 4. LAI inversion results for ZY-3 MUX (**a,d**), GF-1 WFV (**b,e**), and HJ-1 CCD (**c,f**) in Huailai, Hebei Province.

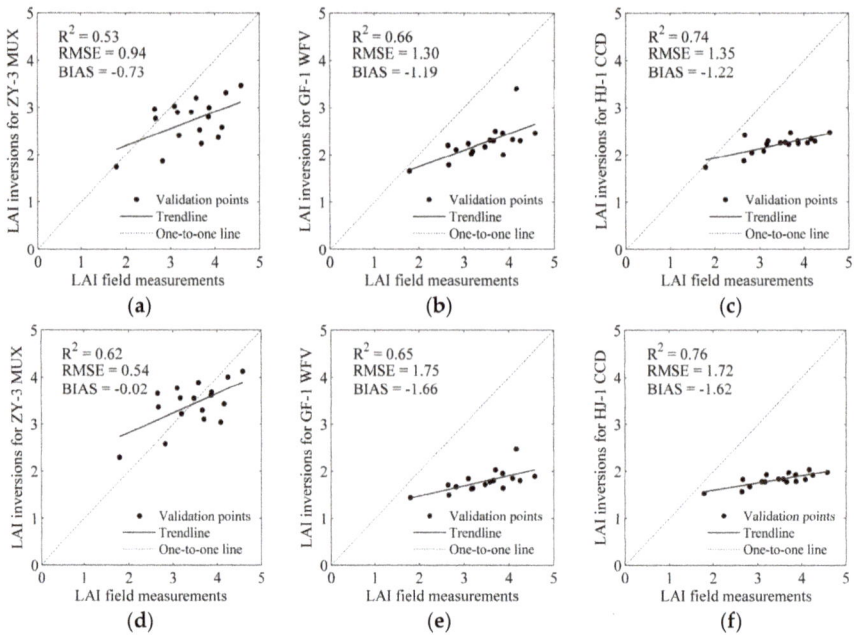

Figure 5. Comparisons of the LAI inversions with the field measurements for ZY-3 MUX (**a,d**), GF-1 WFV (**b,e**), and HJ-1 CCD (**c,f**).

3.2. Influence of Spatial Resolution on LAI Inversion

Spatial resolution influenced the LAI inversions from all the different sensors: the value ranges and accuracies of the LAI inversions both decreased with decreasing spatial resolution (Figures 4 and 5). The spatial representativeness of the LAI field measurements was first assessed by the RAE and CS, according to the methods in Section 3.3 to determine the differences between the LAI inversions from the NDVI-LAI relationships for ZY-3 MUX, GF-1 WFV, and HJ-1 CCD with different spatial resolutions. The LAI inversions from ZY-3 MUX, GF-1 WFV, and HJ-1 CCD were compared with the different spatial representativeness of LAI field measurements, and the results are shown in Figure 6. The performance of the LAI inversions from ZY-3 MUX, GF-1 WFV, and HJ-1 CCD using the LAI field measurements with high spatial representativeness was higher than that obtained using measurements with low spatial representativeness. The LAI inversions using measurements of low spatial resolution exhibited various degrees of underestimation. However, although the performance of the LAI inversions using the LAI field measurements with high spatial representativeness appears to be better in Figure 6a, the LAI inversions for GF-1 WFV and HJ-1 CCD were systematically more underestimated than were those for ZY-3 MUX. It is possible that the size of the pixels of the ZY-3 MUX data was closer to the actual surface. However, the lower spatial resolution of the GF-1 WFV and HJ-1 CCD data recorded more comprehensive information about the surface objects; thus, the vegetation signal was weakened. Therefore, the accuracies of the LAI inversions from the lower-spatial-resolution data were generally lower than those of the higher- spatial-resolution data.

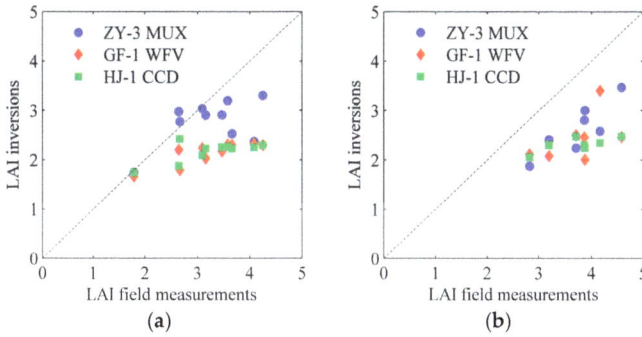

Figure 6. Validation of the LAI inversions from ZY-3 MUX, GF-1 WFV, and HJ-1 CCD using field data with high spatial representativeness (**a**) and low spatial representativeness (**b**).

In addition, this study extracted a 3×3 km^2 range of LAI inversions from the NDVI-LAI relationships to compare the accuracy of LAI inversions among ZY-3 MUX, GF-1 WFV, and HJ-1 CCD at a consistent spatial resolution. The LAI result of ZY-3 MUX was up-scaled to 16 m and 30 m spatial resolution, and the LAI result of GF-1 WFV was up-scaled to 30 m spatial resolution using the upscaling method described in Section 3.3. The up-scaled LAI inversions with 30 m spatial resolution from ZY-3 MUX and GF-1 WFV were compared with the LAI inversion from HJ-1 CCD (Figure 7). The results indicated that the distribution patterns of the LAI inversions were consistent, and that there were higher LAI values in the top left corner of the image near cropland and lower LAI values near roads and in residential areas. However, the LAI inversions that were up-scaled from higher spatial resolution (e.g., ZY-3 MUX and GF-1 WFV) reflected more detail than did the HJ-1 CCD inversion at the same 30 m spatial resolution.

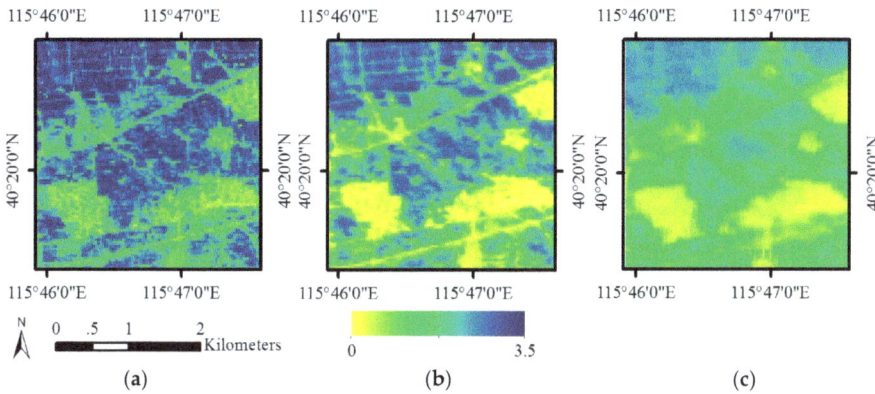

Figure 7. LAI inversions up-scaled to 30 m spatial resolution from ZY-3 MUX (**a**), GF-1 WFV (**b**), and HJ-1 CCD (**c**).

The pixels that were extracted from the LAI inversions in the 3×3 km^2 area were used to further analyze the differences among the three LAI inversions at the same spatial resolution. The pixels in the subregion were separated into pure pixels with uniform surface types and mixed pixels with different surface types, including different vegetation types, roads, or residential areas. The relationships between the LAI inversions from ZY-3 MUX data that were up-scaled to 16 m spatial resolution and those from GF-1 WFV data and the relationships between the LAI inversions from the ZY-3 MUX or

GF-1 WFV data that were up-scaled to 30 m spatial resolution and the LAI inversions from HJ-1 CCD data are shown in Figure 8, for both the pure (blue dots) and mixed (red dots) pixels. Generally, the performances of the LAI inversions between the HJ-1 CCD data and the up-scaled results of the ZY-3 MUX or GF-1 WFV data were preferable for both pure and mixed pixels. The accuracies of the LAI inversions for pure pixels are shown in Figure 8b,c, and these results presented a better agreement between the up-scaled LAI inversions from ZY-3 MUX (R^2 = 0.87, RMSE = 0.42 and BIAS = 0.12) or GF-1 WFV (R^2 = 0.89, RMSE = 0.66 and BIAS = −0.30) and HJ-1 CCD. In contrast, the up-scaled LAI inversions from ZY-3 MUX and GF-1 WFV, as shown in Figure 8a, did not perform as well for the pure (R^2 = 0.58, RMSE = 0.67 and BIAS = −0.45) or the mixed (R^2 = 0.69, RMSE = 0.66 and BIAS = −0.48) pixels. The up-scaled LAI inversions for mixed pixels from ZY-3 MUX (R^2 = 0.52, RMSE = 0.52 and BIAS = 0.06) and GF-1 WFV (R^2 = 0.63, RMSE = 0.58 and BIAS = −0.24) at 30 m spatial resolution were both substantially different from the HJ-1 CCD inversion (Figure 8b,c). In particular, the ZY-3 MUX sensor had much higher performance than the HJ-1 CCD, which demonstrated that the differences between the LAI inversions increased with the increasing spatial resolution between the two sensors.

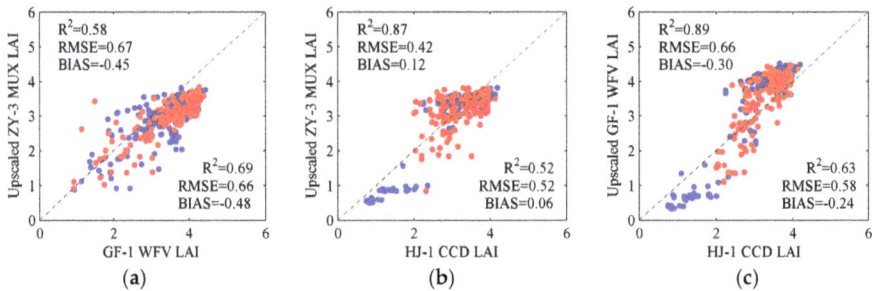

Figure 8. LAI inversions up-scaled to 16 m spatial resolution from ZY-3 MUX plotted against GF-1 WFV LAI inversions (**a**) and LAI inversions up-scaled to 30 m spatial resolution from ZY-3 MUX (**b**) or GF-1 WFV (**c**) plotted against HJ-1 CCD LAI inversions for pure (blue dots) and mixed (red dots) pixels.

3.3. Comparison of Reflectance among Different Sensors

A correlation analysis was used to compare the accuracy of the reflectance among the three different sensors at different LAI values. The reflectance values of ZY-3 MUX, GF-1 WFV, and HJ-1 CCD in the red and NIR bands were extracted, according to the coordinates of the LAI field measurements. The relationships between the LAI field measurements and the corresponding reflectance in the red and NIR bands are shown in Figure 9. The reflectance values of ZY-3 MUX, GF-1 WFV, and HJ-1 CCD decreased with increasing LAI, and the reflectance values were much more scattered in the red band than in the NIR band (Figure 9a). The R^2 values between the reflectance and the LAI were 0.49, 0.50, and 0.59 for ZY-3 MUX, GF-1 WFV, and HJ-1 CCD, respectively. However, the reflectance of the red band for HJ-1 CCD easily reached saturation when the LAI was greater than 3. The reflectance in the NIR band increased with LAI (Figure 9b) in all sensors, and the R^2 values between the reflectance and the LAI were higher than in the red band, which were 0.66, 0.45, and 0.56 for ZY-3 MUX, GF-1 WFV, and HJ-1 CCD, respectively. The degree of variation of the NIR reflectance was higher than the variation in the red reflectance. This was especially true for ZY-3 MUX, which had the highest reflectance values. However, the trends of the NIR reflectance variations were more coincident among ZY-3 MUX, GF-1 WFV, and HJ-1 CCD.

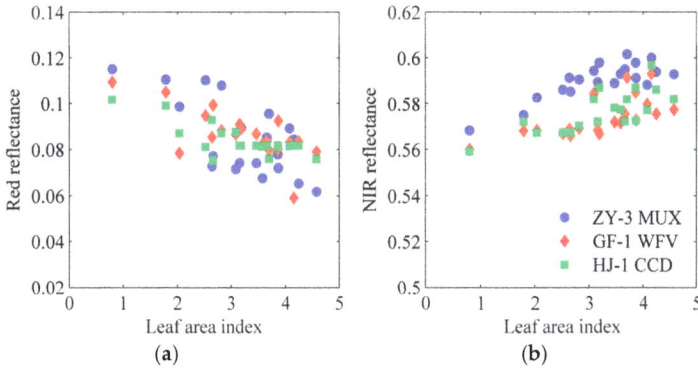

Figure 9. Reflectance in the red band (**a**) and the NIR band (**b**) varied with LAI field measurements for ZY-3 MUX, GF-1 WFV, and HJ-1 CCD.

To determine the influences of different sensor spectral response functions, the reflectance in the red and NIR bands for Landsat-8 OLI and the three Chinese satellite sensors (i.e., ZY-3 MUX, GF-1 WFV, and HJ-1 CCD) were simulated from the PROSAIL model based on the same values of input variables presented in Table 4, with SZA at 30°, VZA at 0°, and LAI varying from 0 to 8 at intervals of 0.1. The theoretical differences in reflectance in the red or NIR band were compared between the simulations of Landsat-8 OLI and those of ZY-3 MUX, GF-1 WFV, or HJ-1 CCD (Figure 10). The results showed that the individual reflectances of ZY-3 MUX, GF-1 WFV, and HJ-1 CCD were higher than those of Landsat-8 OLI in the red band (Figure 10b) and lower than those of Landsat-8 OLI in the NIR band (Figure 10c). The reflectance in the red band for HJ-1 CCD and that in the NIR band for GF-1 WFV were each close to the corresponding bands of Landsat-8 OLI. The differences in reflectance in the NIR band were larger than those in the red band. These differences occurred because the NIR spectral response function of Landsat-8 OLI was much narrower than those of the Chinese satellite sensors for ZY-3 MUX, GF-1 WFV, and HJ-1 CCD, as shown in Figure 2. In addition, the difference in reflectance between Landsat-8 OLI and each of the Chinese satellite sensors was greatly influenced by the sensor's spatial resolution. The reflectance that was theoretically closest to that of Landsat-8 OLI was that of HJ-1 CCD with 30 m spatial resolution, followed by that of GF-1 WFV.

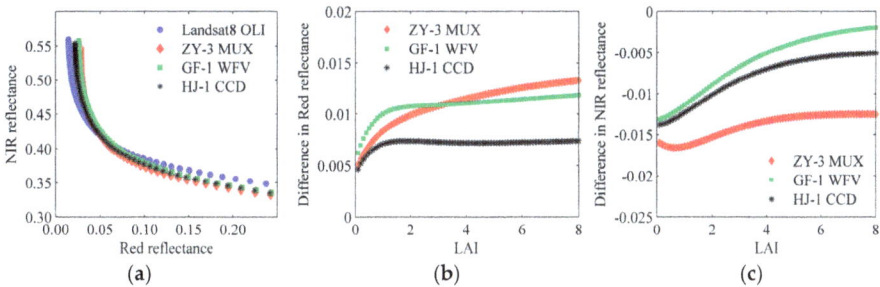

Figure 10. Reflectance in the red and NIR bands with LAI from 0 to 8 at 0.1 intervals for ZY-3 MUX, GF-1 WFV, HJ-1 CCD, and Landsat-8 OLI (**a**), and difference in reflectance in the red (**b**) and NIR (**c**) bands between Landsat-8 OLI and each of ZY-3 MUX, GF-1 WFV, and HJ-1 CCD for maize simulations.

To analyze the stability of the different sensors, the actual differences of reflectance in the red and NIR bands were compared between Landsat-8 OLI and each of ZY-3 MUX, GF-1 WFV, and HJ-1

CCD. The reflectance data of the four sensors were compared at the same spatial resolution of 30 m; thus, the mean values of reflectance for ZY-3 MUX and GF-1 WFV were compared with Landsat-8 OLI reflectance values. Approximately 7000 pixels of uniform cropland were extracted from Landsat-8 OLI, ZY-3 MUX, GF-1 WFV, and HJ-1 CCD images. The percentage density plots of reflectance differences between Landsat-8 OLI and each of ZY-3 MUX, GF-1 WFV and HJ-1 CCD in the red and NIR bands are shown in Figure 11. The majority of reflectance values of ZY-3 MUX and HJ-1 CCD were higher than those of Landsat-8 OLI in both the red and NIR bands (Figure 11a,c). Most of the reflectance values of GF-1 WFV were lower than those of Landsat-8 OLI in the red band and higher than those of Landsat-8 OLI in the NIR band (Figure 11b). However, due to the similar observation geometry conditions between ZY-3 MUX and Landsat-8 OLI, with near nadir observations of Landsat-8 OLI and an SZA of 27.67°, the reflectance differences between ZY-3 MUX and Landsat-8 OLI were lower than those between GF-1 WFV or HJ-1 CCD and Landsat-8 OLI (Figure 11a). In addition, the sensor stability in the NIR band was higher than that in the red band for all three Chinese satellite sensors, with lower reflectance differences from Landsat-8 OLI.

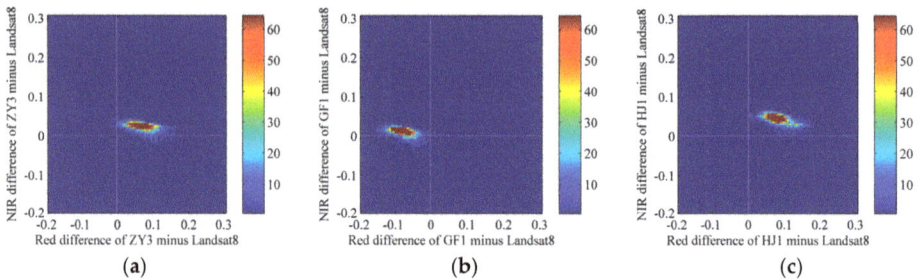

Figure 11. Density scatter plots of the reflectance difference in the red and NIR bands between Landsat-8 OLI and each of ZY-3 MUX (**a**), GF-1 WFV (**b**), and HJ-1 CCD (**c**) for croplands.

4. Discussion

The discrepancies for different instruments or platforms greatly influenced the accuracy of LAI inversion. The factors that influenced LAI inversion included the sensor spectral response function, sun-view geometry, and data preprocessing errors [16,51]. The maximal NDVI difference among ZY-3 MUX, GF-1 WFV, and HJ-1 CCD in theory reached 2.62% due to the different sensor spectral response functions (Figure 10). Because of the similar satellite transit period in the study area, the SZA values of these three sensors were approximately equal, but the VZA difference among these three sensors was approximately 30°. Based on the reflectance in the red and NIR bands in Figure 9, the maximum NDVI difference between ZY-3 MUX, GF-1 WFV, and HJ-1 CCD was 9.74%. The error of geometric correction was less than one pixel for each sensor image in this study. For different sensor resolutions, the one pixel error of geometric correction was 5.8 m for ZY-3 MUX, 16 m for GF-1 WFV and 30 m for HJ-1 CCD. In addition, because of the unstable data quality of HJ-1 CCD, cross-radiometric calibration was performed. In this case, atmospheric correction was achieved based on the method proposed by Zhong et al. [57,59]. After all of the data preprocessing steps, there is still a systematically discrepancies of reflectance (with maximum value of 0.1), both in red and NIR bands when compared with those of Landsat-8 OLI in Figure 11. Therefore, the accuracy of data preprocessing for multiple sensors was the dominant factor that influence the LAI inversion difference among ZY-3 MUX, GF-1 WFV, and HJ-1 CCD.

NIRv, which combined NIR reflectance with the NDVI, was proposed to accurately estimate the global terrestrial gross primary production (GPP) [78]. NIRv is not easily saturated when compared with the NDVI; therefore, it can be applied to improve the LAI estimation. In this study, NIRv was used to estimate LAI from ZY-3 MUX, GF-1 WFV, and HJ-1 CCD. The performance of LAI inversion based on the NIRv-LAI exponential relationship for ZY-3 MUX was best, as shown in Figure 5d. Notably, NIRv depends largely on NIR reflectance. If the NIR reflectance contains noise, the NIRv will deviate more than the NDVI from actual values. NDVI, as a normalized index that can eliminate the fluctuations in the red and NIR bands, and the performance of LAI inversion from the NDVI-LAI relationship was better than that from NIRv-LAI relationship (Figure 5b,c,e,f). In addition, because of the low saturation of NIRv, the superiority of NIRv in LAI inversion will be more apparent when applied for the inversion of high LAI values, such as those of the forests, but not for small values, such as those of the crops as analyzed in this paper.

Currently, the LAI estimation methods at the regional scale are generally based on a single sensor. The primary restriction for generating regional LAI products with moderate to high spatial resolution is the limited number of sensor observations during a specific period. The multi-sensor data during a specific period can greatly increase the number of observations and improve the accuracy of LAI inversion. The sensors with similar spatial resolution, such as ZY-3 MUX, GF-1 WFV, HJ-1 CCD, Landsat-8 OLI, and Sentinel-2, provide a combined multi-sensor dataset for generating LAI products with moderate to high resolutions. The multi-sensor data have more VZA, and the multi-angular observations from multiple sensors are helpful for improving the accuracy of LAI inversion. However, due to the differences in sensor characteristics, geometric and radiometric normalization between different sensors are necessary.

5. Conclusions

This study analyzed the application of LAI inversed from ZY-3 MUX, GF-1 WFV, and HJ-1 CCD data. The method of LAI extraction was based on the SVI-LAI relationship for ZY-3 MUX, GF-1 WFV, and HJ-1 CCD, which was simulated from the PROSAIL model. The LAI inversions were validated using LAI field measurements of maize in Huailai, Hebei Province, China. Regarding the sensor band settings of three Chinese satellite sensors (ZY-3 MUX, GF-1 WFV, and HJ-1 CCD), the performances of these satellite sensors were comparable to that of Landsat-8 OLI. However, the reflectance of the ZY-3 MUX, GF-1 WFV, and HJ-1 CCD images, which was influenced by SZA, VZA, and data processing methods, differed to varying degrees from the reflectance of the Landsat-8 OLI image. The ZY-3 MUX with similar observation geometry conditions as those of Landsat-8 OLI, showed better performance than did GF-1 WFV and HJ-1 CCD in the study area. When compared with the LAI field measurements, the results showed that the performances of the LAI that was inversed from the NIRv-LAI exponential relationships for ZY-3 MUX ($R^2 = 0.62$, RMSE = 0.54 and BIAS = -0.02) were better than the others. However, the performances of the LAI that was inversed from the NIRv-LAI relationships for GF-1 WFV and HJ-1 CCD did not perform as well because of the larger variations in the NIR reflectance. In contrast, the LAI inversions from the NDVI-LAI relationships for ZY-3 MUX, GF-1 WFV, and HJ-1 CCD were much more stable because the NDVI is a normalized index that can eliminate the fluctuations in the reflectance in the red and NIR bands. Overall, LAI inversions tended to be systematically underestimated, especially for the higher LAI values. The scaling effects of the different spatial resolutions could not be ignored, which demonstrated that the LAI inversion differences increased with larger variations in the spatial resolution between the two sensors, especially between ZY-3 MUX and HJ-1 CCD, for mixed pixels. However, more vegetation types and multi-temporal data at different spatial resolutions in LAI inversions need further study.

Acknowledgments: This work was supported by the GF6 Project under Grant 30-Y20A03-9003017/18, the National Natural Science Foundation of China (No. 41401393 and No. 41671374).

Author Contributions: Jing Zhao, Jing Li and Qinhuo Liu conceived, designed and produced the method of LAI estimation from ZY-3 MUX, GF-1 WFV and HJ-1 CCD based on the PROSAIL model; Hongyan Wang, Chen Chen and Shanlong Wu accomplished the multi-sensor data preprocessing; Baodong Xu provided the LAI spatial representativeness evaluation.

Conflicts of Interest: The authors declare no conflict of interest.

References

1. Chen, J.M.; Black, T.A. Defining leaf area index for non-flat leaves. *Plant Cell Environ.* **1992**, *15*, 421–429. [CrossRef]
2. Duveiller, G.; Weiss, M.; Baret, F.; Defourny, P. Retrieving wheat green area index during the growing season from optical time series measurements based on neural network radiative transfer inversion. *Remote Sens. Environ.* **2011**, *115*, 887–896. [CrossRef]
3. Kobayashi, H.; Suzuki, R.; Kobayashi, S. Reflectance seasonality and its relation to the canopy leaf area index in an eastern siberian larch forest: Multi-satellite data and radiative transfer analyses. *Remote Sens. Environ.* **2007**, *106*, 238–252. [CrossRef]
4. Baret, F.; Weiss, M.; Lacaze, R.; Camacho, F.; Makhmara, H.; Pacholcyzk, P.; Smets, B. Geov1: LAI and FAPAR essential climate variables and fcover global time series capitalizing over existing products. Part1: Principles of development and production. *Remote Sens. Environ.* **2013**, *137*, 299–309. [CrossRef]
5. Peng, J.; Dan, L.; Dong, W. Estimate of extended long-term LAI data set derived from AVHRR and MODIS based on the correlations between LAI and key variables of the climate system from 1982 to 2009. *Int. J. Remote Sens.* **2013**, *34*, 7761–7778. [CrossRef]
6. Clark, D.B.; Olivas, P.C.; Oberbauer, S.F.; Clark, D.A.; Ryan, M.G. First direct landscape-scale measurement of tropical rain forest leaf area index, a key driver of global primary productivity. *Ecol. Lett.* **2008**, *11*, 163–172. [CrossRef] [PubMed]
7. Coyne, P.I.; Aiken, R.M.; Maas, S.J.; Lamm, F.R. Evaluating yieldtracker forecasts for maize in western kansas. *Agron. J.* **2009**, *101*, 671–680. [CrossRef]
8. Sharma, L.K.; Bali, S.K.; Dwyer, J.D.; Plant, A.B.; Bhowmik, A. A case study of improving yield prediction and sulfur deficiency detection using optical sensors and relationship of historical potato yield with weather data in maine. *Sensors* **2017**, *17*, 1095. [CrossRef] [PubMed]
9. Asner, G.P.; Scurlock, J.M.O.; Hicke, J.A. Global synthesis of leaf area index observations: Implications for ecological and remote sensing studies. *Glob. Ecol. Biogeogr.* **2003**, *12*, 191–205. [CrossRef]
10. Li, Z.; Tang, H.; Xin, X.; Zhang, B.; Wang, D. Assessment of the MODIS LAI product using ground measurement data and HJ-1A/1B imagery in the meadow steppe of hulunber, China. *Remote Sens.* **2014**, *6*, 6242–6265. [CrossRef]
11. Tian, Y.; Woodcock, C.E.; Wang, Y.; Privette, J.L.; Shabanov, N.V.; Zhou, L.; Zhang, Y.; Buermann, W.; Dong, J.; Veikkanen, B.; et al. Multiscale analysis and validation of the MODIS LAI product : II. Sampling strategy. *Remote Sens. Environ.* **2002**, *83*, 431–441. [CrossRef]
12. Yang, F.; Sun, J.; Zhang, B.; Yao, Z.; Wang, Z.; Wang, J.; Yue, X. Assessment of MODIS LAI product accuracy based on the PROSAIL model, TM and field measurements. *Trans. Chin. Soc. Agric. Eng.* **2010**, *26*, 192–197.
13. Cohen, W.B.; Maiersperger, T.K.; Gower, S.T.; Turner, D.P. An improved strategy for regression of biophysical variables and Landsat ETM+ data. *Remote Sens. Environ.* **2003**, *84*, 561–571. [CrossRef]
14. Chen, J.M.; Cihlar, J. Retrieving leaf area index of boreal conifer forests using landsat tm images. *Remote Sens. Environ.* **1996**, *55*, 153–162. [CrossRef]
15. Colombo, R.; Bellingeri, D.; Fasolini, D.; Marino, C.M. Retrieval of leaf area index in different vegetation types using high resolution satellite data. *Remote Sens. Environ.* **2003**, *86*, 120–131. [CrossRef]
16. Soudani, K.; François, C.; Maire, G.L.; Dantec, V.L.; Dufrêne, E. Comparative analysis of IKONOS, SPOT, and ETM+ data for leaf area index estimation in temperate coniferous and deciduous forest stands. *Remote Sens. Environ.* **2006**, *102*, 161–175. [CrossRef]
17. Turner, D.P.; Cohen, W.B.; Kennedy, R.E.; Fassnacht, K.S.; Briggs, J.M. Relationships between leaf area index and Landsat TM spectral vegetation indices across three temperate zone sites. *Remote Sens. Environ.* **1999**, *70*, 52–68. [CrossRef]
18. Heiskanen, J. Estimating aboveground tree biomass and leaf area index in a mountain birch forest using aster satellite data. *Int. J. Remote Sens.* **2006**, *27*, 1135–1158. [CrossRef]
19. Korhonen, L.; Hadi; Packalen, P.; Rautiainen, M. Comparison of Sentinel-2 and Landsat 8 in the estimation of boreal forest canopy cover and leaf area index. *Remote Sens. Environ.* **2017**, *195*, 259–274. [CrossRef]
20. Fassnacht, K.S.; Gower, S.T.; Mackenzie, M.D.; Nordheim, E.V.; Lillesand, T.M. Estimating the leaf area index of north central wisconsin forests using the landsat thematic mapper. *Remote Sens. Environ.* **1997**, *61*, 229–245. [CrossRef]

21. He, L.I.; Chen, Z.X.; Jiang, Z.W.; Wen-Bin, W.U.; Ren, J.Q.; Liu, B.; Hasi, T. Comparative analysis of GF-1, HJ-1, and Landsat-8 data for estimating the leaf area index of winter wheat. *J. Integr. Agric.* **2017**, *16*, 266–285.

22. Thomas, V.; Noland, T.; Treitz, P.; Mccaughey, J.H. Leaf area and clumping indices for a boreal mixed-wood forest: Lidar, hyperspectral, and landsat models. *Int. J. Remote Sens.* **2011**, *32*, 8271–8297. [CrossRef]

23. Wu, M.; Wu, C.; Huang, W.; Niu, Z.; Wang, C. High-resolution leaf area index estimation from synthetic landsat data generated by a spatial and temporal data fusion model. *Comput. Electron. Agric.* **2015**, *115*, 1–11. [CrossRef]

24. Ilangakoon, N.T.; Gorsevski, P.V.; Milas, A.S. Estimating leaf area index by bayesian linear regression using terrestrial Lidar, LAI-2200 plant canopy analyzer, and landsat tm spectral indices. *Can. J. Remote Sens.* **2015**, *41*, 315–333. [CrossRef]

25. Pu, R. Mapping leaf area index over a mixed natural forest area in the flooding season using ground-based measurements and landsat tm imagery. *Int. J. Remote Sens.* **2012**, *33*, 6600–6622. [CrossRef]

26. Zhang, H.; Chen, J.M.; Huang, B.; Song, H.; Li, Y. Reconstructing seasonal variation of landsat vegetation index related to leaf area index by fusing with modis data. *IEEE J. Sel. Top. Appl. Earth Obs. Remote Sens.* **2014**, *7*, 950–960. [CrossRef]

27. Nguy-Robertson, A.L.; Gitelson, A.A. Algorithms for estimating green leaf area index in C3 and C4 crops for MODIS, Landsat TM/ETM+, MERIS, Sentinel MSI/OLCI, and venμs sensors. *Remote Sens. Lett.* **2015**, *6*, 360–369. [CrossRef]

28. Tang, S.; Chen, J.M.; Zhu, Q.; Li, X.; Chen, M.; Sun, R.; Zhou, Y.; Deng, F.; Xie, D. Lai inversion algorithm based on directional reflectance kernels. *J. Environ. Manag.* **2007**, *85*, 638–648. [CrossRef] [PubMed]

29. Eklundh, L. Estimating leaf area index in coniferous and deciduous forests in sweden using landsat optical sensor data. *Proc. SPIE* **2003**, *4879*, 379–390.

30. Qu, Y.; Zhang, Y.; Xue, H. Retrieval of 30-m-resolution leaf area index from china HJ-1 CCD data and MODIS products through a dynamic bayesian network. *IEEE J. Sel. Top. Appl. Earth Obs. Remote Sens.* **2014**, *7*, 222–228.

31. Walthall, C.; Dulaney, W.; Anderson, M.; Norman, J.; Fang, H.; Liang, S. A comparison of empirical and neural network approaches for estimating corn and soybean leaf area index from landsat ETM+ imagery. *Remote Sens. Environ.* **2004**, *92*, 465–474. [CrossRef]

32. González-Sanpedro, M.C.; Toan, T.L.; Moreno, J.; Kergoat, L.; Rubio, E. Seasonal variations of leaf area index of agricultural fields retrieved from landsat data. *Remote Sens. Environ.* **2008**, *112*, 810–824. [CrossRef]

33. Chen, B.; Wu, Z.; Wang, J.; Dong, J.; Guan, L.; Chen, J.; Yang, K.; Xie, G. Spatio-temporal prediction of leaf area index of rubber plantation using HJ-1A/1B CCD images and recurrent neural network. *ISPRS Photogramm. Remote Sens.* **2015**, *102*, 148–160. [CrossRef]

34. Chen, W.; Cao, C.X.; He, Q.S.; Guo, H.D.; Zhang, H.; Li, R.Q.; Sheng, Z.; Xu, M.; Gao, M.X.; Zhao, J.; et al. Quantitative estimation of the shrub canopy LAI from atmosphere-corrected HJ-1 CCD data in mu us sandland. *Sci. China Earth Sci.* **2010**, *53*, 26–33. [CrossRef]

35. Fernandes, R.; Weiss, M.; Camacho, F.; Berthelot, B.; Baret, F.; Duca, R. Development and assessment of leaf area index algorithms for the sentinel-2 multispectral imager. In Proceedings of the Geoscience and Remote Sensing Symposium, Quebec City, QC, Canada, 13–18 July 2014; pp. 3922–3925.

36. Richter, K.; Hank, T.B.; Vuolo, F.; Mauser, W.; D'Urso, G. Optimal exploitation of the Sentinel-2 spectral capabilities for crop leaf area index mapping. *Remote Sens.* **2012**, *4*, 561–582. [CrossRef]

37. Verrelst, J.; Rivera, J.P.; Veroustraete, F.; Muñoz-Marí, J.; Clevers, J.G.P.W.; Camps-Valls, G.; Moreno, J. Experimental Sentinel-2 LAI estimation using parametric, non-parametric and physical retrieval methods—A comparison. *ISPRS Photogramm. Remote Sens.* **2015**, *108*, 260–272. [CrossRef]

38. Houborg, R.; Mccabe, M.; Cescatti, A.; Gao, F.; Schull, M.; Gitelson, A. Joint leaf chlorophyll content and leaf area index retrieval from landsat data using a regularized model inversion system (regflec). *Remote Sens. Environ.* **2015**, *159*, 203–221. [CrossRef]

39. Verrelst, J.; Camps-Valls, G.; Muñoz-Marí, J.; Rivera, J.P.; Veroustraete, F.; Clevers, J.G.P.W.; Moreno, J. Optical remote sensing and the retrieval of terrestrial vegetation bio-geophysical properties—A review. *ISPRS Photogramm. Remote Sens.* **2015**, *108*, 273–290. [CrossRef]

40. Jonckheere, I.; Fleck, S.; Nackaerts, K.; Muys, B.; Coppin, P.; Weiss, M.; Baret, F. Methods for leaf area index determination. Part I: Theories, sensors and hemispherical photography. *Agric. For. Meteorol.* **2004**, *121*, 19–35. [CrossRef]

41. Price, E. Lai-2200c plant canopy analyzer. In Proceedings of the American Society for Horticultural Science Conference, Orlando, FL, USA, 28–31 July 2014.

42. Leblanc, S.G.; Chen, J.M.; Kwong, M. Tracing radiation and architecture of canopies. Trac manual version 2.1.3. *Gastroenterology* **2002**, *78*, 722–727.

43. Decagon Devices, I. *Accupar Par/Lai Ceptometer Model Lp-80*; Decagon Devices, Inc.: Pullman, WA, USA, 2015.

44. Campos-Taberner, M.; García-Haro, F.J.; Moreno, Á.; Gilabert, M.A.; Sánchez-Ruiz, S.; Martínez, B.; Camps-Valls, G. Mapping leaf area index with a smartphone and gaussian processes. *IEEE Geosci. Remote Sens. Lett.* **2015**, *12*, 2501–2505. [CrossRef]

45. Confalonieri, R.; Foi, M.; Casa, R.; Aquaro, S.; Tona, E.; Peterle, M.; Boldini, A.; Carli, G.D.; Ferrari, A.; Finotto, G. Development of an app for estimating leaf area index using a smartphone. Trueness and precision determination and comparison with other indirect methods. *Comput. Electron. Agric.* **2013**, *96*, 67–74. [CrossRef]

46. Francone, C.; Pagani, V.; Foi, M.; Cappelli, G.; Confalonieri, R. Comparison of leaf area index estimates by ceptometer and pocketlai smart app in canopies with different structures. *Field Crop. Res.* **2014**, *155*, 38–41. [CrossRef]

47. Zhao, J.; Li, J.; Liu, Q.; Fan, W.; Zhong, B.; Wu, S.; Yang, L.; Zeng, Y.; Xu, B.; Yin, G. Leaf area index retrieval combining HJ1/CCD and Landsat 8/OLI data in the heihe river basin, China. *Remote Sens.* **2015**, *7*, 6862–6885. [CrossRef]

48. Zhan, Y.; Menga, Q.; Wanga, C.; Li, J.; Zhouab, K.; Li, D. Fractional vegetation cover estimation over large regions using GF-1 satellite data. In Proceedings of the SPIE Asia-Pacific Remote Sensing, Beijing, China, 27–31 October 2014.

49. Jia, K.; Liang, S.; Gu, X.; Baret, F.; Wei, X.; Wang, X.; Yao, Y.; Yang, L.; Li, Y. Fractional vegetation cover estimation algorithm for chinese GF-1 wide field view data. *Remote Sens. Environ.* **2016**, *177*, 184–191. [CrossRef]

50. Wang, L.; Niu, Z.; Wang, C.; Huang, W.; Chen, H.; Gao, S.; Li, D.; Muhammad, S. Combined use of airborne lidar and satellite GF-1 data to estimate leaf area index, height, and aboveground biomass of maize during peak growing season. *IEEE J. Sel. Top. Appl. Earth Obs. Remote Sens.* **2015**, *8*, 4489–4501.

51. Wang, L.; Yang, R.; Tian, Q.; Yang, Y.; Zhou, Y.; Sun, Y.; Mi, X. Comparative analysis of GF-1 WFV, ZY-3 MUX, and HJ-1 CCD sensor data for grassland monitoring applications. *Remote Sens.* **2015**, *7*, 2089–2108. [CrossRef]

52. Zeng, Y.; Li, J.; Liu, Q.; Li, L.; Xu, B.; Yin, G.; Peng, J. A sampling strategy for remotely sensed lai product validation over heterogeneous land surfaces. *IEEE J. Sel. Top. Appl. Earth Obs. Remote Sens.* **2014**, *7*, 3128–3142. [CrossRef]

53. Satellite Surveying and Mapping Application Center (SASMAC) of National Administration of Surveying Mapping and Geo-Information of China (NASG). Available online: http://sjfw.sasmac.cn/en.html (accessed on 14 December 2017).

54. Gaofen Satellite Data and Information Service System (GFDIS). Available online: http://210.72.27.32:8080/SNFFWeb/WebUI/Main/HomePage.jsp (accessed on 14 December 2017).

55. China Centre for Resources Satellite Data and Application (CRESDA). Available online: http://www.cresda.com/CN/index.shtml (accessed on 14 December 2017).

56. United States Geological Survey (USGS) Earthexplorer. Available online: https://earthexplorer.usgs.gov/ (accessed on 14 December 2017).

57. Zhong, B.; Zhang, Y.; Du, T.; Yang, A.; Lv, W.; Liu, Q. Cross-calibration of hj-1/ccd over a desert site using landsat etm imagery and aster gdem product. *IEEE Trans. Geosci. Remote Sens.* **2014**, 1–17.

58. Zhang, Y.H. HJ1-CCD Cross Radiometric Calibration. Master's Thesis, Shandong University of Science and Technology, Shandong, China, 2011.

59. Zhong, B. Improved estimation of aerosol optical depth from Landsat TM/ETM+ imagery over land. In Proceedings of the IEEE International Geoscience and Remote Sensing Symposium (IGARSS), Vancouver, BC, Canada, 24–29 July 2011; pp. 3304–3307.

60. Kuusk, A. The hot spot effect in plant canopy reflectance. In *Photon-Vegetation Interactions: Applications in Optical Remote Sensing and Plant Ecology*; Myneni, R., Ross, J., Eds.; Springer: New York, NY, USA, 1991; pp. 139–159.

61. Verhoef, W. Light scattering by leaf layers with application to canopy reflectance modeling: The sail model. *Remote Sens. Environ.* **1984**, *16*, 125–141. [CrossRef]

62. Jacquemoud, S.; Baret, F. Prospect: A model of leaf optical properties spectra. *Remote Sens. Environ.* **1990**, *34*, 75–91. [CrossRef]

63. Jacquemoud, S.; Verhoef, W.; Baret, F.; Bacour, C.; Zarco-Tejada, P.J.; Asner, G.P.; François, C.; Ustin, S.L. PROSPECT + SAIL models: A review of use for vegetation characterization. *Remote Sens. Environ.* **2009**, *113*, S56–S66. [CrossRef]

64. Jacquemoud, S.; Verhoef, W.; Baret, F.; Zarco-Tejada, P.J.; Asner, G.P.; Francois, C.; Ustin, S.L. PROSPECT+SAIL: 15 years of use for land surface characterization. In Proceedings of the IEEE International Conference on Geoscience and Remote Sensing Symposium (IGARSS), Denver, CO, USA, 31 July–4 August 2006; pp. 1992–1995.

65. Xiao, Y.; Zhao, W.; Zhou, D.; Gong, H. Sensitivity analysis of vegetation reflectance to biochemical and biophysical variables at leaf, canopy, and regional scales. *IEEE Trans. Geosci. Remote Sens.* **2014**, *52*, 4014–4024. [CrossRef]

66. Zeng, X.; Dickinson, R.E.; Walker, A.; Shaikh, M.; Defries, R.S.; Qi, J. Derivation and evaluation of global 1-km fractional vegetation cover data for land modeling. *J. Appl. Meteorol.* **2000**, *39*, 826–839. [CrossRef]

67. Li, X.; Zhang, J. Derivation of the green vegetation fraction of the whole china from 2000 to 2010 from modis data. *Earth Interact.* **2016**, *20*, 1–8. [CrossRef]

68. Sharma, L.K.; Bu, H.; Franzen, D. *Active Optical Sensor Algorithm for Corn Yield Prediction and Corn Side Dress Nitrogen Rate Aid*; North Dakota State University: Fargo, ND, USA, 2014.

69. Sharma, L.K.; Bu, H.; Denton, A.; Franzen, D.W. Active-optical sensors using red ndvi compared to red edge ndvi for prediction of corn grain yield in North Dakota, U.S.A. *Sensors* **2015**, *15*, 27832–27853. [CrossRef] [PubMed]

70. Bu, H.; Sharma, L.K.; Denton, A.; Franzen, D.W. Comparison of satellite imagery and ground-based active optical sensors as yield predictors in sugar beet, spring wheat, corn, and sunflower. *Agron. J.* **2017**, *109*, 1–10. [CrossRef]

71. Sharma, L.K.; Bu, H.; Franzen, D.W. Comparison of two ground-based active-optical sensors for in-season estimation of corn (*Zea mays*, L.) yield. *J. Plant Nutr.* **2016**, *39*, 957–966. [CrossRef]

72. Sharma, L.K. Evaluation of Active Optical Ground-Based Sensors to Detect Early Nitrogen Deficiencies in Corn. Ph.D. Thesis, North Dakota State University of Agriculture and Applied Sciences, Fargo, ND, USA, April 2014.

73. Franzen, D.W.; Sharma, L.K.; Bu, H.; Denton, A. Evidence for the ability of active-optical sensors to detect sulfur deficiency in corn. *Agron. J.* **2016**, *108*, 1–5. [CrossRef]

74. Bu, H.; Sharma, L.K.; Denton, A.; Franzen, D.W. Sugar beet yield and quality prediction at multiple harvest dates using active-optical sensors. *Agron. J.* **2016**, *108*, 273–284. [CrossRef]

75. Sharma, L.K.; Bu, H.; Franzen, D.W.; Denton, A. Use of corn height measured with an acoustic sensor improves yield estimation with ground based active optical sensors. *Comput. Electron. Agric.* **2016**, *124*, 254–262. [CrossRef]

76. Sharma, L.K.; Franzen, D.W. Use of corn height to improve the relationship between active optical sensor readings and yield estimates. *Precis. Agric.* **2012**, *15*, 331–345. [CrossRef]

77. Tian, Q.; Luo, Z.; Chen, J.M.; Chen, M.; Hui, F. Retrieving leaf area index for coniferous forest in xingguo county, China with Landsat ETM+ images. *J. Environ. Manag.* **2007**, *85*, 624–627. [CrossRef] [PubMed]

78. Badgley, G.; Field, C.B.; Berry, J.A. Canopy near-infrared reflectance and terrestrial photosynthesis. *Sci. Adv.* **2017**, *3*, 1–5. [CrossRef] [PubMed]

79. Xu, B.; Li, J.; Liu, Q.; Huete, A.R.; Yu, Q.; Zeng, Y.; Yin, G.; Zhao, J.; Yang, L. Evaluating spatial representativeness of station observations for remotely sensed leaf area index products. *IEEE J. Sel. Top. Appl. Earth Obs. Remote Sens.* **2016**, *9*, 3267–3282. [CrossRef]

80. Zhu, X.; Feng, X.; Zhao, Y. Scale effect and error analysis of crop LAI inversion. *J. Remote Sens.* **2010**, *14*, 579–592.

remote sensing

MDPI

Article

The Retrieval of 30-m Resolution LAI from Landsat Data by Combining MODIS Products

Jianmin Zhou [1], Shan Zhang [1], Hua Yang [1,*], Zhiqiang Xiao [1] and Feng Gao [2]

1   State Key Laboratory of Remote Sensing Science, Faculty of Geographical Science,
     Beijing Normal University, Beijing 100875, China; 201521170070@mail.bnu.edu.cn (J.Z.);
     201621170033@mail.bnu.edu.cn (S.Z.); zhqxiao@bnu.edu.cn (Z.X.)
2   USDA (United States Department of Agriculture), Agricultural Research Service, Hydrology and Remote
     Sensing Laboratory, 10300 Baltimore Avenue, Beltsville, MD 20705, USA; Feng.Gao@ars.usda.gov
*   Correspondence: yh_crs@bnu.edu.cn; Tel.: +86-10-588-05452

Received: 11 June 2018; Accepted: 24 July 2018; Published: 27 July 2018

Abstract: Leaf area index (LAI) is a critical vegetation structural parameter in biogeochemical and biophysical ecosystems. High-resolution LAI products play an essential role in regional studies. Empirical methods, which normally use field measurements as their training samples and have been identified as the most commonly used approaches to retrieve structural parameters of vegetation from high-resolution remote-sensing data, are limited by the quality of training samples. Few efforts have been made to generate training samples from existing global LAI products. In this study, two methods (a homogeneous and pure pixel filter method (method A) and a pixel unmixing method (method B)) were developed to extract training samples from moderate-resolution imaging spectroradiometer (MODIS) surface reflectance and LAI products, and a support vector regression (SVR) algorithm trained by the samples was used to retrieve the high-resolution LAI from Landsat data at Baoding, situated in the Hebei Province in China, and Des Moines, situated in Iowa, United States. For the homogeneous and pure pixel filter method, two different sets of training samples were designed. One was composed of upscaled Landsat reflectance at the 500-m resolution and MODIS LAI products (dataset A1); the other was composed of MODIS reflectance and LAI products (dataset A2). With them, two inversion models were developed using SVR. For the pixel unmixing method, the training samples (dataset B) were extracted from unmixed MODIS surface reflectance and LAI products at 30-m resolution, and the third inversion model was obtained with them. LAI inversion results showed that good agreement with field measurements was achieved using these three inversion models. The R^2 (coefficient of determination) value and the root mean square error (RMSE) value were computed to assess the results. For all tests, the R^2 values are higher than 0.74 and RMSE values are less than 0.73. These tests showed that three models for the two methods combined with MODIS products can retrieve 30-m resolution LAI from Landsat data. The results of the pixel unmixing method was slightly better than that of the homogeneous and pure pixel filter method.

Keywords: leaf area index; MODIS products; Landsat; high resolution; homogeneous and pure pixel filter; pixel unmixing

1. Introduction

Leaf area index (LAI) is defined as half of the total leaf area per ground area [1]. It is an important input parameter in land biogeochemical and biophysical ecosystems [2,3]. A variety of global LAI products have been produced from satellite data acquired by the advanced very high resolution radiometer (AVHRR) [4,5], the moderate-resolution imaging spectroradiometer (MODIS) [6,7], VEGETATION [8,9], and the multiangle imaging spectroradiometer (MISR) [10], and so forth. However, the existing global LAI products' spatial resolutions are medium or low and need to

Remote Sens. **2018**, *10*, 1187; doi:10.3390/rs10081187   326   www.mdpi.com/journal/remotesensing

be improved. For example, the spatial resolution of the CYCLOPES LAI derived from VEGETATION and the global land surface satellite LAI (GLASS LAI) derived from MODIS are 1/112° and 1000-m, respectively. The spatial resolution of the MODIS LAI is 500-m (MCD15A2H, Version 6). Although these LAI products are globally available, the spatial resolutions of the global LAI products are coarse. This makes them mostly appropriate to global and regional studies [11–13]. High-resolution LAI products are needed to monitor crop growth and study vegetation parameters and canopy structure at small scales.

Quantitative inversion methods for high-resolution LAI can be classified as the empirical approach, physical approach, and the mixed-model methods [14,15]. For these quantitative methods, the training samples and the inversion algorithms are the most important parts. Chen et al. [16] established the relationships between the vegetation indices from high-resolution satellite imagery and ground-based LAI to improve the LAI inversion algorithms and validate Canada-wide LAI map products. Collecting training samples from field measurements is difficult and the measurements are often not sufficient. Duan et al. [17] retrieved the LAI of three typical row crops (maize, potato, and sunflower) using a look-up table (LUT) based on the inversion of the PROSAIL model; the results indicated that the LUT-based inversion of the PROSAIL model was suitable for LAI estimation of these three crops. Some studies obtained training samples through radiative transfer model simulation, which needs many input parameters and the inverse problem is ill-posed [18]. Currently, some global LAI products can provide more information regarding land surface vegetation. Many researchers have carried out studies based on these global LAI products [19,20]. At the same time, some research works show that various global LAI products can be used to choose representative training samples. For example, Xiao et al. [7] developed a method to estimate GLASS LAI based on the MODIS and CYCLOPES LAI products using general regression neural networks (GRNNs). In this method, the MODIS and CYCLOPES LAI products were selected as training samples to train the GRNNs. However, the GLASS LAI products still have low spatial resolution, at 1000-m. In order to retrieve high-resolution LAI, Gao et al. [21] used the MODIS LAI products as a reference to retrieve LAI from Landsat imagery. Results showed that the approach could produce accurate estimates of Landsat LAI for major crops. Nowadays, similar studies on generating training samples from existing global LAI products number too few, but it is a meaningful and important research for high-resolution LAI retrieval.

On the other hand, the inversion algorithm is also an important focus. There are many inversion algorithms for the inversion of LAI, such as the traditional regression model algorithms, which describe the linear or nonlinear relationships of LAI with surface reflectance or the derived vegetation indices (VIs). Vina et al. [22] and Wang et al. [23] both used mathematical statistical approaches to build the LAI–VI relationships to estimate LAI. However, mathematical statistical approaches have poor fitting ability and encounter difficulty in solving multidimensional problems. Chai et al. [24] constructed recurrent neural networks by fusing the MODIS and VEGETATION products to estimate time series LAI. Although the results showed that the method can be helpful to improve the quality of LAI products of the typical vegetation types, the neural networks are too sensitive to the parameters in models [25] and cannot overcome the phenomena of "over-learning" and "local minimum" [26], which would affect the precision of the estimation. Support vector regression (SVR) is a machine-learning method which has a strong nonlinear fitting capability, and the kernel function can solve high-dimensional problems [27]. Durbha et al. [18] adopted a one-dimensional canopy reflectance model (PROSAIL) to retrieve LAI from MISR data using an SVR algorithm, and proposed a kernel-based regularization method to improve the SVR algorithm to solve the ill-posed problem.

The objective of this paper is to improve the high-resolution LAI retrieval approach using the SVR algorithm, especially focusing on the methods whereby to obtain high-quality training samples from the MODIS products. Three different sampling datasets for two methods were compared over two experimental sites. One site is located in the Heibei Province, China. The other site is located in central Iowa, U.S.

The second section introduces the SVR inversion algorithm and different high-quality training sample-acquiring methods. The third section introduces the research area and test data. The fourth section describes the experimental process and the results, followed by analysis and discussion.

2. Methodology

The empirical approach built by the SVR algorithm was developed in this research. Two training sample-selecting methods were developed to extract three different sampling datasets for the study. Namely, the MODIS reflectance products with MODIS LAI, the upscaled Landsat reflectance at the 500-m resolution with MODIS LAI, and unmixed MODIS reflectance and LAI at 30-m resolution were used as SVR training samples to retrieve Landsat LAI at the 30-m resolution from different periods.

Figure 1 shows the data processing framework. The high-quality training samples of SVR were selected by two methods: the homogeneous and pure pixel filter method (method A) and the pixel unmixing method (method B). For the homogeneous and pure pixel filter method, after the homogeneous and pure MODIS LAI pixels were selected, the upscaled Landsat reflectance (aggregate from 30-m to the 500-m resolution in the ENVI software, using a simple average method) of the corresponding pixels (the same location with the above homogeneous and pure MODIS LAI pixels) together with the LAI were selected as the training samples (dataset A1). Next, the corresponding MODIS reflectance of the pixels together with the LAI were also selected as the training samples (dataset A2). The pixel unmixing method took the unmixed LAI after quality control and reflectance of agricultural land pixels at the 30-m resolution as training samples (dataset B), which were obtained through the unmixing of MODIS LAI and reflectance products by linear models. The SVR models trained by the above three sets of training samples for the two methods were then applied to the Landsat surface reflectance to generate the 30-m resolution LAI. The retrieved LAI maps at the 30-m resolution and the temporal trends curves were generated and analyzed. The retrieval LAI results were then compared to the field measurements.

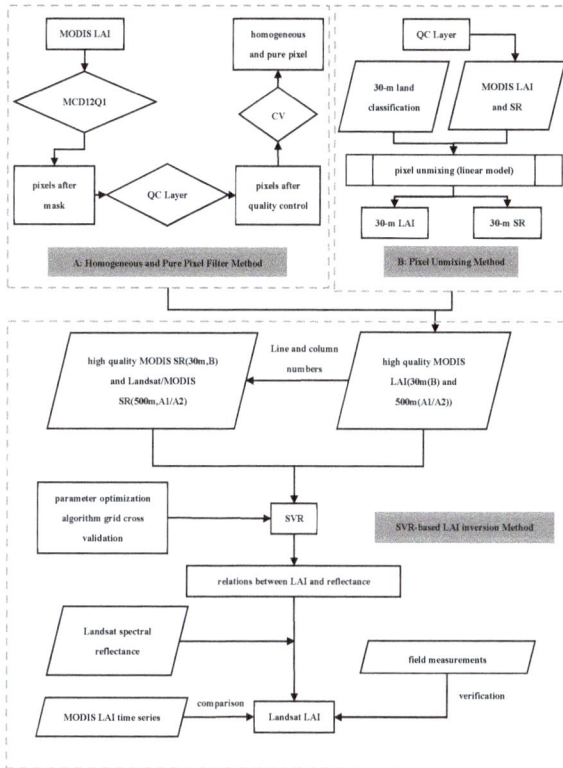

Figure 1. Data processing flowchart of the support vector regression (SVR) inversion method with two training sample-selecting methods: (**A**) the homogeneous and pure pixel filter method and (**B**) the pixel unmixing method.

2.1. Methods of Training Sample Selection

2.1.1. The Homogeneous and Pure Pixel Filter Method

The homogeneous and pure pixel filter method chose the high-quality MODIS LAI pixels by three steps: (1) masking the LAI product images of the research area using MODIS land-cover products (MCD12Q1) to obtain the agricultural land area. According to the International Geosphere—Biosphere Programme (IGBP) global vegetation type classification dataset (Land Cover Type 1), the pixels with the successive agricultural land type in every year during the research period are considered as the agricultural land area; (2) selecting the high-quality LAI pixels using quality control files (MODIS LAI QC layer) to ensure that the pixels' LAI are retrieved from the main algorithm, namely, the look-up table (LUT) algorithm. Through the first two steps, we obtain high-quality LAI pixels of agricultural land, and these pixels are used as the input pixels of the third step: (3) filtering the remaining LAI pixels by the coefficients of variation (CV) to select homogeneous and pure pixels.

It is known that the relationship between the LAI and the spectral reflectance or vegetation index is nonlinear. The relationships vary for pure and mixed pixels at low spatial resolution. To model the relationship between the LAI and the spectral reflectance for different vegetation types, pure pixels should be selected as training samples. The spectral reflectance of different objects in a mixed pixel is different. We use the CV (ratio of standard deviation to the mean value) in a statistical model to represent this difference, similarly to in [21].

The CV formula is as follows:

$$CV = \frac{\sigma}{\mu} \tag{1}$$

where CV is the coefficient of variation and σ and μ are the standard deviation and mean of the surface reflectance of the Landsat pixels within a MODIS pixel, respectively.

The CV of the reflectance of each band was calculated for all Landsat pixels at the MODIS pixel scale, and a threshold was determined based on the pixels' quality and quantity. It is assumed that the MODIS pixels are pure pixels when the CV is small.

When the homogeneous and pure pixels of the MODIS LAI products were selected, the corresponding MODIS surface reflectances were also chosen together to act as the training samples (dataset A2). The MODIS reflectance products have a good correspondence with the MODIS LAI products because of having the same temporal and spatial resolution, with no geometric deviation. In addition, the upscaled Landsat reflectances at the 500-m resolution were aggregated to create another training sample (dataset A1), which allows comparison of these retrieved high-resolution LAI of the training samples for the different selection methods.

2.1.2. The Pixel Unmixing Method

At 500-m resolution, different land cover types may be mixed within a MODIS pixel. The surface reflectance is determined by the spectral characteristics of the different types. Pixel unmixing methods can decompose the pixels of coarse-resolution satellite imagery into different endmembers using high-resolution satellite imagery classification data. These methods can thus obtain the ratio of each type to the coarse-resolution pixel (as a proportion). Ichoku et al. summarized five mixture models. These models are the linear model, the probabilistic model, the geometric-optical model, the stochastic geometric model, and the fuzzy model [28]. In this study, the linear model was used to unmix the MODIS surface reflectance and MODIS LAI. The linear model is a special case of a nonlinear model that ignores multiple scattering [29]. In the linear model, it is assumed that the reflectance of a pixel is a linear combination of the reflectance of each endmember [30]. The weight of the reflectance of the feature type is determined by the ratio of each type to the area of the pixel:

$$R = \sum_{j=1}^{n} \left(f_j \times r_j \right) + \varepsilon \tag{2}$$

where R is the reflectance of the mixed pixel; r_j is the reflectance of the jth endmember; f_j is the proportion of the jth endmember in the mixed pixel, which can be calculated based on the 30-m land cover map from the Landsat image; and ε is the error. In Equation (2), R and f_j are known, and the unknown variable is the r_j. In the study, we assume that the neighboring pixels with the same land cover type have similar surface reflectance or LAI [31]. When the number of endmembers is determined, the number of equations must be greater than or equal to the number of endmembers (j) in order to solve the equation. By using other pixels adjacent to the pixel being processed, one can avoid the ill-conditioning problem caused by too many unknown values, and this can be achieved by sliding the 3 × 3 window [32]. For each mixed pixel, a maximum of nine equations can be derived from the 3 × 3 window, which traverses the entire study area. Each equation group was solved by the constrained least squares method.

The unmixed pixels of MODIS LAI and surface reflectance products at the 30-m resolution using the above method was obtained. Then, choose the pixels of agricultural land cover type and the LAI retrieved from the main algorithm using quality control. Finally, the unmixed MODIS surface reflectance and LAI of the above chosen corresponding pixels were taken as the training samples for SVR.

2.2. Support Vector Regression

Support vector regression (SVR) is a supervised machine-learning method based on the principles of the Vapnik–Chervonenkis dimension theory and structural risk minimization [33]. SVR is more suitable for small samples and nonlinear, high-dimensional problems. SVR aims to construct an optimal super-pipe so that the pipeline can provide as much data as possible under a given accuracy ε, and so that the distance from the sample point to the pipe edge is not larger than ε [34]. This can be expressed by Equation (3):

$$f(x) = (\omega, \varphi(x, x_i)) + b \tag{3}$$

where ω is the normal vector (or the "support vector"), x_i stands for the dataset, and b is the bias.

The loss function in ε-SVR can be expressed as Equation (4):

$$L_\varepsilon = \begin{cases} 0 \; |y - f(x)| \le \varepsilon \\ |y - f(x)| - \varepsilon \text{ otherwise} \end{cases} \tag{4}$$

Operating on the basis of the structural risk minimization theory, SVR finally evolves into a convex optimization problem. Here, the slack variables ξ^*_i and ξ_i are introduced to represent the fitting error:

$$\text{Min} \frac{1}{2} \|\omega\|^2 + C \sum_{i=1}^{n} (\xi_i + \xi^*_i) \tag{5}$$

$$\text{subject to} \begin{cases} y_i - f(x_i) \le \varepsilon + \xi^*_i \\ f(x_i) - y_i \le \varepsilon + \xi_i \\ \xi_i, \xi^*_i \ge 0, i = 1 \ldots \ldots n \end{cases}$$

where C is the margin parameter.

SVR has the advantage of solving the nonlinear problem by introducing a kernel function, which is a good solution to the problem in high-dimensional space and is the core of SVR. The radial basis function (RBF) has been found to be superior to other kernel functions for reflecting the nonlinear relationship between LAI and reflectance. Therefore, the RBF was used in this research. After that, the choice of hyperparameters determines the quality of the model, which affects the quality of the SVR algorithm. The range of hyperparameters and kernel parameters was [−10, 10], and the six-fold cross-validation method was used to find the optimal parameters. As is well known, the higher the proportion of training samples, the closer the inversion results will be to the measured values. Next, the proportion of training samples was fixed at 80% for optimal parameter determination.

Three inversion models were built through training the SVR algorithm using the three training samples datasets obtained by the homogeneous and pure pixel filter method (datasets A1 and A2, method A) and the pixel unmixing method (dataset B, method B), both of which were described in detail in Section 2.1. In the inversion process, Landsat surface reflectance was used as the input to the three SVR inversion models to retrieve high-resolution LAI at 30-m resolution.

3. Study Area and Data Description

3.1. Study Area

Two study sites were selected in this study. The first study area was located at the boundary between Baoding and Shijiazhuang in the Hebei Province, China (Figure 2a). This area is flat, with about a 40-m altitude. Crops can be harvested twice a year and consist mainly of winter wheat and summer maize. The winter wheat-growing season is from October to June of the following year. The second study area was located near the capital of Iowa in Des Moines, Iowa, United States (Figure 2b), at a 300-m altitude, where the highest temperature occurs in July and the main crops are corn and soybean. The two research areas include three different crops and have different crop cultivation (crops can be harvested once or twice a year). Moreover, the results of the two different study sites could be used to test the robustness of the algorithm and different sampling strategies.

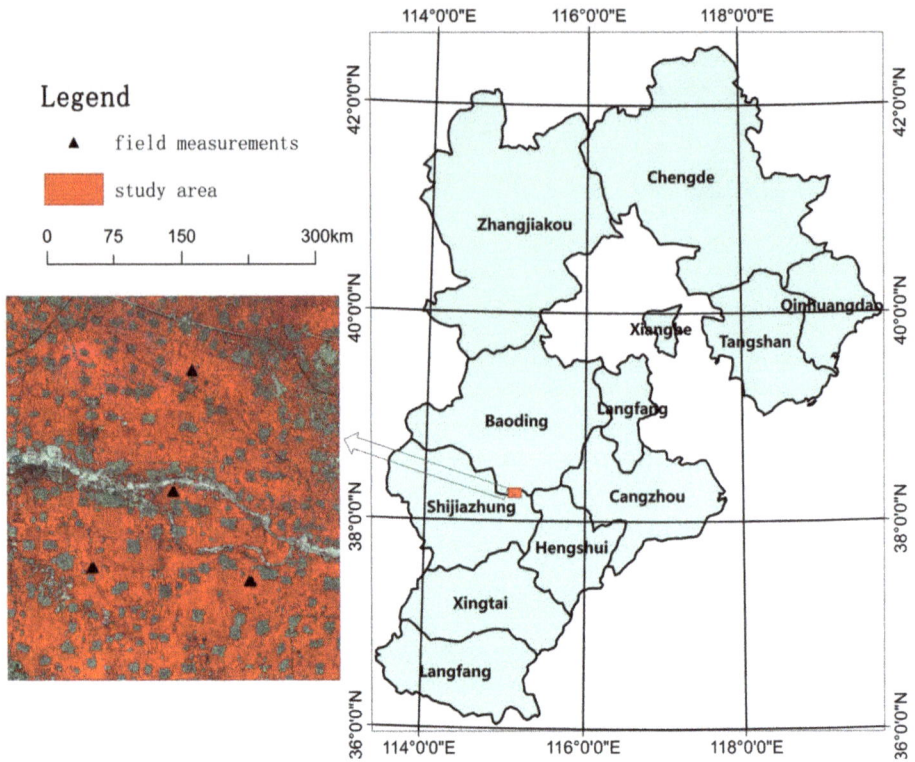

Figure 2. *Cont.*

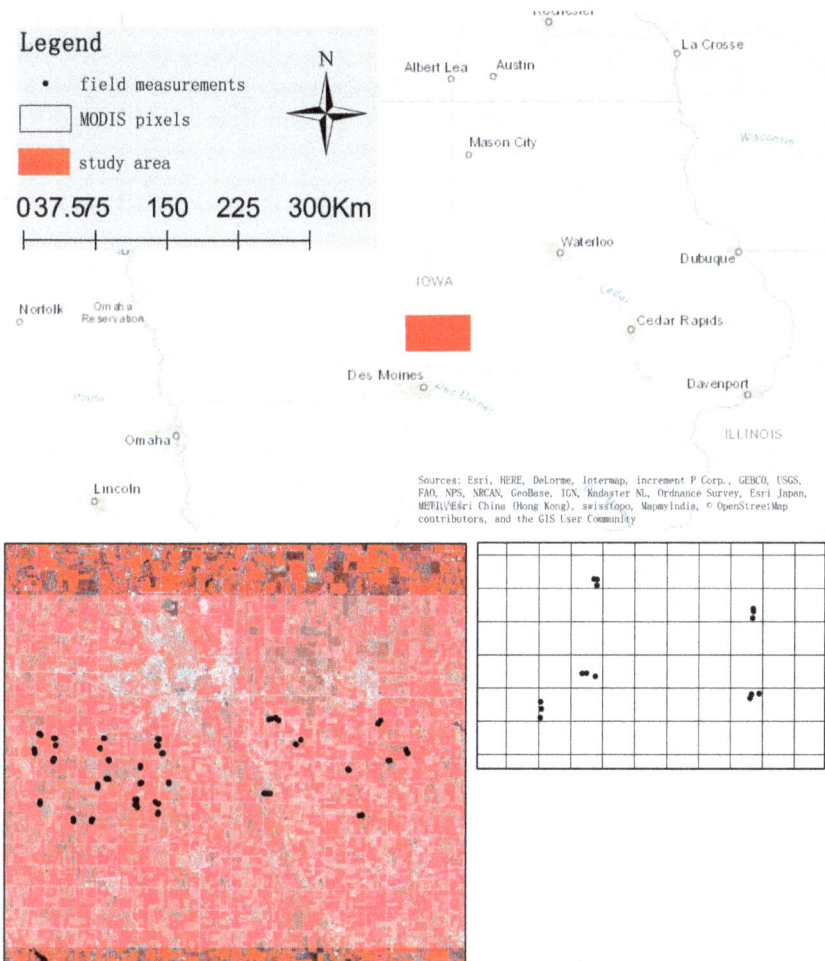

Figure 2. The two study areas: (**a**) Baoding, Hebei Province, China; (**b**) Des Moines, Iowa, United States. The grid in (**b**) shows the MODIS pixel cells at the 500-m resolution.

3.2. Data and Preprocessing

The satellite image datasets consisted mainly of MODIS products and Landsat scenes. MODIS products include a land-cover type product (MCD12Q1) (Collection 5 version), LAI products (MCD15A2H), and reflectance products (MOD09A1) (Collection 6 version). Landsat8 operational land imager (OLI) and Landsat5 thematic mapper (TM) and enhanced thematic mapper plus (ETM+) surface reflectance products at the 30-m spatial resolution were also collected as input data. At the Baoding study area, 29 Landsat8 scenes (paths 123–124 and row 33) from April 2013 to May 2017 were selected. In Des Moines, 29 Landsat5 scenes (paths 26–27 and row 31) from May 2003 to August 2007 were chosen. All data were downloaded from https://earthexplorer.usgs.gov/. The blue band in Landsat imagery is susceptible to atmospheric scattering, and the shortwave infrared (SWIR) band differs greatly in the Landsat and MODIS bands (Table 1). To obtain more band information, this study used the green, red, and near-infrared (NIR) bands.

Table 1. The comparison of band wavelengths between Landsat and MODIS.

Band	TM (μm)	ETM+ (μm)	OLI (μm)	MODIS (μm)
Blue	0.45–0.53	0.45–0.53	0.45–0.51	0.459–0.479
Green	0.52–0.60	0.52–0.60	0.53–0.59	0.545–0.565
Red	0.63–0.69	0.63–0.69	0.64–0.67	0.62–0.67
NIR	0.76–0.90	0.76–0.90	0.85–0.88	0.841–0.876
SWIR	1.55–1.75	1.55–1.75	1.57–1.65	1.628–1.652

For classification products, the IGBP global vegetation type classification dataset (Land Cover Type 1) of the MODIS land-cover product (MCD12Q1) is obtained from Terra and Aqua observations over one year, and has 17 main land-cover types [35]. The spatial overlay analysis of the multiyear data to yield the continuous planted area of crops was done to separate the study region into crop and noncrop areas. The spatial overlay analysis method means image binarization, in which a crop pixel is replaced by 1 and a noncrop pixel is replaced by 0, based on MCD12Q1. Then, the binary value of each pixel for every year is multiplied, so 1 stands for the continuous planted area of crops and 0 stands for noncrop areas. The GlobeLand30 product is the global 30-m resolution land-surface cover product generated by the National Geomatics Center of China with the support of '863' key projects, and has 10 main land-cover types. However, it only produced global land cover maps in 2000 and 2010. (The data set is provided by the National Geomatics Center of China. (DOI: 10.11769/GlobeLand30.2000.db; DOI: 10.11769/GlobeLand30.2010.db) [36,37]).

The MODIS LAI was produced by two algorithms: the LUT main algorithm and the backup algorithm, which is an empirical model relating the LAI and the vegetation index [38]. The quality of MODIS LAI for each pixel is described by the quality control flags of the QC layer, which is generally reflected by the SCF_QC bit. The latter three bits of that describe the LAI inversion algorithm [39]. When the SCF_QC is "000", the main algorithm is used with no saturation, which represents the best-quality LAI.

Field measurements from Des Moines, Iowa, United States were obtained from the 2002 soil moisture experiment (SMEX02) conducted by the U.S. Department of Agriculture (USDA). The purpose of the experiment was to research land–gas interaction, verify ground parameters' accuracy, and evaluate the new method of monitoring soil moisture by remote sensing. Samples of LAI were obtained using LAI-2000 instruments in the inter-row region at least 1 m away from where the biomass sample was taken [40]. The LAI measurements campaign went through four sampling rounds, conducted over the periods: 15–19 June, 27–30 June, 2–3 July, and 5–9 July. Twelve large areas went through the four rounds. Each observation area covered approximately one MODIS pixel, and each region was spaced to small areas measured three or four times (Figure 2). To match the time of the satellite imagery acquisition, the measured LAI were linearly interpolated to the Landsat data-acquisition time. In situ LAI measurements that had been taken four times were interpolated to the Landsat data-acquisition time. In this study, the LAI measurements on 30 June and 2 July constrained or corresponded to the 1 July inversion results, and the LAI measurements on 7 and 9 July constrained or corresponded to the 8 July inversion results.

Field measurements in Baoding, Hebei Province, China were collected from 27 April to 3 May 2016 using LAI-2000, containing ten sampling points. Measurements from each site were collected at least twice. The corresponding Landsat data-acquisition times were 25 April and 4 May, and the inversion LAI were linearly interpolated to each sampling-point time.

4. Experiments and Analysis

4.1. SVR Training Samples

4.1.1. Homogeneous and Pure Pixel Filter

In this part of the experiment, different experimental areas and different input data (upscaled Landsat reflectance or MODIS reflectance) made the size of the study area different. For example, a 143 × 140 (68,640 m × 67,200 m) pixel region in Des Moines, Iowa, United States was initially selected, and 971 noncrop land pixels were removed after masking using the MODIS land-cover product. Using the SCF_QC layer, the MODIS LAI pixels with the highest quality (SCF_QC = 000, main algorithm and no saturation) were extracted. This process removed 1858 pixels.

The reflectances of the Landsat bands were used in the CV calculation (Equation (1)). The CVs of the Landsat bands were calculated at the MODIS pixel scale using Landsat pixels, followed by the QC checking area, and a threshold of CV for pure pixels was determined based on the pixel quality and quantity. In this paper, a threshold of CV of 0.15 for pure pixels was determined for SMEX02 in the Des Moines area. In order to compare the inversion results of these two sites in the paper and conduct validation of the threshold, the threshold for the Baoding area was the same as that in the Des Moines area. The pure pixel filtering step removed 6090 pixels for one tile after using CV filtering from the NIR of Landsat5. Figure 3 shows the results of each step using the MODIS LAI product on the day 201 of 2005 as an example.

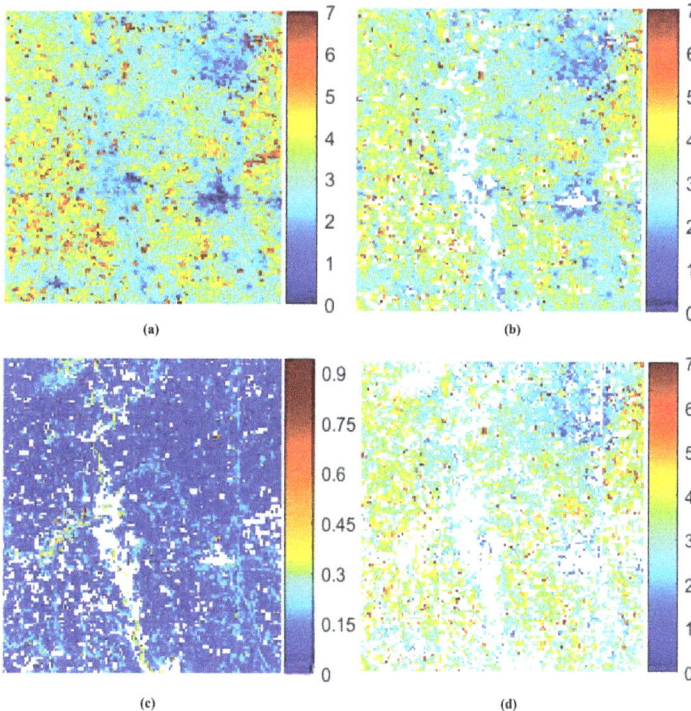

Figure 3. The MODIS leaf area index (LAI) pixel filtering process. From left to right: (**a**) original MODIS LAI; (**b**) MODIS LAI pixels for cropland with the highest retrieval quality (main algorithm and not saturated); (**c**) CV map of NIR band4 from Landsat5; and (**d**) final selected homogeneous and pure MODIS LAI pixels.

4.1.2. Pixel Unmixing

Taking the SMEX02 site area as an example, the area was mainly composed of natural vegetation and crops. Corn and soybean accounted for >80% of the total area. Figure 4 shows the MODIS classification map (Figure 4a) and the 30-m resolution GlobeLand30 surface covering products (Figure 4b,c). The study area included three categories from the MCD12Q1_IGBP classification system: number 12: agricultural land; number 13: urban and construction area; and number 14: junction of agricultural land and natural vegetation. The following classification of the GlobeLand30 product is used: arable land (10), forest (20), grassland (30) (accounting for 5% of the total area), wetland (50) (accounting for ~1% of the total area), water body (60), artificial surfaces (80), and bare land (90).

In this paper, the study area was divided into arable land, artificial surface, forest, grassland, and water bodies. Based on the Landsat scenes from 14 May to 2 August in 2002, it was obvious that the arable land had a regular shape and the spectral characteristics were different from other types. In addition, roads (within the artificial surface category) showed a regular rectangle, especially the roads between arable lands. This study used an object-oriented segmentation approach to classify the Landsat imagery. The eCognition software was used for this. Finally, the classification results were corrected by visual interpretation. Figure 4d shows the final results.

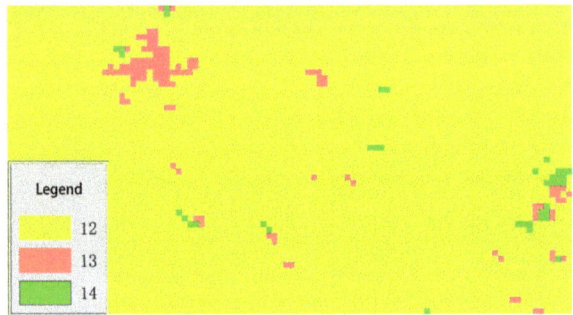

(a)

(b)

Figure 4. *Cont.*

(c)

(d)

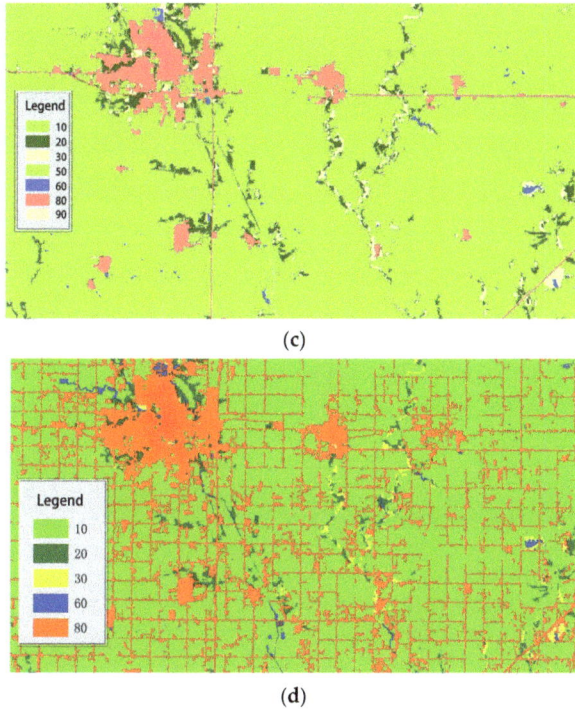

Figure 4. Classification products of the Soil Moisture Experiment 2002 (SMEX02) near Des Moines, Iowa, United States: (**a**) MCD12Q1_IGBP; (**b**) GlobeLand30 in 2000; (**c**) GlobeLand30 in 2010; (**d**) 30-m classification product. Notes: (**a**) 12: agricultural land; 13: urban and construction area; 14: junction of agricultural land and natural vegetation. (**b–d**) 10: arable land; 20: forest; 30: grassland; 50: wetland; 60: water body; 80: artificial surfaces; 90: bare land.

In this paper, a linear model was used to solve the mixed pixels, using the algorithm presented in Equation (3). A 3 × 3 sliding window was used to obtain the optimal solutions to the sets of equations. The MODIS surface reflectance and LAI products were unmixed by this method, described in Section 2. For example, in the SMEX02 study area, 5300 MODIS pixels were used to obtain the unmixing pixels and 2902 MODIS pixels were removed after QC checking. Finally, fewer than 400 arable land pixels were chosen in one tile (the day 137 of 2003).

4.2. Comparison of Regions

4.2.1. Retrieved LAI on SMEX02

In this part of the experiment, a comparison of the 30-m resolution LAI retrieved from Landsat surface reflectance scenes for three periods by different SVR inversion models and the field measurements was made. Figure 5 shows the results.

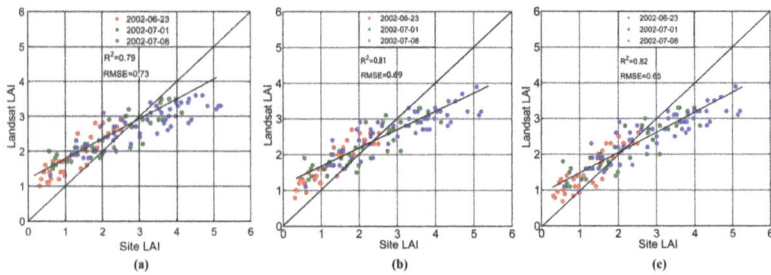

Figure 5. The comparison between the retrieved LAI by three approaches and field measurements: (**a**) the results of dataset A1 (upscaled Landsat reflectance at the 500-m resolution and MODIS LAI products); (**b**) the results of dataset A2 (MODIS reflectance and LAI products); (**c**) the results of dataset B (unmixed MODIS surface reflectance and LAI products at 30-m resolution).

Figure 5 shows that the retrieved LAI agrees well with the field measurements. At the field scale, the R^2 (coefficient of determination) between retrieved Landsat LAI and field measurements is 0.79 for the homogeneous and pure pixel filter method with upscaled Landsat reflectance as the training samples (dataset A1) (Figure 5a). The R^2 value is 0.81 for the homogeneous and pure pixel filter method with MODIS reflectance as the training samples (dataset A2) (Figure 5b), slightly better than dataset A1. The correlation coefficients of dataset A2 and dataset B was similar (Figure 5b,c), where the retrievals of unmixed MODIS reflectance and LAI using the pixel unmixing method (method B) shown in Figure 5c was slightly higher, with the R^2 value of 0.82 and a root mean square error (RMSE) value of 0.65. Although we had obtained high-quality pixels through two methods, it was inevitable that the MODIS LAI was small and underestimated due to mixing pixels at the small LAI stage. When the Landsat reflectance at 30-m resolution (the cover area of vegetation categories of the pixel usually is larger than that of MODIS at the small LAI stage) was used as the input for the SVR models, the results were overestimated when LAI was 0–2. We found that the results of approach B were improved when LAI was <2. However, when the LAI was greater than 3, the inversion results were significantly underestimated, especially on 8 July (Figure 6) [21]. One reason for this could be that, generally, the relationship between LAI and reflectance gradually becomes saturated. Another reason for the underestimation may have been that when the LAI was >3, especially when the LAI was much greater, the MODIS inversion algorithm was mainly using the backup algorithm on SMEX02, and the training samples obtained using the LUT method contained fewer pixels of LAI > 3. The result was that the training sample representativeness with LAI greater than 3 was poor. It is also clear that the mixed-pixel decomposition method had higher accuracy on 1 and 8 July due to the removal of other vegetation types (mainly forest and grassland, which grew better from June–July), and LAI also showed an increasing trend.

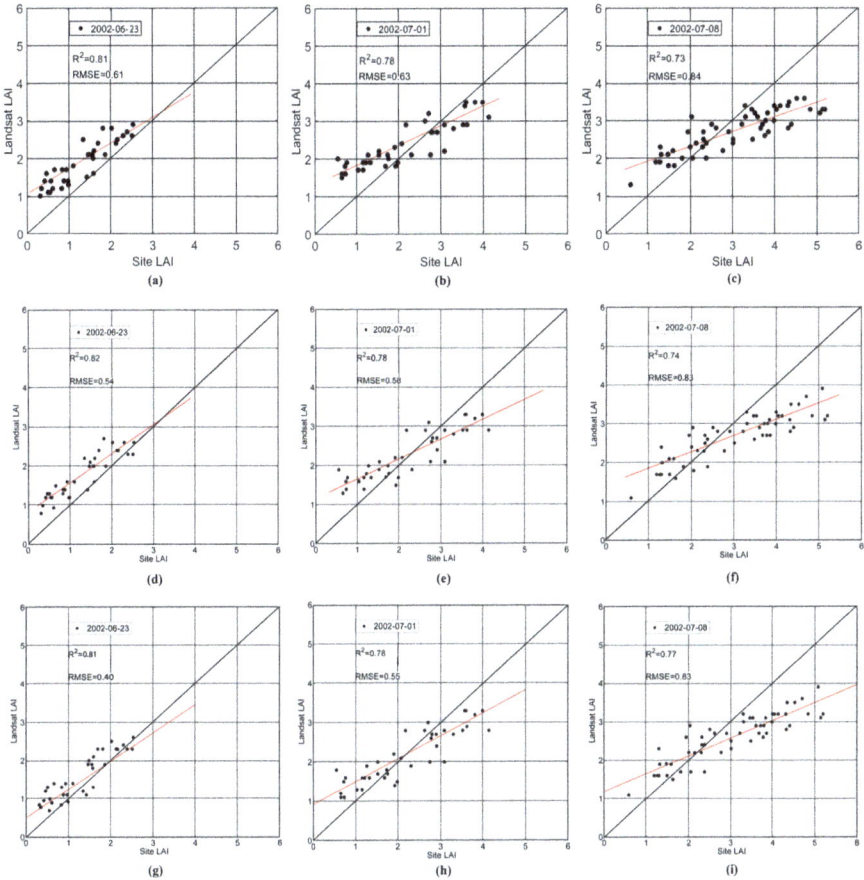

Figure 6. Comparison of inversion LAI in time series with field measurements. (**a–c**) The results of dataset A1; (**d–f**) the results of dataset A2; (**g–i**) the results of dataset B.

4.2.2. Retrieved LAI at the Baoding Study Area

In the present study, the red, near-infrared, and green bands of the Landsat scene were extracted to retrieve 30-m resolution LAI. Figure 7 shows the results of a comparison with measured values. The results of approach A1 and A2 compared with the measured values is shown in Figure 7a: the R^2 is 0.77 and the RMSE is 0.41 of A in general, and the inversion and measured values showed good correlation. The retrievals of approach B shown in Figure 7b were slightly higher, with the R^2 value of 0.79 and an RMSE value of 0.49. However, as the LAI value becomes larger, the inversion value is gradually more severely underestimated, and large deviations from the measured values occurred for the same reason as with SMEX02.

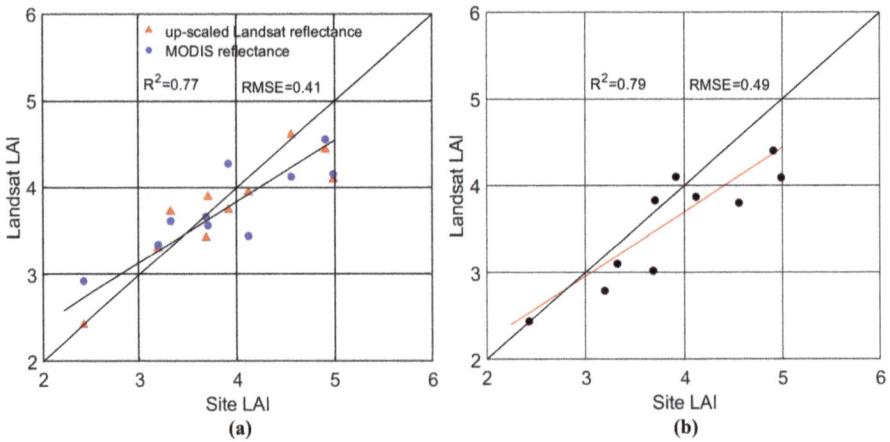

Figure 7. Comparison of inversion LAI with field measurements at the Baoding study area; (**a**) the results of datasets A1 and A2; (**b**) the results of dataset B.

4.2.3. Temporal Trends of LAI at the 30-m Resolution

In this research, Landsat LAI maps at the 30-m resolution at the SMEX02 site were retrieved for three periods, namely, 23 June, 1 July, and 8 July. The three SVR inversion models of datasets A1, A2, and B were used and compared. In Figure 8, no matter which inversion method was used, the LAI inversion values increased over time. Furthermore, they all had a good agreement with the distribution of vegetation in space. The Landsat LAI maps showed that different inversion methods had similar inversion results, and that it was also consistent with the comparison between retrievals and field measurements in Section 4.2.1.

In the Baoding area, ten measurements were located in seven valid MODIS pixels. The Landsat acquisition time extended from 2 April to 4 May, with a total of five measurement dates (2 April, 9 April, 18 April, 25 April, and 4 May). The temporal trend of LAI at the 30-m resolution was compared with that of the MODIS LAI (Figure 9).

The results show that the LAI time series retrieved from the Landsat data was consistent with the trend of the MODIS LAI time series (taking six pixels as an example). On the whole, the LAI curve showed a tendency to rise first and then decline, and reached the maximum near day of year (DOY) = 113. For Figure 9c, the CV of the pixel was >0.15, so this pixel is not pure and homogenous. In most situations, the LAI values of Landsat were higher than the MODIS LAI values, possibly mainly due to mixed pixels, among others. In the MODIS scale, the pixel contains nonvegetation categories and affects the MODIS LAI, but in the Landsat scale, the cover area of vegetation categories of the pixel usually is larger than that of MODIS. Of course, there are other errors, such as the scale effect, and so forth.

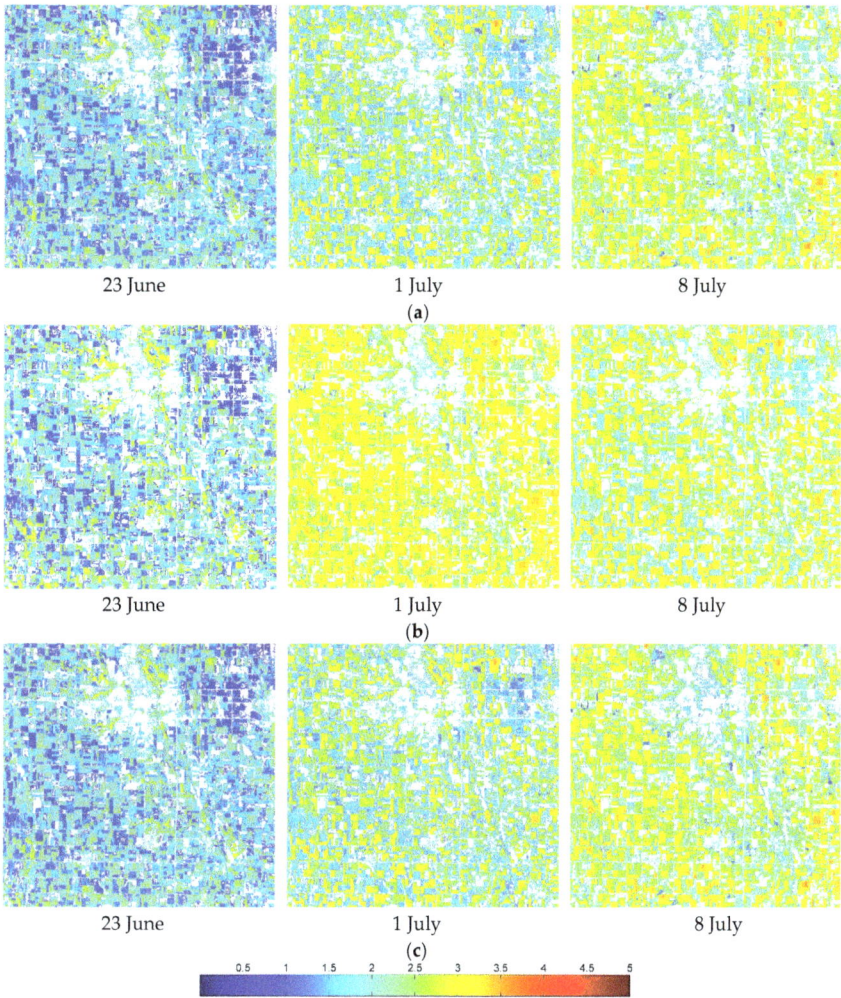

Figure 8. The LAI inversion in time series at the SMEX02 site (23 June, 1 July, and 8 July): (**a**) the results of dataset A1; (**b**) the results of dataset A2; (**c**) the results of dataset B.

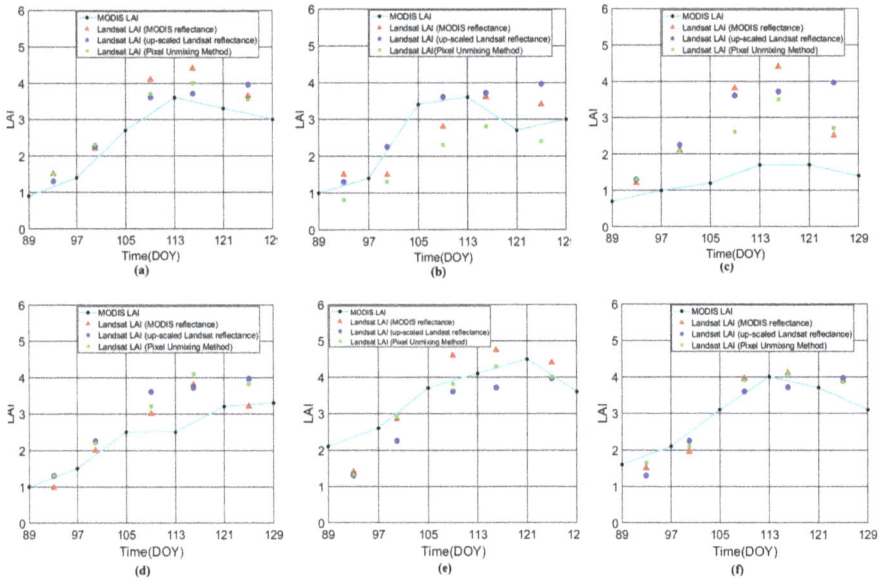

Figure 9. LAI inversion in time series at the Baoding area (MODIS DOY: 89–129), based on the homogeneous and pure pixel filter method and the pixel unmixing method. (**a**) pixel one; (**b**) pixel two; (**c**) pixel three; (**d**) pixel four; (**e**) pixel five and (**f**) pixel six.

5. Discussion

A comparative analysis of accuracy between the LAI retrievals and the field measurements (or MODIS LAI) was performed for the homogeneous and pure pixel filter method (method A) and the pixel unmixing method (method B). The results demonstrated that inversion of high-resolution LAI combining MODIS products based on the SVR algorithm was feasible. Compared to physical modeling methods, the empirical model using the SVR algorithm did not need to consider the physiological characteristics of vegetation and was easy to implement. Compared to other methods of obtaining training samples, our methods can obtain a greater number of high-quality samples based on global LAI products.

Gao et al. [21] used the decision tree with the upscaled Landsat reflectance at the 500-m resolution and MODIS LAI to retrieve LAI at the 30-m resolution. The comparison (R^2 = 0.79, mean bias error = −0.18, mean absolute difference = 0.58) between Landsat retrievals (30-m) of Gao's and field measurements at the field scale (10-m) showed that there was a good agreement with low to moderate LAI (0–3), but retrievals were underestimated for high LAI (3–5). In this paper, the R^2 was 0.79 and the RMSE was 0.73 for the homogeneous and pure pixel filter method when the MODIS LAI and upscaled Landsat reflectance at the 500-m resolution were used as the training samples (dataset A1). The R^2 was 0.81 and RMSE was 0.69 for the homogeneous and pure pixel filter method when the MODIS LAI and reflectance were used as the training samples (dataset A2). In addition, the retrievals of the unmixing method with unmixed MODIS reflectance and LAI at the 30-m resolution as the training samples (dataset B) were slightly higher, with an R^2 value of 0.82 and an RMSE value of 0.65. Compared with Gao et al., we selected the multiyear MODIS products for obtaining training samples to ensure the richness and representativeness of the samples. In addition, the MODIS LAI and reflectance products were also used as training samples to build the SVR inversion model, and yielded a good result. Moreover, the pixel unmixing method was used to obtain the SVR inversion model, and resulted in the highest accuracy.

In particular, the SVR algorithm had an advantage in solving the nonlinear problem because it overcomes the phenomena of "over-learning" and "under-learning" [41]. After the kernel function and the hyperparameters were chosen, the relationship between the MODIS LAI and surface reflectance was fitted. It was shown that the SVR algorithm can represent the relationship between LAI and reflectance and had good generalization ability.

Moreover, the quality of the training samples, including sample distribution, can seriously affect the quality of all empirical approaches, including the SVR. In this study, the training samples spanned multiple years and included the whole crop-growing season. Therefore, the data quality was good and representative. The homogeneous and pure pixel filter method took the quality control file (QC layer) of the MODIS LAI products into consideration and ensured the quality of the MODIS LAI. This was the precondition for taking the MODIS LAI as training samples. The ratio of the standard deviation to the mean value (CV) in a statistical model can represent the difference of spectral reflectance of different objects in a mixed pixel. The MODIS pixels can be considered homogeneous and pure pixels when the thresholds of the QC and CV settle within a certain range. The pixel unmixing method decomposed the pixels of coarse-resolution satellite imagery into different endmembers and obtained the high-resolution LAI (or reflectance). It can obtain higher-quality training samples than other methods.

However, there were also a few limitations to be improved upon. First, when the LAI was greater than 3, the inversion results were underestimated during July. A main reason for this was that the relationship between the surface reflectance and LAI tended to become saturated when the LAI was greater than 3. In this paper, when the MODIS LAI is high, the probability of using the main algorithm is low. In other words, if the training samples were selected by the main algorithm, many samples with LAI > 3 were eliminated, resulting in fewer training samples with high LAI values. Figure 10 shows the histogram of LAI from the LAI-SR samples of SMEX02. As we can see from the histogram, there are fewer large LAI values (>3.0). In the SMEX02, there are 21,500 LAI-SR samples for dataset A1; 2028 for dataset A2, using the homogeneous and pure pixel filter method; and 17,889 for the pixel unmixing method. In Baoding, the LAI-SR samples are 6909, 8797, and 8838, respectively. In addition, the scaling effect may cause the under-representation of high LAI values. The high values can be smoothed out in the coarse-resolution image. For both training sample-acquiring methods, we used the MODIS quality control flags to select the highest-quality retrievals derived from the MODIS LAI main algorithm. We relied on the MODIS LAI data quality flags and have not considered the effect of noise associated with the main algorithm. The noise from the main algorithm retrieval may need to be considered for other regions, such as the tropical area, where clouds are always present.

Furthermore, when the upscaled Landsat reflectance and MODIS LAI were used to build the relational model, geometrical registration between Landsat and MODIS was not performed in this study. This led to a bias when calculating the CV of the Landsat surface reflectance in a MODIS pixel. The CV threshold of the homogeneous MODIS pixel was determined based on LAI sample quality and quantity from subjective experience at present. The threshold may vary with different study sites or landscapes. In order to compare the inversion results from these two sites in the paper, the same threshold of 0.15 was used. Additional study and analysis are needed to quantify the threshold. In this paper, we used the eight-day MODIS LAI products. This means that the MODIS LAI product for the period was made up from eight independent days. However, the Landsat imagery reflected the instantaneous optical characteristics of vegetation for the Landsat acquisition date. Therefore, differences between the Landsat and MODIS data products were observed in time and space.

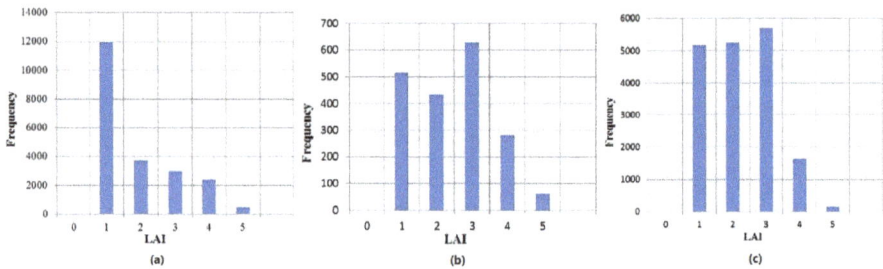

Figure 10. Histogram of LAI from the LAI-SR samples of SMEX02. (**a**) dataset A1 (upscaled Landsat reflectance at the 500-m resolution and MODIS LAI products); (**b**) dataset A2 (MODIS reflectance and LAI products); (**c**) dataset B (unmixed MODIS surface reflectance and LAI products at 30-m resolution).

Note that the spectral response function of the sensor is different, and the wavelengths of the corresponding bands are also different. The spectral response function of the sensor is a function of the wavelength. It is the ratio of the radiance received by the sensor at each wavelength to the radiance of the incident. In this study, we compared the bands of Landsat and MODIS and chose the three bands with small difference. Although the relational model was built using MODIS reflectance and MODIS LAI, the Landsat reflectance was used as the input parameter to retrieve the LAI. There exists a large gap in the wavelength range, particularly in the near-infrared band4 of Landsat5. Thus, the use of a spectral response function to convert the Landsat reflectance and MODIS reflectance will be the next key work.

Finally, an empirical model has advantages, but also limitations. In this study, the verification experiments were conducted over two study areas: the Hebei Province, China and Des Moines, Iowa, United States. Crops included corn, soybean, and winter wheat. Although the inversion model proposed here achieved good results in the study area, applications in other areas need to be further verified. To apply the model to large regions, not only must the method of screening high-quality data be improved, but the biophysical mechanisms of vegetation must also be studied. In the present study, the relationship between LAI and reflectance had a certain scope of application. Moreover, inversion accuracy was high in the main growth period, but was reduced in other periods, especially in the late stages of crop growth. Climate change may impact vegetation growth. The accuracy of LAI retrieval may be decreased if LAI-SR samples cannot cover this variation. In our study sites, climate change from the study period is not significant to enable testing of this hypothesis. Future research may consider different inversion models at different times.

6. Conclusions

In this study, a support vector regression algorithm combined with MODIS LAI and reflectance products which were used to obtain training samples was developed to retrieve high-resolution LAI from Landsat data. The homogeneous and pure pixel filter method and the pixel unmixing method were applied to select high-quality training samples to retrieve high-resolution LAI. Among them, the principle of selecting training samples for the homogeneous and pure pixel filter was simple and easy to operate. Additionally, the pixel unmixing method took into account the problem of mixed pixels in the large scale. The results of the two main methods were in good agreement with the field-measured values. Inversion accuracy of the pixel unmixing method was slightly higher than the homogeneous and pure pixel filter method, but the pixel unmixing method was more complex to implement. Using the homogeneous and pure pixel filter method, the retrievals using MODIS reflectance as the training samples was better than upscaled Landsat reflectance as the training samples. These tests showed that both methods combined with MODIS products can retrieve 30-m resolution

LAI from Landsat imagery. In future research, additional validations need to be done to assess the accuracy of the approach over other regions.

Author Contributions: H.Y. provided the idea for this paper and edited the manuscript. J.Z. wrote the paper and performed the data processing and analysis. S.Z. partly analyzed the data and reviewed the manuscript. Z.X. provided the idea and some suggestions for this paper and reviewed the manuscript. F.G. provided some suggestions and revised the manuscript.

Funding: This research received no external funding. This research was supported by the National Key R&D Plan (2017YFA0603001).

Acknowledgments: Thanks go to the surveyors of the soil moisture experiment of 2002 (SMEX02) for providing the LAI field measurements. Thanks to associate professor Lina Chai from Beijing Normal University for the LAI data. Thanks also to Shan Xu and Aolin Jia from Beijing Normal University for help in writing this paper.

Conflicts of Interest: The authors declare no conflicts of interest.

References

1. Chen, J.M.; Black, T.A. Measuring leaf area index of plant canopies with branch architecture. *Agric. For. Meteorol.* **1991**, *57*, 1–12. [CrossRef]
2. Barbu, A.L.; Calvet, J.C.; Mahfouf, J.F.; Albergel, C.; Lafont, S. Assimilation of soil wetness index and leaf area index into the ISBA-A-gs land surface model: Grassland case study. *Biogeosci. Discuss.* **2011**, *8*, 1971–1986. [CrossRef]
3. Anav, A.; Murray-Tortarolo, G.; Friedlingstein, P.; Sitch, S.; Piao, S.; Zhu, Z. Evaluation of land surface models in reproducing satellite derived leaf area index over the high-latitude northern hemisphere. Part II: Earth system models. *Remote Sens.* **2013**, *5*, 3637–3661. [CrossRef]
4. Zhu, Z.; Bi, J.; Pan, Y.; Ganguly, S.; Anav, A.; Xu, L.; Samanta, A.; Piao, S.; Nemani, R.; Myneni, R. Global data sets of vegetation leaf area index (LAI)3g and fraction of photosynthetically active radiation (FPAR)3g derived from global inventory modeling and mapping studies (GIMMS) normalized difference vegetation index (NDVI3g) for the period 1981 to 2011. *Remote Sens.* **2013**, *5*, 927–948. [CrossRef]
5. Claverie, M.; Matthews, J.; Vermote, E.; Justice, C. A 30+ year AVHRR LAI and FPAR climate data record: Algorithm description and validation. *Remote Sens.* **2016**, *8*, 263. [CrossRef]
6. Myneni, R.B.; Hoffman, S.; Knyazikhin, Y.; Privette, J.L.; Glassy, J.; Tian, Y. Global products of vegetation leaf area and fraction absorbed par from year one of MODIS data. *Remote Sens. Environ.* **2002**, *83*, 214–231. [CrossRef]
7. Xiao, Z.; Liang, S.; Wang, J.; Chen, P.; Yin, X.; Zhang, L.; Song, J. Use of general regression neural networks for generating the GLASS leaf area index product from time-series MODIS surface reflectance. *IEEE Trans. Geosci. Remote Sens.* **2014**, *52*, 209–223. [CrossRef]
8. Baret, F.; Hagolle, O.; Geiger, B.; Bicheron, P.; Miras, B.; Huc, M.; Berthelot, B.; Niño, F.; Weiss, M.; Samain, O.; et al. LAI, FPAR and fCover CYCLOPES global products derived from VEGETATION. Part 1: Principles of the algorithm. *Remote Sens. Environ.* **2007**, *110*, 275–286. [CrossRef]
9. Baret, F.; Weiss, M.; Lacaze, R.; Camacho, F.; Makhmara, H.; Pacholcyzk, P.; Smets, B. Geov1: LAI and FPAR essential climate variables and fCover global time series capitalizing over existing products. Part 1: Principles of development and production. *Remote Sens. Environ.* **2013**, *137*, 299–309. [CrossRef]
10. Hu, J.; Tan, B.; Shabanov, N.; Crean, K.A.; Martonchik, J.V.; Diner, D.J.; Knyazikhin, Y.; Myneni, R.B. Performance of the MISR LAI and FPAR algorithm: A case study in africa. *Remote Sens. Environ.* **2003**, *88*, 324–340. [CrossRef]
11. Valderrama-Landeros, L.H.; España-Boquera, M.L.; Baret, F. Deforestation in Michoacan, Mexico, from CYCLOPES-LAI time series (2000–2006). *IEEE J. Sel. Top. Appl. Earth Obs. Remote Sens.* **2017**, *1–8*, 1–8. [CrossRef]
12. Wang, J.; Wang, J.; Zhou, H.; Xiao, Z. Detecting forest disturbance in northeast China from GLASS LAI time series data using a dynamic model. *Remote Sens.* **2017**, *9*, 1293. [CrossRef]
13. Wang, Q.; Tenhunen, J.; Dinh, N.; Reichstein, M.; Otieno, D.; Granier, A.; Pilegarrd, K. Evaluation of seasonal variation of MODIS derived leaf area index at two European deciduous broadleaf forest sites. *Remote Sens. Environ.* **2005**, *96*, 475–484. [CrossRef]

14. Verrelst, J.; Rivera, J.P.; Veroustraete, F.; Muñoz-Marí, J.; Clevers, J.G.P.W.; Camps-Valls, G.; Moreno, J. Experimental sentinel-2 LAI estimation using parametric, non-parametric and physical retrieval methods—A comparison. *ISPRS J. Photogramm. Remote Sens.* **2015**, *108*, 260–272. [CrossRef]

15. Zhang, W.; Chen, Y.; Hu, S. Retrieving LAI in the heihe and the hanjiang river basins using Landsat images for accuracy evaluation on MODIS LAI product. *Int. Geosci. Remote Sens. Symp. (IGARSS)* **2007**, *3417–3421*, 3417–3421. [CrossRef]

16. Chen, J.M.; Pavlic, G.; Brown, L.; Cihlar, J.; Leblanc, S.G. Derivation and validation of Canada-wide coarse-resolution leaf area index maps using high-resolution satellite imagery and ground measurements. *Remote Sens. Environ.* **2002**, *80*, 165–184. [CrossRef]

17. Duan, S.B.; Li, Z.L.; Wu, H.; Tang, B.H.; Ma, L.; Zhao, E.; Li, C. Inversion of the PROSAIL model to estimate leaf area index of maize, potato, and sunflower fields from unmanned aerial vehicle hyperspectral data. *Int. J. Appl. Earth Obs. Geoinf.* **2014**, *26*, 12–20. [CrossRef]

18. Durbha, S.S.; King, R.L.; Younan, N.H. Support vector machines regression for retrieval of leaf area index from multiangle imaging spectroradiometer. *Remote Sens. Environ.* **2007**, *107*, 348–361. [CrossRef]

19. Huang, J.; Li, X.; Liu, D.; Ma, H.; Tian, L.; Wei, S. Comparison of winter wheat yield estimation by sequential assimilation of different spatio-temporal resolution remotely sensed LAI datasets. *Trans. Chin. Soc. Agric. Mach.* **2015**, *46*, 240–248. [CrossRef]

20. Chen, S.; Zhao, Y.; Shen, S. Applicability of pywofost model based on ensemble kalman filter in simulating maize yield in northeast China. *Chin. J. Agrometeorol.* **2012**, *33*, 245–253.

21. Gao, F.; Anderson, M.C.; Kustas, W.P.; Wang, Y. Simple method for retrieving leaf area index from Landsat using MODIS leaf area index products as reference. *J. Appl. Remote Sens.* **2012**, *6*, 063554. [CrossRef]

22. Viña, A.; Gitelson, A.A.; Nguy-Robertson, A.L.; Peng, Y. Comparison of different vegetation indices for the remote assessment of green leaf area index of crops. *Remote Sens. Environ.* **2011**, *115*, 3468–3478. [CrossRef]

23. Wang Qing, L.J.; Kun-Yong, Y. Inversion of masson pine forest LAI by multiple-perspective vegetation index. *Plant Sci. J.* **2017**, *35*, 48–55. [CrossRef]

24. Linna, C.; Yonghua, Q.; Lixin, Z.; Shunlin, L.; Jindi, W. Estimating time series leaf area index based on recurrent neural networks. *Adv. Earth Sci.* **2009**, *24*, 756–768. [CrossRef]

25. Liang, S.L.; Cheng, J.; Jia, K.; Jiang, B.; Liu, Q.; Liu, S.H.; Xiao, Z.Q.; Xie, X.H.; Yao, Y.J.; Yuan, W.P.; et al. Recent progress in land surface quantitative remote sensing. *J. Remote Sens.* **2016**, *20*, 875–898. [CrossRef]

26. Lv, J. *Hyperspectral Remote Sensing Inversion Models of Crop Chlorophy | | Content Based on Machine Learning and Radiative Transfer Models*; China University of Geosciences: Beijing, China, 2012.

27. Awad, M.; Mariette, R.; Khanna, R. Support vector regression. *Neural Inf. Process. Lett. Rev.* **2007**, *11*, 203–224.

28. Ichoku, C.; Karnieli, A. A review of mixture modeling techniques for sub-pixel land cover estimation. *Remote Sens. Rev.* **1996**, *13*, 161–186. [CrossRef]

29. Chang chun, L.V. A review of pixel unmixing models. *Remote Sens. Inf.* **2003**, *55–58*, 55–58. [CrossRef]

30. Settle, J.J.; Drake, N.A. Linear mixing and the estimation of ground cover proportions. *Int. J. Remote Sens.* **1993**, *14*, 1159–1177. [CrossRef]

31. Tobler, W.R. A computer movie simulating urban growth in the detroit region. *Econ. Geogr.* **1970**, *46*, 234–240. [CrossRef]

32. Liao, L.; Song, J.; Wang, J.; Xiao, Z.; Wang, J. Bayesian method for building frequent Landsat-like NDVI datasets by integrating MODIS and Landsat NDVI. *Remote Sens.* **2016**, *8*, 452. [CrossRef]

33. Vapnik, V.; Golowich, S.E.; Smola, A. Support vector method for function approximation, regression estimation, and signal processing. *Adv. Neural Inf. Process. Syst.* **1997**, *9*, 281–287.

34. Chang, C.C.; Lin, C.J. Libsvm: A library for support vector machines. *ACM Trans. Intell. Syst. Technol. (TIST)* **2011**, *2*, 27. [CrossRef]

35. Friedl, M.A.; Sulla-Menashe, D.; Tan, B.; Schneider, A.; Ramankutty, N.; Sibley, A.; Huang, X. MODIS collection 5 global land cover: Algorithm refinements and characterization of new datasets. *Remote Sens. Environ.* **2010**, *114*, 168–182. [CrossRef]

36. Chen, J.; Ban, Y.; Li, S. China: Open access to earth land-cover map. *Nature* **2014**, *514*, 434–434.

37. Chen, J.; Liao, A.; Chen, J.; Peng, S.; Chen, L.; Zhang, H. *30-Meter Global Land Cover Data Product-Globe Land30*; Geomatics World: Beijing, China, 2017.

38. Knyazikhin, Y.; Glassy, J.; Privette, J.L.; Tian, Y.; Lotsch, A.; Zhang, Y.; Wang, Y.; Morisette, J.T.; Votava, P.; Myneni, R.B.; et al. MODIS Leaf Area Index (LAI) and Fractionof Photosynthetically Active Radiation Absorbed by Vegetation (FPAR) Product (MOD15) Algorithm Theoretical Basis Document. 1999. Available online: http://eospso.gsfc.nasa.gov/atbd/modistables.html (accessed on 11 May 2018).

39. Yang, W.; Huang, D.; Tan, B.; Stroeve, J.C.; Shabanov, N.V.; Knyazikhin, Y.; Nemani, R.R.; Myneni, R.B. Analysis of leaf area index and fraction of par absorbed by vegetation products from the terra modis sensor 2000–2005. *IEEE Trans. Geosci. Remote Sens.* **2006**, *44*, 1829–1842. [CrossRef]

40. Anderson, M.C.; Neale, C.M.U.; Li, F.; Norman, J.M.; Kustas, W.P.; Jayanthi, H.; Chavez, J. Upscaling ground observations of vegetation water content, canopy height, and leaf area index during SMEX02 using aircraft and Landsat imagery. *Remote Sens. Environ.* **2004**, *92*, 447–464. [CrossRef]

41. Liang, D.; Yang, Q.; Huang, W.; Peng, D.; Zhao, J.; Huang, L.; Zhang, D.; Song, X. Estimation of leaf area index based on wavelet transform and support vector machine regression in winter wheat. *Infrared Laser Eng.* **2015**, *44*, 335–340. [CrossRef]

remote sensing

MDPI

Article

Estimating Leaf Area Density of Individual Trees Using the Point Cloud Segmentation of Terrestrial LiDAR Data and a Voxel-Based Model

Shihua Li [1,2,*], Leiyu Dai [1], Hongshu Wang [3], Yong Wang [4], Ze He [1] and Sen Lin [1]

[1] School of Resources and Environment, University of Electronic Science and Technology of China, No. 2006, Xiyuan Ave, West Hi-Tech Zone, Chengdu 611731, China; uestc_dly@163.com (L.D.); 18200113725@163.com (Z.H.); 18200112978@163.com (S.L.)
[2] Center for Information Geoscience, University of Electronic Science and Technology of China, No. 2006, Xiyuan Ave, West Hi-Tech Zone, Chengdu 611731, China
[3] Department of Surveying and Mapping Engineering, Sichuan Water Conservancy Vocational College, Chongzhou 611231, China; wanghongshu5013@sina.com
[4] Department of Geography, Planning, and Environment, East Carolina University, Greenville, NC 27858, USA; wangy@ecu.edu
* Correspondence: lishihua@uestc.edu.cn

Received: 19 September 2017; Accepted: 20 November 2017; Published: 22 November 2017

Abstract: The leaf area density (LAD) within a tree canopy is very important for the understanding and modeling of photosynthetic studies of the tree. Terrestrial light detection and ranging (LiDAR) has been applied to obtain the three-dimensional structural properties of vegetation and estimate the LAD. However, there is concern about the efficiency of available approaches. Thus, the objective of this study was to develop an effective means for the LAD estimation of the canopy of individual magnolia trees using high-resolution terrestrial LiDAR data. The normal difference method based on the differences in the structures of the leaf and non-leaf components of trees was proposed and used to segment leaf point clouds. The vertical LAD profiles were estimated using the voxel-based canopy profiling (VCP) model. The influence of voxel size on the LAD estimation was analyzed. The leaf point cloud's extraction accuracy for two magnolia trees was 86.53% and 84.63%, respectively. Compared with the ground measured leaf area index (LAI), the retrieved accuracy was 99.9% and 90.7%, respectively. The LAD (as well as LAI) was highly sensitive to the voxel size. The spatial resolution of point clouds should be the appropriate estimator for the voxel size in the VCP model.

Keywords: leaf area density; terrestrial LiDAR; tree canopy; vertical structure; voxel

1. Introduction

Foliage plays an important role in the energy budget through photosynthesis, transpiration, respiration, and the maintenance of the plant microclimate [1]. The spatial distribution of leaves is critical for describing the transmission and interception of solar radiation for wood production, species competition, ecosystem dynamics, and biodiversity [2]. The leaf area index (LAI) is generally used for expressing the amount of leaves in a tree canopy, and has been successfully retrieved by using remotely sensed data at different scales [3]. The determination of LAI is common. However, LAI can be difficult to use for characterizing the structure of a heterogeneous canopy, and may be less effective or more complicated to use in cases where leaves have irregular shapes and forms [2,4].

As one of the canopy vertical structure parameters, the leaf area density (LAD) in each horizontal layer is generally used for the quantification of the leaves in the canopy [2]. LAD is defined as the total one-sided leaf area per unit volume [5]. Integrating the LAD profile data vertically, one can calculate the LAI [6]. LAD can be estimated in situ using direct, semi-direct, or indirect approaches [2]. The direct method involves the counting and measurement of leaves, but this application is limited, because it is destructive and time consuming. One of the representative semi-direct methods is Wilson's inclined point quadrat method, which counts the number of contacts of a leaf with probes inserted into the vegetation canopy [7]. Indirect techniques mainly involve the use of passive optical devices based on a gap fraction method, such as hemispherical photography, which relies on the Beer–Lambert law of light transmission through a turbid medium adapted to canopies [8]. However, these methods are limited in the spatial explicitness of their estimates, as well as in their accuracy [2,5].

Light detection and ranging (LiDAR) sensors have recently been applied to obtain the three-dimensional (3D) structural properties of plants [9–12]. A terrestrial LiDAR sensor emitting small-footprint laser pulses at a high pulse repetitive frequency and with small angular steps between consecutive pulses provides a fine spatial resolution, which allows the inner canopies of trees to be assessed from the ground and makes the accurate estimation of LAD profiles possible [1,13]. One of the important prerequisites in the estimation of canopy LAD is the ability to describe the spatial distribution of leaves separately from that of wood, because the point clouds include not only leaves, but also non-leaf components (such as twigs, branches, and the stem/trunk) of the plant, which will affect the estimation accuracy of LAD [13]. The LAD estimation is greatly affected by whether the leaf and non-leaf components are well separated in the point clouds. Hosoi and Omasa showed that the LAD of *Zelkova serrata* was overestimated by 19% if the LiDAR points of the woody components were not eliminated [6]. To classify leaf and non-leaf components in a laser point cloud, many studies have used manual techniques, where laser points associated with different canopy components are visually identified [1,6]. However, these methods are labor intensive and time consuming, which limits the use of LiDAR data at relatively broad spatial scales for estimating LAD [14]. Distinguishing a leaf from a trunk or branch by using the intensity of the reflected pulse relies on the differences in their optical properties at the wavelength of the LiDAR sensor [2,15]. However, the laser return intensity is affected by the distance and incidence angle. The radiometric calibration of the intensity is not easy [14]. The geometric method was also used to separate the photosynthetic and non-photosynthetic components in the terrestrial LiDAR data of forest canopies [14,16]. Unfortunately, the geometric information is not easy to obtain, either. The reflectance values associated with a digital camera can be useful for automatically classifying the structural parameters of the canopy [17]. However, the dimensionless and uncalibrated reflectance values are highly variable [14]. In recent years, a non-destructive and rapid object extraction method called point cloud segmentation has been used for ground object extraction and classification from airborne LiDAR data [18–21]. However, the segmentation method has been seldom used to extract leaf point clouds using high-resolution terrestrial LiDAR data. The leaf is significantly different from the other parts of the tree, such as the stem and trunk, both in shape and size. Consequently, the segmentation of high density unorganized 3D LiDAR point clouds should have the potential to distinguish leaves from the other parts of the tree.

Recently, many researchers have attempted to develop various models for the estimation of LAD using LiDAR sensors [1,15,22–24]. Among the models, the voxel-based method has been commonly used for describing the computation of a 3D matrix of voxels from terrestrial LiDAR point clouds. The method has the characteristic that no assumption about the spatial distribution, size, or shape of canopy components is made. This method is also easy to operate. The vegetation density of a voxel can be computed using the number of echoes inside the voxel [25]. The voxel-based method has been successfully used in individual trees and woody canopy LAD estimation [2,4,13,15,25–28]. Hosio and Omasa developed a voxel-based canopy profiling (VCP) method to express the laser trace information as a voxel that serves as an attribute of a 3D array [1]. Based on each voxel, both LAD profiles and the LAI of an individual tree can be accurately estimated by counting the frequency of contact between laser beams and the foliage of the canopy in each horizontal layer. The same group of researchers applied this method to a natural forest stand [6] and woody materials [11,26]. Wang et al. estimated the LAD of a magnolia canopy using the VCP method based on terrestrial LiDAR and true color images [17]. Therefore, the voxel-based method is a promising way to estimate LAD. However, the voxel size needs to be chosen carefully, because it can significantly influence the estimation accuracy of the LAD [2]. The calculation accuracy and efficiency depend on the voxel size. An assessment of the effect of the voxel size on the LAD estimation model is needed. The objectives of this study are: (1) to develop an effective workflow to estimate LAD for individual magnolia trees on the basis of high-resolution terrestrial LiDAR measurements; (2) to propose a point cloud segmentation method for leaf extraction; and (3) to quantify the impact of voxel size on the LAD estimations for individual trees.

2. Study Site and Field Measurements

The study site was located on the campus of University of Electronic Science and Technology of China, Chengdu, Sichuan, China. The ground area was almost flat. For this study, two individual magnolia trees (Table 1 and Figure 1) were selected and scanned with Leica ScanStation C10, which has a full 360° × 270° field of view, long range (300 m@90% reflectivity), and high scan frequency (50,000 points/s). The laser wavelength is 532 nm.

Table 1. Description variables for the scanned magnolia trees (m).

Tree	Height	Canopy Depth	Crown Size	Average Leaf Length	Average Leaf Width
Magnolia A	6.1	4.1	2.80 × 2.83	0.144	0.075
Magnolia B	6.4	4.5	2.81 × 3.29	0.156	0.078

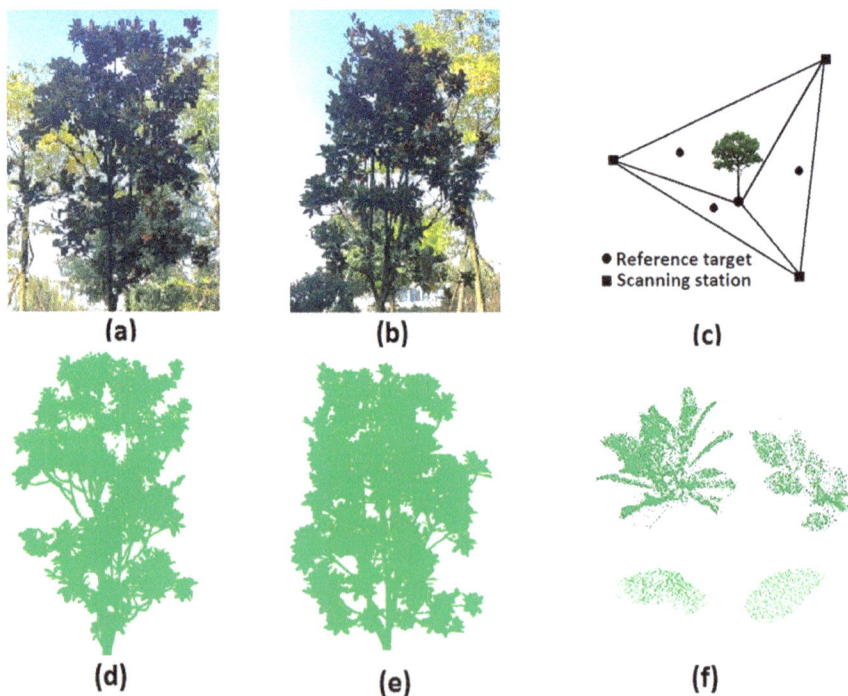

Figure 1. Field measurements. (a) Magnolia A; (b) Magnolia B; (c) The location of scanning stations and reference targets, with the dots representing reference targets that are used to establish the correspondences between different scanning stations (the squares); light detection and ranging LiDAR point clouds of Magnolia A (d); and Magnolia B (e); (f) LiDAR point clouds of leaves.

The tree was scanned from three scan locations, and three reference targets were placed on the ground to establish correspondences between different scanning stations (Figure 1c). For all of the scans, ScanStation C10 was placed on a survey tripod approximately 1.5 m above the ground, and at a distance of about 5–8 m from the tree to ensure that the entire crown was well within the view window. The scans were all performed under very low wind conditions. The high scanning resolution or point spacing for every site was approximately 0.05 m at a distance of 100 m, because this scanning resolution was sufficient to identify a small leaf. More than 600 points exist on one leaf on average, and the mean point spacing (spatial resolution) is approximately 2.5 mm (Figure 1f).

Point clouds from the three scan locations with their individual coordinate systems were registered into one point cloud dataset under a common coordinate system using Leica Cyclone v9.1® and an improved iterative closest point algorithm based on k-dimensional tree [29,30]. The registration error was within 2 mm.

3. Materials and Methods

Figure 2 illustrates the developed LAD estimation workflow, which consists of leaf point cloud extraction and LAD estimation. Details are presented next.

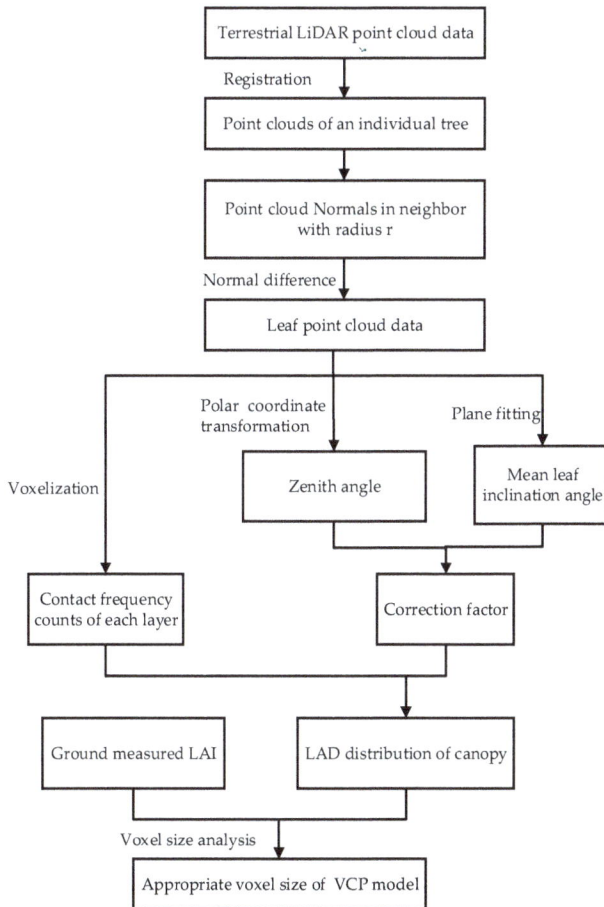

Figure 2. Flowchart of leaf area density (LAD) estimation.

3.1. Extraction of Leaf Point Cloud

In the segmentation of point clouds, a point is partitioned into subregions to extract important features from the cloud data. The segmentation methods can be roughly classified into three categories: edge-detection, region-growth, and hybrid methods. The edge-detection methods detect discontinuities in the surfaces that form the closed boundaries of components in the point data. The region-growth methods detect continuous surfaces that are homogeneous or similar in geometrical properties. The hybrid approaches combine the edge-detection and region-based methods [31].

The normal of the point cloud is a very important characteristic parameter for unorganized 3D point cloud segmentation. If the direction of the two normals is almost identical, the surface structure does not change significantly. If the structure around a center point is significantly different from the other points, the direction of the two estimated normals is likely to vary by a relatively large margin [18]. Since most of the magnolia leaf surfaces are nearly flat, the normal vectors of these point clouds have similar directions. The non-leaf parts are composed from cylindrical segments. The point clouds of the non-leaf components located on the cylindrical surface and the normal vector of these point clouds have different directions (Figure 3). The normal difference of the leaf point cloud is generally smaller than that of the non-leaf component in the neighborhood. The normal difference method was proposed to segment leaf point clouds, which counts the normal difference between each point and the other points in the neighborhood.

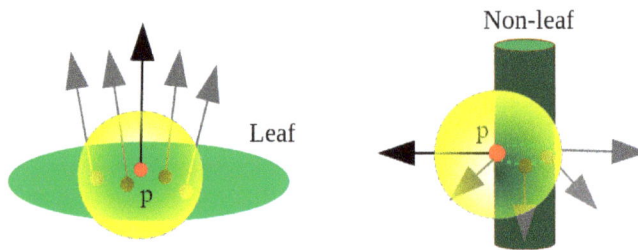

Figure 3. Normals of a point cloud of the leaf and trunk or branch. The red dot is point p, the gray dots are the other points in the neighborhood (yellow sphere). The arrows are the normals of these points.

There are many methods for estimating the normals of point clouds. However, only those using a fixed support radius or a fixed number of neighbors are suitable for point clouds. The normals were estimated by finding the tangent plane and using the principal components of a local neighborhood of fixed support radius around each point [18]. For point p_i in the point cloud P, the normal difference operator in the neighbors are determined by

$$\Delta_{\hat{n}}(p,r) = \frac{1}{N} \sum_{i=1}^{N} (\hat{n}(p) - \hat{n}(p_i)) \tag{1}$$

where $\hat{n}(p)$ is the normal vector of point p in the neighbors, while radius r is the average spatial distance between two leaves. $\hat{n}(p_i)$ is the normal vector of point p_i. $\Delta_{\hat{n}}(p,r)$ is the normal difference operator. The leaf point cloud is determined using the magnitude of $\|\Delta_{\hat{n}}(p,r)\|$ as the threshold. The Otsu algorithm was applied to estimate the threshold [32]. The normal difference results of the point cloud data were viewed as the grey values of images. The appropriate threshold value was calculated through iteration to ensure the maximum variance between the leaf point cloud (foreground) and the non-leaf point cloud (background), as well as the minimal classification error.

Three cubes were chosen to manually evaluate the segmentation accuracy in the upper, middle, and bottom of the canopy, respectively. The non-leaf components were manually deleted. Then, the point number was counted. The leaves' extraction accuracy was calculated from the ratio of the number of segmented leaf points to the number of manually classified leaf points. The overall segmentation accuracy was the average accuracy values of three test cubes.

3.2. Voxel-Based LAD Estimation Method

The voxel-based LAD estimation method mainly includes the voxelization and computation of the contact frequency of the laser beams in each horizontal layer [1].

3.2.1. Voxelization

A bounding box is first constructed to represent the domain of the registered leaf point clouds. The boundaries of voxel coordinates are determined by the minimum and maximum values of X, Y, and Z coordinates from the Cartesian coordinates of the point cloud region. During voxelization, a voxel is defined as a volume element in a 3D array. The range and scan resolution of the LiDAR determine the voxel size, which was set to 2.5 mm in this study. With the voxelization method, the registered leaf point cloud datasets can be grouped into individual voxels [1]. Therefore, voxelization reduces the number of points in a cloud, and improves the computational efficiency of the point cloud contact frequency. Defining the width (w), length (l), and height (h) for each voxel, we grouped the point clouds into $Nl \times Nw \times Nh$ voxels, where $Nl = (Xmax - Xmin)/l$, $Nw = (Ymax - Ymin)/w$, and $Nh = (Zmax - Zmin)/h$ [3]. In addition, the voxel coordinates can be calculated as

$$\begin{cases} i = X_{min} + (int(X - X_{min})/l) \times l \\ j = Y_{min} + (int(Y - Y_{min})/w) \times w \\ k = Z_{min} + (int(Z - Z_{min})/h) \times h \end{cases} \quad (2)$$

where (i, j, k) denote the voxel coordinates in the voxel array. Int is an integer operator. (X, Y, Z) represent the point coordinates of the registered LiDAR point data [1]. The voxel attribute determines the presence of a laser point in the voxel. A voxel with attribute 1 implies that the laser beam is intercepted inside the voxel. A voxel with attribute 0 indicates that there was no interception of the laser beam inside the voxel [1]. The attribute assignment of voxels within a horizontal layer, and a schematic map of the voxel-based model, are shown in Figure 4.

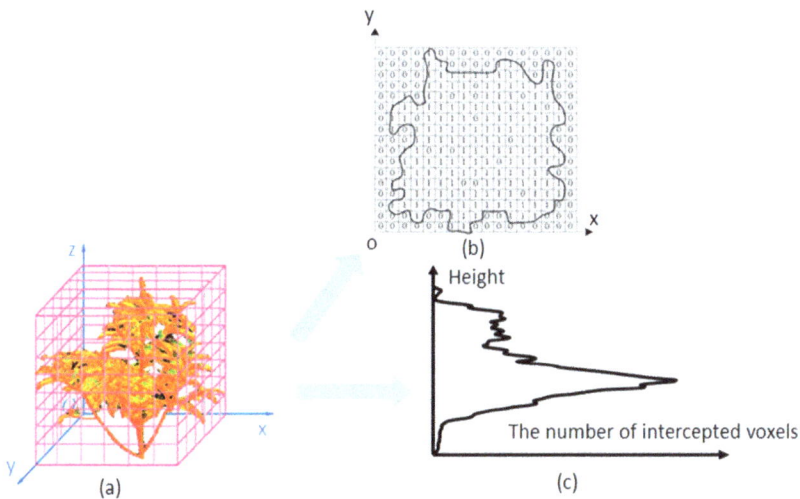

Figure 4. The schematic diagram of a Voxel-based model. (**a**) Illustration of the canopy voxelization; (**b**) the interception (1) and non-interception (0) of the laser beam within a horizontal layer; and (**c**) the vertical distribution of intercepted voxels.

3.2.2. LAD Estimation Model Description

In the LAD computation, a plant region defined as the region above an area covered by a projection of the canopy on the horizontal plane was set in a voxel array. The area covered by the projection of the canopy was produced in the array by projecting all of the voxels with attribute 1 onto the horizontal plane at $k = 0$. Voxels above the area were regarded as voxels within the plant region, and used for the LAD computation. Thus, LAD was calculated in each horizontal layer of the canopy using the voxel-based canopy profiling (VCP) method [1,6]. In particular, LAD(h, ΔH), which is the LAD between heights h and $h + \Delta H$ above the ground, can be calculated as [1,6]

$$\text{LAD}(h, \Delta H) = \alpha(\theta) \frac{1}{\Delta H} \sum_{k=m_h}^{m_h + \Delta H} \frac{n_l(k)}{n_l(k) + n_p(k)} \tag{3}$$

where θ denotes the zenith angle of a laser beam; ΔH represents the horizontal layer thickness (0.5 m in this study); m_h and $m_h + \Delta H$ indicate the voxel coordinates on the vertical axis equivalent to height h and $h + \Delta H$ in orthogonal coordinates ($h = \Delta k \times m_h$); and $n_l(k)$ and $n_P(k)$ denote the numbers of voxels with the attribute values of 1 and 0 in the k-th horizontal layer of the voxel array, respectively. Thus, $(n_l(k) + n_P(k))$ is the total number of incident laser beams that reach the k-th layer. $n_l(k)$ is obtained by counting the number of voxels with attribute 1 in the k-th layer of the voxel-based tree model. $n_P(k)$ is obtained by counting the number of voxels with attribute 0 in the k-th layer. $\alpha(\theta)$ is defined as

$$\alpha(\theta) = \frac{\cos \theta}{G(\theta)} \tag{4}$$

which represents a correction factor that affects the leaf inclination angle at the laser incident zenith angle of θ. $G(\theta)$ stands for the mean projection of a unit leaf area on a plane perpendicular to the direction of the laser beam. $G(\theta)$ determined with the assumption that leaves are azimuthally symmetrical is

$$G(\theta) = \frac{1}{2\pi} \int_0^{2\pi} \int_0^{\pi/2} g(\theta_L) |\cos(\vec{n}_B, \vec{n}_L)| d\theta_L d\varphi_L = \int_0^{\pi/2} g(\theta_L) S(\theta, \theta_L) d\theta_L \tag{5}$$

with

$$S(\theta, \theta_L) = \begin{cases} \cos \theta \cos \theta_L, & \text{for } \theta \leq \frac{\pi}{2} - \theta_L \\ \cos \theta \cos \theta_L \left[1 + \frac{2(\tan x - x)}{\pi} \right], & \text{for } \theta > \frac{\pi}{2} - \theta_L \end{cases} \tag{6}$$

$$x = \cos^{-1}(\cot \theta \cot \theta_L) \tag{7}$$

where θ_L denotes the leaf inclination angle; φ_L is the azimuth angle of the normal to the leaf surface; and $\vec{n}_B = (\sin \theta \cos \theta \cos \varphi, \sin \theta \sin \varphi, \cos \theta)$ and $\vec{n}_L = (\sin \theta_L \cos \theta_L, \sin \theta_L \sin \varphi_L, \cos \theta_L)$ are two unit vectors corresponding to the direction of the laser beam and the direction of the normal to the leaf surface, respectively. To use the field-measured distribution of leaf-inclination angles, one can rewrite Equation (5) as

$$G(\theta) = \sum_{q=1}^{T_q} g(q) S(\theta, \theta_L(q)) \tag{8}$$

where q denotes the leaf inclination angle class, and T_q represents the total number of leaf inclination angle classes. Of each class, the range of the inclination angles is typically the same. Thus, if $T_q = 18$ is the number of leaf inclination angle classes existing from 0° to 90°, each class is 5° in range, or the interval is 5°. $g(q)$ denotes the distribution of the leaf inclination angle for class q, which is the ratio of the leaf area belonging to class q to the total leaf area. $\theta_L(q)$ stands for the midpoint angle of class q, which is the leaf inclination angle used for representing class q. With the eigenvalue method, leaves at different tree heights were randomly selected to fit the leaf planes and estimate the leaf inclination angles [1,6].

Contact frequency is calculated from the values of $n_I(k)$ and $n_P(k)$ in each horizontal thickness layer. Void voxels outside of the canopy exist because of the irregularity of the canopy structure and the 3D voxel model constructed using the maximum and the minimum coordinate values of the point cloud data. The voids should not be regarded as $n_P(k)$, and are not used in the calculation for contact frequency. Thus, the canopy boundary determination and exclusion of invalid elements are important for the contact frequency calculation. Here, the simple and efficient two-dimensional convex hull algorithm, or Graham scan [33], is used to identify the canopy contour range in each horizontal layer. The algorithm can be described as follows. First, in one scan over the list of points, the point with the minimum y-coordinate is found and called p_0. Next, the other points are sorted by their polar angle about the origin p_0. If two points form the same angle with p_0, then the one that is closer to p_0 precedes the other in the ordering. Finally, starting from p_0, the sorted points are sequentially scanned. If these points are on the convex hull polygon, each of the three successive points p_{i-1}, p_i, p_{i+1} should satisfy the following properties: p_{i+1} is located the left side of the vector $<p_{i-1}, p_i>$. If the properties are not met, p_i must not be the apex on the convex hull, and should be deleted [34].

3.3. Validation

The validation method of LAD can be divided into a direct method and an indirect method. A direct method collects leaves hierarchically, and then measures the single leaf area of each layer. The workload of the method is heavy, and the method is destructive. It is almost important (to use the method) if trees are tall and study areas are large. An indirect method measures the LAI of the canopy by using LAI instrument. The sum of each layer's LAD value of the total canopy height of h is called the canopy LAI. The relationship between LAD and LAI can be expressed as follows [35]

$$\text{LAI} = \int_0^h \text{LAD}(z)dz \tag{9}$$

In the assessment of the estimated LAI, the LAI-2200™ instrument was chosen as validation data source provider. The LAI of a tree was measured using a 90° view cap, with the sensor placed near the base of the trunk (Figure 5). The view cap can prevent the sensor from seeing the trunk of the tree. LAI is computed by averaging the LAIs that are measured five times per tree.

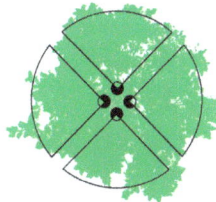

Figure 5. Leaf area index (LAI) measurement of a tree using LAI 2200™ sensor [36]. The sensor is placed near the base of a trunk with a 90° view cap.

3.4. Voxel Size Effects Analysis

Voxel size can influence the layer of detail of the structural information of the extracted canopy, and the computational accuracy of the contact frequency in the VCP model. Thus, voxel size is the key parameter for acquiring the structural information of the vegetation, and influences the LAD estimation accuracy. To understand the effect of the voxel size on the contact frequency, we use seven voxel sizes (2.5 mm, 5 mm, 10 mm, 25 mm, 50 mm, 62.5 mm, and 100 mm). The contact frequencies of each horizontal thickness layer (height interval = 0.5 m) for different voxel sizes will be calculated. The appropriate voxel size is analyzed based on measured LAI data.

4. Results and Discussion

4.1. Extraction of Leaf Point Cloud Data for Two Magnolia Trees

The average spacing of two leaves was 20 mm. Thus, the neighborhood radius was set up as 20 mm. The segmented leaf point clouds from Magnolia A, as shown in Figure 6d, were derived using the normal difference method. The segmentation threshold of 0.5 was chosen from the normal difference of the Otsu algorithm. Compared with the canopy's original point clouds (Figure 6a), the majority of the points were related to leaves. To describe the segmentation results clearly, we showed the point clouds of cube 1 in Figure 6b. Segmented leaf point clouds based on the normal difference method of cube 1 are shown in Figure 6e. All of the leaves were successfully segmented. The obvious erroneous non-leaf parts that were segmented can be deleted manually, and the effect should be minimal. The number of all of the leaf point clouds extracted in test cube 1 (Figure 6e) was 98,042. The extraction accuracy was 86.84% in test cube 1. Similarly, the extraction accuracy levels of test cubes 2 and 3 were 87.22% and 85.54%, respectively. Therefore, the averaged accuracy level for Magnolia A was 86.53%. Close-up views of four leaves are shown in Figure 6c. The detailed segmented leaves' point clouds are shown in Figure 6f. Noise points near the leaf were excluded. It indicated that all of the points located on leaf 1 and leaf 3 were segmented, but a few of the points on leaf 2 and leaf 4 were removed. These points were located on the curved surface of the leaves, and it is difficult to segment the points on a curly leaf since the normal directions of these points were different, and the normal differences of the curly leaf points were similar to the non-leaf components. A visual inspection of the leaves suggested that the shape and edges were kept well, although a few points were incorrectly eliminated.

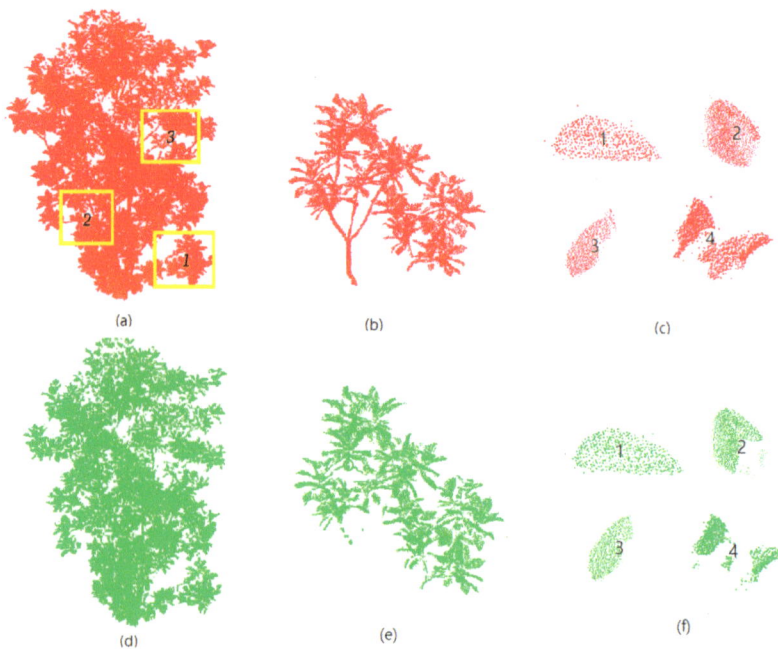

Figure 6. The leaf point clouds extraction of Magnolia A. (**a**) The point clouds of the whole canopy. The yellow rectangle is the tested cube; (**b**) the point clouds of test cube 1; (**c**) the point clouds of leaves; (**d**) the segmented leaves of Magnolia A; (**e**) the segmented leaves of test cube 1; (**f**) the segmented leaves.

Similarly, the leaf point clouds of Magnolia B (Figure 7) were extracted using the same steps and parameter settings (as Magnolia A). The leaves segmentation effects were similar with Magnolia A (Figure 7e,f). The accuracy of three test cubes were 83.23%, 82.81%, and 87.86%, respectively. The averaged accuracy value was 84.63%.

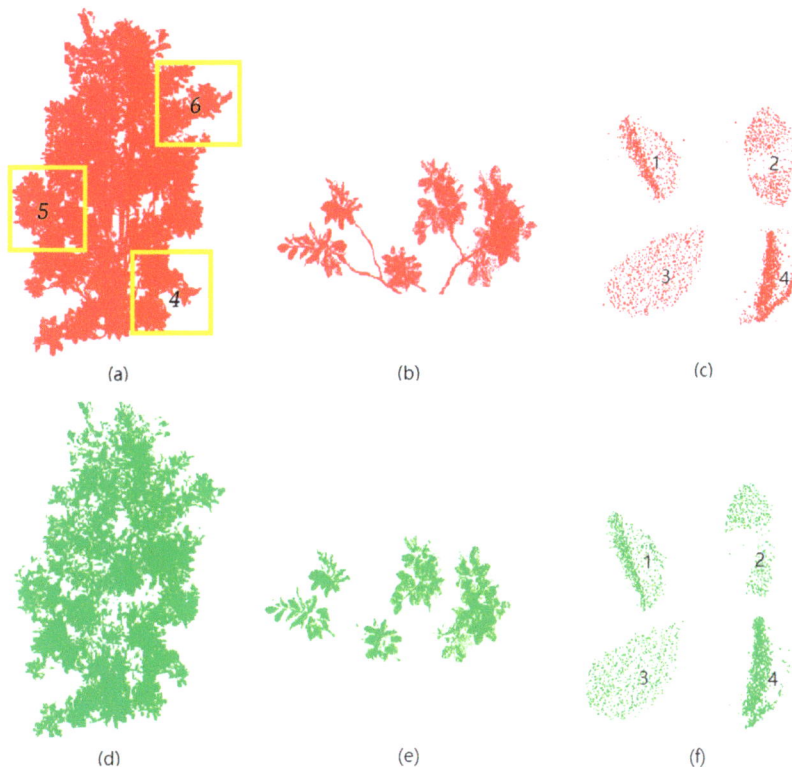

Figure 7. The leaf point clouds extraction of Magnolia B. (**a**) The point clouds of the whole canopy. The yellow rectangle is the tested cube; (**b**) the point clouds of test cube 4; (**c**) the point clouds of leaves; (**d**) the segmented leaves of the tree; (**e**) the segmented leaves of test cube 4; (**f**) the segmented leaves.

4.2. LAD Estimation

4.2.1. LAD Estimation in the Voxel Size of 2.5 mm

The canopy boundary contour in each horizontal layer was determined by the Graham scan algorithm using the location of the intercepted voxels. Figure 8 showed the outline of one horizontal layer that reflects the leaf coverage condition. The large red dots comprise the horizontal thickness of the boundary layer of the tree canopy.

Sixty leaves were randomly selected from each horizontal layer. The distribution probability of the leaf inclination angle was calculated using the probability of the leaf inclination angle at an interval of $10°$, with the range from $0°$ to $90°$. The inclination angle values of Magnolia A ranged between $15°$ and $75°$, with a mean value of $46°$ (Figure 9). The inclination angle values of Magnolia B ranged between $10°$ and $70°$, with a mean value of $37°$ (Figure 10). The mean zenith angle, correction coefficient, and contact frequency in each horizontal layer were calculated. The correction coefficients were mainly determined by mean zenith angle, and were near 1.10. The contact frequency distribution agreed with the leaves distribution. The maximum contact frequency of Magnolia A was 0.19, which occurred in layer 2. For Magnolia B, the maximum contact frequency was 0.17, which occurred in layer 5.

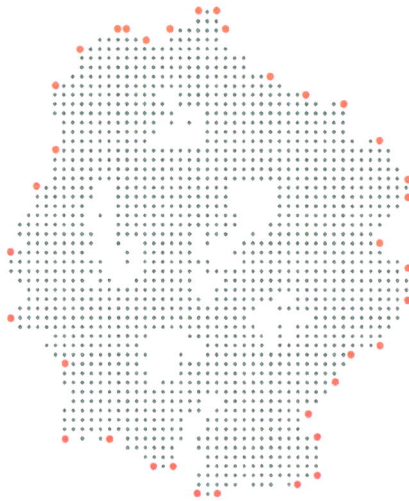

Figure 8. Outline of the horizontal layer. Each small black point represents the interception of the laser beam at a voxel location. A large red dot denotes the convex hull polygon vertex of the point sets, that is, the boundary layer of the tree canopy vertices.

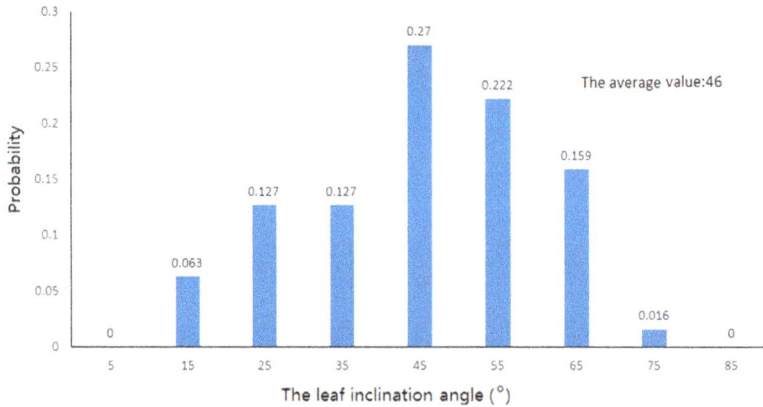

Figure 9. Probability distribution of the leaf inclination angle of Magnolia A.

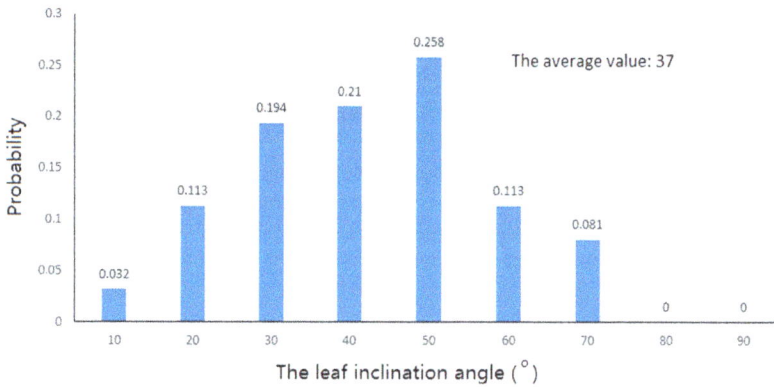

Figure 10. Probability distribution of the leaf inclination angle of Magnolia B.

The vertical distribution of the LAD (Figure 11) was consistent with the vertical leaf distribution of the magnolia canopy (Figure 1). In each horizontal layer, the higher the leaf density, the larger the LAD. The maximum LAD of Magnolia A happened at 1.0 m of the canopy. Then, the LAD was positively related to height until 3.0 m. In the middle to higher layers of the canopy, the LAD of Magnolia A was inversely related to height. The LAD of Magnolia B at 1.0 m is much higher than 1.5 m. The LAD of Magnolia B reached the peak at 2.5 m, and then decreased gradually as the height continuously increased.

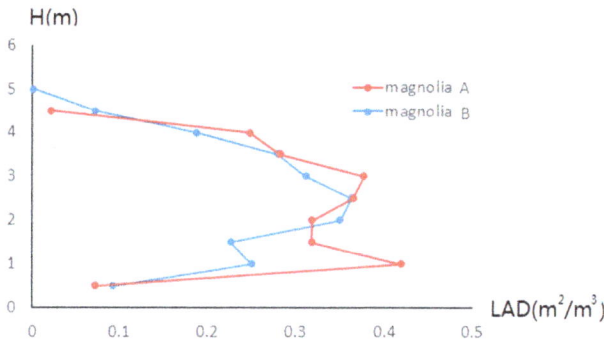

Figure 11. Leaf area density (LAD) profile of the individual broadleaf trees.

The cumulative LAD (LAI) was validated by the ground measured LAI (Table 2). The level of accuracy was 90.7% or higher.

Table 2. The LAI accuracy.

	Magnolia A	Magnolia B
Estimated LAI	1.21	1.07
LAI2200 Measured LAI	1.20	1.18
Accuracy	99.9%	90.7%

4.2.2. Voxel Size Effects

Variations in the contact frequency of different horizontal thickness layers for different voxel sizes are shown in Figure 12 in alphabetical order. It was clear that the contact frequency increased with an increase in the voxel size. They were highly correlated logarithmically. As contact frequency increased, the coefficient of determination decreased. The maximum coefficient of determination (0.99) had layer 9, where the LAD was at its lowest. In contrast, the minimum coefficient of determination (0.89) had layer 2, where the LAD was at its highest. This implied that the estimated contact frequency of a single horizontal thickness layer was very sensitive to the voxel size. A small voxel could express the internal structure of a canopy more carefully, and intercept a laser canopy. Thus, the result could be more accurate (than that derived from a large voxel). However, noise points in the point cloud data could inevitably get involved in the calculation, resulting in unreasonable calculated contact frequencies. In contrast, a large voxel size suppressed the internal structure of the canopy. Given the same thickness of the horizontal layer, the estimated LAD varied at different voxel sizes. Therefore, the expression of the canopy structure and the calculation of the contact frequency by using the appropriate voxel size were very important to the retrieval of reasonable LAD.

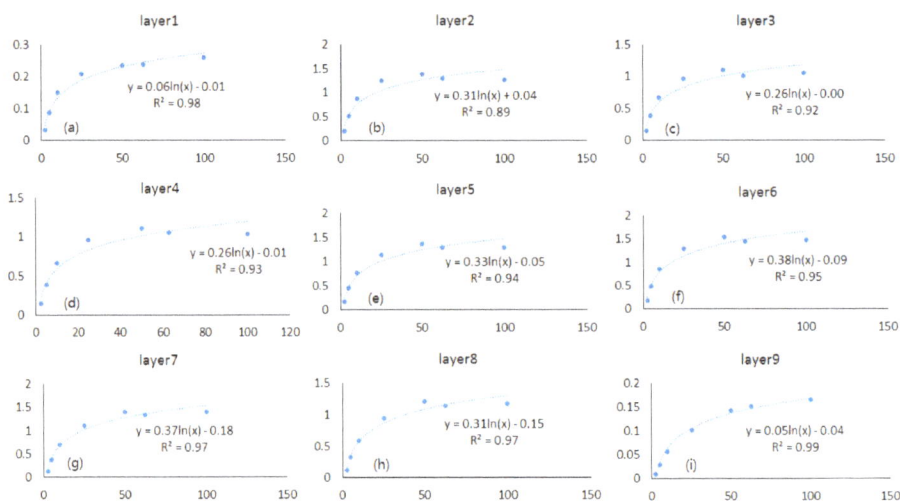

Figure 12. Variation for the contact frequency with the voxel size in different horizontal thickness layers of Magnolia A. (**a**–**i**) indicate layers 1 (the lowest) to 9 (the highest), respectively.

The estimated cumulative LAD (or LAI) of all of the horizontal layers was also positively related to the voxel size (Figure 13). Meanwhile, the estimation efficiency of the VCP model was also positively related to the voxel size. If the voxel size increased two times, the total voxel amount would decrease eight times. The voxels quantity of Magnolia A produced by the VCP model is 2.1 billion when the voxel size is 2.5 mm; thus, it will take too much time to process this model. When voxel size increased to 50 mm, the voxels quantity of Magnolia A produced by the VCP model would decrease to 0.26 million. So, in order to improve the efficiency of the VCP model calculation, increasing the voxel size is necessary. The high correlation between the ratio contact frequency, LAI and voxel size, give the potential of LAD estimation using big voxel size. Improvements to the estimation efficiency of the VCP model through using a big voxel size, which has less voxels, can be further studied in the future. The LAI increased continuously from 1.07 m^2/m^2 to 10.74 m^2/m^2 when the voxel size increased from 2.5 mm to 100 mm (Figure 13). There is a high correlation between LAI and voxel size, and the coefficient of determination reached 0.95 and 0.97. Thus, a large error in the LAI estimation based on the VCP model could occur if the appropriate voxel size was not determined properly. In this analysis, the LAI varied in one order of magnitude. Compared with the ground-measured LAI, the appropriate voxel size of 2.5 mm, which is the spatial resolution of the point clouds in this study, should be chosen. Thus, the spatial resolution of the point cloud data should be the key factor in determining the voxel size.

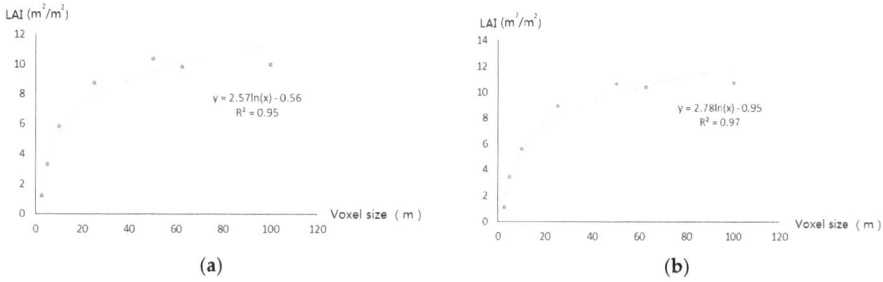

(a)   (b)

Figure 13. LAI of magnolia trees for different voxel sizes. (**a**) Magnolia A; (**b**) Magnolia B.

5. Conclusions

To estimate the LAD of a tree canopy in a timely fashion, we developed an algorithm based on the LiDAR point cloud segmentation algorithm coupled with the voxel-based model. The proposed normal difference method was used to remove non-photosynthetic components from the photosynthetic components in the data. The normal difference method was proposed to calculate leaf point cloud segmentation, because the normal vector difference of the leaf point cloud is different to the non-leaf components in the neighborhood. From the three chosen test cubes, the extraction accuracy was 86.53% and 84.63% for Magnolia A and Magnolia B, respectively. This shows that the normal difference method has big potential for leaf point cloud segmentation in magnolia canopies, because this method mainly considers the leaf structure, and is non-destructive.

The results also suggested that the LAD/LAI estimated by the VCP model were highly sensitive to the voxel size. The estimated LAD/LAI would increase with an increase in voxel size. When the voxel size was larger than 10 times the mean points spacing, the LAD/LAI remained a constant value. Thus, an appropriate voxel size should be identified for the VCP model with the consideration of the density of LiDAR points.

The canopy LAD was estimated by computing the contact frequency for each thickness layer, and the leaf inclination correction factor. The individual magnolia tree LAD and the vertical distribution of the leaf point cloud exhibited an overall agreement when the voxel size was 2.5 mm and the horizontal layer thickness was 0.5 m. The cumulated LAD (LAI) had little difference with the ground measurement LAI when the voxel size was 2.5 mm. Thus, the LAD distribution of individual magnolia trees could be retrieved accurately using terrestrial LiDAR data, which could overcome the limitations of field measurements obtained using traditional methods.

Acknowledgments: This work was supported by the National Natural Science Foundation of China (41471294), China's Special Funds for Major State Basic Research Project (2013CB733405), and Fundamental Research Funds for the Central Universities (ZYGX2015J112).

Author Contributions: Shihua Li, Leiyu Dai and Hongshu Wang conceived and conducted the study, Leiyu Dai acquired and processed the LiDAR data, Shihua Li and Yong Wang wrote the paper, Ze He and Sen Lin acquired and processed the validation data.

Conflicts of Interest: The authors declare no conflict of interest.

References

1. Hosoi, F.; Omasa, K. Voxel-based 3-D modeling of individual trees for estimating leaf area density using high-resolution portable scanning LiDAR. *IEEE Trans. Geosci. Remote Sens.* **2006**, *44*, 3610–3618. [CrossRef]
2. Béland, M.; Baldocchi, D.D.; Widlowski, J.L.; Fournier, R.A.; Verstraete, M.M. On seeing the wood from the leaves and the role of voxel size in determining leaf area distribution of forests with terrestrial LiDAR. *Agric. For. Meteorol.* **2014**, *184*, 82–97. [CrossRef]
3. Zheng, G.; Moskal, L.M. Computational-geometry-based retrieval of effective leaf area index using terrestrial laser scanning. *IEEE Trans. Geosci. Remote Sens.* **2012**, *50*, 3958–3969. [CrossRef]
4. Huang, P.; Pretzsch, H. Using terrestrial laser scanner for estimating leaf areas of individual trees in a conifer forest. *Trees* **2010**, *24*, 609–619. [CrossRef]
5. Weiss, M.; Baret, F.; Smith, G.J.; Jonckheere, I.; Coppin, P. Review of methods for in situ leaf area index (LAI) determination: Part II. Estimation of LAI, errors and sampling. *Agric. For. Meteorol.* **2004**, *121*, 37–53. [CrossRef]
6. Hosoi, F.; Omasa, K. Factors contributing to accuracy in the estimation of the woody canopy leaf area density profile using 3D portable LiDAR imaging. *J. Exp. Bot.* **2007**, *58*, 3463–3473. [CrossRef] [PubMed]
7. Wilson, J.W. Estimation of foliage denseness and foliage angle by inclined point quadrats. *Aust. J. Bot.* **1963**, *11*, 95–105. [CrossRef]
8. Monsi, M.; Saeki, T. On the factor light in plant communities and its importance for matter production. *Ann. Bot.* **2005**, *95*, 549–567. [CrossRef] [PubMed]
9. Qin, Y.; Vu, T.T.; Ban, Y. Toward an optimal algorithm for LiDAR waveform decomposition. *IEEE Geosci. Remote Sens. Lett.* **2012**, *9*, 482–486. [CrossRef]
10. Qin, Y.; Li, S.; Vu, T.T.; Niu, Z.; Ban, Y. Synergistic application of geometric and radiometric features of LiDAR data for urban land cover mapping. *Opt. Express* **2015**, *23*, 13761–13775. [CrossRef] [PubMed]
11. Hosoi, F.; Nakai, Y.; Omasa, K. 3-D voxel-based solid modeling of a broad-leaved tree for accurate volume estimation using portable scanning LiDAR. *ISPRS J. Photogramm. Remote Sens.* **2013**, *82*, 41–48. [CrossRef]
12. Li, W.; Niu, Z.; Li, J.; Chen, H.; Gao, S.; Wu, M.; Li, D. Generating pseudo large footprint waveforms from small footprint full-waveform airborne LiDAR data for the layered retrieval of LAI in orchards. *Opt. Express* **2016**, *24*, 10142–10156. [CrossRef] [PubMed]
13. Béland, M.; Widlowski, J.L.; Fournier, R.A. A model for deriving voxel-level tree leaf area density estimates from ground-based LiDAR. *Environ. Model. Softw.* **2014**, *51*, 184–189. [CrossRef]
14. Ma, L.; Zheng, G.; Eitel, J.U.H.; Moskal, L.M.; He, W.; Huang, H. Improved salient feature-based approach for automatically separating photosynthetic and nonphotosynthetic components within terrestrial LiDAR point cloud data of forest canopies. *IEEE Trans. Geosci. Remote Sens.* **2016**, *54*, 679–696. [CrossRef]
15. Béland, M.; Widlowski, J.L.; Fournier, R.A.; Côté, J.F.; Verstraete, M.M. Estimating leaf area distribution in savanna trees from terrestrial LiDAR measurements. *Agric. For. Meteorol.* **2011**, *151*, 1252–1266. [CrossRef]
16. Tao, S.; Guo, Q.; Su, Y.; Xu, S.; Li, Y.; Wu, F. A geometric method for wood-leaf separation using terrestrial and simulated LiDAR data. *Photogramm. Eng. Remote Sen.* **2015**, *81*, 767–776. [CrossRef]
17. Wang, H.; Li, S.; Guo, J.; Liang, Z. Retrieval of the leaf area density of Magnolia woody canopy with terrestrial Laser-scanning data. *J. Remote Sens.* **2016**, *20*, 570–578. [CrossRef]
18. Ioannou, Y.; Taati, B.; Harrap, R.; Greenspan, M. Difference of normals as a multi-scale operator in unorganized point clouds. In Proceedings of the IEEE Second International Conference on 3D Imaging, Modeling, Processing, Visualization and Transmission, Zurich, Switzerland, 13–15 October 2012.
19. Awrangjeb, M.; Fraser, C.S. Automatic segmentation of raw LiDAR data for extraction of building roofs. *Remote Sens.* **2014**, *6*, 3716–3751. [CrossRef]
20. Li, W.; Guo, Q.; Jakubowski, M.K.; Kelly, M. A new method for segmenting individual trees from the LiDAR point cloud. *Eng. Remote Sens.* **2012**, *78*, 75–84. [CrossRef]
21. Li, Z.; Zhang, L.; Tong, X.; Du, B.; Wang, Y.; Zhang, L.; Zhang, Z.; Liu, H.; Mei, J.; Xing, X.; et al. A three-step approach for TLS point cloud classification. *IEEE Trans. Geosci. Remote Sens.* **2016**, *54*, 5412–5424. [CrossRef]
22. Radtke, P.J.; Bolstad, P.V. Laser point-quadrat sampling for estimating foliage-height profiles in broad-leaved forests. *Can. J. For. Res.* **2001**, *31*, 410–418. [CrossRef]
23. Parker, G.G.; Harding, D.J.; Berger, M.L. A portable LIDAR system for rapid determination of forest canopy structure. *J. Appl. Ecol.* **2004**, *41*, 755–767. [CrossRef]

Remote Sens. **2017**, *9*, 1202

24. Sumida, A.; Nakai, T.; Yamada, M.; Ono, K.; Uemura, S.; Hara, T. Ground-based estimation of leaf area index and vertical distribution of leaf area density in a Betula ermanii forest. *Silva Fenn.* **2009**, *43*, 799–816. [CrossRef]

25. Grau, E.; Durrieu, S.; Fournier, R.; Gastellu-Etchegorry, J.P.; Yin, T. Estimation of 3D vegetation density with Terrestrial Laser Scanning data using voxels. A sensitivity analysis of influencing parameters. *Remote Sens. Environ.* **2017**, *191*, 373–388. [CrossRef]

26. Hosoi, F.; Nakai, Y.; Omasa, K. Estimation and error analysis of woody canopy leaf area density profiles using 3-D airborne and ground-based scanning LiDAR remote-sensing techniques. *IEEE Trans. Geosci. Remote Sens.* **2010**, *48*, 2215–2223. [CrossRef]

27. Van der Zande, D.; Stuckens, J.; Verstraeten, W.W.; Mereu, S.; Muys, B.; Coppin, P. 3D modeling of light interception in heterogeneous forest canopies using ground-based LiDAR data. *Int. J. Appl. Earth Obs. Geoinf.* **2011**, *13*, 792–800. [CrossRef]

28. Iio, A.; Kakubari, Y.; Mizunaga, H. A three-dimensional light transfer model based on the vertical point-quadrant method and Monte-Carlo simulation in a Fagus crenata forest canopy on Mount Naeba in Japan. *Agric. For. Meteorol.* **2011**, *151*, 461–479. [CrossRef]

29. Leica Cyclone 3D Point Cloud Processing Software. Available online: http://hds.leica-geosystems.com/en/Leica-Cyclone_6515.htm (accessed on 16 September 2017).

30. Li, S.; Wang, J.; Liang, Z.; Su, L. Tree point cloud registration using an improved ICP algorithm based on KD-tree. In Proceedings of the IGARSS 2016—2016 IEEE International Geoscience and Remote Sensing Symposium, Beijing, China, 10–15 July 2016.

31. Woo, H.; Kang, E.; Wang, S.; Kelly, M. A new segmentation method for point cloud data. *Int. J. Mach. Tools Manuf.* **2002**, *42*, 167–178. [CrossRef]

32. Yao, C.; Chen, H. Automated retinal blood vessels segmentation based on simplified PCNN and fast 2D-Otsu algorithm. *J. Cent. South Univ. Technol.* **2009**, *16*, 640–646. [CrossRef]

33. Graham, R.L. An efficient algorith for determining the convex hull of a finite planar set. *Inf. Process. Lett.* **1972**, *1*, 132–133. [CrossRef]

34. Alsuwaiyel, M.H. *Algorithms: Design Techniques and Analysis*; World Scientific Publishing Company: Singapore, 1998; pp. 471–474.

35. Zhao, J.; Li, J.; Liu, Q. Review of forest vertical structure parameter inversion based on remote sensing technology. *J. Remote Sens.* **2013**, *17*, 697–716. [CrossRef]

36. LI-COR. *LAI-2200 Plant Canopy Analyzer Instruction Manual*; LI-COR: Lincoln, NE, USA, 2010.

remote sensing

MDPI

Article

Detecting Forest Disturbance in Northeast China from GLASS LAI Time Series Data Using a Dynamic Model

Jian Wang [1,2], Jindi Wang [1,2,*], Hongmin Zhou [1,2] and Zhiqiang Xiao [1,2]

[1] State Key Laboratory of Remote Sensing Science, Jointly Sponsored by Beijing Normal University and Institute of Remote Sensing and Digital Earth of Chinese Academy of Sciences, Beijing 100875, China; leonw63@mail.bnu.edu.cn (J.W.); zhouhm@bnu.edu.cn (H.Z.); zhqxiao@bnu.edu.cn (Z.X.)

[2] Beijing Engineering Research Center for Global Land Remote Sensing Products, Institute of Remote Sensing Science and Engineering, Faculty of Geographical Science, Beijing Normal University, Beijing 100875, China

* Correspondence: wangjd@bnu.edu.cn; Tel.: +86-10-5880-9966; Fax: +86-10-5880-5274

Received: 27 October 2017; Accepted: 9 December 2017; Published: 12 December 2017

Abstract: Large-scale forest disturbance often leads to changes in forest cover and structure, which imposes a great uncertainty in the estimation of the forest carbon cycle and biomass and affects other applications. In northeastern China, the Daxinganling region has abundant forest resources, where the forest coverage is about 30%. The Global LAnd Surface Satellite (GLASS) leaf area index (LAI) time series data provide important information to monitor the possible change of forests. In this study, we developed a new method to detect forest disturbances using GLASS LAI data over the Daxinganling region of Northeast China. As a dynamic model, the season-trend model has a higher sensitivity toward a seasonal change in LAI. Based on the accumulation of multi-year GLASS LAI products from 1997 to 2002, the dynamic model of LAI time series for each pixel is established first. The time-stepping modeling (TSM) process was designed by using the season-trend method, and sequential tests for detecting disturbances from a time series of pixels. Significant changes in the model parameters were captured as disturbance signals. Then, the near-infrared and shortwave-infrared bands of Moderate Resolution Imaging Spectroradiometer (MODIS) surface reflectance are used as auxiliary information to distinguish the types of forest disturbances. Here, the algorithm led to the detection of two different types of disturbances: fire and other (e.g., insect, drought, deforestation). In this study, we took the forest region as the study area, used the 8-day composite GLASS LAI data at 1000-m spatial resolution to identify each pixel as a fire disturbance, other disturbance, or non-disturbance. Validation was performed using reference burned area data derived from Landsat 30 m imagery. Results were also compared with the MCD64 product. The validation results were based on confusion matrices showing the overall accuracy (OA) exceeded 92% for our method and the MCD64 product. Statistical tests identified that TSM's product accuracy is higher than that of MCD64. This study demonstrated that the TSM algorithm using a season-trend model provides a simple and automated approach to identify and map forest disturbance.

Keywords: GLASS LAI time series; forest disturbance; disturbance index

1. Introduction

Forest disturbances are discrete events that cause tree mortality and destruction of plant biomass. An effective method to detect the temporal and spatial distribution of forest disturbance in a large area is to use remote sensing time series data. In many ecological models, accurate monitoring of forest disturbances has a crucial role in Gross Primary Production (GPP) and Net Primary Production (NPP) estimation accuracy and aerosol and biomass estimation [1–10]. The mapping of precise forest disturbances in large areas provides basic data for wildlife protection and strong support for large-scale vegetation biophysical and structural change monitoring [11]. Based on the needs of various applications, the use of remote sensing data for detecting forest disturbances in a large area has been carried out widely.

Many efficient methods have been proposed to detect changes from image time series. The methods include detecting forest disturbance and recovery [3,12,13], detecting trend and seasonal changes [14–16], and extracting seasonality metrics from satellite time series [17,18]. Current remote sensing approaches in monitoring forest disturbance detection are mainly based on vegetation indices (VI) and other vegetation parameters [19–24], such as the normalized difference vegetation index (NDVI) and Fraction of Absorbed Photosynthetically Active Radiation (FAPAR) [11], to determine the disturbance by analyzing the changes in a long time series. VI is a common indicator of vegetation disturbance monitoring. Vegetation is relatively fragile in terrestrial ecosystems, and it is also sensitive to the occurrence of disturbances. It is possible to detect the disturbance events effectively by observing the change of vegetation using remote sensing data. NDVI is the most widely used VI; it can show the growth curve of vegetation over time. It is suitable for medium and long-term vegetation growth monitoring, phenophase monitoring, and crop yield estimation. The hotspot and NDVI differencing synergy (HANDS) [19] algorithm was applied to Canadian forests during the 1995–2000 fire seasons using the annual hotspot masks and differencing of the anniversary date of September NDVI composites. HANDS is designed to produce annual maps of burned forests by combining the active fire detection product with NDVI differencing, a common change detection technique. The HANDS method substantially reduces noise levels by requiring that burned areas identified through NDVI differencing be co-located with hotspots. In 2009, Mildrexler used the Moderate Resolution Imaging Spectroradiometer (MODIS) global disturbance index (MGDI) [25] with satellite data to operationally detect large-scale ecosystem disturbances at 1-km resolution. The MGDI algorithm was designed to contrast annual changes in vegetation density and land surface temperature (LST) following disturbance by enhancing the signal to effectively detect the location and intensity of disturbances.

Disturbances from agents such as fire, insect damage, or strong winds are common throughout the world's forests. Fire is the dominant driver of forest disturbance at the global scale, and fire is one of the most dominant disturbance agents in Northeast China. Currently, the most widely used burned area products are three MODIS data products and three products developed within the fire disturbance project (fire_cci) [26–31]. There is a significant difference between the detection results of different products; through a comparison of a variety of products, statistical tests identified that MCD64 was the most accurate, followed by MCD45 [32]. Biomass density varies through time as a result of disturbances; accurate estimates of biomass emissions focuses on identifying disturbances (whether anthropogenic or natural), but it is difficult to meet the application requirements with most of the forest disturbance products.

In this study, we used The Global LAnd Surface Satellite (GLASS) leaf area index (LAI) time series data to detect forest disturbances in Northeast China. We developed a method for disturbance detection in forest land. Based on the season-trend model [15,16], the GLASS LAI data from 1997 to 2002 was taken as the background dataset to model phenological changes of vegetation in forest land, which represents the annual periodic variation curve of LAI without disturbance. The model parameters were used as the reference parameters for identifying disturbance pixels. Every 8 days, GLASS LAI data acquired after 2002 was iteratively added to the background dataset to model the phenological change step by step. The model parameters could be different in each modeling step and were taken as a signal of whether there was a disturbance. These procedures were iterated until the end of a time series or until a change was detected. We call this algorithm the time-stepping modeling (TSM) process. During the monitoring period, the disturbance was detected by the difference between the criteria parameters and the new parameters which were re-simulated by adding a new LAI. The forest disturbance and the natural growth were distinguished by the disturbance index (DI), which is defined in this paper, and the forest disturbance was automatically identified. Finally, the normalized burn ratio (NBR) was used to reclassify the disturbance of vegetation to determine whether it was a burned area. In order to verify the accuracy of the approach, the types and spatial ranges of forest disturbance were detected in a forest region of Northeast China. The detected result was compared with the burned area of MCD64 and the burned area of Landsat, which were calculated using the differenced NBR (dNBR) method. The results demonstrated the validity of our method through experiments in the Daxinganling Mountains, including the effective detection of disturbances based on the TSM process using a dynamic model, and the discrimination of types of disturbances.

2. Materials

2.1. Study Area

We selected the northeast forest area of China as our study area. The region is characterized by a continental monsoon climate. The forests in the region are horizontally divided into four regions of vegetation: the cool temperate deciduous coniferous forest region, the temperate mixed evergreen coniferous-deciduous broad-leaved forest region, the warm temperate deciduous broad-leaved forest region, and the temperate steppe region. Forests of these types account for about 30% of forest in China. Northeastern China has abundant tree species and a variety of forest types, including evergreen needleleaf forest, deciduous needleleaf forest, deciduous broadleaf forest, and mixed forests. Figure 1 shows the location of the study area and the land classification results of 30 m resolution (Figure 1A) and 1000 m resolution (Figure 1B). In subsequent steps, we use 1000 m land classification results.

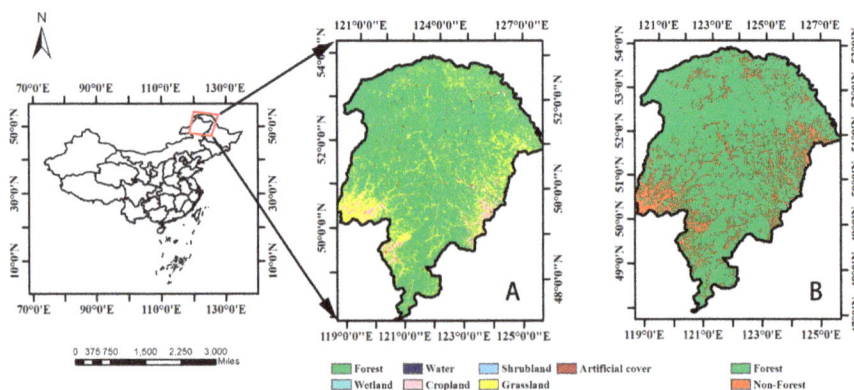

Figure 1. Map showing the location of the study area. (**A**) Land cover map of study area in 2003 produced by Landsat images of 30 m resolution; (**B**) Land cover mapping with 1000 m resolution.

2.2. Data Processing

The remote sensed data we used in this study include the GLASS LAI product from 1997 to 2003, the "MOD09A1" 500-m MODIS atmospherically-corrected Level 3 8-day surface reflectance products, and the Level 3 MODIS "MCD64A1" monthly burned area products. The MOD09A1 and MCD64A1 data were reprojected from the MODIS sinusoidal projection into the Albers equal area projection at 1000 m resolution. In order to test the performance of the forest disturbance detection approach, Landsat TM images of 2003 were used to create high resolution burned area maps. The maps were taken as a reference to assess our disturbance detection results at 1 km spatial resolution.

The GLASS LAI product was retrieved from time series MODIS and Advanced very-high-resolution radiometer (AVHRR) surface reflectance data using general regression neural networks (GRNNs), and the inter-comparison of GLASS LAI products and other existing operational global LAI products, including MODIS and CYCLOPES, indicate that the GLASS LAI product is the most spatially complete and temporally continuous [33].

The mean phenological change of the LAI was computed for 5 years (1997–2001) on a pixel-by-pixel basis. The mean LAI reflects the undisturbed plant growth phenology and provides a background to assess a departure from the LAI variability. The vegetation in the northeast forest area was dominated by deciduous forests; therefore, LAI at the beginning and end of the year were at a very low level. In order to improve the simulation of the LAI model accuracy by excluding non-growth period data, the input parameters were taken from the 96th day to the 306th day of each year.

The study focuses on forest land. We used land classification data to mask non-forest areas. Land use classification is the result of supervised classification using Landsat data to derive seven categories: forestland, artificial cover, water, shrubland, wetland, cropland, grassland, as shown in Figure 1A. We resampled the spatial resolution of the classifications from 30 m to 1000 m using a mode resampling method, and the land use was divided into two categories: forest land area and non-forest land. The classifications data was reprojected from the UTM projection into the Albers equal area projection. Land use classification results are shown in Figure 1B. The remote sensing data used in the present paper are in Table 1.

Table 1. Remote sensing data parameter list. GLASS = Global LAnd Surface Satellite.

Data	Resolution (m)	Temporal Resolution (day)	Date
GLASS	1000	8	1 January 1997–31 December 2003
MCD64	500	Monthly	1 January 2003–31 December 2003
MOD09	500	8	1 January 2001–31 December 2003
Classification	30	Year	2003
Landsat	30	16	10 April 2003 23 April 2003 26 May 2003 13 June 2003

3. Methods

LAI as an indicator of vegetation coverage and vegetation growth status can directly reflect the growth status of vegetation. In the normal growth period of forests, LAI has a similar annual periodic variation. The annual LAI curve also has good continuity, so there is no obvious fluctuations in the curve. When a forest disturbance occurs, such as drought, pest, fire, deforestation, etc., these effects directly affect the LAI, which will be reduced significantly. Through analysis of a long time series of forest LAI data from remote sensing images using our developed dynamic model and the TSM process, we can detect the location and range of the forest disturbances effectively. Figure 2 shows the flow chart of the approach.

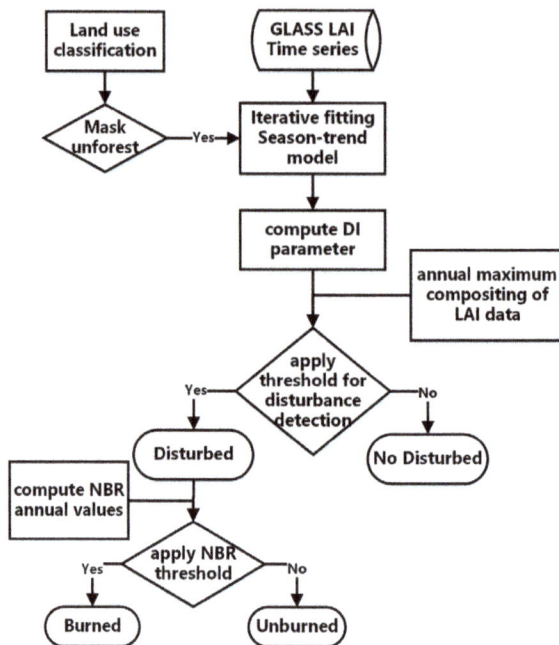

Figure 2. Flow chart of the forest disturbance detection. LAI = leaf area index; DI = disturbance index; NBR = normalized burn ratio.

3.1. Iterative Algorithm to Detect Forest Disturbance

The season-trend method is a linear regression model proposed by Verbesselt to account for seasonal and trend changes typically occurring within climate-driven biophysical indicators derived from satellite data [15,16]. The model is an additive decomposition model, which is based on the characteristics of the periodic variation of remote sensing data from the vegetation-covered surface. It is assumed that the time series is mainly composed of seasons and trends; thus, the modeling results are obtained using the trends and seasons in the model sequence to be decomposed, as follows

$$Y_t = a_1 + a_2 t + \sum_{j=1}^{k} r_j \sin\left(\frac{2\pi jt}{f} + \sigma_j\right) + \varepsilon_t \tag{1}$$

where Y_t is the time series observation at time t, the intercept α_1, slope α_2, amplitudes γ_j, and phases δ_j are the unknown simulation parameters, k is the number of harmonic terms that should be specified manually, f is the frequency of the time series observations, and ε_t is the error term at time t.

We used three harmonic terms to robustly simulate phenological changes within GLASS LAI time series [34,35]. The formula for LAI time series simulation can be expressed as follows

$$Y(t) = a + b * t + c * \sin\left(\frac{2\pi t}{f} + \sigma_1\right) + e * \sin\left(\frac{4\pi t}{f} + \sigma_2\right) + g * \sin\left(\frac{6\pi t}{f} + \sigma_3\right) \tag{2}$$

In a large area of forest disturbance detection, the idea for monitoring techniques is simple. In this paper, the estimation of parameters was performed by iterating through the following steps until reaching the last value of the monitoring period or until a disturbance is detected:

Step 1: Based on a given LAI time series (there are two parts of data: the mean phenological change of 1997–2001 and the year 2002, thus we call it background data), the Equation (2) was used to simulate the LAI time series, the initial simulation parameters (i.e., a and c in Equation (2)) were used as the reference parameters for detection.

Step 2: The iterative procedure began with a simulation of Y (2003) of Equation (2) by using the season-trend method. The modeling data was a combination of background data and monitoring data. The first LAI data in 2003 was the starting point of the monitoring data for detection, and the parameters of the fitting model were calculated.

Step 3: The size of the window was the same as the background data, the moving width of the window for monitoring data was equal to 1. When we add monitor LAI data to the end of the background data, the first of the background data is removed. The fitting model parameters obtained after adding the monitoring data were compared with the reference parameters to determine whether any change was detected.

The mean LAI can represent the normal phenological pattern of forests. The LAI of the prior year shows whether forest growth has changed. Through the simulation of background data, the growth structure of the forest can be determined. It can be considered, as the phenology of the previous year occupied a higher weight in the simulation process. These time-stepping modeling processes were iterated until reaching the last value of the monitoring period or until disturbances are detected. Through analysis of the model parameters, the amplitude parameter was found to be the most sensitive to structural changes in the time series, as shown in Figure 3. For the detection of forest disturbances, we defined the DI_c and DI_a using the following formulas

$$\begin{cases} DI_c = \frac{c_n - c_0}{c_0} < -0.1(c_0 \neq 0) \\ DI_a = a_n - a_0 > 0.15 \end{cases} \tag{3}$$

where c_0 (a_0) is the value of the parameter c in the initial simulation and c_n (a_n) is the value of the parameter c (a) on the n $(1 < n < 27)$ time series used in the monitoring period.

Using a set of simulated LAI data of three years, Figure 3 shows an example of this analysis; four subsets were established within varying degrees of growth. We computed the DI values at each single step length during the disturbance in 2003. Figure 3A illustrates the change in DI in the case of forest growth. With the increase of fitting data, the value of DI_c gradually increases, and the value of DI_a decreases gradually. Figure 3B illustrates the change in DI value without disturbance. The time series of DI basically did not fluctuate, hovering around zero. Figure 3C illustrates the change in DI value with a certain degree of disturbance. As the disturbance continues, the value of DI_a gradually increases, and the value of DI_c decreases gradually. Figure 3D illustrates that with the increase in the degree of disturbance, the value of DI_c appeared to have a greater degree of reduction, while the value of DI_a appeared to increase to a much greater degree. Based on the analysis of the disturbed area, we selected $DI_c < -0.1$ and $DI_a > 0.15$ as our thresholds for detecting those disturbances. In the iterative modeling, when the amplitude of the two parameter changes do not exceed the threshold, forest disturbances will be identified as not occurring. Otherwise, forest disturbances will be detected.

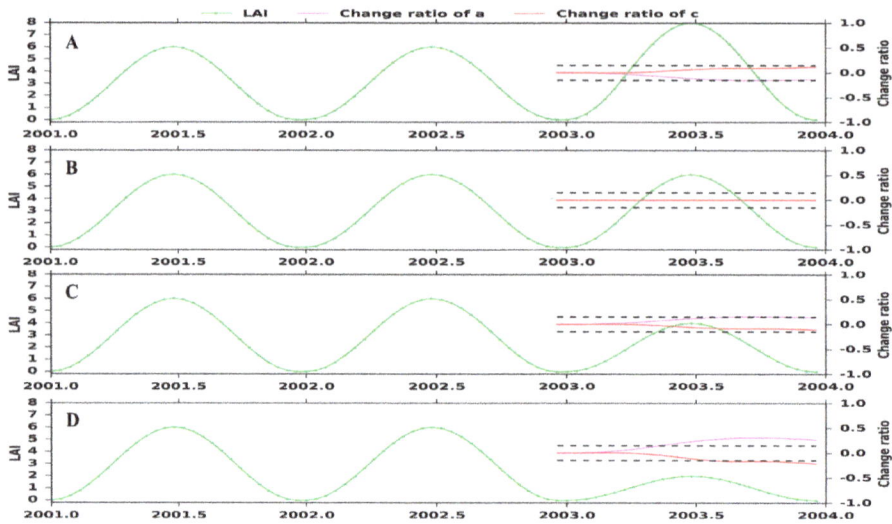

Figure 3. Using the season-trend model to simulate GLASS LAI time series compared to the previous two years LAI values in 2003. (**A**) multiplied the coefficient 4/3, (**B**) is 1, (**C**) is 2/3, and (**D**) is 1/3. Where the red line is the curve of the amplitude of the parameter c, the purple line is the amplitude curve of b.

The LAI value of some of the pixels may have been at a low level; therefore, if there is a slight change in LAI it will lead to a disturbance index that appears to have relatively large fluctuations, resulting in the detection of an error. To avoid this problem, we applied annual maximum value compositing to the LAI data. The LAI data were combined into one image representing the maximum LAI detected at every pixel throughout an annual period, and then this was compared with the annual maximum of the previous year's LAI.

$$LAI\_MAX_{2002} > LAI\_MAX_{2003} + 1.5 \tag{4}$$

If the LAI maximum of the same pixel had a decrease of 1.5 for two years, it was detected as a disturbance in the previous step. It is possible to be confident that the pixel was disturbed, otherwise, no disturbance had occurred. Any pixels above this threshold were flagged as a disturbance.

3.2. Distinguish the Types of Disturbances

The normalized burn ratio (NBR) [36,37] in form is a modification of the NDVI, except that it uses near-infrared (NIR) and a shortwave-infrared (SWIR) bands. The NIR band is sensitive to the chlorophyll content of the vegetation, while the SWIR band responds to the soil moisture and vegetation water content. Healthy vegetation has very high NIR reflectance and low reflectance in the SWIR portion of the spectrum. NBR is calculated as follows

$$NBR = \frac{B5 - B7}{B5 + B7} \tag{5}$$

A high NBR value generally indicates healthy vegetation, while a low value indicates bare ground and recently burned areas. This ratio spectral index shows a significant decrease following a burn and provides good burned–unburned discrimination using MODIS data.

On 5 May 2003 there was a large forest fire across the Jinhe forestry region, and the NBR was calculated using the surface reflectance data of the fire pixels. The time series curve of the NBR is shown in Figure 4A. It can be seen at the end of May 2003 that the NBR experienced a sharp decline from around 0.4 to −0.3, and the NBR of the non-fire pixels were always above 0.2 (Figure 4B). This characteristic abrupt decrease in NBR is the primary indicator used within the algorithm to identify burned areas. Combined with the first four steps of the detection results, when the pixel is labeled as disturbance and during this period of time NBR calculation results are less than 0, the candidate pixel was identified and labeled as burned. This algorithm attempts to distinguish the types of forest disturbances.

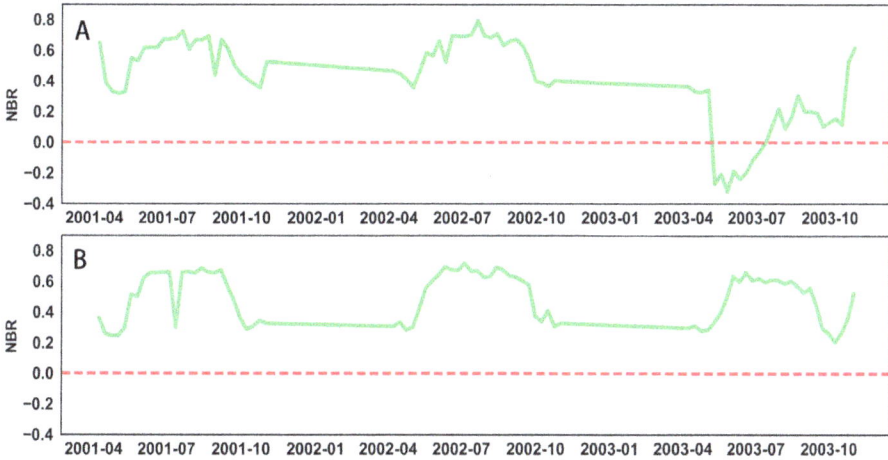

Figure 4. Difference between the NBR anomalies for burned (**A**) and unburned (**B**) areas in 2003. The green curve is the value of NBR, and the red dotted line is the horizontal line when NBR is zero.

4. Results

We used the TSM processing to model the GLASS LAI time series (1997–2003) to detect the Daxinganling forest fire that occurred in 2003. The disturbance detection result using the TSM algorithm was compared with the reference burned area data created using high-resolution Landsat imagery and the MCD64 products, respectively.

4.1. Comparing Results with MCD64 Products

Figure 5 presents the two burned area products over the study area in the 2003 period. There were 6082 burned pixels in the MCD64 data, 4022 pixels were burned in the GLASS data, and the difference between the two results was more than 2000 pixels. The differences were mainly concentrated on the left and right sides of the study area.

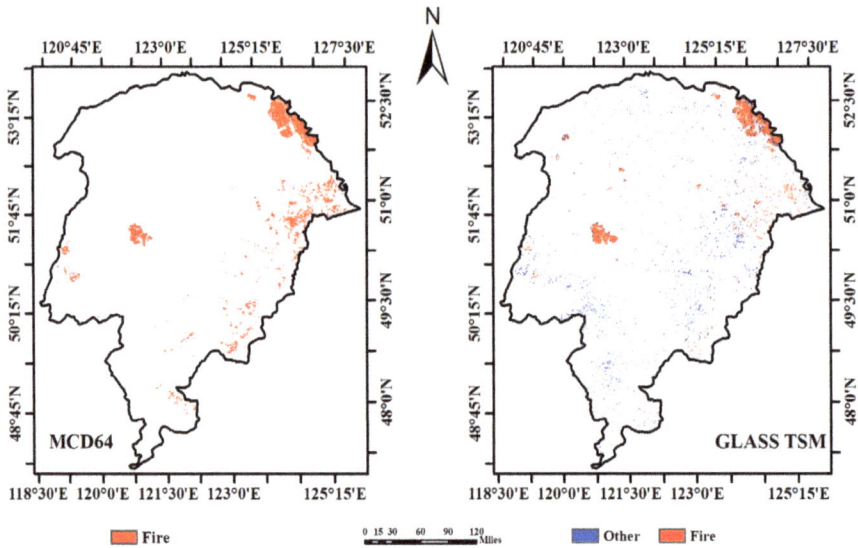

Figure 5. Comparison of GLASS burned area data and the MCD64A1 product data.

In order to compare the detection accuracy, the differences between the two products were selected in the study area. There were 978 pixels that MCD64 identified as unburned, but the TSM algorithm detected them as burned pixels; the LAI and NBR time series are shown in Figure 6A,B. The LAI in 2003 compared to the previous two years had a large degree of reduction; at the same time, the value of NBR in 2003 was generally at a low level compared with that of the previous two years. In addition, there were 2904 unburned pixels in the TSM algorithm detection that were marked as burned pixels in the MCD64, as shown in Figure 6C,D. In 2003, the LAI value did not decrease significantly, and the value of NBR was also at a high level. There were a large number of missed and wrong detection phenomenon in MCD64 product, and the TSM algorithm effectively avoided the phenomenon.

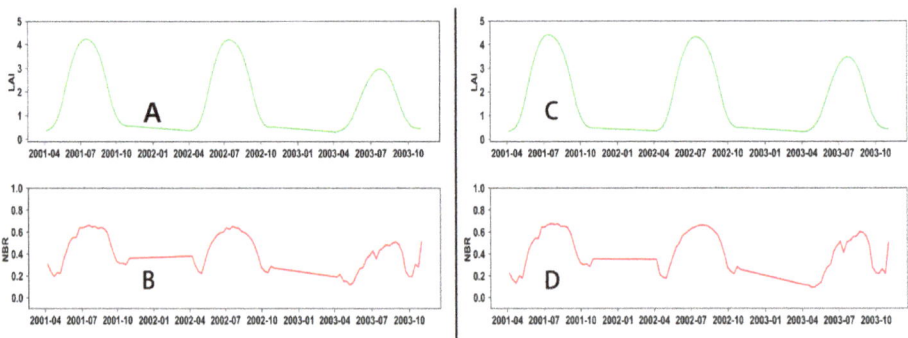

Figure 6. The LAI (green) and NBR (red) time series in the study area. (**A,B**) are detected as burned pixels in the TSM algorithm, and (**C,D**) are detected as unburned.

4.2. Results of Other Disturbances

Fires, pests, hurricanes, deforestation, etc., can result in a disturbance of forest growth. Different types of forest disturbances influence forest ecosystems in different ways; we need to distinguish between them. In Figure 7, the left side is the pixels with other disturbance types, which are marked in blue, with a total of 3457 pixels, accounting for 1.762% of the study area. For these non-fire disturbance pixels, two regions (A and B) were selected at random. On the right side of Figure 7 are the mean LAI and NBR time series curves for the two regions. There were 209 pixels presumed to have been disturbed in region A and 1248 in region B. Compared with the previous two years, the LAI was significantly reduced in 2003. The NBR time series curve is above 0.2 in 2001 to 2003; the reduction of LAI can exclude the cause as a fire disturbance. Our algorithm can effectively detect the occurrence of non-fire disturbances.

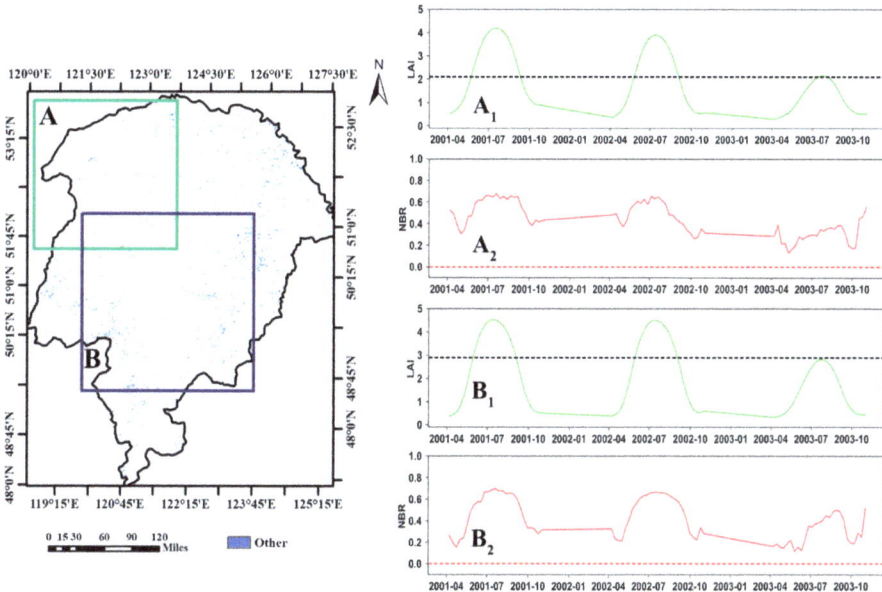

Figure 7. Other types of forest disturbances and mean LAI (subscript is 1) and NBR (subscript is 2) time series of two verification regions. Region **A**: The maximum mean LAI in 2003 at a value of 2.1; maximum mean LAI was 4.1 for 2000 and 2001. Region **B**: The maximum mean LAI in 2003 at a value of 2.8; maximum mean LAI was 4.5 for 2000 and 2001.

5. Discussion

5.1. Comparing Results with TM dNBR Map

We assessed the accuracy of burned area maps produced with the TSM algorithm in two different regions. On 5 May 2003, there was a large forest fire in the Jinhe forestry region in Inner Mongolia, and on 17 May 2003, there was a large forest fire in the Shibazhan forestry region in Heilongjiang Province. The results of the GLASS and the MCD64 product were compared with the results of the fire using high spatial resolution Landsat images. Landsat TM data resolution is 30 m; its spatial detail features were more obvious compared to those of the 1000 m spatial resolution of the MODIS data. The green leaves of the fire area were reduced, and compared with the reflectance of normal growing forest, there was an increase in the SWIR spectral region and a NIR reflectance drop. Bi-temporal image differencing is frequently applied on pre-and post-fire NBR images, resulting in a differenced NBR (dNBR). Through the dNBR calculation results, statistics of the area and location of the fire and the fire condition in the subpixel of GLASS can be clearly determined. The results were sampled to 1000 m, and the detection accuracy of GLASS was evaluated on the basis of sampled results.

$$dNBR = \left(\frac{B4 - B7}{B4 + B7}\right)_{pre-fire} - \left(\frac{B4 - B7}{B4 + B7}\right)_{post-fire} \tag{6}$$

Landsat datasets for the study area were downloaded from the United States Geological Survey (USGS) satellite data download site. Data related to the May 2003 Jinhe fires were downloaded for 8 April 2003 (pre-fire) and 26 May 2003 (post-fire). Data related to the May 2003 Shibazhan fires were downloaded for 10 April 2003 (pre-fire) and 13 June 2003 (post-fire). The images were subjected to geometric, radiometric, and atmospheric correction.

In Figure 8, the dNBR results are sampled to 1000 m spatial resolution using a mode resampling method. The statistical analysis of the sizes of the burned area before and after resampling and the burned area of the two study regions were reduced by different degrees after resampling. The burned area of the Jinhe was reduced by about 4000 hectares and the burned area of the Shibazhan was reduced by about 2000 hectares. Cloud and cloud shadow contamination will result in a dNBR calculation error. We identified cloud and cloud shadows in the Landsat images using visual interpretation, and a mask was applied to the GLASS and MCD64 to subset the image graphically to mask contaminated pixels.

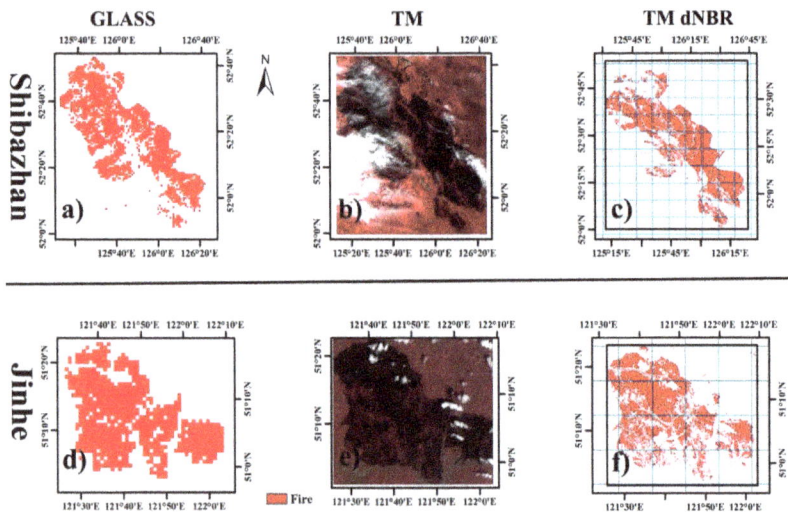

Figure 8. Regional images of the 2003 GLASS and Landsat fire disturbance in Shibazhan and Jinhe. The comparison results of two algorithms in Shibazhan are shown in the upper row: (**a**) TSM results; (**b**) Landsat image; (**c**) Landsat dNBR results. The comparison results of two algorithms in Jinhe are shown in the bottom row: (**d**) TSM results; (**e**) Landsat image; (**f**) Landsat dNBR results. The Landsat image is displayed in band 4 (red), 3 (green), 2 (blue) color composite. The cyan boxes in the differenced NBR (dNBR) represent 10 × 10 GLASS pixels.

A reference of the burned area data for each region was compiled using high-resolution Landsat imagery. The results of the burned area of the two regions were compared using a confusion matrix to evaluate the spatial fidelity of mapping on a per-pixel basis. Table 2 provides the producer accuracy and user accuracies obtained when comparing GLASS and MCD64 versus reference maps of burned areas derived from Landsat TM over Jinhe and Shibazhan in the 2003 period. The overall accuracy is the percentage of all validation pixels correctly classified. The burned region map was validated using Landsat data. The TSM algorithm mapped the spatial distribution of the burns in Jinhe with an overall kappa coefficient of 0.776 and an accuracy of 96.5%; in Shibazhan, the overall accuracy was 98%, and the kappa coefficient was 0.838. The overall accuracy of the MCD64 product, when compared in Jinhe and Shibazhan, was 92.52% and 97.14%, respectively, and the kappa coefficient was 0.758 and 0.634, respectively. The results indicate that the GLASS-derived burned map of our study area has a high accuracy.

Table 2. Regional confusion matrices and kappa coefficient (*k*) from geographic accuracy assessment. MODIS = Moderate Resolution Imaging Spectroradiometer; TSM = time-stepping modeling.

Landsat	MODIS		Producer's Accuracy	GLASS TSM		Producer's Accuracy
	Burned	Unburned		Burned	Unburned	
Jinhe (*k* = 0.758)				*k* = 0.776		
Burned	473	107	81.5%	521	59	89.8%
Unburned	150	3955	96.3%	103	4002	97.5%
User's accuracy	75.9%	97.4%		83.5%	98.5%	
Shibazhan (*k* = 0.634)				*k* = 0.838		
Burned	1407	265	94.7%	1579	93	94.4%
Unburned	1115	17377	97.3%	321	18171	98.3%
User's accuracy	80.8%	99.3%		83.1%	99.5%	

5.2. Why Use GLASS LAI Data?

There are many remote sensing products that can be used as input parameters for the detection of disturbances, such as NDVI, enhanced vegetation index (EVI), and LAI. GLASS LAI [33] was selected as the input data for the following reasons: (1) the time series of GLASS LAI products have good continuity; (2) GLASS LAI in the vegetation cycle is relatively smooth. If the time series data has large fluctuations, it needs to be smoothed before use; In addition, (3) the LAI is more sensitive to detecting disturbances in forested land, whereas NDVI is prone to moderate to high saturation.

After extracting the fire area coordinates in the Jinhe large forest fire, a time series curve from 2001 to 2005 was drawn from burn-related data using MODIS NDVI. In 2003, NDVI data showed a rapid rise after a brief decline; compared to the previous data with no fire, the NDVI changes were not obvious. One of the consequences of fire is that a burning forest will leave a wealth of organic matter. Fire is necessary to cycle nutrients, especially on sites with deep organic soils. After a fire, adequate nutrients play a significant role in weed and understory growth [38].

The satellite observes forest land from a high altitude; therefore, both the canopy and understorey will be reflected in the results of NDVI. LAI can be regarded as a three-dimensional characteristic parameter of forests, which is superior to NDVI as an input parameter. The time series of MODIS LAI data of the same burned pixel were plotted. Compared with NDVI, the LAI appears to be significantly reduced after May 2003, but MODIS LAI in the time series has a very large jump; the occurrence of jump points will affect the modeling parameters so that the detection results will be in error. If we use MODIS LAI data to model, we need to smooth the LAI in accordance with the growth pattern and weaken the detection error caused by the jump point.

As can be seen from Figure 9, the curves using GLASS LAI data are smoother than those using MODIS NDVI and MODIS LAI and are, therefore, more suitable for season-trend modeling and the detection of disturbances.

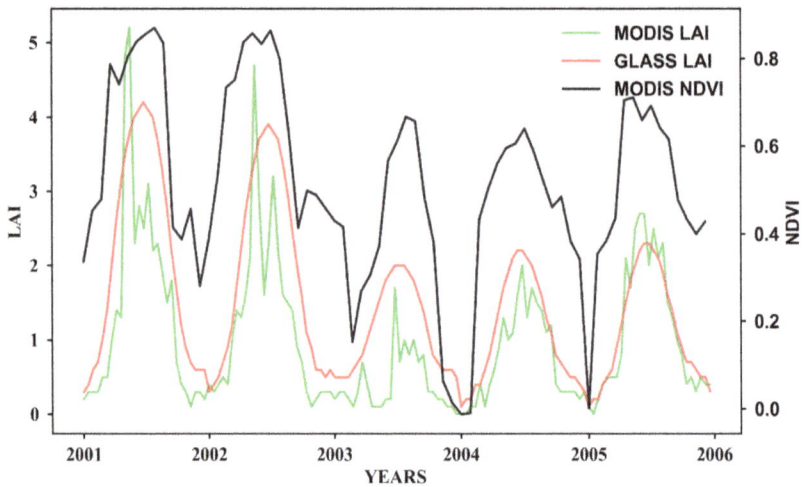

Figure 9. The time series curve of different products at the same location (3 × 3 km); GLASS LAI (red), MODIS LAI (green) and MODIS normalized difference vegetation index (NDVI) (black).

As with any remote sensing method, our developed algorithm has a limit and cannot distinguish some non-fire types of forest disturbances. In a future analysis we will use auxiliary data to achieve a more detailed classification of the results in order to enhance the versatility of the algorithm and distinguish the different types of forest disturbances. The LAI of forests will be at a lower level in winter and spring, the variations of the model parameters that occurred during this period were not as sensitive as that of the growing season. Although we can use subsequent data to detect the occurrence of disturbances by changes in LAI time series, the time and range of the disturbance in the detection results will be affected. We will improve the LAI's simulation method to allow it to detect those sedondary disturbance too.

As it can be seen in Figure 3, the season-trend model can detect forest recovery. We will extend the application of the model and the TSM process method to estimate the restoration of the forest after the disturbance according to its parameter variation characteristics and evaluate the ecological restoration quantitatively.

6. Conclusions

In this study, we describe a new algorithm for detecting forest disturbances using 8-day composite LAI data at 1000 m spatial resolution from the GLASS product, which uses the season-trend model and the time-stepping modeling (TSM) process, monitors the variation in the amplitude parameter of the dynamic model over a long period, and detects disturbance signals by capturing structural changes in the time series data. The resultant forest burn map had an overall accuracy greater than 96% and kappa coefficient greater than 0.77 based on the validation data derived from Landsat images. The results show that the TSM algorithm can improve the application of remote sensing data in forest disturbance detection and provide effective support for long-term monitoring of forest ecosystems. The results from this study also demonstrate that the TSM algorithm is automatic and robust and can be used to map forest disturbances in forest land. In the present study, by using the SWIR band as auxiliary data, the algorithm achieved good performance in Daxinganling and distinguished between burned and unburned areas. To distinguish other types of forest disturbances, future studies need to include more auxiliary information and more complicated scenarios.

Acknowledgments: This research was supported by the National Basic Research Program of China under grant No. 2013CB733403.

Author Contributions: Jindi Wang conceived and designed the study. Jian Wang contributed to the conception of the study, performed the data analysis, and wrote the paper. Hongmin Zhou and Zhiqiang Xiao aided with the discussion and the manuscript revision. Jindi Wang reviewed and edited the manuscript. All authors read and approved the manuscript.

Conflicts of Interest: The authors declare no conflict of interest.

References

1. Bonan, G.B. Forests and climate change: Forcings, feedbacks, and the climate benefits of forests. *Science* **2008**, *320*, 1444–1449. [CrossRef] [PubMed]
2. Dixon, R.K.; Andrasko, K.J.; Sussman, F.G.; Lavinson, M.A.; Trexler, M.C.; Vinson, T.S. Forest sector carbon offset projects: Near-term opportunities to mitigate greenhouse gas emissions. *Water Air Soil Pollut.* **1993**, *70*, 561–577. [CrossRef]
3. Kennedy, R.E.; Yang, Z.; Cohen, W.B. Detecting trends in forest disturbance and recovery using yearly landsat time series: 1. Landtrendr—Temporal segmentation algorithms. *Remote Sens. Environ.* **2010**, *114*, 2897–2910. [CrossRef]
4. Foster, D.; Swanson, F.; Aber, J.; Burke, I.; Brokaw, N.; Tilman, D.; Knapp, A. The importance of land-use legacies to ecology and conservation. *Bioscience* **2003**, *53*, 77–88. [CrossRef]
5. Mladenoff, D.J.; White, M.A.; Pastor, J.; Crow, T.R. Comparing spatial pattern in unaltered old-growth and disturbed forest landscapes. *Ecol. Appl.* **1993**, *3*, 294–306. [CrossRef] [PubMed]
6. Nielsen, S.E.; Boyce, M.S.; Stenhouse, G.B. Grizzly bears and forestry. I. Selection of clearcuts by grizzly bears in west-central Alberta, Canada. *For. Ecol. Manag.* **2004**, *199*, 51–65. [CrossRef]
7. Linke, J.; Franklin, S.E.; Huettmann, F.; Stenhouse, G.B. Seismic cutlines, changing landscape metrics and grizzly bear landscape use in alberta. *Landsc. Ecol.* **2005**, *20*, 811–826. [CrossRef]
8. Rollins, M.G. Landfire: A nationally consistent vegetation, wildland fire, and fuel assessment. *Int. J. Wildland Fire* **2009**, *18*, 235–249. [CrossRef]
9. David, P.; Wang, J. A sub-pixel-based calculate of fire radiative power from MODIS observations: 2. Sensitivity analysis and potential fire weather application. *Remote Sens. Environ.* **2013**, *129*, 231–249.
10. Wooster, M.J.; Zhang, Y.H. Boreal forest fires burn less intensely in Russia than in north America. *Geophys. Res. Lett.* **2004**, *31*, 183–213. [CrossRef]
11. Potter, C.; Tan, P.N.; Steinbach, M.; Klooster, S.; Kumar, V.; Myneni, R.; Genovese, V. Major disturbance events in terrestrial ecosystems detected using global satellite data sets. *Glob. Chang. Biol.* **2010**, *9*, 1005–1021. [CrossRef]
12. Zhu, Z.; Woodcock, C.E.; Olofsson, P. Continuous monitoring of forest disturbance using all available landsat imagery. *Remote Sens. Environ.* **2012**, *122*, 75–91. [CrossRef]
13. Devries, B.; Verbesselt, J.; Kooistra, L.; Herold, M. Robust monitoring of small-scale forest disturbances in a tropical montane forest using landsat time series. *Remote Sens. Environ.* **2015**, *161*, 107–121. [CrossRef]
14. Lunetta, R.S.; Knight, J.F.; Ediriwickrema, J. Land-cover characterization and change detection using multitemporal MODIS NDVI data. *Remote Sens. Environ.* **2006**, *105*, 142–154. [CrossRef]
15. Verbesselt, J.; Hyndman, R.; Newnham, G.; Culvenor, D. Detecting trend and seasonal changes in satellite image time series. *Remote Sens. Environ.* **2010**, *114*, 106–115. [CrossRef]
16. Verbesselt, J.; Zeileis, A.; Herold, M. Near real-time disturbance detection using satellite image time series. *Remote Sens. Environ.* **2012**, *123*, 98–108. [CrossRef]
17. Jonsson, P.; Eklundh, L. Seasonality extraction by function fitting to time-series of satellite sensor data. *IEEE Trans. Geosci. Remote Sens.* **2002**, *40*, 1824–1832. [CrossRef]
18. Jönsson, P.; Eklundh, L. Timesat—A program for analyzing time-series of satellite sensor data. *Comput. Geosci.* **2004**, *30*, 833–845. [CrossRef]
19. Fraser, R.H.; Li, Z.; Cihlar, J. Hotspot and NDVI differencing synergy (hands): A new technique for burned area mapping over boreal forest. *Remote Sens. Environ.* **2000**, *74*, 362–376. [CrossRef]

20. Hilker, T.; Wulder, M.A.; Coops, N.C.; Linke, J.; Mcdermid, G.; Masek, J.G.; Gao, F.; White, J.C. A new data fusion model for high spatial- and temporal-resolution mapping of forest disturbance based on Landsat and MODIS. *Remote Sens. Environ.* **2009**, *113*, 1613–1627. [CrossRef]

21. Barbosa, P.M.; Grégoire, J.M.; Pereira, J.M.C. An algorithm for extracting burned areas from time series of AVHRR GAC data applied at a continental scale. *Remote Sens. Environ.* **1999**, *69*, 253–263. [CrossRef]

22. Masek, J.G.; Cohen, W.B.; Leckie, D.; Wulder, M.A.; Vargas, R.; De Jong, B.; Healey, S.; Law, B.; Birdsey, R.; Houghton, R.A. Recent rates of forest harvest and conversion in north America. *J. Geophys. Res. Biogeosci.* **2015**, *116*, 1451–1453. [CrossRef]

23. Hansen, M.C.; Stehman, S.V.; Potapov, P.V. Quantification of global gross forest cover loss. *Proc. Natl. Acad. Sci. USA* **2010**, *107*, 8650–8655. [CrossRef] [PubMed]

24. Mildrexler, D.J.; Zhao, M.; Heinsch, F.A.; Running, S.W. A new satellite-based methodology for continental-scale disturbance detection. *Ecol. Appl.* **2007**, *17*, 235–250. [CrossRef]

25. Mildrexler, D.J.; Zhao, M.S.; Running, S.W. Testing a MODIS global disturbance index across north America. *Remote Sens. Environ.* **2009**, *113*, 2103–2117. [CrossRef]

26. Giglio, L.; Loboda, T.; Roy, D.P.; Quayle, B.; Justice, C.O. An active-fire based burned area mapping algorithm for the MODIS sensor. *Remote Sens. Environ.* **2009**, *113*, 408–420. [CrossRef]

27. Giglio, L.; Descloitres, J.; Justice, C.O.; Kaufman, Y.J. An enhanced contextual fire detection algorithm for MODIS. *Remote Sens. Environ.* **2003**, *87*, 273–282. [CrossRef]

28. Boschetti, L.; Roy, D.; Hoffmann, A.A. *MODIS Collection 5 Burned Area Product-MCD45*; Users Guide Ver 2008; University of Maryland: College Park, MD, USA, 2008.

29. George, C.; Rowland, C.; Gerard, F.; Balzter, H. Retrospective mapping of burnt areas in Central Siberia using a modification of the normalised difference water index. *Remote Sens. Environ.* **2006**, *104*, 346–359. [CrossRef]

30. Loboda, T.; O'Neal, K.J.; Csiszar, I. Regionally adaptable dnbr-based algorithm for burned area mapping from MODIS data. *Remote Sens. Environ.* **2007**, *109*, 429–442. [CrossRef]

31. Roy, D.; Descloitres, J.; Alleaume, S. The MODIS fire products. *Remote Sens. Environ.* **2002**, *83*, 244–262.

32. Padilla, M.; Stehman, S.V.; Ramo, R.; Corti, D.; Hantson, S.; Oliva, P.; Alonso-Canas, I.; Bradley, A.V.; Tansey, K.; Mota, B. Comparing the accuracies of remote sensing global burned area products using stratified random sampling and estimation. *Remote Sens. Environ.* **2015**, *160*, 114–121. [CrossRef]

33. Xiao, Z.; Liang, S.; Wang, J.; Chen, P.; Yin, X.; Zhang, L.; Song, J. Use of general regression neural networks for generating the glass leaf area index product from time-series MODIS surface reflectance. *IEEE Trans. Geosci. Remote Sens.* **2013**, *52*, 209–223. [CrossRef]

34. Geerken, R.A. An algorithm to classify and monitor seasonal variations in vegetation phenologies and their inter-annual change. *ISPRS J. Photogramm. Remote Sens.* **2009**, *64*, 422–431. [CrossRef]

35. Julien, Y.; Sobrino, J.A. Comparison of cloud-reconstruction methods for time series of composite NDVI data. *Remote Sens. Environ.* **2010**, *114*, 618–625. [CrossRef]

36. Key, C.H.; Benson, N.C. Landscape assessment: Ground measure of severity, the composite burn index; and remote sensing of severity, the normalized burn ratio. In *FIREMON: Fire Effects Monitoring and Inventory System*; Rocky Mountain Research Station, USDA Forest Service: Fort Collins, CO, USA, 2006; pp. LA8–LA51.

37. Cocke, A.E.; Crouse, J.E. Comparison of burn severity assessments using differenced normalized burn ratio and ground data. *Int. J. Wildland Fire* **2005**, *14*, 189–198. [CrossRef]

38. Smith, J.K.; Lyon, L.J.; Huff, M.H.; Hooper, R.G.; Telfer, E.S.; Schreiner, D.S.; Smith, J.K. *Wildland Fire in Ecosystems. Effects of Fire on Fauna*; RMRS-GTR-42; USDA Forest Service: Fort Collins, CO, USA, 2000.

remote sensing

MDPI

Article

Spatio-Temporal Analysis and Uncertainty of Fractional Vegetation Cover Change over Northern China during 2001–2012 Based on Multiple Vegetation Data Sets

Linqing Yang [1,2], Kun Jia [1,2,*], Shunlin Liang [3], Meng Liu [4,5], Xiangqin Wei [6], Yunjun Yao [1,2], Xiaotong Zhang [1,2] and Duanyang Liu [1,2]

[1]   State Key Laboratory of Remote Sensing Science, Faculty of Geographical Science, Beijing Normal University, Beijing 100875, China; linqingyang@mail.bnu.edu.cn (L.Y.); boyyunjun@163.com (Y.Y.); xtngzhang@bnu.edu.cn (X.Z.); 201721170045@mail.bnu.edu.cn (D.L.)
[2]   Beijing Engineering Research Center for Global Land Remote Sensing Products, Faculty of Geographical Science, Beijing Normal University, Beijing 100875, China
[3]   Department of Geographical Sciences, University of Maryland, College Park, MD 20742, USA; sliang@bnu.edu.cn
[4]   State Key Laboratory of Earth Surface Processes and Resource Ecology, Beijing Normal University, Beijing 100875, China; mengliu@mail.bnu.edu.cn
[5]   College of Global Change and Earth System Science, Beijing Normal University, Beijing 100875, China
[6]   Institute of Remote Sensing and Digital Earth, Chinese Academy of Sciences, Beijing 100101, China; weixq@radi.ac.cn
*   Correspondence: jiakun@bnu.edu.cn; Tel.: +86-10-5880-0152

Received: 7 January 2018; Accepted: 2 April 2018; Published: 3 April 2018

Abstract: Northern China is one of the most sensitive and vulnerable regions in the country. To combat environmental degradation in northern China, a series of vegetation protection programs, such as the Three-North Shelter Forest Program (TNFSP), have been implemented. Whether the implementation of these programs in northern China has improved the vegetation conditions has merited global attention. Therefore, quantifying vegetation changes in northern China is essential for meteorological, hydrological, ecological, and societal implications. Fractional vegetation cover (FVC) is a crucial biophysical parameter which describes land surface vegetation conditions. In this study, four FVC data sets derived from remote sensing data over northern China are employed for a spatio-temporal analysis to determine the uncertainty of fractional vegetation cover change from 2001 to 2012. Trend analysis of these data sets (including an annually varying estimate of error) reveals that FVC has increased at the rate of $0.26 \pm 0.13\%$, $0.30 \pm 0.25\%$, $0.12 \pm 0.03\%$, $0.49 \pm 0.21\%$ per year in northern China, Northeast China, Northwest China, and North China during the period 2001–2012, respectively. In all of northern China, only 33.03% of pixels showed a significant increase in vegetation cover whereas approximately 16.81% of pixels showed a significant decrease and 50.16% remained relatively stable.

Keywords: fractional vegetation cover (FVC); multi-data set; northern China; spatio-temporal; inter-annual variation; uncertainty; standard error of the mean

1. Introduction

Northern China has typical characteristics of fragile ecological situations and is one of the most sensitive and vulnerable regions in China. For historical reasons, farming practices, grazing and other reasons, northern China is suffering long-term land degeneration, a lack of fresh water, drought, and other extreme weather, etc. [1–4]. Therefore, northern China is a key area deserving of scholarly attention. Faced with these problems, the Chinese government has realized the seriousness of the situations and has implemented a series of policy measures to ease the environmental crisis.

Vegetation, which is bonded soil, climatic, hydrologic, and other elements in the whole ecosystem, is a sensitive indicator of climate change and human activities and thus influences climate by affecting the energy, water, and carbon cycle [5–7]. Vegetation is also a positive factor in the prevention of soil and water loss as well as in the control of sandstorms; it also is an important factor for soil erosion prediction. Vegetation absorbs CO_2 through photosynthesis from the atmosphere to mitigate global warming. Meanwhile, vegetation can increase precipitation, runoff regulation, reduce flood and drought, reduce pollution, and improve the ecological environment. Vegetation change can also affect the energy balance as well as biochemical and biophysical processes [8].

Therefore, many ecological engineering programs have been implemented in northern China to improve the regional ecological environment [5,9] such as the Three-North Shelterbelt Forest Program (TNSFP), the Green to Grain Program (GTGP), and the Natural Forest Conservation Program (NFCP) [10–13]. In these ecological programs, TNFSP is the oldest, invested in the most, and affected the widest range. The TNSFP was officially launched in 1979 and involves 13 provinces, autonomous regions, and municipalities in the Three-North region with a total planned area of more than 4 million km^2, i.e., nearly 42% of the total area of China. To date, a total of 30.6 million ha of afforestation has been carried out at a total cost of ¥4 billion [1,6]. Because of its huge geographic extent and complexity, the project will extend to 2050 and will provide important information through the monitoring of its long-term progress [14]. The main purpose of such projects is to prevent land desertification, control sandstorms, and improve both local water resources and the natural environment. Whether the implementation of these programs throughout northern China has improved the vegetation conditions has merited global attention.

Therefore, quantifying vegetation changes in northern China is essential. Fractional Vegetation Cover (FVC), which is an important variable describing land surface vegetation, is generally defined as the fraction of green vegetation as seen from the nadir of the statistical area. FVC is also a crucial biophysical parameter for studying the atmosphere, pedosphere, hydrosphere, and biosphere as well as their interactions [15–18]. Reliable information on FVC change over northern China is needed for environment and ecological monitoring, environmental assessment, and the evaluation of vegetation change feedbacks in climate. For example, Su et al. [19] used MODIS data to detect vegetation changes in the agricultural-pastoral areas of northern China from 2001 to 2013. Liu et al. [20] used SPOT-VGT data from 1998 to 2007 to detect vegetation change throughout northern China. Zhang et al. [21] analyzed the spatio-temporal vegetation changes of northern China from 2000 to 2012. Li et al. [22] analyzed the spatial-temporal pattern and change of FVC in northern China during 2001–2012. Li et al. [23] conducted a comparison of multiple forest cover data sets to monitor forest cover changes across China.

However, most of the studies used a single data set in which may exist large uncertainties. The accuracy of the FVC data set is unclear in northern China resulting from a lack of ground measurements. As a result, its attributions and any response to climate change generated from a single source may lead to large uncertainties. A multi-data set approach to analyze the vegetation change is a logical response to the challenges mentioned above as it fuses the strengths of the various platforms and methodologies as well as provides an estimate of the uncertainty. Therefore, the main object of this study is to develop a multi-data set estimate of FVC change throughout northern China for the period of 2001–2012. It is also expected to provide reliable and accurate information for regional sustainable development, ecological restoration project planning, and ecological environmental protection.

2. Data and Methods

2.1. Study Area

The study area (Figure 1), based on provincial boundaries, contains 13 provinces, cities, autonomous regions, and municipalities in northern China. Reflecting on the diversity of the natural environment, northern China has been further subdivided into northeastern China, northwestern China, and northern China when conducting statistical analysis at a regional scale.

Northeast China contains Heilongjiang, Liaoning, Jilin province as well as the eastern part of the Inner Mongolia Autonomous Region. Northeast China is characterized by temperate monsoon climate with a low mean annual temperature (5.2 °C) and annual precipitation reaching 300~1000 mm. Northeast China contains almost all major forest types in northern Eastern Asia which include cold-temperate conifer mixed forests, temperate conifer forests, broadleaf mixed forests, and warm-temperate deciduous broadleaf mixed forests and covers the largest area of natural forest in China [24].

North China contains Beijing, Tianjin, Hebei, Shandong, Shanxi, Shaanxi as well as the middle part of the Inner Mongolia Autonomous Region. North China is defined by plains and a warm sub-humid continental climate and has a large annual range of temperature; the annual precipitation reaches 400–800 mm, mainly in the summer.

Northwest China contains the western part of the Inner Mongolia Autonomous Region, the Xinjiang Uygur autonomous region, the Ningxia Hui autonomous region, along with Gansu province. The climate of Northwest China varies and includes a temperate continental monsoon climate, arid and semi-arid climates, and a warm temperate continental arid climate. It is distinguished by a low annual mean temperature, a large annual range of temperature, and low precipitation (50–200 mm/year). It has the biggest desert in China, the Taklimakan desert, and the main vegetation type is grassland, shrubland, etc.

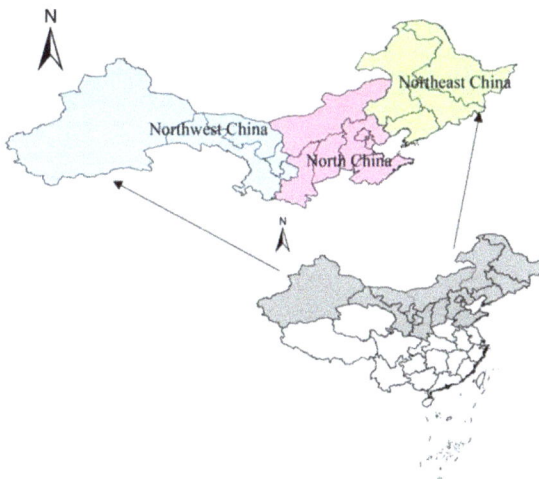

Figure 1. Location of the research area (The green area represents northeast China, the pink area for North China, the blue area for Northwest China, respectively).

2.2. Data Sets

Four FVC data sets were used to estimate vegetation change over northern China in this study, including the Global LAnd Surface Satellite (GLASS) FVC product, GEOV1 FVC product, TRAGL FVC product, and Li product, which are summarized in Table 1.

Table 1. The summary of fractional vegetation cover (FVC) data sets used in the analysis.

Product Name	Sensor	Available Time	Temporal Resolution	Spatial Coverage	Spatial Resolution	Reference
GLASS-MODIS	MODIS	2001–now	8 days	Global	500 m	[16]
GEOV1	SPOT VGT	2001–now	10 days	Global	1 km	[25]
TRAGL	MOIDS	2001–2012	8 days	Global	1 km	[26]
Li	MODIS	2001–2012	8 days	Northern China	0.011°	[16]

2.2.1. GLASS MODIS-FVC Product

The GLASS FVC product [16] is one of the new products in the GLASS product suite, which is supported by China's National High Technology Research and Development Program to generate long-term global land surface parameters. The GLASS MODIS-FVC product is generated using the generalized regression neural networks (GRNNs) with training data derived from MODIS Version 5 surface reflectance data (MOD09A1) and FVC values obtained from Landsat data using the dimidiate pixel model. The temporal and spatial resolution of GLASS MODIS-FVC are 8-day and 0.5 km with a sinusoidal grid projection, respectively. Jia et al. compared the GLASS FVC with GEOV1 FVC which was the best global FVC product and results indicated GLASS FVC presented a much better spatial and temporal continuity and marginally better accuracy with over 44 validation of land European remote sensing instruments (VALERI) validation sites.

2.2.2. GEOV1 FVC Product

The GEOV1 FVC product (http://land.copernicus.eu/global/products/FCover) that derived from SPOT/VEGETATION data from 1999 to the present is an improvement of CYCLOPES FVC product [25]. The product is provided in a Plate Carrée projection at 1/112° spatial resolution and a 10-day frequency. The GEOV1 FCover product was derived from SPOT/VEGETATION sensor data using back-propagation neural networks. The CYCLOPES FCover product was scaled to train the back-propagation neural networks with the SPOT/VEGETATION top-of-canopy directionally normalized reflectance values over the BELMANIP (Benchmark Land Multisite Analysis and Intercomparison of Products) network of sites [25]. The GEOV1 FVC product corrects the underestimate problem of CYCLOPES FVC product and is closer to the real value [27].

2.2.3. TRAGL FVC Product

The TRAGL FVC product was retrieved from GLASS LAI product using physical relations between FVC and LAI [26]. The GLASS LAI product was retrieved using general regression neural networks (GRNNs) from MODIS Version 5 surface reflectance data (MOD09A1)/AVHRR reflectance data [28]. Unlike existing neural network methods that use remote sensing data acquired only at a specific time to retrieve LAI, the GRNNs were trained using fused time series LAI values from MODIS and CYCLOPES LAI products and reprocessed time series MODIS. The temporal and spatial resolution is 8-day and 1 km with geographic projection. The TRAGL FVC product is spatially and temporally complete. A comparison with GEOV1 FVC product showed that both FVC products were generally consistent in their spatial patterns.

2.2.4. Li FVC Product

Li et al. [22] estimated the FVC of northern China from MODIS Version 5 surface reflectance data (MOD09A1) using the dimidiate pixel model, which is one of the most widely used FVC estimation methods [29,30]. It assumed that a pixel consisted of only vegetation and non-vegetation components and its value was a linear combination of these two components. If normalized differential vegetation index (NDVI) was used to represent the spectral response, the mathematical expression of the mixed pixel model would be

$$NDVI = f * NDVI_v + (1 - f) * NDVI_s \tag{1}$$

then,

$$f = \frac{NDVI - NDVI_s}{NDVI_v - NDVI_s} \tag{2}$$

where f was the proportion of vegetation area in the mixed pixel (FVC), $NDVI$ was the NDVI of the mixed pixel, and $NDVI_v$ and $NDVI_s$ were the $NDVI$ of the fully vegetated and bare soil pixel, respectively. The value of $NDVI_v$ and $NDVI_s$ was 0.848 and 0.0133, respectively. The result showed a good performance in the change trend of both inter-annual and within the year. The temporal and spatial resolution is 8-day and 0.011° with geographic projection. In the following sections, this data source is called Li FVC.

Because of the inconsistent spatial resolution and projection between the four data sets, the data sets were processed to be spatially matched with geographic projection and the spatial resolution was converted to 0.01°. Then, annual maximum FVC images of four data sets on a pixel-by-pixel basis from 2001 to 2012 were calculated, respectively. In the following sectors, FVC data sets refer to the maximum FVC images of four data sets.

2.3. Methodology

2.3.1. Inter-Annual Change Trend of FVC

(1) Mann–Kendall Methods

The Mann–Kendall test [31,32] is a nonparametric method for testing the significance of time series data in hydrological processes and other related physical variables [33–35]. The advantage of this method is that the data does not need to conform to any particular distribution and it has a low sensitivity to abrupt breaks due to the inhomogeneous time series [36]. For a time series, $X = \{x_1, x_2, \cdots, x_3\}$, the Mann–Kendall test statistic is given as follows:

$$S = \sum_{i=1}^{n-1} \sum_{j=i+1}^{n} \text{sgn}(x_j - x_i) \tag{3}$$

where n is the number of data points, x_i and x_j are the data values in time series i and j ($j > i$), respectively, and $\text{sgn}(x_j - x_i)$ is the sign function as follows:

$$\text{sgn}(x_j - x_i) = \begin{cases} 1 & x_j - x_i > 0 \\ 0 & x_j - x_i = 0 \\ -1 & x_j - x_i < 0 \end{cases} \tag{4}$$

In cases where the sample size is more than 10, the standard normal test statistic, Z, is computed by:

$$Z = \begin{cases} \frac{S-1}{\sqrt{Var(S)}} & S > 0 \\ 0 & S = 0 \\ \frac{S+1}{\sqrt{Var(S)}} & S < 0 \end{cases} \tag{5}$$

The variance is computed by:

$$Var(S) = \frac{n(n+1)(2n+5) - \sum_{i=1}^{m} t_i(t_i - 1)(2t_i + 5)}{18} \tag{6}$$

where m is the number of tied groups and t_i denotes the number of ties of extent i. A tied group is a set of sample data with the same value.

Positive values of Z indicate increasing trends while negative Z values show decreasing trends [36]. The trend's significance is assessed by comparing the Z value with the standard normal variance at the pre-specified level of statistical significance [37]. The null hypothesis is rejected and a significant time series trend exists when $|Z| > Z_{1-\alpha/2}|$. $Z_{1-\alpha/2}$ is obtained from the standard normal distribution table. In this study, significance levels $\alpha = 0.05$ was used which correspond to $Z_{1-\alpha/2}$ values of 1.960.

(2) Sen's Slope Estimator

True slope can be estimated by using a non-parametric method developed by Sen [37] if a linear trend is presented by the Mann–Kendall method. Sen's slope estimator can be computed efficiently and is insensitive to outliers. The slope estimator of N pairs of data are first computed by:

$$\beta = Median(\frac{x_j - x_i}{j - i})i = 1, 2, \cdots N \tag{7}$$

where $N = n(n-1)/2$ when there is only one datum in each time period, while $N < n(n-1)/2$ when there are multiple observations in one or more time periods; where n is the total number of observations [38]. If β is positive that indicates an increase in X; if negative, then the X decreases or 0 remains constant.

In this study, the temporal change trends of FVC based on the four FVC products during 2001–2012 were calculated using the Mann–Kendall Method and Sen's slope estimator. The time series X was FVC_i which denotes the annual maximum FVC value of the ith year. n is the number of years (equal to 12 in this study) and i represents the year number ($i = 1, 2, 3, \ldots, 12$).

2.3.2. Multi-Data FVC Retrieval and Uncertainty Analysis

A multi-data approach [39] was employed to develop an integrated FVC and reduce the uncertainty from individual data sets. The consistency of each data set was evaluated by computing the correlation and the root-mean-square error (RMSE) of the multi-data set mean, excluding the data set being verified. This method was applied at pixel level to help remove the individual FVC data set with poor data quality from the final averaged FVC. An estimate of the uncertainty in FVC in each year is obtained from the standard error of the mean (*SEM*):

$$SEM = s/\sqrt{n} \tag{8}$$

which depends on the standard deviation s of the n data sets. First, the uncertainty analysis was carried out at the pixel scale. Then, the uncertainty analysis was calculated using the annual mean maximum FVC of the four data sets.

3. Results

3.1. Results of Single FVC Data Sets

3.1.1. Spatial Patterns of Each Single FVC Data Set

Figure 2 shows the spatial patterns of mean annual maximum FVC over the period 2001–2012 in northern China derived from each four data sets respectively. The individual source of FVC values differed in mean amplitude and spatial distribution and the disparity was mainly distributed in Northeast China and North China. Although all the four data sets had the biggest FVC values in Northeast China, the FVC values derived from TRAGL FVC was lower than those of other three data sets by about 0.15 and Li FVC sometimes existed in a saturation phenomenon. Meanwhile, GLASS FVC and TRAGL FVC was almost zero in the southern part of the Xinjiang Uygur autonomous region, the western part of the Inner Mongolia Autonomous Region, and the northwestern part of Ningxia Hui autonomous region; however, FVC values of GEOV1 FVC and Li FVC in those regions were greater than zero. This was mainly because Li FVC utilized the dimidiate pixel model in which the choice of $NDVI_v$ and $NDVI_s$ led to higher values. The GEOV1 FVC product demonstrated that the FVC values were higher than those from SEVIRI in the Validation Report of Land Surface Analysis Vegetation Products (2008). Mu et al. noted that GEOV1 FVC product was generally overestimated for crops by up to 0.20 in the Heihe Basin. GLASS FVC and TRAGL FVC considered the terrestrial ecoregion and land cover type when produced, which made the estimation results more accurate [16].

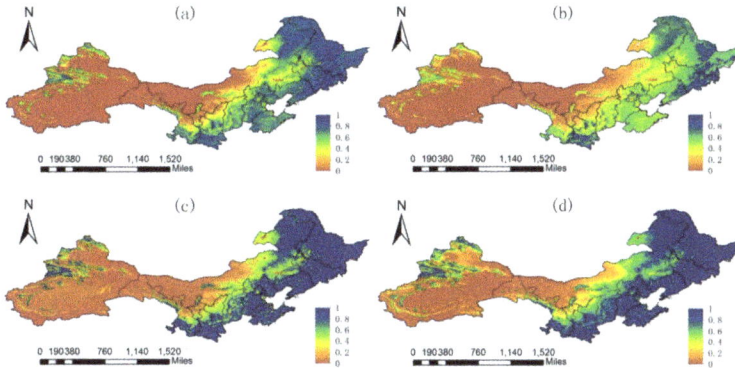

Figure 2. The spatial patterns of maximum FVC over the period 2001–2012 in northern China. (**a**) GLASS FVC; (**b**) TRAGL FVC; (**c**) GEOV1 FVC; (**d**) Li FVC.

3.1.2. Variation Trends of Each FVC Data Set

To evaluate the spatial heterogeneity of FVC change trends, overall linear trends and the linear trends that passed significance level ($p < 0.05$) were calculated at pixel scale shown in Figure 3, respectively. The spatial distribution difference of linear trends was small between the data sets and most of the area had increased in FVC but not significantly. One significant difference among the four data sets was detected in North China, the south part of Xinjiang Uygur autonomous region, the western part of Inner Mongolia Autonomous Region, as well as the northwestern part of Ningxia Hui autonomous region. Of these regions, North China had a significant increase in the four data sets whereas FVC values of the south part of Xinjiang Uygur autonomous region, the western part of Inner Mongolia Autonomous Region, as well as the northwest part of Ningxia Hui autonomous region did not change. Those of GEOV1 FVC showed a significant decrease and those of Li FVC presented a significant increase in some areas of these regions.

Figure 4 shows the inter-annual variations of annual maximum FVC of four data sets in northern China during the period 2001–2012. The mean amplitude, mean variations, and mean variation trend of the four FVC data sets varied among the individual sources. In terms of mean amplitude, the annual mean values of Li FVC were the highest, significantly higher than the other three data sets. This was largely because Li FVC utilizes the dimidiate pixel model in which the choice of $NDVI_v$ and $NDVI_s$ may lead to overestimates or even saturation. In contrast, those of TRAGL FVC were the lowest and GLASS FVC and GEOV1 FVC were closer to the mean values generated from the four data sets in the study area. As for mean variations and mean variation trends, GEOV1 was significantly higher than the mean values and GLASS FVC was closer to the mean values. Summaries of mean variation trends of FVC estimates over northern China, Northeast China, Northwest China, and North China from the four data sets are shown in Table 2. FVC increased at the rate of 0.26%, 0.30%, 0.12%, and 0.49% per year in northern China, North China, Northeast China and Northwest China during the period 2001–2012, respectively. By contrast, GLASS FVC and GEOV1 FVC was closer to the mean values of the four data sets, while the performance of GEOV1 FVC was not better than that of GLASS FVC in terms of mean variations and mean variation trends.

Remote Sens. **2018**, *10*, 549

Table 2. Mean variation trend values of annual average maximum FVC seen by four data sets for 2001–2012.

Region	GLASS FVC	GEOV1 FVC	TRAGL FVC	Li FVC	Average
Northern china	0.0020	0.0048	0.0016	0.0019	0.0026
Northeast china	0.0017	0.0072	0.0021	0.0010	0.0030
Northwest china	0.0012	0.0016	0.0008	0.0013	0.0012
North China	0.0040	0.0084	0.0029	0.0041	0.0049

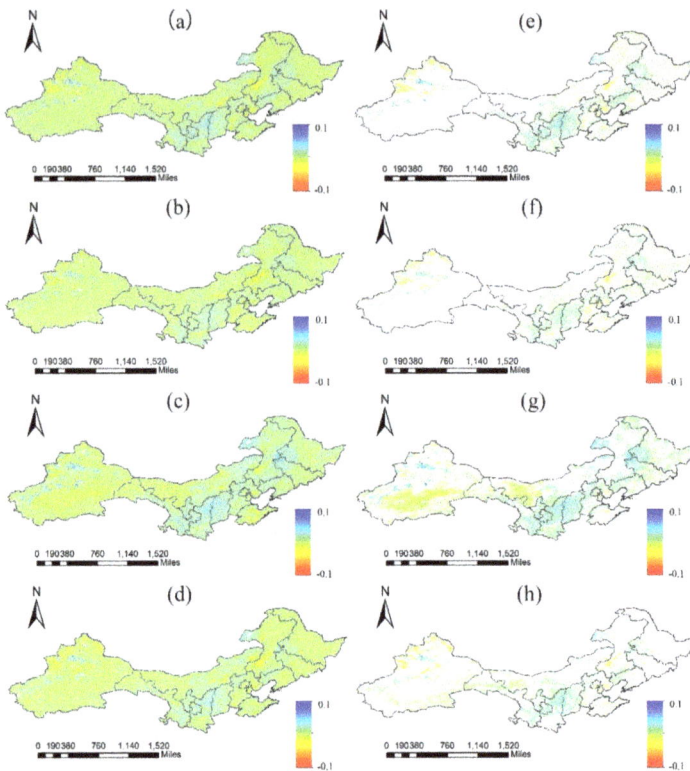

Figure 3. The temporal trends of annual maximum FVC in northern China during the periods 2001–2012. Left column is the temporal trends of (**a**) GLASS FVC, (**b**) TRAGL FVC, (**c**) GEOV1 FVC, (**d**) Li FVC. Right column is the temporal trends of annual maximum FVC that passed the significant test in northern China during the periods 2001–2012. (**e**) GLASS FVC; (**f**) TRAGL FVC; (**g**) GEOV1 FVC; (**h**) Li FVC.

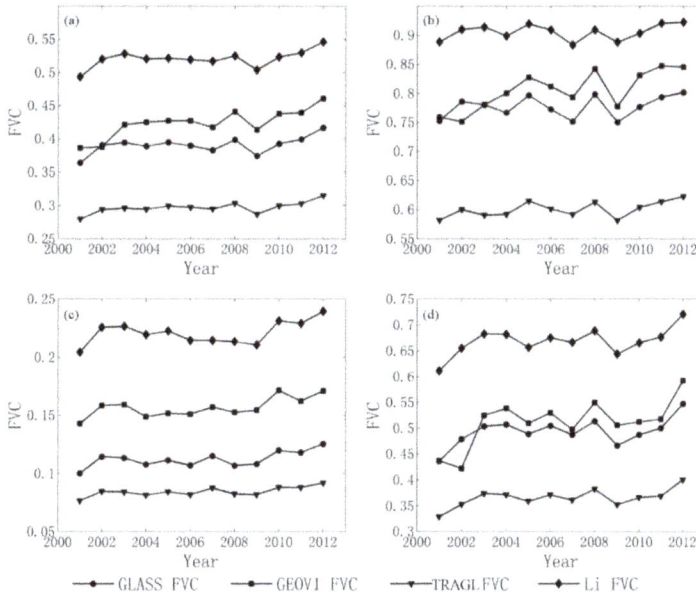

Figure 4. The averages of annual mean maximum FVC of four data sets over the period 2001–2012. (**a**) northern China; (**b**) Northeast China; (**c**) Northwest China; (**d**) North China.

3.2. Analysis of Multi-Source FVC Data Set

3.2.1. Data Set Evaluation

The consistency of each data set was evaluated by computing the correlation of coefficient and root mean square error (RMSE) between each data set and averaged FVC from the three other FVC data sets. The evaluation was intended to remove the poorer performing data sets from the average FVC series [39]. This approach was also applied at pixel scale to remove the poorer performing individual FVC data set from the final average data set in each pixel. The correlation and RMSE of each FVC data set and averaged FVC from the three other FVC data sets above the 95% confidence interval over northern China from 2001 to 2012 are displayed in Figure 5.

The evaluation had been intended to remove the poorer performing data sets from the average FVC series [39]. However, multiple regression analysis revealed that all of the data sets were statistically significant (0.05 level) variables in explaining the variance in the multi-data set series. Accordingly, there was no reason to eliminate any of the four data sets. The stratification of the evaluation results revealed that the GLASS FVC had the highest correlation and lowest RMSE compared to the multi-data set average. In contrast, the TRAGL FVC showed higher agreement with other data sets with correlation of a greater coefficient.

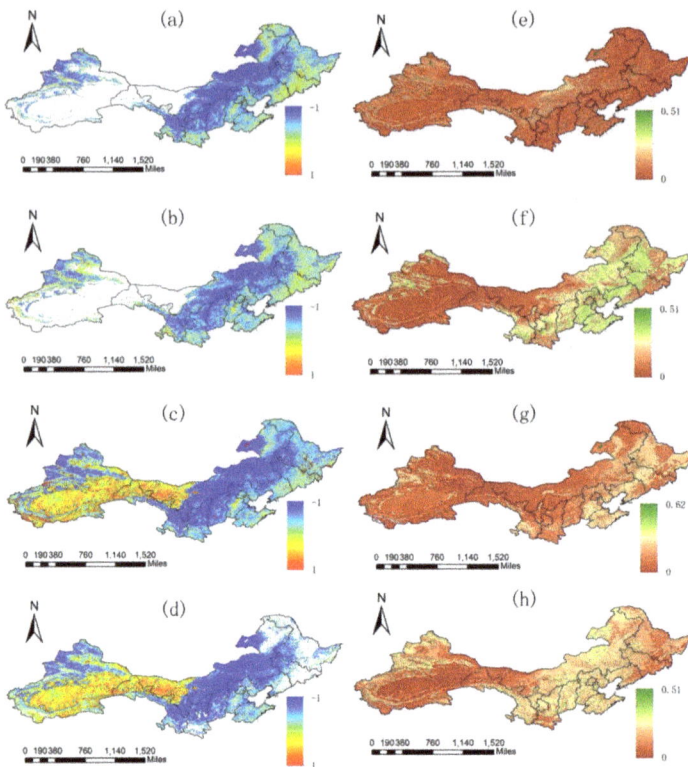

Figure 5. Correlation and root mean square error (RMSE) of each FVC data set with the averaged FVC from the three other FVC data sets used in this study. Left column is the correlation of (**a**) GLASS FVC, (**b**) TRAGL FVC, (**c**) GEOV1 FVC, (**d**) Li FVC with the average from the other FVC data sets. Right column is the corresponding RMSE value of (**e**) GLASS FVC, (**f**) TRAGL FVC, (**g**) GEOV1 FVC, (**h**) Li FVC with the average from the other FVC data sets.

3.2.2. Change Trends of Multi-Source FVC Data Sets

Figure 6a shows the slope values calculated from the regression analysis, Figure 6b shows the linear trends that passed significance level ($p < 0.05$) by Mann–Kendall test, and Table 3 shows the percentage of significant levels over three parts of northern China. The slope estimate results show that the FVC increased over most areas. Increased regions were mainly distributed in Northeast China and North China. In contrast, the northwestern part of the Xinjiang Uygur Autonomous Region (including Toli, Tacheng, Ili, and Yining counties) and the southern part of the Xinjiang Uygur Autonomous Region, Hulunbeir Plateau, as well as most part of Shandong Province showed various degrees of vegetation decline.

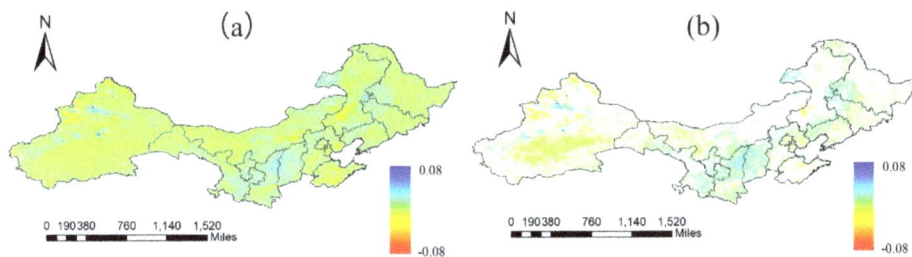

Figure 6. (**a**) The temporal trends of annual averaged FVC of four data sets in northern China during the period 2001–2012. (**b**) The temporal trends of annual averaged FVC of four data sets that passed the significant test in northern China during the period 2001–2012.

Table 3. Statistics for the percentage of significant level of FVC change during the period 2001–2012.

Regions	Significantly Increased	No Significant Change	Significantly Decreased
Northern china	33.03%	50.16%	16.81%
Northeast china	44.88%	50.12%	5.00%
Northwest china	23.34%	51.86%	24.80%
North China	56.05%	35.59%	8.36%

In Northeast China, 44.88% of pixels showed significant increased fractional vegetation cover. 5.00% of Northeast China showed significant decrease and 50.12% remained stable. Significant increases were concentrated in the Northeast Plain and the decreased regions were scattered amongst the Daxing' anling Mountains, the Xiaoxing' anling Mountains, and the Changbai Mountains.

In North China, 50.05% of pixels showed significant increased fractional vegetation cover. 8.36% of the North China showed a significant decrease in FVC. The areas where vegetation cover declined were mainly distributed in Shandong, Hebei, Tianjin, and Beijing. In 35.59% of the pixels, the vegetation cover remained stable and had no significant change.

In Northwest China, 23.34% showed significant increased fractional vegetation cover. Approximately 24.80% of pixels showed a significant decrease and 51.76% remained stable. The increases were mainly distributed in the southern Tianshan Mountains and in northern and southern oasis areas. The unchanged regions were mainly distributed in the northwestern part of Xinjiang Uygur Autonomous Region, the extremely arid desert and Gobi Desert regions, as well as areas that are difficult to use and develop. By contrast, an insignificant decrease occurred in the middle parts of the Kunlun Mountains where the climate is very dry.

3.2.3. Results of Multi-Data FVC and Uncertainty Analysis

Figure 7a,b shows the spatial distribution of multi-data FVC and uncertainty results generated from the four data sets over northern China during the period 2001–2012, respectively. From the spatial distribution, the FVC values were high over Northeast China while those from Northwest China were almost 0 and those of North China fell in between the two. The uncertainty results of much of Northwest China were almost 0 which indicated low uncertainties. The uncertainty results of North China were around 0.15 while that of Northeast China was about 0.25. In general, in contrast to the low FVC area, the high FVC area of northern China had a high degree of uncertainty.

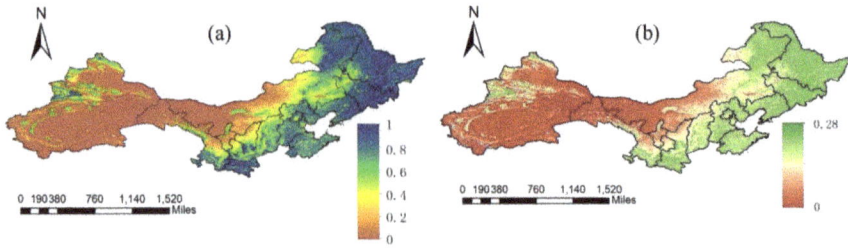

Figure 7. Spatial pattern of mean maximum FVC (**a**) and standard error (**b**) of multi-data sets over the northern China during the period 2001–2012.

Annual-averaged maximum FVC from 2001 to 2012 with error bars (standard error) generated from multi-data sets over northern China, Northeast China, Northwest China, and North China during the period 2001–2012, is shown in Figure 8, respectively. The multi-data set shows a significant increase in FVC over northern China from 2001 to 2012 and the estimates show a more linear increase in FVC than the previous single data set. Low FVC can be seen in 2001 and 2009, especially in Northeast China. 2001 was a year of extreme drought and the precipitation was lower than normal in most regions of northern China [40]. Different degrees of drought occurred in all seasons. In 2009, low temperatures, snowfall in winter, as well as drought in spring and autumn [41] inhibited vegetation growth in this region, causing the annual mean FVC to decrease.

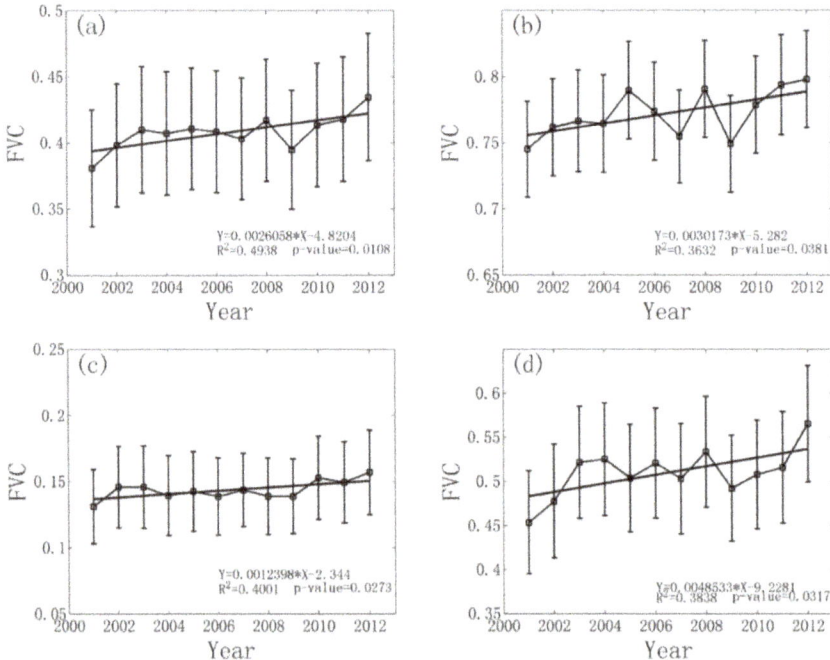

Figure 8. The averages of annual mean maximum FVC and trend in FVC with the error bars showing the standard error of multi-data set average. (**a**) northern China; (**b**) Northeast China; (**c**) Northwest China; (**d**) North China.

From the regional scale, the FVC over Northeast China significantly increased at a rate of 0.3%/year from 2001 to 2012 (R^2 = 0.3632, *p*-value = 0.0381). Similarly, Northwest China had a significant increase at the rate of 0.12%/year in FVC from 2001 to 2012 (R^2 = 0.4001, *p*-value = 0.0273). In general, the growth trend was relatively flat. The FVC over North China significantly increased at a rate of 0.49%/year during the period 2001–2012. Similarly, North China underwent an intense and prolonged drought episode in 2009.

Comparing the three parts, North China had a more obvious increasing trend, while that of Northeast and Northwest China was not so significant. In all of northern China, 33.03% of pixels showed significant increase fractional vegetation cover, approximately 16.81% of pixels showed significant decrease, and 50.16% remained stable. In general, vegetation in northern China has increased during the period 2001–2012.

4. Discussions

The FVC dynamics from 2001–2012 in northern China has been analyzed from four satellite data products in this study. We found a more linear increase in FVC than did previous studies using a single data set. The four data sets show a consistent agreement. The spatial patterns of temporal trends of annual mean are similar among the data sets; however, the GEOV1 FVC product shows relatively bigger variations in trend. From the points of temporal trends and inter-annual variability of annual mean FVC, the GLASS FVC is closer to the mean values of the four data sets.

To date, analysis of vegetation cover change over northern China has relied on single sources of information that can be affected by both estimation methods and satellite sensors. The multi-data set approach taken in this study can reduce the impact of inconsistencies and provide a more reliable estimate of the uncertainty of FVC in each year. Trend analysis of the multi-data set (including an annually varying estimate of error) reveals that FVC has increased at a rate of 0.26 ± 0.13%, 0.30 ± 0.25%, 0.12 ± 0.03%, 0.49 ± 0.21% per year in northern China, Northeast China, Northwest China, and North China during the period 2001–2012, respectively. Most areas of northern China have increased in vegetation, especially in the Northeast Plain, the central part of North China, and the Hulunbuir prairie. In all of northern China, 33.03% of pixels showed significantly increased fractional vegetation cover, approximately 16.81% of pixels showed significant decrease, and 50.16% remained stable. Meanwhile, North China had a more obvious increasing trend while that of Northeast and Northwest China was not so significant.

In this study, the multi-data FVC was retrieved from four FVC data set from 2000 to 2012. However, the ecological programs were implemented in the 1980s. Therefore, the method which considered different FVC data sets can be used to provide a long-term multi-data set analysis. For example, the GEOV1 FVC can be obtained from 1982 to the present while the GLASS FVC, which provided the FVC estimates from 1982 to 2016, is going to be released in the near future. In addition, other land surface products, such as tree cover products and land use products, can also be used to provide evidence about the condition of vegetation change in northern China.

5. Conclusions

This study conducted a spatio-temporal analysis of fractional vegetation cover change in northern China during 2001–2012 based on multiple data sets. Results indicated that fractional vegetation cover increased in northern China from 2000 to 2012 but not significantly. In addition, this study also provides an estimate of uncertainty in FVC at pixel and regional scale. However, this study, which covered the period from 2000 to 2012, is limited in its ability to provide direct evidence for the effects of ecological programs on vegetation change of northern China since the ecological programs were implemented in the 1980s. Although the methodology used in this study was not inherently complex, it is very effective and has been used in many related studies and can be used for large areas or even globally. Further work will focus on the evaluation of vegetation changes by using more land surface products, such as land cover type data and/or tree cover data.

Acknowledgments: This study was partially supported by the National Key Research and Development Program of China (NO. 2016YFA0600103) and the National Natural Science Foundation of China (No. 41671332).

Author Contributions: L.Y., K.J. and S.L. conceived and designed the experiments; L.Y. performed the experiments; K.J. and L.Y. conducted the analysis of the results; and all of the authors contributed towards writing the manuscript.

Conflicts of Interest: The authors declare no conflict of interest.

References

1. Cao, S. Why Large-Scale Afforestation Efforts in China Have Failed To Solve the Desertification Problem. *Environ. Sci. Technol.* **2008**, *42*, 1826–1831. [CrossRef] [PubMed]
2. Brown, L.R.; Halweil, B. China's water shortage could shake world food security. *World Watch* **1998**, *11*, 10–16. [PubMed]
3. Zhai, P.; Chao, Q.; Zou, X. Progress in China's climate change study in the 20th century. *J. Geogr. Sci.* **2004**, *14*, 3–11.
4. Wang, W.; Shao, Q.; Peng, S.; Zhang, Z.; Xing, W.; An, G.; Yong, B. Spatial and temporal characteristics of changes in precipitation during 1957–2007 in the Haihe River basin, China. *Stoch. Environ. Res. Risk Assess.* **2011**, *25*, 881–895. [CrossRef]
5. Bonan, G.B. Forests and climate change: Forcings, feedbacks, and the climate benefits of forests. *Science* **2008**, *320*, 1444–1449. [CrossRef] [PubMed]
6. Jiang, B.; Liang, S.; Yuan, W. Observational evidence for impacts of vegetation change on local surface climate over northern China using the Granger causality test. *J. Geophys. Res. Biogeosci.* **2015**, *120*, 1–12. [CrossRef]
7. Anderson, R.G.; Canadell, J.G.; Randerson, J.T.; Jackson, R.B.; Hungate, B.A.; Baldocchi, D.D.; Ban-Weiss, G.A.; Bonan, G.B.; Caldeira, K.; Cao, L.; et al. Biophysical considerations in forestry for climate protection. *Front. Ecol. Environ.* **2011**, *9*, 174–182. [CrossRef]
8. Duan, H.; Yan, C.; Tsunekawa, A.; Song, X.; Li, S.; Xie, J. Assessing vegetation dynamics in the Three-North Shelter Forest region of China using AVHRR NDVI data. *Environ. Earth Sci.* **2011**, *64*, 1011–1020. [CrossRef]
9. Yin, R.; Rothstein, D.; Qi, J.; Liu, S. Methodology for an Integrative Assessment of China's Ecological Restoration Programs. In *An Integrated Assessment of China's Ecological Restoration Programs*; Springer: Dordrecht, The Netherlands, 2009; pp. 39–54.
10. Uchida, E.; Xu, J.; Rozelle, S. Grain for Green: Cost-Effectiveness and Sustainability of China's Conservation Set-Aside Program. *Land Econ.* **2005**, *81*, 247–264. [CrossRef]
11. Pimentel, D.; Harvey, C.; Resosudarmo, P.; Sinclair, K.; Kurz, D.; McNair, M.; Crist, S.; Shpritz, L.; Fitton, L.; Saffouri, R.; et al. Environmental and economic costs of soil erosion and conservation benefits. *Science* **1995**, *267*, 1117–1123. [CrossRef] [PubMed]
12. Jacinthe, P.A.; Lal, R. A mass balance approach to assess carbon dioxide evolution during erosional events. *Land Degrad. Dev.* **2001**, *12*, 329–339. [CrossRef]
13. Wu, Y.; Zeng, Y.; Wu, B.; Li, X.; Wu, W.B. Retrieval and analysis of vegetation cover in the Three-North Regions of China based on MODIS data. *Chin. J. Ecol.* **2009**, *28*, 1712–1718.
14. Dong, L.S.; Bo, Z.H. The Comparison Study on Forestry Ecological Projects in the World. *Acta Ecol. Sin.* **2002**, *22*, 1976–1982.
15. Liang, S.; Li, X.; Wang, J. *Advanced Remote Sensing: Terrestrial Information Extraction and Applications*; Academic Press: Oxford, UK, 2012; p. 800.
16. Jia, K.; Liang, S.; Liu, S.; Li, Y.; Xiao, Z.; Yao, Y.; Jiang, B.; Zhao, X.; Wang, X.; Xu, S.; et al. Global Land Surface Fractional Vegetation Cover Estimation Using General Regression Neural Networks From MODIS Surface Reflectance. *IEEE Trans. Geosci. Remote Sens.* **2015**, *53*, 4787–4796. [CrossRef]
17. Jia, K.; Liang, S.; Wei, X.; Li, Q.; Du, X.; Jiang, B.; Yao, Y.; Zhao, X.; Li, Y. Fractional Forest Cover Changes in Northeast China From 1982 to 2011 and Its Relationship With Climatic Variations. *IEEE J. Sel. Top. Appl. Earth Obs. Remote Sens.* **2015**, *8*, 775–783. [CrossRef]
18. Yang, L.; Jia, K.; Liang, S.; Liu, J.; Wang, X. Comparison of Four Machine Learning Methods for Generating the GLASS Fractional Vegetation Cover Product from MODIS Data. *Remote Sens.* **2016**, *8*, 682. [CrossRef]
19. Su, W.; Yu, D.-Y.; Sun, Z.-P.; Zhan, J.-G.; Liu, X.-X.; Luo, Q. Vegetation changes in the agricultural-pastoral areas of northern China from 2001 to 2013. *J. Integr. Agric.* **2016**, *15*, 1145–1156. [CrossRef]

20. Liu, S.; Wang, T.; Guo, J.; Qu, J.; An, P. Vegetation change based on SPOT-VGT data from 1998 to 2007, northern China. *Environ. Earth Sci.* **2009**, *60*, 1459–1466. [CrossRef]

21. Zhang, R.; Feng, Q.; Guo, J.; Shang, Z.; Liang, T. Spatio-temporal Changes of NDVI and Climatic Factors of Grassland in Northern China from 2000 to 2012. *J. Desert Res.* **2015**, *35*, 1403–1412.

22. Li, Y.; Jia, K.; Wei, X.; Yao, Y.; Sun, J.; Mou, L. Fractional vegetation cover estimation in northern China and its change analysis. *Remote Sens. Land Resour.* **2015**, *27*, 112–117.

23. Li, Y.; Sulla-Menashe, D.; Motesharrei, S.; Song, X.P.; Kalnay, E.; Ying, Q.; Li, S.; Ma, Z. Inconsistent estimates of forest cover change in China between 2000 and 2013 from multiple datasets: Differences in parameters, spatial resolution, and definitions. *Sci. Rep.* **2017**, *7*, 8748. [CrossRef] [PubMed]

24. Wang, X.; Fang, J.; Tang, Z.; Zhu, B. Climatic control of primary forest structure and DBH–height allometry in Northeast China. *For. Ecol. Manag.* **2006**, *234*, 264–274. [CrossRef]

25. Baret, F.; Weiss, M.; Lacaze, R.; Camacho, F.; Makhmara, H.; Pacholcyzk, P.; Smets, B. GEOV1: LAI and FAPAR essential climate variables and FCOVER global time series capitalizing over existing products. Part1: Principles of development and production. *Remote Sens. Environ.* **2013**, *137*, 299–309. [CrossRef]

26. Xiao, Z.; Wang, T.; Liang, S.; Sun, R. Estimating the Fractional Vegetation Cover from GLASS Leaf Area Index Product. *Remote Sens.* **2016**, *8*, 337. [CrossRef]

27. Camacho, F.; Cernicharo, J.; Lacaze, R.; Baret, F.; Weiss, M. GEOV1: LAI, FAPAR essential climate variables and FCOVER global time series capitalizing over existing products. Part 2: Validation and intercomparison with reference products. *Remote Sens. Environ.* **2013**, *137*, 310–329. [CrossRef]

28. Xiao, Z.; Liang, S.; Wang, J.; Chen, P.; Yin, X.; Zhang, L.; Song, J. Use of General Regression Neural Networks for Generating the GLASS Leaf Area Index Product From Time-Series MODIS Surface Reflectance. *IEEE Trans. Geosci. Remote Sens.* **2014**, *52*, 209–223. [CrossRef]

29. Gutman, G.; Ignatov, A. The derivation of the green vegetation fraction from NOAA/AVHRR data for use in numerical weather prediction models. *Int. J. Remote Sens.* **1998**, *19*, 1533–1543. [CrossRef]

30. Qi, J. Spatial and temporal dynamics of vegetation in the San Pedro River basin area. *Agric. For. Meteorol.* **2000**, *105*, 55–68. [CrossRef]

31. Mann, H.B. Nonparametric Tests against Trend. *Econometrica* **1945**, *13*, 245. [CrossRef]

32. Forthofer, R.N.; Lehnen, R.G. *Rank Correlation Methods*; Springer: New York, NY, USA, 1981; pp. 146–163.

33. Zhao, X.; Tan, K.; Zhao, S.; Fang, J. Changing climate affects vegetation growth in the arid region of the northwestern China. *J. Arid Environ.* **2011**, *75*, 946–952. [CrossRef]

34. Tabari, H.; Marofi, S. Changes of Pan Evaporation in the West of Iran. *Water Resour. Manag.* **2010**, *25*, 97–111. [CrossRef]

35. Hamed, K.H. Exact distribution of the Mann–Kendall trend test statistic for persistent data. *J. Hydrol.* **2009**, *365*, 86–94. [CrossRef]

36. Jaagus, J. Climatic changes in Estonia during the second half of the 20th century in relationship with changes in large-scale atmospheric circulation. *Theor. Appl. Climatol.* **2005**, *83*, 77–88. [CrossRef]

37. Sen, P.K. Estimates of the Regression Coefficient Based on Kendall's Tau. *J. Am. Stat. Assoc.* **1968**, *63*, 1379–1389. [CrossRef]

38. Tabari, H.; Hosseinzadeh Talaee, P. Analysis of trends in temperature data in arid and semi-arid regions of Iran. *Glob. Planet. Chang.* **2011**, *79*, 1–10. [CrossRef]

39. Brown, R.; Derksen, C.; Wang, L. A multi-data set analysis of variability and change in Arctic spring snow cover extent, 1967–2008. *J. Geophys. Res.* **2010**, *115*. [CrossRef]

40. Lu, J. Features of Weather/Climate over China in 2001. *Meteorol. Mon.* **2002**, *29*, 32–36.

41. Chen, H.; Fan, X. Some Extreme Events of Weather, Climate and Related Phenomena in 2009. *Clim. Environ. Res.* **2010**, *15*, 322–336.

MDPI

St. Alban-Anlage 66

4052 Basel

Switzerland

Tel. +41 61 683 77 34

Fax +41 61 302 89 18

www.mdpi.com

Remote Sensing Editorial Office

E-mail: remotesens@mdpi.com

www.mdpi.com/journal/remotesens

www.ingramcontent.com/pod-product-compliance
Lightning Source LLC
Chambersburg PA
CBHW051707210326
41597CB00032B/5397